Selected Titles in This Series

273 A. V. Kelarev, R. Göbel, K. M. Rangaswamy, P. Schultz, and C. Vinsonhaler, Editors, Abelian groups, rings and modules, 2001

272 Eva Bayer-Fluckiger, David Lewis, and Andrew Ranicki, Editors, Quadratic forms and their applications, 2000

271 J. P. C. Greenlees, Robert R. Bruner, and Nicholas Kuhn, Editors, Homotopy methods in algebraic topology, 2001

270 Jan Denef, Leonard Lipshitz, Thanases Pheidas, and Jan Van Geel, Editors, Hilbert's tenth problem: Relations with arithmetic and algebraic geometry, 2000

269 Mikhail Lyubich, John W. Milnor, and Yair N. Minsky, Editors, Laminations and foliations in geometry, topology and dynamics, 2001

268 Robert Gulliver, Walter Littman, and Roberto Triggiani, Editors, Differential geometric methods in the control of partial differential equations, 2000

267 Nicolás Andruskiewitsch, Walter Ricardo Ferrer Santos, and Hans-Jürgen Schneider, Editors, New trends in Hopf algebra theory, 2000

266 Caroline Grant Melles and Ruth I. Michler, Editors, Singularities in algebraic and analytic geometry, 2000

265 Dominique Arlettaz and Kathryn Hess, Editors, Une dégustation topologique: Homotopy theory in the Swiss Alps, 2000

264 Kai Yuen Chan, Alexander A. Mikhalev, Man-Keung Siu, Jie-Tai Yu, and Efim I. Zelmanov, Editors, Combinatorial and computational algebra, 2000

263 Yan Guo, Editor, Nonlinear wave equations, 2000

262 Paul Igodt, Herbert Abels, Yves Félix, and Fritz Grunewald, Editors, Crystallographic groups and their generalizations, 2000

261 Gregory Budzban, Philip Feinsilver, and Arun Mukherjea, Editors, Probability on algebraic structures, 2000

260 Salvador Pérez-Esteva and Carlos Villegas-Blas, Editors, First summer school in analysis and mathematical physics: Quantization, the Segal-Bargmann transform and semiclassical analysis, 2000

259 D. V. Huynh, S. K. Jain, and S. R. López-Permouth, Editors, Algebra and its applications, 2000

258 Karsten Grove, Ib Henning Madsen, and Erik Kjær Pedersen, Editors, Geometry and topology: Aarhus, 2000

257 Peter A. Cholak, Steffen Lempp, Manuel Lerman, and Richard A. Shore, Editors, Computability theory and its applications: Current trends and open problems, 2000

256 Irwin Kra and Bernard Maskit, Editors, In the tradition of Ahlfors and Bers: Proceedings of the first Ahlfors-Bers colloquium, 2000

255 Jerry Bona, Katarzyna Saxton, and Ralph Saxton, Editors, Nonlinear PDE's, dynamics and continuum physics, 2000

254 Mourad E. H. Ismail and Dennis W. Stanton, Editors, q-series from a contemporary perspective, 2000

253 Charles N. Delzell and James J. Madden, Editors, Real algebraic geometry and ordered structures, 2000

252 Nathaniel Dean, Cassandra M. McZeal, and Pamela J. Williams, Editors, African Americans in Mathematics II, 1999

251 Eric L. Grinberg, Shiferaw Berhanu, Marvin I. Knopp, Gerardo A. Mendoza, and Eric Todd Quinto, Editors, Analysis, geometry, number theory: The Mathematics of Leon Ehrenpreis, 2000

For a complete list of titles in this series, visit the AMS Bookstore at **www.ams.org/bookstore/**.

Abelian Groups, Rings and Modules

László Fuchs, November 2000

CONTEMPORARY MATHEMATICS

273

Abelian Groups, Rings and Modules

AGRAM 2000 Conference
July 9–15, 2000
Perth, Western Australia

A. V. Kelarev
R. Göbel
K. M. Rangaswamy
P. Schultz
C. Vinsonhaler
Editors

American Mathematical Society
Providence, Rhode Island

Editorial Board

Dennis DeTurck, managing editor

Andreas Blass   Andy R. Magid   Michael Vogelius

The Proceedings of the AGRAM 2000 Conference on Abelian Groups, Rings and Modules, held at the University of Western Australia, Perth, Western Australia, July 9–15, 2000.

2000 *Mathematics Subject Classification.* Primary 05Axx, 06Bxx, 06Fxx, 13Cxx, 15–XX, 16–XX, 20Kxx, 20Exx.

Library of Congress Cataloging-in-Publication Data
AGRAM 2000 Conference (2000 : Perth, Western Australia)
  Abelian groups, rings, and modules : AGRAM 2000 Conference July 9–15, 2000, Perth, Western Australia / A. V. Kelarev... [et. al.], editors.
    p. cm. — (Contemporary mathematics, ISSN 0271-4132 ; 273)
  Includes bibliographical references.
  ISBN 0-8218-2751-0 (alk. paper)
  1. Abelian groups—Congresses. I. Kelarev, A. V. (Andrei V.) II. Contemporary mathematics (American Mathematical Society) ; v. 273.

QA251.3.A35 2000
512′.2—dc21   2001016047

Copying and reprinting. Material in this book may be reproduced by any means for educational and scientific purposes without fee or permission with the exception of reproduction by services that collect fees for delivery of documents and provided that the customary acknowledgment of the source is given. This consent does not extend to other kinds of copying for general distribution, for advertising or promotional purposes, or for resale. Requests for permission for commercial use of material should be addressed to the Assistant to the Publisher, American Mathematical Society, P. O. Box 6248, Providence, Rhode Island 02940-6248. Requests can also be made by e-mail to reprint-permission@ams.org.

Excluded from these provisions is material in articles for which the author holds copyright. In such cases, requests for permission to use or reprint should be addressed directly to the author(s). (Copyright ownership is indicated in the notice in the lower right-hand corner of the first page of each article.)

© 2001 by the American Mathematical Society. All rights reserved.
The American Mathematical Society retains all rights
except those granted to the United States Government.
Printed in the United States of America.

∞ The paper used in this book is acid-free and falls within the guidelines
established to ensure permanence and durability.
Visit the AMS home page at URL: http://www.ams.org/
10 9 8 7 6 5 4 3 2 1   06 05 04 03 02 01

Contents

Preface ... ix

Conference Participants ... xi

Part I. Introduction

László Fuchs and his "moddom" work
   L. Salce ... 3

Part II. Survey Articles

Error-correcting codes as ideals in group rings
   A. V. Kelarev and P. Solé ... 11

Homomorphisms and duality for torsion–free modules
   B. Olberding ... 19

Generalizations of isomorphism in torsion–free abelian groups
   K. C. O'Meara and C. Vinsonhaler ... 39

Automorphism groups of abelian groups
   P. Schultz ... 51

Part III. Contributed Papers

Direct sum decompositions of torsion–free abelian groups of finite rank
   D. M. Arnold ... 65

The endomorphism ring of a bounded abelian p–group
   M. A. Aviñó and P. Schultz ... 75

The Baer–Kaplansky theorem for almost completely decomposable groups
   E. Blagoveshchenskaya, G. Ivanov, and P. Schultz ... 85

Maximal pure independent sets
   A. Blass and J. Irwin ... 95

Characterization of the tori via density of the solution set of linear equations
   D. Dikranjan and M. Tkachenko ... 107

Quotient divisible mixed groups
  A. A. FOMIN   117

Stacked bases over h–local Prüfer domains
  L. FUCHS AND S. B. LEE   129

Groups with locally defined heights and products of $\mathfrak{R}^*$ groups
  A. J. GIOVANNITTI   135

Reflexive subgroups of the Baer–Specker group and Martin's axiom
  R. GÖBEL AND S. SHELAH   145

Σ–isotype subgroups of local k-groups
  P. HILL, C. MEGIBBEN, AND W. ULLERY   159

Character modules and endomorphism rings of modules over Artinian serial rings
  G. IVANOV   177

Topologically pure extensions
  P. LOTH   191

Rings having simple adjoint semigroup
  N. R. MCCONNELL AND T. STOKES   203

Invariants of global crq-groups
  A. MADER, L. G. NONGXA, AND M. A. OULD–BEDDI   209

On varieties of groups generated by wreath products of abelian groups
  V. H. MIKAELIAN   223

Existence of rigid indecomposable almost completely decomposable groups
  O. MUTZBAUER   239

C2–rings and the FGF–conjecture
  W. K. NICHOLSON AND M. F. YOUSIF   245

Lifting direct sum decompositions of bounded abelian p–groups
  B. L. OSOFSKY   253

On modules and submodules with finite projective dimension
  K. M. RANGASWAMY   261

On the torsion groups in cotorsion classes
  L. STRÜNGMANN AND S. L. WALLUTIS   269

Cotorsion theories induced by tilting and cotilting modules
  J. TRLIFAJ   285

Steadiness is tested by a single module
  J. ŽEMLIČKA   301

Preface

This volume contains the Proceedings of the AGRAM 2000 Conference on Abelian Groups, Rings and Modules held at The University of Western Australia, in Perth, Australia from July 9 to July 15, 2000. In addition to talks presented at the Conference a few specially invited papers make up this volume.

AGRAM 2000 brought together 50 researchers and graduate students from 5 continents and 17 countries.

The AGRAM Conference was another link in a chain of meetings dealing with abelian groups, related structures and applications that have taken place in recent years in the United States, Italy, Germany, Curaçao and Ireland. The next link will be forged in 2001 in Honolulu, Hawaii.

These Proceedings are dedicated to Professor László Fuchs on the occasion of his 75th birthday, and are headed by a tribute and a review of his recent research by his long time collaborator Professor Luigi Salce.

Then follow four surveys in which experts in the field present recent results in active research areas:

- Error correcting codes as ideals in group rings
- Duality in module categories
- Automorphism groups of abelian groups
- Generalizations of isomorphism in torsion-free abelian groups.

The volume contains also 22 research articles in diverse areas connected with the themes of the conference, including various aspects of abelian groups and their endomorphism rings, modules over various rings, commutative and non–commutative ring theory, varieties of groups and topological aspects of algebra.

The papers in this volume have not been published before and underwent a rigorous refereeing procedure. The Editors wish to thank Professor Fuchs for his invaluable assistance in the production of this volume.

We would like to thank the Australian Mathematical Society for a generous grant and the Department of Mathematics and Statistics of The University of Western Australia for assistance in kind and for hosting the Conference. Phill Schultz did all of the organizational work single-handedly: he deserves the lion's share of the credit for the success of the conference.

Also, we would like to thank graduate students Sandra Maria Pereira and John Bamberg for their efforts to ensure the smooth administration of the Conference. Special thanks are due to Beth Schultz for many activities which made the Conference more enjoyable both for the participants and for their accompanying spouses.

Conference Participants

Dave Arnold	Baylor University	arnoldd@baylor.edu
Robert Bashir	Charles University	bashir@karlin.mff.cuni.cz
Kostia Beidar	Taiwan	beidar@mail.ncku.edu.tw
Howard Bell	Brock University	hbell@spartan.ac.brocku.ca
Dikran Dikranjan	University of Udine	dikranjan@dimi.uniud.it
Alberto Facchini	University of Padua	facchini@math.unipd.it
Laszlo Fuchs	Tulane University	fuchs@mailhost.tcs.tulane.edu
Tony Giovanitti	West Georgia State	agiovann@westga.edu
Ruediger Goebel	University of Essen	R.Goebel@uni-essen.de
Brendan Goldsmith	Dublin Institute of Technology	bgoldsmith@dit.ie
Vishal Goundar	University of the South Pacific	vi_go@excite.com
Ola Helenius	University of Goteburg	olahe@math.chalmers.se
George Ivanov	Macquarie University	ivanov@ics.mq.edu.au
Yasuo Iwanaga	Shinshu University	iwanaga@gipnc.shinshu-u.ac.jp
Ming-chang Kang	National Taiwan University	kang@math.ntu.edu.tw
Andrei Kelarev	University of Tasmania	kelarev@hilbert.maths.utas.edu.au
Toshiko Koyama	Tokyo	koyama@is.ocha.ac.jp
Fang Li	Zhehang University	lifang3@sohu.com.cn
Peter Loth	Sacred Heart University	lothp@sacredheart.edu
Neil McConnell	Department of Defence	nickm@defcen.gov.au
Adolf Mader	University of Hawaii	adolf@math.hawaii.edu
Claudia Metelli	University of Naples	cmetelli@math.unipd.it
Ray Mines	New Mexico State University	ray@nmsu.edu
Bernhard Neumann	Australian National University	Bernhard.Neumann@wintermute.anu.edu.au
W. Keith Nicholson	University of Calgary	wmkeith@home.com
Loyiso Nongxa	University of the Western Cape	gnongxa@uwc.ac.za
Kevin O'Meara	University of Canterbury	K.OMeara@math.canterbury.ac.nz
Takashi Okuyama	Toba National College	okuyamat@toba-cmt.ac.jp
K. M. Rangaswamy	University of Colorado	ranga@math.uccs.edu
Phill Schultz	University of Western Australia	schultz@maths.uwa.edu.au
Mike Siddoway	Colorado College	MSiddoway@ColoradoCollege.edu
Tim Stokes	Murdoch University	stokes@prodigal.murdoch.edu.au
Alexander Stolin	University of Goteborg	astolin@math.chalmers.se
Lutz Struengmann	Hebrew University	lutz@math.huji.ac.il
Jan Trlifaj	Charles University	trlifaj@karlin.mff.cuni.cz
Chuck Vinsonhaler	University of Connecticut	vinson@math.uconn.edu
Carol Walker	New Mexico State University	hardy@NMSU.Edu
Elbert Walker	New Mexico State University	elbert@nmsu.edu
Simone Wallutis	University of Essen	simone.wallutis@uni-essen.de
Tsai-Lien Wong	Kaohsiung	tlwong@ibm7.math.nsysu.edu.tw
Paolo Zanardo	University of Padua	pzanardo@math.unipd.it
Jan Zemlicka	Charles University	zemlicka@karlin.mff.cuni.cz

Part I

Introduction

László Fuchs and his "moddom" work

Luigi Salce

On the occasion of the festivities surrounding László Fuchs' 75th birthday it is not an easy task to describe in an original way his contribution to mathematics. The reasons are twofold: the spectrum of his interests is too wide to be captured by a short note; furthermore, a few years ago, on the occasion of László's 70th birthday, Rüdiger Göbel wrote a very nice paper [Periodica Math. Hung., 32 (1996), 13–29] illustrating his personal – but comprehensive and illuminating – evaluation of László's contribution to mathematics. Göbel's paper concentrated on László's work between 1973 (when his second volume on infinite abelian groups appeared) and 1996; it also contained an up-to-date complete list of László's publications. The bibliography at the end of this paper is a continuation of that in Göbel's paper; the references given below are to these bibliographies.

Since the appearance of Göbel's paper, the scientific activity of László Fuchs has continued in a very intensive way, focusing mainly on modules over commutative integral domains. He has also written a book on this matter, in a "joint venture" with me; the preparation of this book took a significant portion of our lives in the last four years. In view of these facts, I would like to discuss here only Fuchs' contribution to the subject of his main interest in recent years: modules over commutative integral domains. The title of this paper derives from the short name: "moddom", that we gave to our recent book when it was in progress. One of the goals of this paper, surely shared by László, is to encourage people working in abelian groups to extend their interests and research activity to this area of algebra.

I would also like to offer at the end of the paper a personal tribute to the mathematician that I consider "un grande maestro ed amico".

The "domains" of interest

The word "domain" can be understood both in the usual meaning of "area" or "sector", and in the technical meaning of "commutative integral domain".

Naturally, the first domain of interest for Fuchs' activity is the ring $\mathbb{Z}$ of rational integers; in fact, he is universally considered as the "father" of the theory of $\mathbb{Z}$-modules, i.e., of abelian groups. The persistence of his passion for abelian groups is testified by his recent production on Butler groups, a subject of great interest for abelian group theorists: in the 90's he wrote more then 10 papers on this topic, and in 1999 [189] he considered a generalization to valuated Butler groups, with G. Viljoen.

It is not far from the truth to say that the second domain of interest for Fuchs' activity are valuation domains. I remember that the project of starting a systematic investigation of modules over valuation domains was proposed to me by László at a picnic during the Séminaire de Mathématique Supérieure, organized by Khalid Benabdallah in Montreal in July 1979. The idea of valuation domains was fruitful, since that area has offered many attractive problems over the last twenty years. In few years a lot of new results were obtained, and we collected old and new results in the monograph [4] published in 1985.

The passage from the local to the global case of Prüfer domains came very late, namely, after about 15 years. From my point of view, this delay was due to the crowded agenda of problems over valuation domains, which left little time to think about the obvious – but often very difficult – generalizations to the global case. But in recent years Prüfer domains played a central role in László's investigation, absorbing a large portion of his mathematical time (which is most of his time, as Shula Fuchs often underlines). So we can say that Prüfer domains now form the third domain of interest for László's activity.

Another domain of interest, in connection with homological and topological properties of modules, are Matlis domains, as we call them today. They are those domains R whose field of quotients Q has projective dimension 1. In case R is a valuation domain, this just means that Q is countably generated. Between 1984 and 1987 László intensively studied divisible modules over domains. Over Matlis domains they coincide with h-divisible modules (i.e., quotients of injective modules). h-exact sequences were introduced in a paper with de la Rosa [136], where the associated relative homological algebra was investigated. László also wrote two papers, [154] and [168], with L.M. Pretorius on topological questions over Matlis domains, which may therefore be viewed as a fourth domain of interest.

Finally, specific classes of modules over general or particular domains also captured László's attention; some of them are natural generalizations of important classes of abelian groups that are milestones of the theory. Other classes of modules have been deeply investigated, which are of real interest only over certain domains, and have no counterpart in abelian group theory, as for example non-standard uniserial modules, divisible modules and modules of projective dimension 1.

In a short note it is impossible to discuss all these contributions, many of which are discussed in Göbel's paper. I will confine myself only to some recent results on the following subjects: modules over valuation and Prüfer domains, and particular classes of modules over domains. Thus this exposition is not intended to be exhaustive. Important recent papers are not considered, as for instance the two papers [185] and [188] on extensions of compact abelian groups by discrete groups with K.H. Hofmann, and the paper [181] with P. Vámos on the Jordan-Zassenhaus theorem.

Some recent results of Fuchs

There are two papers contributing to the theory of valuation domains not discussed in Göbel's report; the first one, [182], is included in Göbel's bibliography, although it only appeared in 1996. The latter, [194], is about to appear.

While Ulm's theorem has been generalized to some particular modules, for instance to TAG-modules by Singh-Asari and by Benabdallah-Singh, the theory of simply presented p-groups had not been generalized to a more general setting, until

László published his paper [182] in the Colorado Spring Proceedings of 1996. A deep knowledge of modules over valuation domains, in particular of heights and tight systems, and the improvement of the notion of proper element, are used by László to obtain a significant generalization to simply presented modules over valuation domains. The second paper [194] is a joint paper with K. Rangaswamy on the global balanced-projective dimension of valuation domains. Under the assumption of GCH, they prove that, if the valuation domain R has finite global dimension n, then its global balanced-projective dimension is either n or $n-1$, according as p.d. $Q = n$ or p.d. $Q \leq n - 1$.

Now I pass to a discussion of the recent papers [186], [187], [190] and [191] dealing with different questions on modules over global Prüfer domains. In the papers [186] and [191], both with S.B. Lee, the two connected problems of characterizing equivalent presentations of modules, and of the existence of stacked bases for presentations of direct sums of cyclic modules, are investigated for modules of projective dimension ≤ 1. The solution of the stacked bases problem involves the existence of primary decompositions of torsion modules, so the question is restricted to h-local Prüfer domains. The method of proof is to reduce to countably generated – and then to finitely generated – modules; in the last case, one can assume the Prüfer domains to be more general, namely, local-global domains.

A completely different topic is investigated in the joint paper [187] with P. Zanardo and myself. We study when the notion of pure-essential extension is transitive. The answer is that, for a Prüfer domain, transitivity occurs if and only if the domain is a DVR. A consequence of this surprising result is that some proofs on the existence of pure-injective envelopes of modules are incorrect.

The fourth paper on Prüfer domains [190] deals with multiplicative ideal theory. It extends some results obtained by S. Bazzoni on the class semigroup of a Prüfer domain of finite character. This vein of research was inaugurated by the discovery, made by S. Bazzoni and myself in 1996, that the class semigroup of a valuation domain is a Clifford semigroup. László extends Bazzoni's results to the subsemigroup $\mathcal{S}^*(R)$ of an arbitrary Prüfer domain R, consisting of the isomorphy classes representable by ideals containing some element of finite character. The results are particularly attractive when the ring is a Bézout domain or a strongly discrete Prüfer domain. The increasing infiltration of Clifford semigroups in the theory of valuation and Prüfer domains, and of their uniserial modules, is an interesting subject illustrated in the "moddom" book.

I mentioned above that in many papers László investigated the possible extensions to modules over general domains of some classes of abelian groups that are milestones of the theory: Baer modules in [147] and [153], Whitehead modules in [150], Butler modules (over valuation domains) in [158], large indecomposable modules in [141] and [144]. Thus he demonstrated his constant desire to transfer the most challenging questions arising in abelian group theory to a wider setting. These papers are discussed in Göbel's report, so I will pass to a discussion of the results obtained in three papers on this subject published after 1996. The first two papers with S.B. Lee, [183] and [184], deal with primary decomposition of torsion modules over arbitrary domains. A generalization of the property of a domain of being h-local is studied in [183]. The two properties defining an h-local domain are imposed on the elements of a multiplicative part S of the domain, and on prime ideals intersecting S. It is shown that S-torsion modules have primary decompositions. One obtains h-locality when $S = R \setminus \{0\}$. In the paper [184], a maximal overring R_0 of an arbitrary domain R is defined (which turns out to be a generalized quotient ring of R in the sense of Richman) such that, if $R \leq A \leq Q$, then A/R has primary decomposition exactly if $A \leq R_0$. The same happens on substituting A/R with an A-torsion module M, i.e., $M \cong \operatorname{Tor}_1^R(A/R, M)$. R is h-local exactly when $R_0 = Q$.

The third paper [193], jointly with Macías-Díaz, deals with a completely different subject: the generalization to modules over arbitrary domains of Hill's criterion for freeness of abelian groups, extending the well-known Pontryagin criterion. It is proved that a module M is projective if and only if it is the union of a countable ascending chain of projective submodules M_n ($n \in \omega$) such that M_n is a relatively divisible submodule of M_{n+1} for all n. The demonstration, based on the $G^*(\kappa)$-families technique, is inspired by Hill's original proof for abelian groups.

A personal point of view of Fuchs' mathematical activity

I would like to touch briefly on three aspects of Fuchs' mathematical activity: the large number of his coauthors, the long and fruitful collaboration across the bridge between Padova and New Orleans; and my twofold experience as coauthor in writing mathematical books.

László has an impressive number of scientific collaborators all over the world; a quick computation shows that, of his 194 research papers, 81 are joint papers with 40 different coauthors. In his collaborations László was able to capture from his coauthors their best mathematical and human capacities, to encourage them in the painful moments, with his friendly sense of humor and appropriate Latin sentences.

The bridge between Padova and New Orleans was inaugurated when I went to Tulane for the first time in 1975. After that, the bridge was often very crowded. László came to Padova so many times that I cannot remember all of them. His coauthors in Padova are: S. Bazzoni, C. Metelli (now in Napoli), E. Monari-Martinez, P. Zanardo and myself. I completely agree with Göbel's comment that László had a strong influence on the growth of the Padova algebra school around Adalberto Orsatti.

Now just a few words on my personal experience as coauthor. I was very happy when László asked me to collaborate with him on the first book on modules over valuation domains. But I was more proud and conscious of the task proposed to me to write our second book on modules over non-Noetherian domains. It soon seemed to be a really hard job, but the experience of László, our close friendship and his strong resolution removed my doubts. In both these experiences I admired the qualities of an outstanding mathematician and of a generous man. László has an extraordinary capacity for assimilating and elaborating large and complex portions of mathematics, for simplifying proofs and for finding new perspectives on old results. Other impressive qualities are his obstinacy in pursuing a result through a full immersion in the problem, and his capacity for identifying the good veins in the great mine of mathematics.

To conclude this note, I'm sure that I represent the common feeling of the abelian community (and its extensions) in wishing that László will continue to produce many more lemmas, theorems, corollaries, theories, lecture notes and books, justly aware that his contribution will long remain in the history of the progress of algebra.

Publications of László Fuchs (1996–2000)

The following list is a continuation of Göbel's list, which stops at item **4** for monographs and books, and at item **183** for papers.

MONOGRAPHS AND BOOKS

5. *Modules over non-Noetherian Domains*, Mathematical Surveys and Monographs, Amer. Math. Soc., Providence, Rhode Islands, (to appear) – coauthor: L. Salce

PAPERS

182. Simply presented torsion modules over valuation domains, *Proceedings of Conference on Abelian Groups and Modules, Colorado Springs, Lecture Notes Pure Appl. Math.*, N. 182, Marcel Dekker (1996), 23–44.

183. Primary decompositions over domains, *Glasgow Math. J.* 38 (1996), 321–326 – coauthor: S.B. Lee

184. Primary decompositions of torsion modules over domains, *Mathematika* 44 (1997), 88–99 – coauthor: S.B. Lee

185. Extensions of compact abelian groups by discrete ones and their duality theory, *J. Algebra* 196 (1997), 578–594 – coauthor: K.H. Hofmann

186. Equivalent presentations of modules over Prüfer domains, *Canad. Math. Bull.* 41 (1998), 151–157 – coauthor: S.B. Lee

187. Note on the transitivity of pure-essential extensions, *Coll. Math.* 78 (1998), 283–291 – coauthors: L. Salce and P. Zanardo

188. Extensions of compact abelian groups by discrete ones and their duality theory II, *Abelian groups, module theory, and topology, Proceedings of the Padova 1997 Conference, Lecture Notes Pure Appl. Math.*, No. 201, Marcel Dekker (1998), 205–225 – coauthor: K.H. Hofmann

189. Valuated Butler groups of special type, *Czech. Math. J.* 49 (124) (1999), 507–516 – coauthor: G. Viljoen

190. On the class semigroups of Prüfer domains, *Abelian Groups and Modules, Proceedings of the Dublin 1998 Conference*, Birkhäuser (1999), 319–326

191. Stacked bases over h-local Prüfer domains, (this Volume) – coauthor: S.B. Lee

192. The fully-invariant-extension property for abelian groups, *Comm. in Algebra* (to appear) – coauthors: G.F. Birkenmeier, G. Călugăreanu and H. P. Goeters

193. A generalization of the Pontryagin-Hill theorem to projective modules, (to appear) – coauthor: J. Macías-Díaz

194. On the global balanced-projective dimension of valuation domains, (to appear) – coauthor: K.M. Rangaswamy

Part II

Survey Articles

Error-correcting Codes as Ideals in Group Rings

A.V. Kelarev and P. Solé

Dedicated to Professor Laszlo Fuchs in honour of his 75th birthday.

ABSTRACT. We present a survey of known results on error-correcting codes which are ideals in group algebras.

1. Introduction

This survey is intended for researchers with background in group and ring theory and some familiarity with the elementary notions of coding as per the first chapter of [26].

All cyclic codes of length n are ideals in the group algebra $F[C_n]$, where C_n is the cyclic group of order n. For several types of well-known cyclic codes it has been shown that they are ideals in group algebras of other groups. This additional algebraic structure helps to find more efficient encoding and decoding algorithms for known codes, and to find new codes in the group algebras. On the other hand, various codes have been originally defined in terms of certain group algebras.

A group-algebra code is a one-sided ideal in a group algebra $F[G]$, where F is a finite field, and G is a finite group. All elements of the field F are letters of the encoding alphabet. The order $n = |G|$ of the group is the length of codewords.

By way of illustration let us include the following easy introductory example. Consider the 1-error-detecting $(4,3)$ binary cyclic code which adds a parity-check digit $c_3 = c_0 + c_1 + c_2$ to each message $m = (c_0, c_1, c_2)$. Denote by g a generator of the cyclic group C_4 in multiplicative notation. If we identify each codeword $c = (c_0, c_1, c_2, c_3)$ with the element $c_0 e + c_1 g + c_2 g^2 + c_3 g^3$ of the group algebra $F_2[C_4]$, then the set of all codewords forms an ideal of $F_2[C_4]$. On the other hand, take the elementary abelian group $G = C_2 \times C_2$. In additive notation $G = \{(0,0), (0,1), (1,0), (1,1)\}$. If we identify the codeword $c = (c_0, c_1, c_2, c_3)$ with the element $c_0(0,0) + c_1(0,1) + c_2(1,0) + c_3(1,1)$ of the group algebra $F_2[G]$, then the same code is embedded into another group algebra, and again it is clear that it forms an ideal in this algebra.

1991 *Mathematics Subject Classification.* Primary: 16P10; Secondary: 94B..

2. Abelian group code theory

Since every finite Abelian group $F[G]$ is a direct product of its primary components, it follows that the group algebra $F[G]$ is a tensor product of the group algebras of these components, and therefore every code in $F[G]$ is a direct sum (i.e., a concatenation) of codes in the group algebras of p-groups, for all prime divisors p of the order of G. Taking this into account, we consider only the case where G is a p-group.

Let K be an extension of F, containing all n-th roots of unity. A character of G is a homomorphism of G into the multiplicative group of a given field K. The set G^* of all characters of G becomes an Abelian group under pointwise multiplication: $(\chi_1 \chi_2)(g) = \chi_1(g)\chi_2(g)$, for $g \in G$ and $\chi_1, \chi_2 \in G^*$. Each element $g \in G$ defines a character g^* of G^* by the rule $g^*(\chi) = \chi(g)$. This gives a homomorphism of G into G^{**}. If we look at every element g as the characteristic function $g : G \to K$ given by

$$g(h) = \begin{cases} 1 & \text{if } g = h, \\ 0 & \text{otherwise}, \end{cases}$$

then each character χ of G can be regarded as an element $\chi = \sum \chi(g) g$ of the group algebra $K[G]$, and in this way extends to a linear function $\chi : F[G] \to K$. Each character χ defines the idempotent

$$e_\chi = \frac{1}{|G|} \sum_{g \in G} \chi(g)^{-1} g$$

in the group algebra $K[G]$.

Let C be an ideal in $F[G]$. The *root set* $R(C)$ is the set of characters χ such that $\chi(C) = 0$. In the special case of cyclic codes, the characters which are roots correspond to the roots of the generator polynomial of the code.

THEOREM 2.1. *Let G be a finite Abelian group, F a finite field with characteristic not dividing $|G|$, and let K be an extension of F, containing all $|G|$-th roots of unity. Then*

(1) *all three groups G, G^* and G^{**} are isomorphic;*
(2) *the homomorphism $g \mapsto g^*$ is an isomorphism of G and G^{**};*
(3) *if H is a subgroup of G, then every character of G/H gives us a character of G, and*

$$(G/H)^* \cong H^0 = \{ \chi \in G^* \mid \chi(g) = 1 \text{ for all } g \in H \}$$

(4) *if we identify G and G^{**}, then $(H^0)^0 = H$;*
(5) *if H is a subgroup of G, then $H^* \cong G^*/H^0$;*
(6) *the characters of G form a basis of $K[G]$;*
(7) *the idempotents e_χ, $\chi \in G^*$, are mutually orthogonal and generate one dimensional ideals in $K[G]$;*
(8) *each idempotent e in $K[G]$ is equal to the sum of all e_χ for all χ with $\chi(e) = 1$;*
(9) *if C is an ideal in $F[G]$, then $e = \sum_{\chi \notin R(C)} e_\chi$ is the idempotent generator of C.*

Thus the identity of the Abelian group algebra $F[G]$ is equal to the sum of central idempotents $e_1, \ldots, e_m$ such that $e_i e_j = 0$ whenever $i \neq j$, and each e_i is a unique nonzero idempotent in the principal ideal it generates. Every ideal of $F[G]$

is generated by the idempotent equal to the sum of all the e_i which belong to this ideal.

There is a nice trace description of codes in $F[G]$ which generalizes the cyclic case [**22**]. Let Tr denote the trace of $F[G]$ downto F. With an ideal I of $F[G]$ we attach a code by the rule

$$C(I) := \{\sum_{g \in G} Tr(X^g a) X^{-g} | \ a \in I\}.$$

In general $I \neq C(I)$. Recall that an ideal is radical if and only if it is equal to its radical. The socle of a ring is the union of its radical ideals. The annihilator of an ideal I in a commutative ring R is defined by

$$\mathrm{ann}(I) = \{x \in R \mid xR = 0\}.$$

THEOREM 2.2. *If I is an ideal in the socle of $K[G]$ then*

$$I = C(K[G]/\mathrm{ann}(I)).$$

3. Examples

3.1. Generalized quadratic residue codes. Let $q = p^m$, where p is an odd prime, m is a positive integer, and let F be a field of characteristic s not equal to p. The generalized quadratic residue codes (or GQR codes) of length q are ideals of the group algebra $F[G]$, where G is the additive group of F_q, defined as follows.

Denote by α a primitive p-th root of unity in the algebraic closure of F. A linear character of G is a homomorphism of G into the multiplicative group of nonzero elements of $F(\alpha)$. If $x \in G$, then put $\mathrm{Tr}(x) = \sum_{i=0}^{r-1} x^{p^i}$. For $g \in G$, define the character χ_g by

$$\chi_g(h) = \alpha^{\mathrm{Tr}(hg)} \text{ for } h \in F_q.$$

The set $\{\chi_g \mid g \in G\}$ is the set of all characters of G.

Let Q be the set of all nonzero squares in F_q, and let N be the set of nonsquares in F_q. The generalized quadratic residue codes $C_Q(F)$ and $C_N(F)$ (or GQR codes) of length q are defined by

$$C_Q(F) = \{c \in F[G] \mid \chi_g(c) = 0 \text{ for all } g \in Q\},$$

$$C_N(F) = \{c \in F[G] \mid \chi_g(c) = 0 \text{ for all } g \in N\}.$$

Boths of these codes have dimension $\frac{1}{2}(q+1)$. If p=q, then these codes are the classical quadratic residue cyclic codes.

For $a \in F_q$, put

$$e(a) = \frac{1}{q} \sum_{g \in G} \alpha^{\mathrm{Tr}(-ag)} g \in F[G].$$

The GQR codes coincide with ideals generated in $F[G]$ by the idempotents

$$\sum_{a \in Q} e(a) \text{ and } \sum_{a \in N} e(a).$$

3.2. Embedding extended cyclic codes into group algebras.

Every cyclic code C of length $n = p^m - 1$ over F_p can be extended by the overall parity-check digit and embedded in the group algebra $F_p[G]$, where G is the additive group of F_{p^m}, as follows.

Choose a primitive element α in F_{p^m}, i.e., a generator of the multiplicative group of nonzero elements of F_{p^m}. Let e be the identity of the group G. Map a codeword $c = (c_0, c_1, \ldots, c_{n-1})$ to the element

$$c \mapsto -\sum_{i=0}^{n-1} c_i e + \sum_{i=0}^{n-1} c_i \alpha^i \in J(F_p[G])$$

of the group algebra $F_p[G]$.

The *defining set* T of C is the set of all integers k such that $0 \leq k < n$ and α^k is a root of the generator polynomial of C.

3.3. Generalized Reed-Muller codes.

Let p be a prime, $q = p^r$. Let ρ be an integer with $0 \leq \rho < m(q-1)$. The generalized Reed-Muller (or GRM) code of order ρ and length q^m over F_q is the linear subspace $\mathrm{GRM}_{F_q}(\rho, m)$ of the space L of all functions $f : F_q^m \to F_q$ spanned by all polynomials in the coordinate variables $x_1, \ldots, x_m$ with degree at most ρ, i.e.,

$$\langle x_1^{i_1} \cdots x_m^{i_m} \mid i_1 + \cdots + i_m \leq \rho \rangle.$$

By $\mathrm{wt}_q(k)$ we denote the number of nonzero digits in the q-ary notation for k.

The following theorem shows that GRM codes over F_p coincide with powers of the Jacobson radical of a group algebra of an elementary abelian p-group (for related results see, for example, [10]).

THEOREM 3.1. *Let G be the direct product of m copies of the cyclic group of order p, and let $0 \leq \rho < m(p-1)$. Then the generalized Reed-Muller code $\mathrm{GRM}_{F_p}(\rho, m)$ is equivalent to the code given by the power $J(F_p[G])^t$ of the radical of the group algebra $F_p[G]$, where $t = m(p-1) - \rho$. It has length p^m, dimension*

$$|\{k \mid 0 \leq k < p^m, \mathrm{wt}_p(k) \leq m(p-1) - t\}|$$

and minimum weight $(b+1)p^a$, where $t = a(p-1) + b$ and $0 \leq b < p-1$. As an extended cyclic code its defining set consists of all numbers k such that $0 \leq k < p^m - 1$ and $\mathrm{wt}_p(k) < t$.

Let F be a field of characteristic p, and let G be an arbitrary group of order p^m. A Jennings basis of G is a sequence $g_1, \ldots, g_m \in G$ such that each element of G can be uniquely written as $g_1^{k_1} \cdots g_m^{k_m}$ with $0 \leq k_i \leq p-1$. Denote by h_i the *height* of g_i, i.e., the minimum number such that $g_i^{p^{h_i}} = e$. We may assume that $1 = h_1 \leq \cdots \leq h_m$. The power $J(F[G])^h$ of the radical $J(F[G])^h$ is spanned by all products $(g_1 - 1)^{t_1} \cdots (g_m - 1)^{t_m}$ such that $0 \leq t_i \leq p-1$ and $\sum_{i=1}^m t_i h_i \geq h$. The product $(g_1 - 1)^{t_1} \cdots (g_m - 1)^{t_m}$ has weight $\prod_{i=1}^m (t_i + 1)$. Ward has proved that they form a *visible* basis of $F[G]$ in the sense that every code spanned by a set B of these products has minimum weight equal to the minimum weight of the products in B.

Descriptions of the Jacobson radical are known also in more general ring constructions. Here we only mention that in [16] (see also [18, 20]) the Jacobson radical of an arbitrary S-graded ring R has been reduced to the radicals of subrings of R graded by subgroups of S, for all locally finite, or completely regular, or algebraic linear semigroups S. We refer to [15, 17] and the survey [19] for descriptions of the radicals of semigroup rings.

3.4. Cauchy codes. Let K be a finite extension of F_q, and let $K[x,y]_m$ denote the set of homogeneous polynomials over K of degree m in commuting variables x and y. Denote by $\overline{F}_q$ the projective line over F_q, i.e., $\overline{F}_q = F_q \cup \{\infty\}$ and put $f(z) = f(z\phi)$, for $z \in \overline{F}_q$, $f \in K[x,y]_m$, where ϕ is the coordinatization of $\overline{F}_q$, i.e.,

$$z\phi = \begin{cases} (z,1) & \text{if } z \in F_q, \\ (1,0) & \text{if } z = \infty. \end{cases}$$

Take any $\alpha = (\alpha_0, \ldots, \alpha_{n-1})$, where $\alpha_0, \ldots, \alpha_{n-1}$ are distinct elements of $\overline{F}_q$, and $v = (v_0, \ldots, v_{n-1})$, where $v_0, \ldots, v_{n-1} \in K \setminus \{0\}$. For $1 \leq k < n$, the *Cauchy code* with *location vector* α, *scaling vector* v, and *location set* $L_\alpha = \{\alpha_0, \ldots, \alpha_{n-1}\}$ is defined by

$$C_k(\alpha, v, q, K) = \{(v_0 f(\alpha_0), \ldots, v_{n-1} f(\alpha_{n-1})) \mid f \in K[x,y]_{k-1}\}.$$

In the special case where the location set is $F_q \setminus \{0\}$, and $v = 1$, then $C_k(\alpha, v, q, K)$ is the Reed-Solomon code; if the location set is F_q and and $v = 1$, then $C_k(\alpha, v, q, K)$ is an extended Reed-Solomon code; and if the location set is a subset of F_q, then $C_k(\alpha, v, q, K)$ is the generalized Reed-Solomon code $GRS_k(\alpha, v)$. Assuming that the polynomial f is nonzero, the code $C_k(\alpha, v, q, K)$ has dimension k and minimum distance $n - k + 1$.

3.5. Bicyclic Codes. A bicyclic code (also called 2-D cyclic code) is an abelian code with respect to the direct product of two cyclic groups of orders, say, n' and n''. If β (resp. γ) denote two elements of respective order n' and n'' in the algebraic closure of F_q then such a code can be defined by the data of a certain subset $\mathcal{A}$ of $[n-1] \times [n-1]$ in the following manner

$$C := \{c(x,y) \mid (j', j'') \in \mathcal{A} \Rightarrow c(\beta^{j'}, \gamma^{j''}) = 0\}.$$

Here $c(x,y)$ is a polynomial representation of a generic codeword of C. Its degree in x (resp. y) is $< n'$ (resp $< n''$.) An important class of bicyclic codes is introduced by specifying

$$\mathcal{A} := \{(j', j'') \mid (j'+1)(j''+1) \leq d\}.$$

These are the so called *hyperbolic codes* of designed distance d. For instance there is such a code over F_8 of parameters $[49, 35, 7]$. These codes can be decoded efficiently by the Sakata Algorithm. See [26, p.1614] for details and references.

4. Non-abelian group codes

If G is non abelian then the ring FG is not commutative and codes will be defined as being one-sided ideals. Representation theory is essential. See [13] for background and undefined terms.

A *representation* T of degree n of a group G over a field F is an homomorphism from G into $GL(n, F)$. The *character* χ *afforded* by T is the map $g \mapsto \det(T(g))$. Let T_i, $i = 1, \cdots, k$ be the irreducible representations of G over K; let n_i be the degree of T_i and χ_i the representation it affords. With each T_i we attach a two-sided ideal of FG generated by the idempotent

$$e_i := \frac{1}{|G|} \sum_{g \in G} \chi_i(\frac{1}{g}) X^g.$$

Thus FG decomposes as a direct sum of minimal two-sided ideals

$$FG := \oplus_{i=1}^{k}(e_i).$$

If $F = L$ is the splitting field of G, then each summand decomposes in turn as a sum of one-sided ideals.

THEOREM 4.1. *If T is an irreducible representation of G over L with attached ideal V, then*

$$V = \oplus_{i=0}^{n-1} W_i,$$

where each W_i is the one-sided ideal generated by

$$\frac{n}{|G|} \sum_{g \in G} T_{ii}(\frac{1}{g}) X^g.$$

4.1. Metabelian codes. Given three integers m, n, r, the metacyclic group $G(m, n, r)$ is defined by the presentation

$$\langle x, y | \ x^m = y^n = 1 \ \& \ yx = x^r y \rangle.$$

Absolutely irreducible representations of such groups are well-understood [13, 333-340] and allow for a straightforward application of the general theory. A binary [125, 20, 44] code is constructed in that way in [38].

4.2. Classical codes. The binary and ternary Golay codes are described as ideals in group rings of non-abelian groups in [21] and extended Golay codes in [4].

4.3. Conclusion. Abelian codes have been studied to some depth and many connections with classical coding theory have been established. Codes in the group algebra of a non-abelian group have been touched upon. In another direction codes which are ideals in the semigroup rings of commutative semigroups have been studied in [5] even in a somewhat more general setting of certain polynomial quotient rings. Without trying to be complete we have included in the bibliography several useful references related to this research direction.

References

[1] S.D. Berman, *On the theory of group codes*, Cybernetics **3** (1967), 25–31.

[2] S.D. Berman and I.I. Grushko, *Code parameters of principal ideals of group algebra of group $(2, 2, \cdots, 2)$ over field of characteristic 2*, Problems Inform. Transmission **14** (1978), 239–246.

[3] S.D. Berman and I.I. Grushko, *Parameters of abelian codes in the group algebra KG of $G = (a) \times (b)$, $a^p = b^p = 1$, p prime, over a finite field K with primitive pth root of unity and related MDS-codes*, "Representation Theory, Group Rings, and Coding Theory", Contemp. Math. **93** (1989), 77–83.

[4] F. Bernhardt, P. Landrock and O. Manz, *The extended Golay codes considered as ideals*, J. Combin. Theory Ser. A **55** (1990), 235–246.

[5] J. Cazaran, A.V. Kelarev, *Generators and weights of polynomial codes*, Arch. Math. (Basel) **69** (1997), 479–486.

[6] P. Charpin, *The Reed-Solomon code as ideals of a modular algebra*, C. R. Acad. Sci. Paris Ser. I Math. **294** (1982), 597–600.

[7] P. Charpin, *Extended cyclic codes and principal ideals of a modular algebra*, C. R. Acad. Sci. Paris Ser. I Math. **295** (1982), 313–315.

[8] P. Charpin, *The extended Reed-Solomon codes considered as ideals of a modular algebra*, Annals Discrete Math. **17** (1983), 171–176.

[9] P. Charpin, *A generalization of Berman's construction of p-ary Reed-Muller codes*, Comm. Algebra **16** (1988), 2231–2246.

[10] P. Charpin, T. Berger, *The automorphism group of generalized Reed-Muller codes*, Discrete Math. **117** (1993), 1–17.

[11] P. Charpin, *Self-dual codes which are principal ideals of the group algebra* $\mathbf{F}_2[\{\mathbf{F}_{2^m}, +\}]$, J. Statist. Plann. Inference **56** (1996), no.1, 79–92.

[12] X. Chen, I.S. Reed, T. Helleseth and T.K. Truong, *Algebraic decoding of cyclic codes: a polynomial ideal point of view*, "Finite Fields: Theory, Applications, and Algorithms", Contemp. Math. **168**, 15–22.

[13] C.W. Curtis, I. Reiner, *"Representation Theory of Finite Groups and Associative Algebras"*, Wiley Interscience, New York, 1962.

[14] V. Drensky and P. Lakatos, *Monomial ideals, group algebras and error correcting codes*, "Applied Algebra, Algebraic Algorithms and Error-Correcting Codes", Lecture Notes in Comput. Sci. **357** (1989), 181–188.

[15] A.V. Kelarev, *On the Jacobson radical of semigroup rings of commutative semigroups*, Math. Proc. Cambridge Philos. Soc. **108** (1990), 429–433.

[16] A.V. Kelarev, *Radicals of graded rings and applications to semigroup rings*, Comm. Algebra **20** (1992), 681–700.

[17] A.V. Kelarev, *The Jacobson radical of commutative semigroup rings*, J. Algebra **150** (1992), 378–387.

[18] A.V. Kelarev, *A general approach to the structure of radicals in some ring constructions*, "Theory of Radicals", Szekszárd, 1991, Coll. Math. Soc. János Bolyai, **61** (1993), 131–144.

[19] A.V. Kelarev, *Radicals of semigroup rings of commutative semigroups*, Semigroup Forum **48** (1994), 1–17.

[20] A.V. Kelarev, *On the structure of the Jacobson radical of graded rings*, Quaestiones Math. **19** (1996), 331–340.

[21] P. Landrock, O.Manz, *Classical codes as ideals in group algebras*, Designs, Codes and Cryptography **2** (1992), 273–285.

[22] P. Langevin, *Weights of abelian codes*, Designs, Codes and Cryptography **14** (1998), 239–245.

[23] F. Laubie, *Ideal codes of some modular algebras and ramification*, Comm. Algebra **15** (1987), 1001–1016.

[24] R. Lidl, G. Pilz, *"Applied Abstract Algebra"*, Springer-Verlag, New York, 1998.

[25] R.L. Miller, *Minimal codes in abelian group algebras*, J. Combin. Theory Ser. A **26** (1979), 166–178.

[26] V.S. Pless, W.C. Huffman, R.A. Brualdi, *"Handbook of Coding Theory"*, Elsevier, New York, 1998.

[27] A. Poli, *Codes dans les algebres de groupes abeliens (codes semi simples, et codes modulaires)*, "Information Theory" (Proc. Internat. CNRS Colloq., Cachan, 1977), Colloq. Internat. CNRS **276** (1978), 261–271.

[28] A. Poli and L. Huguet, *"Error-Correcting Codes: Theory and Applications"*, Prentice-Hall, 1992.

[29] A. Poli and M. Ventou, *Nilpotent principal codes, of maximum dimension, in the $\mathbf{F}_q$ algebra of an elementary abelian p group*, C. R. Acad. Sci. Paris Ser. I Math. **296** (1983), 283–285.

[30] A. Poli and M. Ventou, *Construction de codes autoduaux de profondeur 1 ou 2 dans* $A = F_2[X_1, \ldots, X_n]/(X_1^2 - 1, \ldots, X_n^2 - 1)$, Annals of Discrete Mathematics **17** (1983), 549–557.

[31] P. Rabizzoni, *Binary images of principal ideals of a group algebra*, Rev. CETHEDEC Cahier (1981), no.2, 57–78.

[32] B.S. Rajan and M.U. Siddiqi, *Transform domain characterization of abelian codes*, IEEE Trans. Inform. Theory **38** (1992), 1817–1821.

[33] B.S. Rajan and M.U. Siddiqi, *A generalized DFT for abelian codes over* Z_m, IEEE Trans. Inform. Theory **40** (1994), 2082–2090.

[34] C. Renteria and H. Tapia Recillas, *Some Artin-Schreier codes and ideals*, Congr. Numer. **88** (1992), 51–56.

[35] C. Renteria and H. Tapia Recillas, *Reed-Muller codes: an ideal theory approach*, Comm. Algebra **25** (1997), 401–413.

[36] R.E. Sabin, *On minimum distance bounds for abelian codes*, Appl. Algebra Engrg. Comm. Comput. **3** (1992), 183–197.

[37] R.E. Sabin, *An ideal structure for some quasi-cyclic error-correcting codes*, "Finite Fields, Coding Theory, and Advances in Communications and Computing", Lecture Notes in Pure and Appl. Math. **141** (1993), 183–194.

[38] R.E. Sabin, *On determining all codes in semi-simple group rings*, Lect. Notes in Comp. Sci. **673** (1993), 279–290.

[39] R.E. Sabin, *On row cyclic codes with algebraic structure*, Designs, Codes and Cryptography **4** (1994), no.2, 145–155.

[40] K. Saints and C. Heegard, *Algebraic geometric codes and multidimensional cyclic codes: a unified theory and algorithms for decoding using Grobner bases*, IEEE Trans. Inform. Theory **41** (1995), 1733–1751.

[41] S. Sakata, *Two-dimensional cyclic codes—an algorithm for determining the information bits and its application to codes of even area*, Electron. Comm. Japan **63** (1980), 16–25.

[42] U. Vellbinger, *On the minimum distance of ideals in group algebras*, Acta Math. Inform. Univ. Ostraviensis **4** (1996), 97–103.

[43] P.A. von Kaenel, *Generators of principal left ideals in a noncommutative algebra*, Rocky Mountain J. Math. **11** (1981), 27–30.

[44] H.N. Ward, *Visible codes*, Arch. Math. (Basel) **54** (1990), 307–312.

[45] H.N. Ward, *Quadratic residue codes in their prime*, J. Algebra **150** (1992), 87–100.

[46] S.K. Wasan, *On codes over Z_m*, IEEE Trans. Inform. Theory **28** (1982), 117–120.

[47] J. Wolfmann, *A group algebra construction of binary even self dual codes*, Discrete Math. **65** (1987), 81–89.

[48] J. Wolfmann, *New decoding methods of the Golay code $(24, 12, 8)$*, "Combinatorial Mathematics", North-Holland Math. Stud. **75** (1983), 651–656.

[49] J. Wolfmann, *A new construction of the binary Golay code $(24, 12, 8)$ using a group algebra over a finite field*, Discrete Math. **31** (1980), 337–338.

[50] K.H. Zimmermann, *On generalizations of repeated root cyclic codes*, IEEE Trans. Inform. Theory **42** (1996), 641–649.

DEPARTMENT OF MATHEMATICS, FACULTY OF SCIENCE AND ENGINEERING, UNIVERSITY OF TASMANIA, BOX 252-37, HOBART, TASMANIA 7001, AUSTRALIA
E-mail address: Andrei.Kelarev@utas.edu.au

CENTRE NATIONAL DE LA RECHERCHE SCIENTIFIQUE, LABORATOIRE I3S, ESSI, ROUTE DES COLLES, SOPHIA ANTIPOLIS BP 145, 06 903 FRANCE
E-mail address: ps@essi.fr

Homomorphisms and Duality for Torsion-Free Modules

Bruce Olberding

Dedicated to Professor Laszlo Fuchs in honor of his 75th birthday.

ABSTRACT. In this article we survey and complete some work of the last decade on Warfield duality and Hom and $\otimes$ formulas for torsion-free modules over integral domains. Specific attention is given to the commutative ring theory needed to characterize the rings possessing these formulas. In particular, we touch on stable rings, generalized Dedekind domains, 2-generator rings, strongly discrete and almost maximal Prüfer domains.

Introduction

The title of this article is meant to echo the 1968 paper of R. B. Warfield, Jr., "Homomorphisms and duality for torsion-free abelian groups." This fundamental paper deals with the structure of the abelian groups $G \otimes_{\mathbb{Z}} X$, $Hom_{\mathbb{Z}}(G, X)$, $Hom_{\mathbb{Z}}(X, G)$ and $Hom_{\mathbb{Z}}(\bigwedge^{n-k} G, \bigwedge^n G)$ where G and X are finite rank torsion-free groups with X a rank one group. Warfield settles definitively when a group is isomorphic to one of the above Hom or tensor groups for particular G and X, thus proving a kind of representation theorem for finite rank torsion-free abelian groups.

Warfield approached this result in an interesting way, fixing a rank one group X and studying the functors $Hom_{\mathbb{Z}}(X, -)$, $Hom_{\mathbb{Z}}(-, X)$ and $X \otimes_{\mathbb{Z}} -$. These functors induce a category equivalence and a duality on appropriate full subcategories of the category of finite rank torsion-free abelian groups. From these considerations, he deduces what has ever since been known as "Warfield duality" and an important category equivalence that pairs $Hom_{\mathbb{Z}}(X, -)$ and $X \otimes_{\mathbb{Z}} -$. A variation on these ideas, via exterior powers, yields the representation theorems for groups.

This functorial approach still resonates in the literature. The actual method and devices of Warfield's proofs, however, have become somewhat divorced from more recent studies of these and related functors. One reason for this is that the newer techniques better accomodate generalization of Warfield's results to modules over Dedekind domains, and, suggestively, to modules over much larger classes of integral domains. It is this latter direction of research I wish to review here, as well as fill in some of the gaps in the theory that remain.

1991 *Mathematics Subject Classification.* Primary 20K15, 13G05.

© 2001 American Mathematical Society

The motivation is two-fold. First, the study in this larger context of the various *Hom* and tensor groups that arise in Warfield's paper sheds some light on the complicated structure of torsion-free modules over integral domains. Second, in placing these functors in the setting of modules, one encounters interesting and rather canonical classes of integral domains. To name a few: 2-generator rings, stable rings, generalized Dedekind domains, strongly discrete Prüfer domains and almost maximal Prüfer domains. From a more philosophical point of view, by seeking the validity of Warfield's results for modules over integral domains, one is able to thematize otherwise disparate classes of commutative rings. A priori unrelated classes of rings prove to be unified by properties of their torsion-free modules.

Let R be an integral domain with quotient field Q and let G and H be torsion-free modules. The *rank* of G is the dimension of the Q-vector space $Q \otimes_R G$. We say that H *dominates* G if the canonical homomorphism,

$$Hom_R(G, H) \otimes_R Q \to Hom_R(G, H \otimes_R Q)$$

is an isomorphism. If G and H are both finite rank then H dominates G if and only if $OT(G) \leq IT(H)$, where the inner and outer types of a module are defined as in [**Re**]. A rank one module Y dominates a torsion-free module G of finite rank n if and only if G embeds in a direct sum of n copies of Y. Similarly, G dominates Y if and only if a direct sum of n copies of Y embeds in G. These assertions are well-known for abelian groups and can be found in [**W**]. That the first statement holds for modules was observed in [**BS**, Lemma 3.1]; the second statement can be proved by a dual argument.

The *coefficient ring* of a torsion-free module G is the ring,

$$R(G) = \{q \in Q : qG \subseteq G\}.$$

Sometimes we will regard G as an $R(G)$-module rather than an R-module, especially when flatness is at issue. If $n = \text{rank}(G)$ and $\bigwedge^n G$, the nth exterior power of G, is a torsion-free R-module, then we set $R_1(G) = R(\bigwedge^n G)$. This agrees with Warfield's definition of $R_1(G)$; we must, however, add the caveat that $\bigwedge^n G$ be torsion-free in order to guarantee $R_1(G)$ is well-defined.

We are interested in the four properties described below; the import of Warfield's paper is that all of them hold for $R = \mathbb{Z}$. A 1989 paper of E. L. Lady, using the methods of locally free modules, shows the first three properties hold for Dedekind domains [**La**]. Results in the following sections will show all four properties prove to hold in significantly broader contexts. We've rephrased some of the properties slightly to account for subtleties involving flatness and endomorphism rings. When R is Dedekind, however, our formulation is equivalent to Warfield's.

(**Warfield Domain**): For all torsion-free finite rank modules G, if Y is a rank one module, then the canonical homomorphism

$$G \to Hom_R(Hom_R(G, Y), Y)$$

is an isomorphism if and only if Y dominates G and $R(Y) \subseteq R(G)$.

(**TH**): For all torsion-free modules G, if X is a rank one module, then the canonical homomorphism

$$X \otimes_{R(X)} Hom_{R(X)}(X, G) \to G$$

is an isomorphism if and only if G dominates X and $R(X) \subseteq R(G)$

(HT): For all torsion-free modules G, if X is a rank one module, then the canonical homomorphism
$$G \to Hom_{R(X)}(X, X \otimes_{R(X)} G)$$
is an isomorphism if and only if $R(X) \subseteq R(G)$

(EP): For each torsion-free module G of finite rank n, the canonical homomorphism
$$\bigwedge^k G \to Hom_R(\bigwedge^{n-k} G, \bigwedge^n G)$$
is an isomorphism for $0 < k < n$ if and only if $R(G) = R_1(G)$.

Each property is formulated as a biconditional and it is not hard to see that the implication "$\Rightarrow$" is necessary if the given homomorphism is to hold, with a possible exception of (EP) that will be discussed presently. Thus the force of the assertion that one of these properties holds for an integral domain is that the necessary condition is indeed sufficient for the homomorphism to be an isomorphism.

Property (EP) presents several difficulties; it is not even clear how to state it in our general context. The main problem lies with the definition of $R_1(G)$, since $\bigwedge^k G$ need not be torsion-free. In such a case it is not appropriate to consider elements of Q as acting on $\bigwedge^k G$. We skirt this complication by dealing with (EP) only for *Prüfer domains*, those integral domains for which every finitely generated ideal is invertible. Every torsion-free module G over a Prüfer domain is flat and it follows that $\bigwedge^k G$ is also torsion-free (see Section 5). Thus, for Prüfer domains, we do get that $R(G) = R_1(G)$ is a necessary condition for the canonical homomorphism in (EP) to be an isomorphism.[1]

Only Warfield domains are designated here with a "proper" name; the other three properties are referenced by a temporary shorthand because these properties prove to collapse to previously named classes of integral domains. Thus the acronym (TH) (which stands for Tensor-Hom) is only needed provisionally, as are the acronyms (HT) (= Hom-Tensor) and (EP) (= Exterior Power). In [**G2**], property (TH) was termed "solvable" in analogy with U. Albrecht's solvability condition for abelian groups. Since we will show that the terminology is redundant, that solvability is simply the ideal-theoretic property of "stability" in another guise, we have opted here for (TH) for the sake of symmetry with property (HT). Property (HT) proves to coincide with the notion of a "strongly faithful" domain, a class of rings first introduced in [**GO1**]. Property (EP), as discussed above, is a different matter: we only study it over Prüfer domains and in this case we show that it coincides with (HT). As a consequence, in Theorem 5.4, we obtain: *a Prüfer domain has (EP) if and only if R is a "generalized Dedekind domain."*

Much of the commutative ring theory needed to characterize these four properties has been developed over the past decade. Warfield domains were the first to be studied. J. D. Reid, in his article [**Re**], presented an approach to Warfield duality that freed arguments from PID-based assumptions. He reduced Warfield duality to two properties, one multiplicative in nature, the other homological, and concluded that Warfield duality holds for Dedekind domains. H. P. Goeters, in [**G1**], further investigated these properties, showing that the homological requirement could be translated into a statement about torsion-free *Ext*. By analyzing the Noetherian

---

[1] Perhaps a better definition for $R_1(G)$ in more general contexts would be $R_1(G) = End_R(\bigwedge^n G)$.

case, he was able to link Warfield duality to the extensive literature on reflexive domains (defined in Section 6), and in particular to the 2-generator property for ideals when the ring has module-finite integral closure. In so doing, he found a source of numerous examples of Noetherian Warfield domains that are not Dedekind. In their 1996 paper, [**BS**], S. Bazzoni and L. Salce fully articulated this link to reflexive rings by proving that an integral domain R is a Warfield domain if and only if every overring of R is a reflexive domain. Using this result, they completely describe both the Noetherian and integrally closed classes of Warfield domains. The remaining case, that of non-Noetherian non-integrally closed Warfield domains, was settled in [**O8**] and there is now a complete classification of Warfield domains. These results are summarized in Section 6.

A complete classification of (TH) domains has also been obtained, though this fact has not been explicitly stated in the literature. In Section 2, we collect results from the theory of stable rings to show that the class of (TH) domains is precisely the class of stable domains.

Property (HT) has been characterized in the Noetherian case, where it proves to coincide with (TH) [**GO1**, Main Theorem]. In Section 3, we characterize (HT) for arbitrary domains. Although in the arbitrary case (HT) does not coincide with (TH), we are still able to give a complete description of (HT) domains using the ideal-theoretic analysis of Section 4.

Once all these properties have been characterized, the relationship between them is easily deduced: (HT) is equivalent to (EP) in the Prüfer case and for arbitrary integral domains,

$$\text{Dedekind domain} \Rightarrow \text{Warfield domain} \Rightarrow \text{(TH)} \Rightarrow \text{(HT)}.$$

Examples show that none of these implications can be reversed (see Section 6).

Notation and terminology. Throughout this article, R represents a commutative integral domain; Q is reserved for its quotient field. The integral closure of R in Q is denoted $\bar{R}$. A *quasilocal domain* is an integral domain with a unique maximal ideal; a *local domain* is a Noetherian quasilocal domain. A domain R has *finite character* if every non-zero ideal of R is contained in at most finitely many maximal ideals of R. An R-module G is *locally free* if $G_M = G \otimes_R R_M$ is a free R_M-module for all maximal ideals M of R. A torsion-free module G is a *faithful* R-module if $GM \neq G$ for all maximal ideals M of R. Thus a torsion-free module G is a faithfully flat R-module if and only it is faithful and flat as an R-module. If X is a submodule of Q and G is a torsion-free R-module, then XG denotes the image of $X \otimes G$ in QG, the divisible hull of G.

1. Multiplicative properties of rank one modules

In this section we collect some preliminary results on the canonical homomorphisms that occur in properties (TH) and (HT). In [**GO2**] it is shown that these mappings can be translated into multiplicative properties of rank one modules. This reduction to the rank one case is what provides the link between module theory and the commutative ring theory needed to characterize (HT), (EP) and (TH).

By way of orientation, we first reformulate the canonical mappings in category-theoretic terms. Fix an integral domain R with quotient field Q. Let X be a submodule of Q. Define $\mathcal{E}_X$ to be the category of torsion-free $R(X)$-modules, and let $\mathcal{D}_X$ be the category of torsion-free $R(X)$-modules that dominate X. Define a

pair of functors
$$H_X : \mathcal{D}_X \to \mathcal{E}_X$$
and
$$T_X : \mathcal{E}_X \to \mathcal{D}_X$$
by $H_X(G) = Hom_{R(X)}(X, G)$ for all $G \in \mathcal{D}_X$ and $T_X(H) = X \otimes_{R(X)} H$ for all $H \in \mathcal{E}_X$.

Properties (HT) and (TH) can now be encoded into categorical language.

(HT): For all submodules X of Q, $H_X T_X \cong 1_{\mathcal{E}_X}$.
(TH): For all submodules X of Q, $T_X H_X \cong 1_{\mathcal{D}_X}$.

The behavior of the functors H_X and T_X depends, of course, on X. In order to clarify this dependence, the following terminology was introduced in [**GO2**]. If R is an integral domain and X is a submodule of Q, then X is a *cancellation module* for R if for all submodules Y and W of Q such that $XY = XW$, it must be that $Y = W$. In the case that X is an ideal of R, then the notion of cancellation module coincides with the classical notion of a cancellation ideal. This is a consequence of the following proposition.

PROPOSITION 1.1. [**GO2**, Theorem 2.3] *Let R be an integral domain. The following are equivalent for a submodule X of the quotient field of R.*

(1) *X is a cancellation module for R.*
(2) *X is a locally free R-module.*
(3) *X is a faithfully flat R-module.*
(4) *The canonical homomorphism $G \to Hom_R(X, XG)$ is an isomorphism for all torsion-free R-modules G.*

If X is a cancellation module for $R(X)$, then combining (3) and (4) of Proposition 1.1 shows that $H_X T_X \cong 1_{\mathcal{E}_X}$. The converse, that $H_X T_X \cong 1_{\mathcal{E}_X}$ implies X is a cancellation module for $R(X)$, is probably not true (flatness is the obstacle), although I know of no counterexample.

We say a submodule X of the quotient field Q of an integral domain R is a *divisor module* for R provided that for each submodule Y of Q containing X, there exists a submodule W of Q such that $XW = Y$.

PROPOSITION 1.2. [**GO2**, Corollary 3.4] *If R is an integral domain and X is a submodule of Q, then X is a divisor module for $R(X)$ if and only if, for all torsion-free modules H that dominate X, the canonical homomorphism*
$$X \otimes_{R(X)} Hom_{R(X)}(X, H) \to H$$
is an isomorphism.

Proposition 1.2 can be rephrased as: X is a divisor module for R if and only if $T_X H_X \cong 1_{\mathcal{D}_X}$.

There is a close relationship between divisor modules and cancellation modules. Lemma 1.3 implies these two concepts are the same when the coefficient ring of the module is quasilocal.

LEMMA 1.3. [**GO2**, Theorem 3.9] *Let R be an integral domain of finite character with quotient field Q. If X is a submodule of Q, then X is a cancellation module for R if and only if X is a divisor module for R and $R(X) = R$.*

PROPOSITION 1.4. *Let R be an integral domain of finite character and let X be a submodule of the quotient field of R such that $R(X) = R$. Then the functors $H_X : \mathcal{D}_X \to \mathcal{E}_X$ and $T_X : \mathcal{E}_X \to \mathcal{D}_X$ induce a category equivalence if and only if X is a locally free R-module.*

PROOF. Apply Lemma 1.3, Proposition 1.2 and Proposition 1.1, and use the fact that locally free modules are flat. □

2. Property (TH)

The commutative ring theory needed to characterize (TH) has already been developed elsewhere, so the proofs in this section are mainly editorial in nature. The key notion is that of a *stable ideal*. If R is an integral domain and I is an ideal of R, then I is stable if it is a projective as a module over $End_R(I)$. Thus stability is equivalent to asserting that I is an invertible ideal of $R(I)$, the coefficient ring of I. The domain R is *stable* if every non-zero ideal of R is stable. It is easy to see that Dedekind domains are stable. The converse, however, fails is a strong way: stable domains need not be Noetherian, one-dimensional or integrally closed (see [**O7**]).

We show in this section that a domain R satisfies (TH) if and only if R is stable. In keeping with the focus of this article, we omit a discussion of the more technical properties of stable ideals and their origins. For details, see the survey article [**O5**].

It is easy to see that if a divisor module is an ideal, then it is stable. By Lemma 1.2, an integral domain R has (TH) if and only if each submodule X of Q is a divisor module for $R(X)$. It follows at once that (TH) domains are stable. That the converse is true is more surprising, since in general it is hard to obtain information about rank one modules of integral domains solely from ideal-theoretic conditions.

LEMMA 2.1. *Let R be a stable integral domain. If X is a non-zero submodule of Q, then X is a cancellation module for $R(X)$.*

PROOF. In [**O6**, Theorem 5.4] it is shown that if R is a stable domain and X is a submodule of Q, then X is a locally free $R(X)$-module. Hence the claim follows from Proposition 1.1. □

LEMMA 2.2. [**O6**, Theorem 3.3] *An integral domain R is stable if and only if R has finite character and for all maximal ideals M of R, R_M is a stable domain.*

THEOREM 2.3. *An integral domain R has (TH) if and only if R is stable.*

PROOF. Assume R is stable. Then by Lemma 2.1, each submodule X of the quotient field Q is a cancellation module for $R(X)$. Let X be a submodules of Q. Every overring of R is stable [**O6**, Theorem 5.1], so we assume without loss of generality that $R(X) = R$. By Lemmas 1.3 and Lemma 2.2, X is a divisor module for R. By Proposition 1.2, $X \otimes_R Hom_R(X, H) \to H$ is an isomorphism for all torsion-free R-modules H that dominate X. By Proposition 1.1, we have also that $G \to Hom_R(X, XG)$ is an isomorphism for all torsion-free R-modules G. Finally, by Lemma 2.1, X is flat as an R-module and hence $X \otimes_R G$ can be identified with XG. Thus the pair of functors induces a category equivalence. Conversely, Proposition 1.2 implies that each submodule X of Q is a divisor module for $R(X)$. In particular, if I is an ideal of R, then there exists a submodule J of Q such that $IJ = R(I)$, and hence I is stable. □

COROLLARY 2.4. *An integral domain R is stable if and only if for all submodules X of Q, the pair of functors $H_X : \mathcal{D}_X \to \mathcal{E}_X$ and $T_X : \mathcal{E}_X \to \mathcal{D}_X$ induces a category equivalence between $\mathcal{D}_X$ and $\mathcal{E}_X$.*

PROOF. Apply Theorem 2.3, Lemma 2.1, Lemma 1.3 and the fact that since overrings of a stable domain are stable [**O6**, Theorem 5.1], every overring of a stable domain has finite character (Lemma 2.2). □

3. Property (HT)

In this section we prove that the domains satisfying (HT) are precisely the strongly faithful domains, a class of integral domains introduced in [**GO1**] and characterized in [**GO3**].

An integral domain R is *strongly faithful* if and only if for all torsion-free modules G, if X is a rank one module and $R(X) \subseteq R(G)$, then the canonical homomorphism,

$$G \to Hom_R(X, XG)$$

is an isomorphism. Comparison with the defining criterion of (HT) domains shows that strong faithfulness ostensibly differs from (HT) only with respect to flatness, and indeed the proof of Theorem 3.2, where (HT) is shown to be equivalent to strong faithfulness, does turn on this issue of flatness. By Proposition 1.1, R is strongly faithful if and only if X is a cancellation module of $R(X)$ for all submodules X of Q. This reduction to the rank one case helps make characterizations of strongly faithful domains tractable. Moreover, by Lemma 2.1, it follows that stable domains, and hence (TH) domains, are strongly faithful.

In order to establish the identification of strong faithfulness with (HT), we appeal first to a technical lemma devoted to torsion modules over Prüfer domains. The proof of this lemma relies on a valuation-theoretic notion of discreteness. A *discrete rank one valuation ring (DVR)* is a valuation domain that has value group isomorphic to $\mathbb{Z}$; equivalently, R is a valuation domain of Krull dimension one such that its maximal ideal M is not idempotent, i.e. $M \neq M^2$. (There is a subtlety here: in multiplicative ideal theory, DVR entails something stronger than "discrete valuation ring." DVR requires the ring be of Krull dimension one, or what is the same, rank one; "discrete valuation ring" does not. See [**Gi**].) This observation motivates the following definition: A Prüfer domain is *strongly discrete* if each non-zero prime ideal P is non-idempotent, that is, $P \neq P^2$. A quasilocal Prüfer domain, i.e. a valuation domain, is strongly discrete if and only if it is stable. This implies that for each non–zero prime ideal P of a strongly discrete valuation domain R, $P \cong R_P$ [**BS**, Proposition 7.6], and if R has finite Krull dimension n, then the value group of G is a free abelian group of rank n [**FHuP**, Corollary 5.3.4].

LEMMA 3.1. *Suppose R is an integral domain such that $\bar{R}$ is a strongly discrete Prüfer domain. If T is a non-zero torsion R-module, then there exists $t \in T$ such that $P = Ann_R(t)$ for some non-zero prime ideal P of R.*

PROOF. Consider the class $\mathcal{F}$ of proper ideals A of $\bar{R}$ with $A = \cup_\alpha Ann_{\bar{R}}(x_\alpha)$ for some set of non-zero elements x_α of $\bar{R} \otimes_R T$ of the form $x_\alpha = 1 \otimes_R t_\alpha$, where $t_\alpha \in T$. Clearly $\mathcal{F}$ is non-empty, for if $0 = 1 \otimes_R t$ in $\bar{R} \otimes_R T$ for some $0 \neq t \in T$, then $rt = 0$ for some $r \in R$ such that $r^{-1} \in \bar{R}$. Since r^{-1} is then integral over R, this forces $r^{-1} \in R$ and hence r is a unit in R and $t = 0$, a contradiction. Also, every chain of ideals in $\mathcal{F}$ has a maximal element in $\mathcal{F}$, so by Zorn's Lemma, there exists a maximal element, say L, of $\mathcal{F}$. We claim L is a prime ideal of $\bar{R}$. Write $W = \{t \in T : Ann_{\bar{R}}(1 \otimes_R t) \subseteq L\}$. Suppose $r, s \in \bar{R}$ and $rs \in L$. If $s(1 \otimes_R y) = 0$ for some $y \in W$, then $s \in Ann_{\bar{R}}(1 \otimes_R y) \subseteq L$. Otherwise, $s(1 \otimes_R y) \neq 0$ for all $y \in W$. Define $L' = \cup_{y \in W} Ann_{\bar{R}}(1 \otimes_R sy)$. We claim that L' is an ideal of $\bar{R}$. If $a, b \in L'$, then $a(1 \otimes_R sy) = 0$ and $b(1 \otimes_R sy') = 0$ for some $y, y' \in W$. Since $y \in W$, $as \in Ann_{\bar{R}}(1 \otimes_R y) \subseteq L$. Similarly, since $y' \in W$, $bs \in L$. Thus $as + bs \in L$ and there exists $w \in W$ such that $(as + bs)(1 \otimes_R w) = 0$. Thus $a + b \in Ann_{\bar{R}}(1 \otimes_R sw) \subseteq L'$ and L' is an ideal of R. By the assumption that $1 \otimes_R sy \neq 0$ for all $y \in W$, it follows that $L' \in \mathcal{F}$. Since $L \subseteq L'$, the maximality of L in $\mathcal{F}$ implies $L = L'$. Thus $r \in Ann_{\bar{R}}(1 \otimes sx) \subseteq L' = L$ and hence L is prime. Now since $\bar{R}$ is a strongly discrete Prüfer domain, $L_L = \bar{R}_L a$ for some $a \in L$. Since $a \in L$, $a \in K := Ann_{\bar{R}}(1 \otimes_R x)$ for some $x \in T$ with $K \subseteq L$. Thus $L_L = K_L$ and hence if $T' := (\bar{R}_L \otimes_R T)$, $L_L = Ann_{\bar{R}_L}(1 \otimes_R x)$, with $1 \otimes_R x \in T'$. Thus $1 \otimes rx = 0$ in T' for all $r \in L$, and so $L \subseteq K$, proving $K = L$. Set $P = L \cap R$. Then $P = Ann_R(x)$ and the claim is proved. □

THEOREM 3.2. *An integral domain R has (HT) if and only if R is strongly faithful.*

PROOF. If R is strongly faithful, then the claim follows from the fact that each rank one module X is a flat $R(X)$-module. To prove the converse, it suffices to show each submodule X of the quotient field of R is a flat $R(X)$-module. For then $X \otimes_{R(X)} G \cong XG$ for all torsion-free $R(X)$-modules G, and the claim follows from the definition of strong faithfulness. The proof of this is in three steps.

(i) *$\bar{R}$ is a strongly discrete Prüfer domain.* It is shown in [**O4**, Lemma 2.2] that an integral domain D is a strongly discrete Prüfer domain if and only if for every non-zero prime ideal P of D, P_P is a principal ideal of D_P. We will use this to prove (i). Let P be a prime ideal of $\bar{R}$ and set $S = \bar{R}_P$ and $N = P_P$. We show first that N is a flat ideal of S. To do this it suffices to show $N \otimes_S I$ is torsion-free for all finitely generated ideals I of S. Let I be a finitely generated ideal of S. Observe that since S is integrally closed and I is finitely generated, $R(I) = S$. Set $T = Tor_S(I, S/N)$ and consider the exact sequence,

$$0 \to T \to I \otimes_S N \to IN \to 0.$$

This yields another exact sequence,

$$0 \to Hom_S(I, T) \to Hom_S(I, I \otimes_S N).$$

By (HT), $Hom_S(I, I \otimes_S N) \cong N$. Thus the torsion S-module, $Hom_S(I, T)$, must be the trivial module, and so $Hom_S(I/IN, T) \cong Hom_S(I, T) = 0$. But T is an S/N-vector space and, by Nakayama's Lemma, I/IN is a non-zero S/N-vector space. Hence $T = 0$ and $I \otimes_S N \cong IN$. We conclude N is a flat S-module. In particular, $N \otimes_S N \cong N^2$ and it follows from (HT) that $N^2 \neq N$. Hence, N is a faithfully flat S-module. By Proposition 1.1, N is a cancellation module of S, and since S is quasilocal, this implies N is a principal ideal of S and (i) follows.

(ii) *If $P \neq 0$ is a prime ideal of R and $R(P_P) = R_P$, then P_P is a principal ideal of R_P.* We show first that P_P is a flat ideal of R_P. Set $E = R_P$ and $N = P_P$. Let A be an ideal of E and set $T = Tor_E(A, E/N)$. By assumption $Hom_E(N, N \otimes_E A) \cong A$. There is an exact sequence,

$$0 \to T \to A \otimes_E N \to AN \to 0,$$

which induces an exact sequence,
$$0 \to Hom_E(N,T) \to Hom_E(N, N \otimes_E A).$$

The last entry in the latter sequence is torsion-free, since it is isomorphic to A. Thus the torsion module $Hom_E(N,T)$ must be the trivial module. But T is an E/N-vector space. Moreover, N/N^2 is also an E/N-vector space. Since $Hom_E(N/N^2, T) \cong Hom_E(N,T) = 0$, this forces $N = N^2$ or $T = 0$. The former case is impossible. For if $N = N^2$, then $\bar{E}N = \bar{E}N^2$, contrary to (i), since $\bar{E}$, as an overring of $\bar{R}$, is a strongly discrete Prüfer domain. Thus N is a flat E-module. In the course of verifying flatness, we showed that $N^2 \neq N$ and thus that N is faithfully flat as an E-module. By Proposition 1.1, $N = P_P$ must be a principal ideal of $E = R_P$.

(iii) *If X is submodule of Q, then X is a flat $R(X)$-module.* Since every overring of R inherits our hypotheses we will assume without loss of generality that $R(X) = R$. To show X is flat over R, we show that $X \otimes_R J \cong XJ$ for all ideals J of R. Let T be the kernel of the mapping $X \otimes_R J \to XJ$, and observe T is a torsion module. We will show that $T = 0$. There is an exact sequence,
$$0 \to Hom_R(X,T) \to Hom_R(X, X \otimes_R J).$$

By assumption, $Hom_R(X, X \otimes J) \cong J$, so the torsion module $Hom_R(X,T)$ must be the trivial module. By Lemma 3.1, there exists a non-zero prime ideal P and a non-zero torsion element $t \in T$ such that $P = Ann_R(t)$. We claim that $Hom_R(X,T) \otimes R_P \cong Hom_R(X_P, T_P)$. There is a commutative diagram (with tensor products taken over R),

$$\begin{array}{ccccccc}
0 & \to & Hom_R(X,T) \otimes R_P & \to & Hom_R(X, X \otimes J) \otimes R_P & \to & C \otimes R_P \\
& & \downarrow & & \downarrow & & \downarrow \\
0 & \to & Hom_R(X, T_P) & \to & Hom_R(X, X \otimes J_P) & \to & Hom_R(X, XJ_P)
\end{array}$$

where C is the image of $Hom_R(X, X \otimes J)$ in $Hom_R(X, XJ)$. By (HT) and the flatness of R_P as an R-module, the middle map is an isomorphism. Also, since $Hom_R(X, XJ_P)$ is torsion-free and C is torsion-free of rank one, it follows that the right vertical mapping is an embedding and hence, by the Snake Lemma, $Hom_R(X,T) \otimes_R R_P \cong Hom_R(X, T_P)$, as claimed.

Since P was chosen to be the annihilator of an element of T, T_P contains a simple R_P-module which we will denote by U.

Now $Hom_R(X_P, U) \hookrightarrow Hom_R(X_P, T_P) = 0$. Also, $PX_P \neq X_P$, for otherwise $R(P_P) \subseteq R(X_P)$ and by (HT), $R(X_P) \cong R(X) \otimes_R R_P \cong R_P$. In this case, $R(P_P) = R_P$ and, by (ii), P_P is a principal ideal of R_P. Since $R(X_P) = R_P$, this forces $PX_P \neq X_P$. Thus $Hom_R(X_P/XP_P, U) = Hom_R(X_P/XP_P, R_P/P_P) \neq 0$, contrary to assumption. We conclude $T = 0$ and $X \otimes_R J \cong XJ$. This establishes the flatness of X as an R-module and the claim is proved. □

4. Ideal-theoretic analysis of (TH) and (HT)

In Sections 2 and 3, (TH) and (HT) were linked to stable and strongly faithful domains. The ideal theory of these classes of domains can be extensively described, and in this section we restate some results from [**GO3, O6, O7**] in the present context. For a recent survey of the relevant commutative ring theory, see [**O5**].

In light of Theorems 2.3 and 3.2, stability is interchangeable with (TH), as is strong faithfulness with (HT). The results in this section are quoted from studies of stability and strong faithfulness, but in stating these results, we retain the now redundant terminology of (HT) and (TH) for the reader's convenience.

THEOREM 4.1. ([**O7**, Theorem 2.3] and [**GO3**, Theorem 3.11]) *Consider the following axioms for an integral domain R.*

(i) *Finitely generated ideals of R are stable.*
(ii) *Prime ideals of R are stable.*
(iii) *For all non-zero non-maximal prime ideals P of R, R_P is a valuation domain.*
(iv) *R has finite character.*

Then

(1) *R has (HT) if and only if R satisfies (i) - (iii).*
(2) *R has (TH) if and only if R satisfies (i) - (iv).*

An immediate consequence of Theorem 4.1 is that (HT) and (TH) coincide for finite character domains. For Noetherian domains, each of these conditions implies finite character and hence this identification can be maintained. The integrally closed case is more complicated and there is significant distance between (HT) and (TH). Theorems 4.2 and 4.3 will make this more precise.

The integral closure of a strongly faithful domain is a Prüfer domain. In [**GO3**, Theorem 3.7], it is shown that the integral closure of a strongly faithful domain is in fact a *generalized Dedekind domain*, a Prüfer domain for which every non-zero prime ideal is stable. There are a number of interesting characterizations of generalized Dedekind domains; the definition we have given here is actually a characterization from [**O1**, Theorem 4.7] and [**Ga**, Theorem 5], and not the original definition given by N. Popescu in [**P**]. For more on this class of domains, see Chapter 5 of the monograph [**FHuP**].

The next theorem describes the Noetherian rings with (HT) and (TH). These properties are equivalent to stability, and we have chosen D. Rush's characterization of Noetherian stable domains to illuminate this case.

THEOREM 4.2. (Noetherian Case) *Let R be a Noetherian domain. Then (HT) and (TH) are both equivalent to the property: R has Krull dimension one, each R-submodule of $\bar{R}$ containing R is a ring and there are most two maximal ideals of $\bar{R}$ lying over each maximal ideal of R.*

PROOF. Apply [**GO1**, Main Theorem] and [**R1**, Theorem 2.4]. □

THEOREM 4.3. (Integrally Closed Case) *Let R be an integral domain.*

(i) *R is an integrally closed domain with (TH) if and only if R is a strongly discrete Prüfer domain of finite character.*
(ii) *R is an integrally closed domain with (HT) if and only if R is a generalized Dedekind domain.*

PROOF. The proof of (i) can be found in [**O1**, Theorem 4.6]. Statement (ii) is proved in [**GO3**, Theorem 3.7]. □

The following two theorems give the technical specifications for (TH) and (HT) rings. They can be considered classifications of (TH) and (HT) in that they describe how all (TH) and (HT) rings are assembled from one-dimensional quasilocal stable domains and strongly discrete valuation domains. The ideal theory of the latter rings being very well-understood (see Section 5.3 of [**FHuP**], for example), the only gap in the classification is in the theory of one-dimensional quasilocal stable domains. This gap is rather small: in most cases, quasilocal one-dimensional stable domains are Noetherian and hence there is an extensive literature on their structure. There are, however, non-Noetherian one-dimensional stable domains, but results in [**O7**] show such rings must meet a number of peculiar conditions. See [**O7**, Section 5] for the construction of an example.

THEOREM 4.4. [**O7**, Corollary 2.7] *An integral domain R has (TH) if and only if R has finite character and for all $M \in Max(R)$,*
  (i) *R_M is a stable domain of Krull dimension at most one,*
  (ii) *R_M is a strongly discrete valuation domain, or*
  (iii) *R_M arises in a pullback diagram of the form*

$$\begin{array}{ccc} R = \nu^{-1}(A) & \to & A \\ \downarrow & & \downarrow \\ B & \overset{\nu}{\to} & k \end{array}$$

where (B, M) is a quasilocal domain satisfying (i), $k = B/M$, $\nu : B \to k$ is the canonical projection of B onto k, and A is a ring that satisfies (ii) and has quotient field k.

The prime spectrum, $Spec(R)$, of an integral domain R is *Noetherian* if it is Noetherian as a topological space endowed with the Zariski topology. Thus $Spec(R)$ is Noetherian if and only if the prime ideals of R satisfy ACC. It is shown in [**GO3**, Theorem 3.10] that: *A domain R is strongly faithful if and only if $Spec(R)$ is Noetherian and for every maximal ideal M of R, R_M is a stable domain.* Combining this characterization with Theorem 4.4 yields the classification of (HT) domains:

THEOREM 4.5. *An integral domain R has (HT) if and only if $Spec(R)$ is a Noetherian space and for all $M \in Max(R)$, R_M satisfies (i), (ii) or (iii) of Theorem 4.4.*

Generalized Dedekind domains approximate the behavior of Dedekind domains in some ways– maximal ideals are invertible, for example– but are quite different in many others, especially with respect to the spectrum of prime ideals. This variance is well demonstrated by Proposition 4.6. A *tree* is a partially ordered set $(X, \leq)$ such that if $x, y \leq w$ for $x, y, w \in X$, then $x \leq y$ or $y \leq x$. A tree is *Noetherian* if every ascending chain is eventually stationary.

PROPOSITION 4.6. (Existence of (HT) and (TH) domains) *Let $(X, \leq)$ be a partially ordered set with least element x_0.*
  (i) *There exists an (HT) domain R such that $Spec(R)$ is order-isomorphic to $(X, \leq)$ if and only if X is a Noetherian tree.*
  (ii) *There exists a (TH) domain such that $Spec(R)$ is order-isomorphic to $(X, \leq)$ if and only if X is a Noetherian tree such that every element $x \neq x_0$ of X is contained in at most finitely many maximal elements of X.*

PROOF. If R is an (HT) domain, then by Theorem 4.5, $Spec(R)$ is Noetherian. The converse is due to Facchini: In [**Fa**, Theorem 5.3], he proves (i) when "(HT) domain" is replaced by "generalized Dedekind domain." Thus our version of (i) follows from Facchini's result and Theorem 4.3. Statement (ii) follows from (i) and Theorem 4.1. □

It is clear from this proposition that there exist (TH) domains that do not possess (HT). It is not hard to exhibit such examples. The ring $\mathbb{Z} + X\mathbb{Q}[X]$, with X an indeterminate, is an (HT) domain that is not a (TH) domain. It is, in fact, a generalized Dedekind domain that is not stable [**GO1**, p. 193].

5. Property (EP)

This section is devoted to the Prüfer case of property (EP). We begin by stating some of the rather special traits of exterior powers of modules over Prüfer domains.

LEMMA 5.1. [**O2**, Proposition 2.2] *If R is a Prüfer domain and G is a torsion-free module of rank n, then $\bigwedge^k G = 0$ for all $k > n$ and $\bigwedge^k G$ is a torsion-free R-module of rank $\binom{n}{k}$ for all $0 < k \leq n$.*

Let G be a torsion-free module of finite rank n. A *composition series* for G is a sequence of submodules of G,

$$0 = H_0 \subset H_1 \subset H_2 \subset \cdots \subset H_n = G$$

such that for all $0 < i \leq n$, H_i/H_{i-1} is a torsion-free rank one module. The *chief factors* F_i of the composition series are the rank one modules $F_i = H_i/H_{i-1}$ with $0 < i \leq n$. It is easy to see that in general, torsion-free finite rank modules have infinitely many different composition series. Jordan-Hölder rarely holds for the chief factors of the different composition sequences; the *product* of the chief factors, however, is unique up to isomorphism. This is a consequence of the following lemma.

LEMMA 5.2. [**O2**, Theorem 2.8] *An integral domain R is a Prüfer domain if and only if for each torsion-free R-module G of finite rank n,*

$$\bigwedge^n G \cong F_1 \otimes_R F_2 \otimes_R \cdots \otimes_R F_n$$

for the chief factors $\{F_i\}$ of any composition series of G.

LEMMA 5.3. *If R is a Prüfer domain, then R is a generalized Dedekind domain if and only if for all torsion-free finite rank R-modules G, G is locally free as an $R(G)$-module whenever $R(G) = R_1(G)$.*

PROOF. Suppose R is a generalized Dedekind domain and let G be a torsion-free R-module of finite rank n. Then by Lemma 5.1, $\bigwedge^n G$ is a torsion-free rank one module. By Theorem 4.3, R is strongly faithful. Thus, by Proposition 1.1, $\bigwedge^n G$ is locally free over $R(G) = R_1(G)$. In particular, if M is a maximal ideal of $R(G)$, then $\bigwedge^n G_M \cong R(G)_M$. This last fact implies that, since $R(G)_M$ is a quasilocal ring, G_M is a free R_M-module [**Gar**, Corollary 11.6]. Thus G is a locally free $R(G)$-module. Conversely, we have that each rank one module X is locally free over $R(X)$. Thus, by Proposition 1.1, R is strongly faithful and by Theorem 4.3, R is a generalized Dedekind domain. □

THEOREM 5.4. *An Prüfer domain R has (EP) if and only if R is a generalized Dedekind domain.*

PROOF. Suppose first that R has (EP). In order to prove R is strongly faithful, it is enough, by Proposition 1.1, to show that if X is a submodule of Q, the quotient field of R, then X is flat and faithful as an $R(X)$-module. Since R is a Prüfer domain, every torsion-free R-module is flat and hence we must only verify faithfulness. To this end, let X be a submodule of Q. The fact that R is a Prüfer domain implies every overring of R is a flat extension of R, and so (EP) holds for every overring of R. Thus we may assume without generality that $R(X) = R$. Suppose $MX = X$ for some maximal ideal M of R. Define $H = M \oplus X$ and observe $R(H) = R$. Moreover, $\bigwedge^2 H \cong M \otimes_R X \cong MX$, since X is flat as an R-module. Applying $Hom_R(-, MX)$ to the sequence

$$0 \to M \to H \to X \to 0$$

yields the commutative diagram,

$$\begin{array}{ccccccccc} 0 & \to & M & \to & H & \to & X & \to & 0 \\ & & \downarrow & & \downarrow & & \downarrow & & \\ 0 & \to & Hom_R(X, XM) & \to & Hom_R(H, XM) & \to & Hom_R(M, XM) & \to & 0 \end{array}$$

Now by (EP), the middle map is an isomorphism since $R_1(H) = R(XM) = R(X) = R$. Thus the Snake Lemma implies the left mapping is an ismorphism. Since $XM = X$, this forces $M = R(X) = R$, a contradiction. It follows that X is faithful and hence R is strongly faithful. By Theorems 3.2 and 4.3, R is a generalized Dedekind domain.

Conversely, suppose R is strongly faithful with quotient field Q, and let G be a torsion-free R-module of finite rank $n > 1$ such that $R(G) = R_1(G)$. We lose no generality in assuming that $R(G) = R$, since overrings of strongly faithful domains are strongly faithful and overrings of Prüfer domains are flat extensions, allowing us to identify $\bigwedge_R^k G$ and $\bigwedge_{R(G)}^k G$. By Lemma 5.3, G is a locally free R-module. We may view G as a submodule of QG, the image of G in $Q \otimes_R G$; similarly, we identify $\bigwedge^n G$ with its image in $Q \bigwedge^n G$. Let k be such that $0 < k < n$. We also view $Hom_R(\bigwedge^k G, \bigwedge^n G)$ as a subset of $Hom_R(\bigwedge^k G, Q \bigwedge^n G)$. Now for all maximal ideals M of R, it is easy to check that

$$\bigwedge^{n-k} G_M \to Hom_R(\bigwedge^k G_M, \bigwedge^n G_M)$$

is an isomorphism since G_M is a free R_M-module. Let ϕ denote the canonical homomorphism $\bigwedge^{n-k} G \to Hom_R(\bigwedge^k G, \bigwedge^n G)$. If $f \in Hom_R(\bigwedge^k G, \bigwedge^n G)$, then $f \in Hom_{R_M}(\bigwedge^k G_M, \bigwedge^n G_M)$ and hence $f = \phi(x)$ for some $x \in \bigwedge^{n-k} G_M$. Observe that $\bigwedge^n G$ dominates G, since by Lemma 5.2 and the flatness of torsion-free modules over Prüfer domains, it follows that every image of G in Q embeds in $\bigwedge^n G$. Thus we may extend ϕ to a Q-vector space isomorphism,

$$\phi' : \bigwedge^{n-k} QG \to Hom_Q(\bigwedge^k QG, \bigwedge^n QG).$$

It follows that ϕ is injective. Hence the choice of x must be the same for each maximal ideal M of R. Consequently, $x \in \cap_{M \in Max(R)} \bigwedge^{n-k} G_M = \bigwedge^{n-k} G$ and it follows that ϕ is surjective. $\square$

As noted in the introduction, there are difficulties in treating (EP) over non-Prüfer domains. The statement of (EP) is probably not very well suited for the general case. In addition to the previously discussed problem of the definition of

$R_1(G)$, the exterior powers would likely need to be relativized to $\bigwedge_{R(G)}^k G$ rather than simply $\bigwedge^k G$, if the goal is to link (EP) to strong faithfulness.

6. Warfield domains

From a historical point of view, three articles bear heavily on the development of Warfield duality for modules over integral domains. Each was published in 1968, though the authors were motivated by different problems and were evidently unaware of each other's work. The first, of course, is Warfield's original paper of 1968. Second is Matlis extensive study of *reflexive domains*, those integral domains for which every submodule G of a finite rank free R-module satisfies $G \cong Hom_R(Hom_R(G, R), R)$ via the canonical duality homomorphism [**M2**]. Third, W. Heinzer's study of domains for which every ideal is divisorial [**H**] is crucial to the link between reflexive rings and Warfield domains established in the paper [**BS**] of Bazzoni and Salce. This link is what we wish to describe first; it is key to the rest of the results of the section. We omit the technical devices that enable this link, but one of the main steps of the characterization is easy enough to describe.

The basic strategy in characterizing Warfield, (HT) or (TH) domains is to reduce to the quasilocal case, characterize it in ideal-theoretic terms and then find a necessary and sufficient condition for this characterization to globalize. This strategy works in our context because rank one modules localize well in the following sense. If R is an (HT) domain and X is a rank one module, then the localization of X at a maximal ideal of its coefficient ring is a fractional ideal of some overring of R. Thus, in a loose sense, rank one modules are locally ideals, and as local properties, Warfield duality, (HT) and (TH) are determined by ideal-theoretic properties of *overrings* of the base ring.

In [**G1**], Goeters showed this strategy works for Warfield duality over Noetherian domains with module-finite integral closure. Bazzoni and Salce then proved that, regardless of Noetherian hypotheses, rank one modules *always* localize to ideals for a class of integral domains, the class of "totally divisorial" domains, that is properly situated between the class of (TH) domains and the class of Warfield domains [**BS**]. That this method of localization to ideals continues to work for the larger class of stable domains is proved in [**O6**, Theorem 5.4].

Thus, after analysis of appropriate globalizing conditions, Warfield duality reduces to information about ideals of overrings of R, and we have:

THEOREM 6.1. [**BS**, Theorem 6.6] *An integral domain R is a Warfield domain if and only if R is a totally reflexive domain.*

Theorem 6.1 reduces Warfield duality to ideal-theoretic data of overrings, since reflexive domains are characterized by the property that every ideal is reflexive and the domain has injective dimension one as a module over itself [**M1**, Theorem 29]. However, since overrings of an integral domain can be mysterious, what remains is to make a final reduction: to ideal-theoretic information of the base ring alone. Since every overring of a Warfield domain is clearly Warfield, any characterization in terms of the base ring must travel "upwards" through the overrings. Stability, indeed, is inherited by overrings of the base ring [**O6**, Theorem 5.1] and this fact is what makes the following theorem possible.

THEOREM 6.2. [**O8**, Theorem 4.2] *An integral domain R with quotient field Q is a Warfield domain if and only if R is a stable domain and Q/R is an injective R-module.*

Warfield duality can be loosened a bit to require only that, for a fixed rank one module Y, the mapping $G \to Hom_R(Hom_R(G,Y),Y)$ is an isomorphism whenever Y dominates G and $R(Y) \subseteq R(G)$. If R possesses this duality with respect to Y, then R is said to by Y-*reflexive*. A general approach to Y-reflexive rings is presented in [**BS**]. The article [**G3**] contains Noetherian characterizations of this property, while the integrally closed case is treated in [**O3**]. The papers [**HHuP**] and [**B**] are also relevant to this topic.

Bazzoni and Salce have completely characterized Noetherian and integrally closed Warfield domains. Recall that an *almost maximal ring* is a ring R for which R/I is linearly compact for all non-zero ideals I of R. A Prüfer domain is almost maximal if and only if every (prime) ideal of R has injective dimension at most one [**O3**, Theorem 3.5 and Corollary 3.6].

THEOREM 6.3. [**BS**, Theorems 7.3 and 7.8] *Let R be an integral domain.*
 (i) *R is a Noetherian Warfield domain if and only if every ideal of R can be generated by at most two elements.*
 (ii) *R is an integrally closed Warfield domain if and only if R is an almost maximal strongly discrete Prüfer domain.*

It follows from Theorem 6.2 that Warfield domains are stable and hence have (TH). Theorem 4.1 implies that (TH) domains have (HT). Thus we have substantiated the implications:

$$\text{Dedekind domain} \Rightarrow \text{Warfield domain} \Rightarrow \text{(TH)} \Rightarrow \text{(HT)}$$

Theorem 6.3(i) makes it clear that the first implication cannot be reversed. The existence of (HT) domains that do not have (TH) was remarked on in Section 4; thus the last implication cannot be reversed. Nor can the second be reversed, even for Noetherian domains. Sally and Vasconcelos have constructed an example of a Noetherian stable domain that does not have the 2-generator property and hence cannot be Warfield [**SV**].

In [**O8**], it is shown that there is a sense in which the Noetherian and integrally closed cases determine but do not exhaust the class of Warfield domains. This is the content of the next theorem, which can be found in [**O8**, Theorem 4.7]. We say a domain is *complete* if it is complete in the R-topology (see [**M1**]).

THEOREM 6.4. (Classification of Warfield domains) *An integral domain R is a Warfield domain if and only if R has finite character and for all $M \in Max(R)$, R_M satisfies one of the following statements:*
 (i) *every ideal of R_M can be generated by two elements,*
 (ii) *R_M is an almost maximal strongly discrete valuation domain, or*
 (iii) *R_M occurs in a pullback diaagram of the form*

$$\begin{array}{ccc} R = \nu^{-1}(A) & \to & A \\ \downarrow & & \downarrow \\ B & \stackrel{\nu}{\to} & k \end{array}$$

where (B, M) is a quasilocal domain satisfying (ii), $k = B/M$, $\nu : B \to k$ is the canonical projection of B onto k, and A is a quasilocal complete domain satisfying (i).

Using Theorem 6.4, it is easy to construct Warfield domains that are neither Noetherian nor integrally closed.

COROLLARY 6.5. [**O8**, Corollary 4.8] *Let R be a complete local domain for which every ideal can be generated by two elements but R is not a DVR. Then if Q is the quotient field of R, $R + XQ[X]_{(X)}$ is a non-Noetherian, non-integrally closed Warfield domain.*

An integral domain R is *h-local* if R has finite character and each non-zero prime ideal of R is contained in a unique maximal ideal of R. It is interesting that one is able to omit h-locality from the previous discussion since reflexive, and hence Warfield, domains are h-local. The full strength of h-locality is not needed for globalization because of a remarkable property of reflexive Prüfer domains that was noticed by Facchini in 1994 [**Fa**]: *A Prüfer domain R is reflexive if and only if R has finite character and R_M is a reflexive domain for all maximal ideals M of R.* (Facchini's proof must be slightly modified to prove this statement; see [**O3**, Lemma 4.1].) The fact that finite character suffices where before h-local was needed is a consequence of the valuation theory of maximal valued fields. In particular, if there are two independent maximal valuations on a given field, then these valuations have divisible value groups (see, for example, [**V**, Theorem A and Proposition 9] or [**Br**, Section 14]). This fact is remarked on again at the end of this section. In the context of Warfield domains, finite character suffices even for non-Prüfer rings; this is evidenced by Theorem 6.4. We can draw an interesting consequence from this discussion:

PROPOSITION 6.6. [**O3**, Corollary 4.5] *Let $\{V_1, \ldots, V_n\}$ be a collection of Warfield valuation domains all having the same quotient field. Then $\cap_{i=1}^n V_i$ is a Warfield Prüfer domain.*

Thus, if $\{V_1, \ldots, V_n\}$ is an irredundant collection of Warfield valuation domains all having the same quotient field, then it is an *independent* collection, meaning no two valuation domains share the same prime ideal. This independence property is rather strong and will be remarked on again below.

Warfield domains also capture another aspect of Noetherian rings for which every ideal can be generated by two elements. In his article [**Ba**] of 1963, Bass proves that if every finitely generated torsion-free R-module is isomorphic to a direct sum of ideals, then every ideal of R can be generated by two elements. The converse of this assertion remained open for Noetherian rings until 1991, when Rush extended results of L. Levy and R. Wiegand to prove this decomposition property was equivalent to the 2-generator property for Noetherian Cohen-Macaulay local rings. In fact, he proved that finitely generated torsion-free modules over these rings decompose in a rigid manner [**R1**, Theorem 4.3]. This result is extended to the general case in [**O8**].

THEOREM 6.7. [**O8**, Theorem 5.3] *Let R be an integral domain. The following statements are equivalent for R.*

(1) *R is a Warfield domain.*
(2) *Every submodule G of a finite rank free R-module can be decomposed as*

$$G \cong S_1 \oplus S_2 \oplus \cdots \oplus S_{n-1} \oplus I,$$

where $S_1 \subseteq S_2 \subseteq \cdots \subseteq S_n$ are overrings of R and I is an invertible ideal of S_n.

(3) *Every submodule G of a finite rank free R-module is isomorphic to a direct sum of stable ideals.*

Since for Noetherian domains, Warfield duality is equivalent to the 2-generator properties for ideals, there are numerous interesting examples of Noetherian Warfield domains that occur in "nature." The coordinate ring of an affine curve is a Warfield domain if and only if every singularity of the curve is a double point; every quadratic order $\mathbb{Z}[\sqrt{n}]$, $n \in \mathbb{Z}$, is a Warfield domain; and every integral group ring $\mathbb{Z}G$, with G a finite abelian group of square free order is a Warfield domain. (These examples are cribbed from [**LW**].) It is also worth noting that for analytically unramified Noetherian domains, those integral domains with module-finite integral closure, stability is equivalent to the 2-generated property, and hence equivalent to the existence of Warfield duality. Thus for the "excellent" rings of algebraic and arithmetic geometry, e.g. finitely generated algebras over fields or $\mathbb{Z}$, Warfield duality, (TH) and (HT) *all* coincide.

The existence of interesting examples of integrally closed Warfield domains that are not Dedekind is more troubling, at least if one seeks examples with infinitely many maximal ideals. Classical valuation theory guarantees an abundance of examples of Warfield valuation domains. Simply take a maximal extension field F of the quotient field of a strongly discrete valuation domain V; then there is an extension V' of V to F such that V' is a *maximal valuation domain*, , a valuation domain that cannot be enlarged to extension fields without changing the residue field or value group. A maximal extension preserves value groups, so V' remains strongly discrete and hence is, by Theorem 6.3, a Warfield domain.

How to construct non-quasilocal examples that are not Dedekind is less clear. To my knowledge, the following question remains open: *Does there exist a Prüfer Warfield domain with infinitely many maximal ideals that is not Dedekind?* Presumably the answer is yes, although there is little evidence for it yet. Note that such an example must have Krull dimension greater than one, since one-dimensional Prüfer Warfield domains are Noetherian and hence Dedekind. Thus, at the very least, in order to answer the question affirmatively one must be able to construct *an almost maximal Prüfer domain having infinitely many maximal ideals but with Krull dimension greater than one.* Evidently this has not yet been achieved. In fact, the only example (to my knowledge) of an almost maximal Prüfer domain that has infinitely many maximal ideals but is not Dedekind is due to P. Vamos [**V**, Proposition 12]. This example is one-dimensional. On the other hand, S. Weigand has shown that every finite tree can be realized as the prime spectrum of a locally almost maximal Prüfer domain [**Wi**].

It is doubtful that the constructions of Vamos and Wiegand can be applied directly to the problem of constructing Warfield domains. This is because the localizations of the rings in their examples are maximal valuation domains. Vamos has shown that if the quotient field of a maximal valuation possesses another maximal valuation (i.e. the field is "multiply maximally complete"), then the value group of any maximal valuation must be divisible. It follows that a Prüfer Warfield domain such that all localizations V at maximal ideals are maximal valuation domains must be quasilocal. For V has a principal maximal ideal and hence the value group has a summand isomorphic to $\mathbb{Z}$. This is impossible if the quotient field of V possesses a second maximal valuation domain, since the value group of V must then be divisible.

The preceding discussion suggests it is the homological aspect of Warfield domains that proves difficult to capture in Prüfer examples. The existence of Prüfer domains whose *rank one* modules possess Warfield duality can be established via the construction of Facchini discussed in Section 4. Bazzoni and Salce show in [**BS**] that these domains are simply the h-local strongly discrete Prüfer domains. In [**O1**, Proposition 5.5], it is shown that if a partially ordered set X is a Noetherian tree with least element x_0 and each element x except possibly x_0 is contained in a unique maximal element of X, then X is order-isomorphic to the prime spectrum of some h-local strongly discrete Prüfer domain. Hence there exist Prüfer domains with arbitrarily "large" prime spectra whose rank one modules possess Warfield duality.

References

[Ba] H. Bass, On the ubiquity of Gorenstein rings, Math. Z. **82** (1963), 8-28.

[B] S. Bazzoni, Divisorial domains, preprint.

[BS] S. Bazzoni and L. Salce, Warfield domains, J. Algebra **185** (1996), 836-868.

[Br] W. Brandal, *Commutative rings whose finitely generated modules decompose*, Springer-Verlag, New York, 1979.

[Fa] A. Facchini, Generalized Dedekind domains and their injective modules, J. Pure Appl. Algebra **94** (1994), 159-173.

[FHuP] M. Fontana, J. Huckaba, and I. Papick, *Prüfer Domains*; Marcel Dekker, New York, 1997.

[Ga] S. Gabelli, A class of Prüfer domains with nice divisorial ideals, Commutative Ring Theory (Fès 1995), 313-318, Lecture Notes in Pure and Appl. Math., 185, Dekker, New York, 1997.

[Gar] R. B. Gardener, *Exterior algebras over commutative rings*, University of North Carolina at Chapel Hill, 1972.

[Gi] R. Gilmer, *Multiplicative ideal theory*, Queen's Papers in Pure and Applied Mathematics 90, Queen's University, 1992.

[G1] H. P. Goeters, Warfield duality and extensions of modules over an integral domain, Abelian groups and modules (Padova, 1994), 239-248, Math. Appl., 343, Kluwer Acad. Publ., Dordrecht, 1995.

[G2] H. P. Goeters, Noetherian stable domains, J. Algebra **202** (1998), 343-356.

[G3] H. P. Goeters, Reflexive rings and subrings of a product of Dedekind domains, J. Pure Appl. Algebra **137** (1999), no. 3, 237-252.

[GO1] H. P. Goeters and B. Olberding, Faithfulness and cancellation over Noetherian domains, Rocky Mountain J. Math. **30** (2000), 185-194.

[GO2] H. P. Goeters and B. Olberding, On the mulplicative properties of submodules of the quotient field of an integral domain, Houston J. Math. **26** (2000), 241-254.

[GO3] H. P. Goeters and B. Olberding, Extension of ideal-theoretic properties of a domain to submodules of a quotient field, J. Algebra, to appear.

[H] W. Heinzer, Integral domains in which every non-zero ideal is divisorial, Mathematika **15** (1968), 164-169.

[HHuP] W. Heinzer, J. Huckaba and I. Papick, m-canonical ideals in integral domains, Comm. Algebra **26** (1998), no. 9, 3021-3043.

[La] E. L. Lady, Warfield duality and rank-one quasi-summands of tensor products of finite rank locally free modules over Dedekind domains, J. Algebra **121** (1989), 129-138.

[LW] L. Levy and R. Wiegand, Dedekind-like behavior of rings with 2-generated ideals, J. Pure Appl. Algebra **37** (1985), 41-58.

[M1] E. Matlis, *Torsion-free modules*, The University of Chicago Press, Chicago, 1972.

[M2] E. Matlis, Reflexive domains, J. Algebra **8** (1968), 1-33.

[O1] B. Olberding, Globalizing local properties of Prüfer domains, J. Algebra **205** (1998), 480-504.

[O2] B. Olberding, Composition series of modules over Prüfer domains, Proc. Amer. Math. Soc. **127** (1999), 1917-1921.

[O3]   B. Olberding, Almost maximal Prüfer domains, Comm. Algebra **27** (1999), pp. 4433-4458.
[O4]   B. Olberding, Factorization into prime and invertible ideals, J. London Math. Soc., to appear.
[O5]   B. Olberding, Stability of ideals and its applications, Ideal-theoretic methods in commutative algebra, Marcel Dekker, to appear.
[O6]   B. Olberding, On the structure of stable domains, preprint.
[O7]   B. Olberding, On the classification of stable domains, preprint.
[O8]   B. Olberding, Stability, duality, 2-generated ideals and a canonical decomposition of modules, preprint.
[P]   N. Popescu, On a class of Prüfer domains, Rev. Roumaine Math. Pures Appl. **29** (1984), 777-786.
[Re]   J. D. Reid, Warfield duality and irreducible groups, Contemporary Math. **130** (1992), 361-370.
[R1]   D. Rush, Rings with 2-generated ideals, J. Pure Appl. Algebra **73** (1991), 257-275.
[SV]   J. D. Sally and W. V. Vasconcelos, Stable rings, J. Pure Appl. Algebra **4** (1974), 319-336.
[V]   P. Vamos, Multiply maximally complete fields, J. London Math. Soc. **12** (1975), 103-111.
[W]   R. B. Warfield, Jr., Homomorphisms and duality for abelian groups, Math. Z. **107** (1968), 189-212.
[Wi]   S. Wiegand, Locally maximal Bezout domains, Proc. Amer. Math. Soc. **47** (1975), 10-14.

DEPARTMENT OF MATHEMATICS, UNIVERSITY OF LOUISIANA AT MONROE, MONROE, LA 71209
E-mail address: `maolberding@ulm.edu`

Generalizations of Isomorphism in Torsion–free Abelian Groups

K.C. O'Meara and C. Vinsonhaler

ABSTRACT. We survey the literature on four generalizations of isomorphism in the category of torsion–free abelian groups of finite rank. One relatively new application is a variation on the standard K_0 for this category. A section on parallels in the theory of regular rings is followed by a list of open problems.

1. Introduction.

A major breakthrough in the study of torsion–free abelian groups of finite rank occured in 1957 with Jónnson's theorem on unique decomposition, up to quasi–isomorphism, into strongly indecomposables [24]. Much later, Jónnson's result was put into a categorical setting by Walker [38]. Subsequently other categories were introduced with corresponding definitions of isomorphism that proved useful in studying mixed abelian groups and later, torsion–free abelian groups of finite rank (hereafter, tffr groups). Lady introduced near isomorphism into the study of tffr groups [25]. Stable isomorphism was studied by Goodearl [17] and Warfield [39]. Most recently, the paper [32] placed the notion of multiple isomorphism into the tffr toolchest, bringing the total number of ways to call groups "the same" to five. In (strictly) descending order, they are,

$$\begin{aligned}&\text{isomorphism}\\ \Longrightarrow\ &\text{multiple–isomorphism}\\ \Longrightarrow\ &\text{stable isomorphism}\\ \Longrightarrow\ &\text{near–isomorphism}\\ \Longrightarrow\ &\text{quasi–isomorphism}\end{aligned}$$

Our purpose here is to survey the known results and applications involving these concepts, mention connections to other areas, and indicate directions for future research. After reviewing the definitions, we establish the above hierarchy and give some examples (Section 2). Next we discuss an application of multiple isomorphism to the definition of a new type of Grothendieck group (Section 3). We follow this by giving some parallels in the theory of von Neumann regular rings and exchange rings (Section 4). Finally we list some open problems (Section 5).

1991 *Mathematics Subject Classification.* Primary:20K15, 20K30 .
Key words and phrases. multiple–isomorphism, stable, near, quasi .

© 2001 American Mathematical Society

DEFINITION 1. For A and B torsion–free abelian groups, we call A and B:

- **multiple–isomorphic** (or *multi-isomorphic*) if $A^n \simeq B^n$ for every integer $n \geq 2$.
- **stably isomorphic** ([14] uses "equivalent") if $A \oplus C \simeq B \oplus C$ for some torsion–free C of finite rank.
- **near–isomorphic** if for each prime p there is a homomorphism $A \to B$ that induces an isomorphism $A_p \to B_p$ on the localizations of A and B at the prime p.
- **quasi–isomorphic** if A is isomorphic to a subgroup of finite index in B.

All of these isomorphisms can be put into categorical settings and extended to more general classes of modules, an exercise we leave to the reader.

2. The Isomorphism Hierarchy

Obviously ordinary isomorphism implies multiple–isomorphism. To see that multiple isomorphism implies stable isomorphism, note that if A is multiple–isomorphic to B, then $A \oplus (A \oplus A) \cong B \oplus (B \oplus B)$. But also, $A \oplus A \cong B \oplus B$, so A and B are stably isomorphic.

The notion of near isomorphism was originally defined by Lady [25]. Warfield used instead the term *genus*, which was standard in other areas, and showed that A and B are in the same genus (nearly isomorphic) if and only if $A^n \simeq B^n$ for some positive integer n ([39], Theorem 5.9). The fact that stable isomorphism implies near–isomorphism follows from the work of Goodearl (see [17], Theorem 5.1). The final implication, that near isomorphism implies quasi–isomorphism, is a direct consequence of the definitions.

Before showing that no two of the definitions are equivalent, we look at some examples of classes of groups where some collapsing does occur.

THEOREM 2. *Within the class of tffr groups:*

1. *quasi–isomorphism implies isomorphism for Murley groups ($\dim G/pG \leq 1$ for all primes p), in particular rank-one groups;*
2. *Near isomorphism implies isomorphism for semilocal groups ($pG = G$ for all but a finite number of primes p) and for direct sums of rank-one groups;*
3. *Stable isomorphism implies isomorphism for $End(G)$ a Dedekind domain;*
4. *Multiple isomorphism implies isomorphism for all G such that $QEnd(G)$ modulo its nil radical does not have a factor that is a totally definite quaternion algebra (definition below).*

Part 1 is due to Murley [30], and is a straightforward calculation for the expert. Part 2 is also straightforward, see [6], Corollary 7.19 and [29], Theorem 9.2.7. The last two items of Theorem 2 require some explanation. In particular, we will need two classic theorems from the literature. The first, due to a number of authors such as Dress and Arnold-Lady, is critical in the construction of examples that illustrate the differences between our five isomorphisms (see, for example [6], Corollary 9.6). Recall that for G an abelian group, a group H is G-projective if H is a summand of a finite direct sum of copies of G.

PROPOSITION 3. *Let G be a torsion–free abelian group of finite rank. There is a category equivalence from G-projectives to finitely generated projective $End(G)$-modules given by $H \to Hom(G, H)$. The inverse sends P to $P \otimes G$.*

The next theorem reviews the well-known structure of finitely generated projectives over a Dedekind domain ([**35**], Theorem 4.13).

THEOREM 4. *If I_i and J_j are nonzero (fractional) ideals in a Dedekind domain of rank n, then $I_1 \oplus I_2 \oplus ... \oplus I_m \cong J_1 \oplus J_2 \oplus ... \oplus J_n$ if and only if $m = n$ and $I_1 I_2 ... I_n \cong J_1 J_2 ... J_n$.*

Before returning to the proof of Part 3 in Theorem 2, we need two more standard definitions. If G and C are tffr groups, denote by $C(G)$ the canonical image of $Hom(G,C) \otimes G$ in C; and by $C[G]$ the intersection of the kernels of all maps $C \to G$. Now suppose G, H and C are tffr groups with $G \oplus C \cong H \oplus C$ (G and H are stably isomorphic). It follows readily that $G \oplus (C/C[G]) \cong H \oplus (C/C[G])$, so we may assume $C[G] = 0$. Then also $G \oplus C(G) \cong H \oplus C(G)$, so we may assume $C(G) = C$. A final lemma leads to our goal..

LEMMA 5. *Let G be a tffr group with $End(G)$ Dedekind. If C is a tffr group such that $C[G] = 0$ and $C(G) = C$, then C is G–projective.*

PROOF. Since $C[G] = 0$, we can embed C into a finite direct sum of copies of $G: 0 \to C \to G^r$. Applying $Hom(G, \_)$ we obtain an embedding of right R–modules, $0 \to Hom(G, C) \to R^r$, where $R = End(G)$ is Dedekind. It follows that $Hom(G,C)$, as a submodule of a free module, is projective. Moreover the canonical map, $\theta : Hom(G,C) \otimes_R G \to C$ is onto, because $C(G) = C$. Finally, the rank of $Hom(G,C) \otimes_R G$ is the same as the rank of C. Thus, the map θ is an isomorphism, and C is G–projective. □

Going back to Theorem 2(3) and the isomorphism $G \oplus C \cong H \oplus C$, we can by Lemma 5 and the discussion preceding it assume that C is G-projective. Adding an additional summand to each side if necessary, we may assume $C = G^m$ for some positive integer m, that is, $G \oplus G^m \cong H \oplus G^m$. Apply the equivalence $Hom(G, \_)$ of Proposition 3 to both sides to obtain $R^{m+1} \cong I \oplus R^m$ for $I \cong Hom(G, H)$, an ideal of R. By Theorem 4, $I \cong R$. A reverse application of the equivalence shows $G \cong H$. This completes the proof that stable isomorphism implies ordinary isomorphism when the endomorphism ring of one of the groups is a Dedekind domain.

For item number 4 in Theorem 2, we need some additional machinery. Recall (see [**6**] or [**39**], for example) that *n is in the stable range* of a ring R provided that whenever f_i, g_i are elements of R such that

$$f_1 g_1 + f_2 g_2 + ... + f_{n+1} g_{n+1} = 1$$

in R, then there are elements h_i, k_i in R such that

$$(f_1 + f_{n+1} h_1) k_1 + (f_2 + f_{n+1} h_2) k_2 + ... + (f_n + f_{n+1} h_n) k_n = 1.$$

By a classic result of Evans (1973), if $End(G)$ has 1 in the stable range, then G can be cancelled in the class of abelian groups. That is, $G \oplus C \simeq H \oplus C$ implies $G \simeq H$. The weaker condition of 2 in the stable range gives some cancellation properties, as developed in Warfield [**39**]. Warfield's result that 2 is in the stable range of $End(G)$ for all tffr groups G is integral to our study here. In particular, we make frequent use of the next proposition.

PROPOSITION 6. [**32**]. *Let G, H be right R-modules with 2 in the stable range of $End_R(G)$. Then the following are equivalent:*

1. G is multiple–isomorphic to H;
2. $G \oplus G \simeq G \oplus H \simeq H \oplus H$;
3. $G \oplus G \simeq G \oplus H$.

As noted already, this Proposition applies when G and H are torsion–free abelian groups of finite rank. In general the properties in the Proposition are not equivalent (see [2]). We can now close the book on part 4 of Theorem 2 with the next theorem, which combines Proposition 6 with Theorems 8.18 and 8.19 of [6]:

THEOREM 7. *If G is a tffr group, then multiple isomorphism implies ordinary isomorphism if the quasi-endomorphism ring of G modulo the nil radical does not contain a totally definite quaternion algebra as a direct factor. In particular, this is the case if G is almost completely decomposable.*

Recall [12], [6] that an algebra A is a *totally definite quaternion algebra* if A is a four-dimensional algebra $A = F \cdot 1 \oplus Fi \oplus Fj \oplus Fk$ over a subfield F of the reals such that every embedding of F into the complex numbers actually embeds F in the reals, and multiplication is given by $i^2 = j^2 = -1; ij = -ji = k$. For an example, take the rational quaternions ($F = \mathbb{Q}$).

Our next task is to show that our five isomorphisms are all distinct. It is well-known that quasi–isomorphism is strictly weaker than near–isomorphism - an almost completely decomposable group is quasi–isomorphic to a completely decomposable group, but not nearly isomorphic unless it is itself completely decomposable (see [29]). The work of Lady [25] shows that near–isomorphism is weaker than stable isomorphism; see also [8], Example 5.1. To construct your own example, let R be a ring of integers in an algebraic number field that has a non-free ideal I such that I^n is free for some (minimal) positive integer $n \geq 2$. Use Corner's theorem (or [40]) to construct a tffr G such that $End(G) = R$. Then because $I^n \cong R^n$, the Proposition 3 equivalence used above imples that $G^n \cong (IG)^n$. Hence G is nearly isomorphic to IG. But G and IG are not stably isomorphic by Theorem 2(3), since I and R are not isomorphic.

The reference [6] contains examples (e.g. Example 2.10) of locally completely decomposable groups that are stably isomorphic, but not multiple–isomorphic, the latter a consequence of Theorem 2(4) above and Theorem 8.18 in [6]. Finally, in view of the next theorem, the example of [36] and the examples of [32] show that multiple–isomorphism is weaker than isomorphism.

We close with two other results from the literature surrounding multiple isomorphism that seem worthy of note.

THEOREM 8. [37] *If A is a totally definite quaternion algebra, then there are only finitely many Z-orders R in A with the property that if I is a right ideal of R, then $I \oplus R \cong R \oplus R$ implies $I \cong R$.*

This result, along with those of [22], provides *existence* proofs for the inequality of multiple and ordinary isomorphism. Specifically, for any positive integer d, there *exist* examples of rank $4d$ of tffr groups G and H such that G is muliple-isomorphic, but not isomorphic, to H. This follows from the fact that there are infinitely many nonisomorphic orders in a totally definite quaternion algebra ([35]), so by Theorem 8 there are infinitely many orders R that contain a non-free ideal I such that $I \oplus R \simeq R \oplus R$. To translate back into the abelian group setting is routine, first writing $R = End(G)$ for some torsion-free abelian group of the same rank as R [40], then using the hom-tensor equivalence of Proposition 3. A technique of T. Y. Lam, based on the following elegant result, may be modified to produce concrete examples, as in [**32**].

THEOREM 9. [**26**] *Let M be an ideal in the ring R such that the quotient ring R/M is semisimple (or just unit regular), and let P, Q be right ideals of R containing M. If $P/M \cong Q/M$ as right R/M-modules, then P and Q are multiple isomorphic as R-modules.*

3. Applications and A New Look at K_0.

Quasi–isomorphism was introduced by Jónnson to obtain a Krull–Schmidt theorem for decompositions of tffr groups into strongly indecomposables. The notion has also been used effectively in the study of Butler groups, torsion–free homomorphic images of finite direct sums of subgroups of the additive rationals $\mathbb{Q}$. More recently, near isomorphism has been shown to be the right notion for studying almost completely decomposable groups, those groups quasi–isomorphic to a finite direct sum of subgroups of $\mathbb{Q}$ [**29**]. Stable isomorphism was used by Arnold to construct a K_0 for tffr groups, as well as a K_0 for the class of A-projectives, for a given tffr group A [**6**], Chapter 13.

We next review a recent construction [**32**] of K_0–like objects using multiple isomorphism rather than the traditional stable isomorphism. Recall that one obtains the usual K_0 (of the category of torsion–free abelian groups of finite rank) by starting with the monoid V of isomorphism classes of tffr's (under $\oplus$) and factoring by the relation of stable isomorphism. Then one obtains a cancellative abelian monoid by defining $[A] + [B] = [A \oplus B]$, and an abelian group K_0 by adjoining differences. What happens if we factor by the (much finer) relation of multiple–isomorphism instead of stable isomorphism? In this case, the resulting monoid $(M, +)$ of multiple–isomorphism classes turns out to be *separative*: $a + a = a + b = b + b$ implies $a = b$; or equivalently, $na = nb$ for all $n > 1$ implies $a = b$. See [**11**] or [**2**] for a more complete discussion (in a semigroup setting) of separativity and the related notions we introduce next. Two multiple–isomorphism classes $[A]$ and $[B]$ lie in the same *archimedean component* if and only if A is B-projective (a summand of a finite direct sum of copies of B) and B is A-projective. We denote by Arch(A) the archimedean component of the class $[A]$. The archimedean components of M partition M into *cancellative* abelian semigroups [**11**].

The preceding definitions and facts are well-known to semigroup theorists. The next result spells out the results germane to our new setting.

THEOREM 10. [**32**]. *Let A be a torsion-free abelian group of finite rank. Then:*
1. *Arch(A) is a cancellative abelian semigroup under $[B] + [C] = [B \oplus C]$;*
2. *The abelian group $G(A)$ of differences $[B] - [C]$ of multiple-isomorphism classes in Arch(A) has torsion subgroup $T = \{[A] - [B] : A$ and B are near–isomorphic$\}$;*
3. *The group $G(A)$ is finitely generated.*

Moreover, multiple–isomorphism is the smallest congruence on the monoid V of isomorphism classes for which the factor monoid is separative.

Multi–isomorphism is much finer than stable isomorphism — it nearly always coincides with isomorphism; and Arch(A) uses a restricted subset of the A-projectives. Therefore one could expect that the groups $G(A)$ and T from Theorem

10, together with the subgroup U of T given by $U = \{[A] - [B] : A$ and B are stably isomorphic$\}$, will provide important information about the group A. In particular we can use $U \subseteq T$ to measure the relative coarseness of the relations multiple-isomorphism, stable isomorphism and near isomorphism. Specifically, we have the following numerical invariants for A:

$[near : stable] = |T/U|$;
$[stable : multi] = |U|$;
$[near : multi] = |T|$

For example, if $[stable : multi] = 3$, then there are precisely three multiple-isomorphism classes of torsion–free abelian groups of finite rank that are stably isomorphic to A. The (finite) rank of the free group $G(A)/T$ also is an invariant. It gives the number of generators of $G(A)$ up to near isomorphism. We surmise that the collection of these invariants provides enough useful information about a group A to warrant further study. Corollary 12 below lends support to this view. We are grateful to Ken Goodearl for the next theorem.

THEOREM 11. *Suppose A is a strongly indecomposable tffr group. Then $[A]$ generates a direct summand of $G(A)$.*

PROOF. Choose a basis $[D_i] - [C_i]$, $i = 1, ..., n$, for $G(A)$ modulo $T(A)$, with $[D_i], [C_i]$ in $Arch(A)$. Because C_i is A-projective, there exists X_i such that $C_i \oplus X_i \simeq A^m$ for some integer $m \geq 0$. By adding extra copies of A to X_i if necessary, we may assume $[X_i] \in Arch(A)$ and that the same m works for all i. Then $[D_i] - [C_i] = [D_i \oplus X_i] - [A^m]$, so without loss of generality we may assume $[C_i] = [A^m]$ for all i.

Next write $[A] = \sum_{i=1}^{n} a_i ([D_i] - [A^m]) + t$ for integers a_i and torsion element $t \in T(A)$. Let k be a positive integer such that $kt = 0$, so that $k[A] = \sum_{i=1}^{n} ka_i([D_i] - [A^m])$. This yields $k(1 + \sum_{i=1}^{n} a_i)[A] = \sum_{i=1}^{n} ka_i[D_i]$. Thus, there is a positive integer d such that the sum of $d(1 + \sum_{i=1}^{n} a_i)$ copies of A is isomorphic to $\bigoplus_{i=1}^{n} D_i^{da_i}$. By uniqueness of quasi-decomposition into strongly indecomposables, each D_i is quasi–isomorphic to a direct sum of copies of A, so by counting copies of A we have $d(1 + \sum_{i=1}^{n} a_i) = \sum_{i=1}^{n} da_i b_i$ for some positive integers b_i. Thus, the integers a_i are relatively prime, and $[A]$ generates a summand of $G(A)$ by a standard argument. $\square$

COROLLARY 12. *Let A and B be strongly indecomposable tffr groups such that B is A-projective and A is B-projective. If $G(A)$ has torsion–free rank one, then A and B are nearly isomorphic.*

PROOF. Let $G = G(A) = G(B)$ and assume G has torsion–free rank one. Let T be the torsion subgroup of G. By Theorem 11, $[A]$ and $[B]$ are each generators of the infinite cyclic group G modulo T. Hence $[A] = \pm[B]$ modulo T, so for some positive integer n, we have $n[A] = \pm n[B]$. Clearly we can't have $n[A] = -n[B]$ in $G(A)$ since $n[A] + n[B] = n[A \oplus B]$ is nonzero. Therefore $n[A] = n[B]$, whence A^n is multi-isomorphic to B^n. In particular, $A^{2n} \cong B^{2n}$, so by the discussion at the beginning of Section 2, A and B are nearly isomorphic. $\square$

We close this section with an example to illustrate the ideas.

EXAMPLE 13. Given an integer n such that $2n + 1$ is prime, we construct a collection of nearly isomorphic almost completely decomposable groups of rank two that contains exactly n isomorphism classes.

Let $r = 2n + 1$ and choose two primes p, q congruent to 1 modulo r. There are infinitely many such primes by the well-known result of Dirichlet. Define torsion-free abelian groups of rank two as follows:
$G_0 = Z[p^{-1}] \oplus Z[q^{-1}]$
$G_i = G_0 + \frac{1}{r}Z(1, i)$, $1 \leq i \leq n$.
Here $Z[p^{-1}]$ denotes the subring of the rationals generated by $1/p$. Each G_i is an almost completely decomposable group (contains a direct sum of two rank-one groups as a subgroup of finite index). For almost completely decomposable groups, isomorphism is the same as multiple-isomorphism ([6], Corollary 8.19) and stable isomorphism is the same as near isomorphism ([6], Exercise 12.1). A routine computation shows that all the groups G_i are nearly isomorphic, hence stably isomorphic. Moreover, all the isomorphism classes of groups in this near-isomorphism class are represented by some G_i (see Corollary 12.3.2 of [29]). It is easy to show that $G_i \simeq G_{r-i}$ under the map $(a, b) \to (a, -b)$. Standard arguments, using the fact that p and q are congruent to 1 modulo r, show that G_i and G_j are not isomorphic if $0 \leq i < j \leq n$. Then one needs an application of the decomposition theorem for Butler groups with critical typeset of size two (see, for example, [28]) to see that $\text{Arch}(A)$ contains precisely the classes $[G_i]$ for $1 \leq i \leq n$ and their direct sums. It follows easily that for the group $G = G(G_1)$ of Theorem 10(b), $\text{rank}(G/T) = 1$; $|U| = n$; and $|T/U| = 1$. We assert without proof that the torsion subgroup T is cyclic.

REMARK 14. If a tffr group A has endomorphism ring R that is a Dedekind domain, then by Theorem 2(3) and Proposition 3, it follows that $G(A) \cong K_0(R)$. In turn, for R a Dedekind domain, $K_0(R) \cong \mathbb{Z} \oplus C(R)$, where $C(R)$ is the class group of R. The arguments are similar to those in [6], Theorem 13.13. In particular, the torsion subgroup of $G(A)$ can be as least as general as that of $C(R)$, where R is an integrally closed subring of an algebraic number field. On the other hand, if B is completely decomposable, $G(B)$ is torsion-free by Theorem 2(2), and can be shown to have arbitrary positive finite rank. Combining the above observations, it can be shown that $G(A)$ can have arbitrary positive torsion-free rank and the same generality of torsion subgroup described above.

4. Parallels in Regular Rings and Exchange Rings

In the theory of von Neumann regular rings, stable isomorphism and K_0 have played key roles. See [1], [16], [18], [19], [20], [21]. More recently, considerable activity in the area has been focused on cancellation (most especially separative cancellation) questions for finitely generated projective modules over regular rings [1], [2], [3], [4], [5], [18], [19], [31], [33], [34]. Although the concept of separativity was formulated in semigroup theory over 45 years ago (arising from a 1956 characterisation by Hewitt and Zuckerman of when there are enough characters of a semigroup to separate its elements [23], [11]), its introduction into module theory is quite new. We pause for the relevant definitions.

DEFINITION 15. A class of modules Γ is said to satisfy

1. *Separative Cancellation* if for all A, B in Γ, $A \oplus A \cong A \oplus B \cong B \oplus B$ implies $A \cong B$;
2. *Strong Separative Cancellation* if for all A, B in Γ, $A \oplus A \cong A \oplus B$ implies $A \cong B$.

Strong separative cancellation was called "self–cancellation" in [**6**] and "cancellation of small projectives" in [**5**]. Proposition 6 shows that for a class of tffrs, both separative cancellation and strong separative cancellation are equivalent to the statement that multiple isomorphism implies isomorphism. multiple–isomorphism for tffr groups does not appear to have been studied before [**32**], even though it is closely related to self–cancellation, which was investigated by Warfield and Arnold nearly twenty years ago. Indeed, the fact that the condition $A \oplus A \simeq A \oplus B$ defines an equivalence relation $A \sim B$ in the category of tffr groups, is not at all obvious until one establishes that the condition is the same as saying A and B are multiple–isomorphic.

A ring R is called *separative* if its finitely generated projective (right) modules satisfy separative cancellation. Separativity for exchange rings (defined below) is preserved under finite matrix rings, direct products, direct limits, factor rings and extensions of ideals by factor rings [**2**]. It was first applied in 1993 to a regular ring R through its monoid $V(R)$ (under direct sum) of finitely generated projective modules, by Ara, Goodearl and O'Meara. This was subsequently extended in 1998 to exchange rings by the above and Pardo [**2**]. Although the definition makes sense for general rings, in the exchange setting the property turns out to be particularly robust, principally because there $V(R)$ enjoys the common refinement property. To focus our discussion, we restate the

Fundamental Separativity Problem: Are all regular rings (more generally, exhange rings) separative? (see [**2**], p. 129)

Much research has gone into this question. For example, it has been shown that an affirmative answer for all regular rings is equivalent to the following statements:

- All 2 by 2 matrices over a regular ring can be diagonalized [**3**];
- All 2 by 2 matrices over $M_n(F)$ can be uniformly diagonalized independently of n and for all fields F.

See [**31**] for a more complete discussion of the second statement. It is fair to say that the resolution of the Fundamental Separativity Problem, at least for regular rings, would be a most prized trophy. The "smart money" says non–separative regular rings almost certainly exist. (An aside: one of the initial motivations for the authors to look at separative cancellation in abelian groups was to use this as a guide as to what to expect in regular rings) There are a number of outstanding open problems which have recently been settled within the class of separative exchange rings. Before listing some of these, we recall that a (right) module M over a ring R satisfies the *finite exchange property* if for every R-module A and any decompositions $A = M' \oplus N = A_1 \oplus ... \oplus A_n$ with $M' \cong M$, there exist submodules $A'_i \subseteq A_i$ such that $A = M' \oplus A'_1 \oplus ... \oplus A'_n$. Following Warfield, we call R an *exchange ring* if R_R satisfies the finite exchange property.

Returning to the recently resolved problems, let R be a separative exchange ring. Recall that R is *directly finite* if one-sided inverses are two-sided ($xy = 1$ implies $yx = 1$).

1. If R is directly finite, then so are all its matrix rings $M_n(R)$ [**2**], Proposition 2.3. (Hence a finitely generated projective module is not isomorphic to a proper direct summand of itself.)
2. If R is simple and directly finite, then R has stable range 1 (and so a finitely generated projective module can be cancelled from a direct sum) ([**2**], Theorem 3.4) . For regular rings this means R is unit–regular.

3. The stable range of R can only be 1, 2 or infinity ([**2**], Theorem 3.3).
4. Every regular square matrix A over R is equivalent to a diagonal matrix, i.e. PAQ is diagonal for some invertible matrices P and Q ([**3**], Theorem 2.8).
5. All invertible square matrices over R are products of elementary matrices and an invertible diagonal matrix. Consequently, the natural homomorphism from $GL_1(R)$ to the Whitehead group $K_1(R)$ is surjective ([**4**], Theorem 2.8).

Even for regular rings, all of 1–5 are still open in general. For an exchange ring R, separativity is equivalent to multi–isomorphism implying isomorphism on the finitely generated projective R–modules. However, unlike the situation for tffr abelian groups, separativity is much weaker than strong separativity, because the latter implies that the stable range is at most two (whereas, for example, a full ring of linear transformations on an infinite dimensional vector space has infinite stable range).

Very recently, O'Meara and Raphael [**31**] have shown that the Fundamental Separativity Problem for regular rings is equivalent to a linear algebra problem: for a field F, is there a formula $PAQ = \begin{bmatrix} * & 0 \\ 0 & * \end{bmatrix}$ for diagonalising a 2x2 matrix $A = \begin{bmatrix} a & b \\ c & d \end{bmatrix}$ over $M_n(F)$, independently of n? Here P and Q are required to be invertible matrices whose entries are fixed regular F–algebra expressions in a, b, c and d, relative to a quasi–inverse operation (sending x to its quasi-inverse x' such that $xx'x = x$) on $M_n(F)$, but the formula must work for all choices of such an operation, a myriad of possibilities. (For fixed n such a formula exists. Also there is a formula that works for all unit–quasi–regular operations, independently of n.)

One of the really nice payoffs for working in the setting of exchange rings is that it provides a bridge from von Neumann regular rings to operator algebras. Specifically, the C^*–algebras of real rank zero, introduced in 1991 by Brown and Pedersen [**10**], are exchange rings [**2**]. This has allowed separativity analogues of 1–5 to be carried over to these C^*–algebras. See [**2**], [**4**] and [**34**] . (Here separativity is formulated in terms of orthogonal sums of projections and the Murray–von Neumann equivalence of projections : $p \oplus p \sim p \oplus q \sim q \oplus q$ implies $p \sim q$.)

See the excellent surveys by Goodearl [**19**] and Lam [**27**], and Facchini's book [**13**] for further background and details on the various types of cancellation problems. Recent papers in the area also include [**9**] and [**15**].

5. Open Questions

The following problems arise naturally from our study:

1. Find examples or constructions that show which sets of the new K_0-invariants can be realized: $[near : stable] = |T/U|$; $[stable : multi] = |U|$ $[near : multi] = |T|$.
2. What are the possible finite group structures for T, U, and T/U? Notice that T can have a very general structure by Remark 14.
3. Investigate how the structure of the group $G(A)$ can be used to recover or describe the group A. Corollary 12 is an example of a result in this direction.
4. In particular, what does the torsion–free rank of $G(A)$ say about A?

5. When does A, B indecomposable and $Arch(A) = Arch(B)$ imply A nearly isomorphic to B? Again, see Corollary 12.
6. Are all regular rings (or exchange rings) separative?

References

[1] Ara, P., Stability properties of exchange rings, Proceedings Korea–China–Japan Ring Theory Conference, Birkhauser, 1999.

[2] Ara, P., K. R. Goodearl, K. C. O'Meara and E. Pardo, Separative cancellation for projective modules over exchange rings, Israel J. Math. **105** (1998), 105–137.

[3] Ara, P., K. R. Goodearl, K. C. O'Meara and E. Pardo, Diagonalisation of matrices over regular rings, Linear Algebra and its Applications **265** (1997), 147–163.

[4] Ara, P., K. R. Goodearl, K. C. O'Meara and R. Raphael, K_1 of separative exchange rings and C*–algebras with real rank zero, Pacific J. Math. **951** (2000), 261–275.

[5] Ara, P., K. C. O'Meara and D. V. Tyukavkin, Cancellation of projective modules over regular rings with comparability, J. Pure Appl. Algebra **107** (1996), 19–38.

[6] Arnold, D. M., *Finite Rank Torsion Free Abelian Groups and Rings*, Lecture Notes in Mathematics **931**, Springer–Verlag, Heidelberg, 1992.

[7] Arnold, D. M., *Abelian Groups and Representations of Finite Partially Ordered Sets*, C.M.S. Books in Mathematics, Springer–Verlag, New York, 2000.

[8] Arnold, D. M. and E. L. Lady, Endomorphism rings and direct sums of torsion free Abelian groups, Trans. Amer. Math. Soc. **211** (1975), 225–237.

[9] Brookfield, G., Direct sum cancellation of noetherian modules, J. Algebra **200** (1998), 207–224.

[10] Brown, L .G., and G. K. Pedersen, C*–algebras of real rank zero, J. Func. Anal. **99** (1991), 131–149.

[11] Clifford, A. H. and G. B. Preston, *The Algebraic Theory of Semigroups*, Vol 1, Math Surveys 7, Amer. Math. Soc., Providence, 1961.

[12] Eichler, M., Über die Idealklassenzahl total definiter Quaternionalgebren, Math. Zeit. **43** (1938), 102–109.

[13] Facchini, A., *Module Theory: Endomorphism rings and direct sum decompositions in some classes of modules*, Progress in Mathematics **167**, Birkhauser Verlag, Basel, 1998.

[14] Fuchs, L., *Infinite Abelian Groups II*, Academic Press, New York, 1973.

[15] Fuchs, L., and P. Vamos, The Jordan–Zassenhaus Theorem and direct Decompositions, J. Algebra **230** (2000), 730–748.

[16] Goodearl, K. R., *Partially Ordered Abelian Groups with Interpolation*, Math. Surveys and Monographs 20, Amer. Math. Soc., Providence, 1986.

[17] Goodearl, K. R., Power cancellation of groups and modules, Pacific J. Math. **64** (1976), 387–411.

[18] Goodearl, K. R., *Von Neumann Regular Rings*, Pitman, London, 1979; Second Ed., Krieger, Malabar, Fl. 1991.

[19] Goodearl, K. R., Von Neumann regular rings and direct sum decomposition problems, in Abelian Groups and Modules, Padova 1994, Kluwer, Dordrecht, 1995, 249–255.

[20] Goodearl, K. R., Torsion in K_0 of unit–regular rings, Proc. Edinburgh Math. Soc. **38** (1995), 331–341

[21] Goodearl, K. R., D. E. Handelman and J. W. Lawrence, Affine representations of Grothendieck groups and applications to Rickart C*–algebras and aleph nought continuous regular rings, Memoirs Amer. Math. Soc. **234** (1980).

[22] Guralnick, R., The genus of a module. II. Roiter's theorem, power cancellation and extension of scalars, J. Number Theory **26** (1987), no. 2, 149–165.

[23] Hewitt, E. and H. S. Zuckerman, The ℓ_1–algebra of a commutative semigroup, Trans. Amer. Math. Soc. **83** (1956), 70–97.

[24] Jónnson, B., On direct decompositions of torsion–free abelian groups, Math. Scand. **5** (1957), 230–233.

[25] Lady, E. L., Nearly isomorphic torsion free abelian groups, J. Algebra **35** (1975), 235–238.

[26] Lam, T. Y., A lifting theorem, and rings with isomorphic matrix rings, in Fifty Years as a Mathematician and Educator: On the 80th Birthday of Professor Y.C. Wong, World Scientific Publ. Co., London–Singapore–Hong Kong, 1995, 169–186.

[27] Lam, T. Y., Modules with Isomorphic Multiples and Rings with Isomorphic Matrix Rings, Monographie No. 35 de L' Enseignement Mathématique, Genève 1999 (71 pages).

[28] Lewis, W .S., Almost completely decomposable groups with two critical types, Communications in Algebra **21** (1993), 607–614.

[29] Mader, A., *Almost Completely Decomposable Groups*, Algebra, Logic and Applications, Gordon and Breach, Amsterdam, 2000.

[30] Murley, C., The classification of certain classes of torsion free abelian groups, Pac. J. Math. **40** (1972), 647–665.

[31] O'Meara, K. C. and R. Raphael, Uniform Diagonalization of Matrices over Regular Rings, to appear in Algebra Universalis.

[32] O'Meara, K. C. and C. Vinsonhaler, Separative Cancellation and Multiple Isomorphism in torsion–free Abelian Groups, J. Algebra **221** (1999), 536–550..

[33] Pardo, E., Comparability, separativity, and exchange rings, Comm. Algebra **24** (1996), 2915–2929.

[34] Perera, F., Lifting units modulo exchange ideals and C*–algebras with real rank zero, J. Reine Angew. Math., **522** (2000), 51–62.

[35] Reiner, I., *Maximal Orders*, Academic Press, London, 1975.

[36] Swan, R. G., Projective modules over group rings and maximal orders, Ann. Math.**76** (1962), 55–61.

[37] Vigneras, M.-F., Simplification pour les orders des corps de quaternions totalement définis, J. Reine Angew. Math. **285/287** (1976), 257–277.

[38] Walker, E. A., Quotient categories and quasi–isomorphisms of abelian groups, Proc. Colloq. Abelian Groups, Budapest, 1964, 147–162.

[39] Warfield, R. B., Jr., Cancellation of modules and groups and stable range of endomorphism rings, Pac. J. Math. **91** (1980), 457–485.

[40] Zassenhaus, H., Orders as endomorphism rings of modules of the same rank, J. London Math. Soc. **42** (1967), 180–182.

(O'Meara) DEPARTMENT OF MATHEMATICS, UNIVERSITY OF CANTERBURY, CHRISTCHURCH, NEW ZEALAND

E-mail address: `K.OMeara@math.canterbury.ac.nz`

(Vinsonhaler) DEPARTMENT OF MATHEMATICS, UNIVERSITY OF CONNECTICUT, STORRS, CT 06269

E-mail address: `vinson@uconnvm.uconn.edu`

Automorphism Groups of Abelian Groups

Phill Schultz

This paper is dedicated to Laszlo Fuchs on his 75th birthday.

ABSTRACT. This paper is a survey of the properties, realisation and uniqueness of the automorphism groups of abelian groups.

The topic has three major theorems. Historically, the first is a characterisation of the finite groups which may be realised as the automorphism group of some torsion–free abelian group; the second characterises a large class of groups which may be realised as the automorphism group of some torsion–free abelian group; and the third states that any torsion abelian group is determined by its automorphism group.

In addition, there are many results describing necessary or sufficient conditions for an isomorphism between a pair of subgroups of an abelian group to be extended to an automorphism of the group.

1. Properties of automorphism groups of abelian groups

In view of the brilliant success of applications of automorphism groups in other areas of mathematics, such as finite group theory, representation theory, K–theory, combinatorics and geometry, it is somewhat surprising that automorphism groups have been a neglected backwater of abelian group theory. Perhaps a major reason is the difficulty in obtaining transparent structure theorems for these groups. Nevertheless, in some special cases, a great deal is known about $\mathrm{Aut}(G)$ for certain abelian groups G.

1.1. Abelian p–groups. Leaving aside elementary p–groups, torsion–free divisible groups, free abelian groups and cyclic groups, whose automorphism groups form part of the theory of general linear groups or number theory, the first attempt to characterize the automorphism group of an abelian group was by Shoda [**Sh28**], [**Sh30**] and [**Sh30a**]. These papers dealt with finite abelian p–groups, represented as direct sums of n cyclic groups. Shoda characterized the invertible $n \times n$ integral matrices which represented their automorphism groups in terms of the maximal powers of p which could divide the entries. In [**Sh30**], Shoda was also the first to indicate the close connection between nil ideals of the endomorphism ring and normal subgroups of the automorphism group of an abelian group.

1991 *Mathematics Subject Classification.* Primary 20K30, 20F28.

The first results relating the structure of $\text{Aut}(G)$ to that of G for arbitrary abelian p–groups were obtained by Baer, [**Ba37**], [**Ba55**] and [**Ba64**]. For example, he showed in [**Ba55**] that the subgroups of G satisfy the descending chain condition if and only if all torsion subgroups of $\text{Aut}(G)$ are finite. Along similar lines, Fuchs [**Fu60a**] described ascending chains of normal subgroups in $\text{Aut}(G)$ with elementary abelian factors.

In the 1970's Hausen produced a sequence of papers, [**Ha68**], [**Ha70**], [**Ha71**], [**Ha71a**], [**Ha71b**], [**Ha72**], [**Ha72a**], [**Ha73**], [**Ha74**], [**Ha74a**], [**Ha75**], [**Ha77**] and [**HJ76**], elucidating the normal subgroup structure of the automorphism group of a torsion abelian p–group of infinite rank. For example, she proved that a finite normal subgroup Γ of $\text{Aut}(G)$ is nilpotent, and that if p does not divide the order of Γ or if the index of the centralizer of Γ is finite and not divisible by $p-1$, then Γ is central. Her major themes were the action of $\text{Aut}(G)$ on G (see Section 3 below) and the construction of the maximal normal p–subgroup of $\text{Aut}(G)$.

Brandl [**Br90**] proved that for abelian p–groups G, $\text{Aut}(G)$ is solvable if and only if G is bounded, and Shlayfer [**Shl88**] showed that for an arbitrary abelian group G, $\text{Aut}(G)$ is a finite solvable group if and only if tG is finite with solvable automorphism group and $\text{Aut}(G/tG)$ and $\text{Hom}(G/tG, tG)$ are finite. Vilyatser [**V61**] described some specific examples of $\text{Aut}(G)$ for abelian p–groups G.

1.2. Torsion–free groups.

The structure of automorphism groups of torsion–free abelian groups is not so thoroughly studied. As a by–product of their important work on the realisation of rings as endomorphism rings, Pierce [**P64**], Corner [**Co67**], Shelah [**Sh74**] and Corner and Göbel [**CG85**] found torsion–free abelian groups of large cardinality having trivial automorphism group $\text{Aut}(G) = \{1, -1\}$. In Section 2 below I discuss other situations in which $\text{Aut}(G)$ is finite.

In [**MS00**], Mader and I described the automorphism groups of completely decomposable groups, giving necessary and sufficient conditions on G for $\text{Aut}(G)$ to be torsion, abelian or a direct sum of an elementary 2-group of given rank and a free abelian group of given rank. We showed that an almost completely decomposable (acd) group G is rigid if and only if $\text{Aut}(G)$ is the direct product of an elementary 2-group whose rank is that of G, and a free abelian group whose rank can be calculated from the invariants of G. We also gave sufficient conditions for an arbitrary acd group to have abelian or torsion automorphism group.

2. Realization theorems

A realization theorem is one that states that a given a group Γ is an automorphism group and determines the abelian groups G for which $\text{Aut}(G) \cong \Gamma$.

The most notorious realisation theorem is the classification of finite groups that are the automorphism groups of finite rank torsion–free abelian groups. This problem was first considered in a series of five papers by K. A. Hirsch and his student T. J. Hallett, [**HH65**], [**HH70**], [**HH70a**], [**HH73**], [**HH77**], and a further paper by Hirsch and Zassenhaus, [**HZ66**]. These papers showed that if Γ is a finite group isomorphic to $\text{Aut}(G)$ for some torsion–free group G, then Γ is a subdirect product of finitely many groups of the following six types, called **primordial** by Corner, [**Co88**]:

cyclic groups $\mathbb{Z}(2)$, $\mathbb{Z}(4)$ and $\mathbb{Z}(6)$;
the quaternion group Q of order 8;
the dicyclic group $D = \langle a, b : a^3 = b^2 = (ab)^2 = 1\rangle$ of order 12 and

the binary tetrahedral group $B = \langle a, b : a^3 = b^3 = (ab)^2 = 1\rangle$ of order 24.

The authors found finite rank torsion–free abelian groups whose automorphism groups are realised by each of the primordial groups. They also found a set of four necessary conditions that a finite subdirect product of primordial groups must satisfy in order to realise $\text{Aut}(G)$ for some G, and purported to show that these conditions are in fact sufficient.

However, in unpublished lecture notes, Corner [**Co88**] pointed out errors in the proofs. In fact he constructed a finite group Γ which is a subdirect product of primordial groups satisfying the Hallett–Hirsch conditions but which cannot be the group of units of an order in a rational algebra, and hence cannot be $\text{Aut}(G)$ for a torsion–free group G. The first main theorem mentioned in the Abstract above is Corner's theorem in [**Co88**] giving necessary and sufficient conditions for a subdirect product of primordial groups to be the group of units of an order in a rational algebra. These conditions are too technical to reproduce in this essay, but let us hope that one day the theorem finds its way into print.

Krempa [**Kr95**] extended the problem to the realisation of torsion groups as automorphism groups of torsion–free abelian groups, once again using orders in rational algebras. He showed that for finite rank G, $\text{Aut}(G)$ is torsion if and only if it is finite, whereas in general if $\text{Aut}(G)$ is torsion it is locally finite. Bekker and Kozhukov also [**BK88**] described torsion–free abelian groups with torsion automorphism groups.

In other work related to the realisation problem for finite groups, Kozhukov and Nikiforov [**KN89**] described the torsion–free groups of rank 2 with automorphism group $\mathbb{Z}(2)$ or $\mathbb{Z}(4)$; Kozukhov [**Ko88**] solved several problems on finite automorphism groups of finite rank torsion–free groups; Shlyafer [**Shl85**] found necessary and sufficient conditions on a finite rank torsion free group G for $\text{Aut}(G)$ to be Hamiltonian or torsion abelian. Bekker [**Be86**] found properties of the automorphism groups of the torsion–free groups having no nilpotent endomorphisms.

Kozhukov [**Ko86**] proved that if G is torsion–free of rank $p-1$ for some prime p such that G has an automorphism of order p, then $\text{Aut}(G) \cong \mathbb{Z}(2p) \times \prod_n \mathbb{Z}$ for some $n \leq \aleph_0$ if and only if G is strongly indecomposable.

The most far–reaching result in this area is due to Warren May, [**May99**]. He completely described the abelian groups of finite and countable rank which can be realized as the automorphism groups of torsion–free abelian groups.

On slightly different lines, Nikiforov [**Ni86**] found the number of torsion–free groups of given cardinality with given finite automorphism group.

Another type of realisation theorem was proved by Dixon and Evans [**DE90**] who showed that there are infinitely many non–isomorphic countable torsion–free abelian groups whose automorphism group is a countable divisible abelian group. However, $\text{Aut}(G)$ cannot be a finite rank divisible abelian group for any G.

The second main theorem mentioned in the Abstract is again due to Corner, [**Co67**, Theorem 2.6]. As a by–product of his realisation theorem for endomorphism rings, he showed that for every totally ordered group U of cardinality less than the least strongly inaccessible cardinal, there is a torsion–free abelian group G with $U \times \mathbb{Z}(2) \cong \text{Aut}(G)$.

3. The action of Aut(G) on G

The most fruitful way of studying Aut(G) is to consider it as a group acting on a set, either the set of elements of G or some set of subgroups or factor groups of G. For example, many authors have studied the stabilizers of chains of subgroups of G, defined as follows. Let $\{G_i : i \in I\}$ be a decreasing chain of characteristic subgroups of G, so that Aut(G) acts on each G_i and G_i/G_{i+1}. Let

$$\text{Stab}(\{G_i\}) = \{\alpha \in \text{Aut}(G) : \text{ for all } i \in I, \alpha \text{ fixes } G_i/G_{i+1}\}.$$

Notice that this definition includes for example Fix(H), the fixed group of any characteristic subgroup H of G and Resid(H), the residual of H = the fixed group of G/H.

It is readily checked that Stab($\{G_i\}$) is a normal subgroup of Aut(G). Since the lattice of characteristic subgroups is known if G is a sufficiently well–behaved p–group, it is not surprising that most progress has been made in this case. For example, let $G_i = p^i G[p]$ for all positive integers i. Then Stab($\{G_i\}$) is related to the maximum normal p–subgroup $O_p(\text{Aut}(G))$ of Aut(G). For $p \geq 5$, it was shown by Freedman [**Fr62**], Leptin [**L64**], Hill [**Hi71**] and Hausen [**Ha72**] that $O_p(\text{Aut}(G))$ is the set of elements of finite order in Stab($\{G_i\}$). Jutta Hausen and I [**HS98**] proved the same result for $p = 3$, while for $p = 2$, we showed that $O_p(\text{Aut}(G)) = \text{Stab}(\{G_i\}) \times \{1, -1\}$.

Let A be a finite rank completely decomposable group and e a positive integer. Then $\overline{A} = e^{-1}A/A$ is a finite group whose automorphism group contains a subgroup **TypAut**$\overline{A}$ consisting of the automorphisms which preserve the images of the type subgroups of $e^{-1}A$. In his book [**M00**, Section 8.2], Mader considers the action of TypAut $\overline{A}$ on the set of subgroups of $\overline{A}$ and uses it to classify the isomorphism classes and near isomorphism classes of almost completely decomposable groups between A and $e^{-1}A$.

Whenever one has a group Γ acting on a structure X, one has a Galois Theory, relating substructures of X to the subgroups which fix them and subgroups of Γ to the substructures they fix. The first published Galois Theory for automorphism groups acting on abelian groups is due to Baer, [**Ba35**]. He established a Galois correspondence between characteristic subgroups of an arbitrary abelian group G and certain normal subgroups of Aut(G).

A different type of Galois Theory was produced by Tarwater in [**T67**] and [**T68**]. Mimicking the construction of algebraic extensions in classical field theory, he constructed normal and separable extensions of an abelian group and established a Galois correspondence between them and normal subgroups of Aut(G) as well as a Galois cohomology.

In a sequence of papers [**Sch94**], [**Sch98**] and [**Sch99**] I developed a three–way Galois correspondence among characteristic subgroups H, factor groups G/H and normal subgroups of Aut(G) for an abelian p–group G.

4. Extension Theorems

An extension theorem is one which describes the characteristic subgroups H of G for which any automorphism of H can be extended to an automorphism of G. The importance of such results for the study of automorphism groups is that the map which sends each automorphism of G to its restriction to such a subgroup H is an epimorphism whose kernel Fix(H) is a normal subgroup of Aut(G).

In [**Me66**], Megibben showed that for arbitrary abelian p–groups G, while every automorphism of $p^n G$ can be extended to an automorphism of G, there exist (non–countable) G for which not every automorphism of $p^\omega G$ extends to an automorphism of G. In [**Hi71**], Hill showed that if G is a countable reduced p-group then every automorphism of $p^{\omega^2} G$ can be extended to an automorphism of G inducing the identity map on each $p^{\omega n} G/p^{\omega n+1} G$. Hill and Megibben [**HM83**] as well as Parker and Walker [**PW68**] found conditions for the extendibility to G of automorphisms of $p^\alpha G$ for various ordinals α, while Hill [**Hi68**] and [**Hi69**] considered the possibility of the extension to $\mathrm{Aut}(G)$ of automorphisms of various characteristic subgroups of G.

A dual property to extensibility concerns the characteristic subgroups H of G for which any automorphism of G/H lifts to an automorphism of G. Mishina in [**Mi72**] showed that every subgroup has this property if and only if each primary component of G is homocyclic or rank 1 divisible. Analogously to the subgroup case, if a characteristic subgroup H of G has the property that automorphisms of G/H lift to automorphisms of G, then the epimorphism of $\mathrm{Aut}(G)$ onto $\mathrm{Aut}(G/H)$ which maps each automorphism of G to the induced automorphism of G/H has as kernel the normal subgroup $\mathrm{Resid}(H)$ of $\mathrm{Aut}(G)$.

Another type of extension theorem concerns extending isomorphisms between subgroups of G to automorphisms of G. There is some confusion of notation in the literature about this concept, so I will attempt to standardise by declaring that subgroups H and K are **equivalent** if each isomorphism of H onto K extends to an automorphism of G and **congruent** if some isomorphism of H onto K extends to an automorphism of G. Leptin [**L60**] and Enochs [**E64**] found necessary and sufficient conditions for basic subgroups of an abelian p–group to be equivalent. Tarwater [**T70**] classified the p–groups for which all isomorphic subgroups are congruent. Hill [**Hi70**] found necessary and sufficient conditions for basic subgroups of a countable p–group to be congruent and proved that high subgroups of countable torsion groups are always congruent. The situation for arbitrary isomorphic subgroup pairs in p-groups is more complicated. The strongest theorem presently is due to Hill and Megibben [**HM83**]. They showed that if G is totally projective of limit length and H and K are almost balanced subgroups, then H and K are congruent if and only if they have the same Ulm invariants and $G/H \cong G/K$. It follows as a Corollary that isomorphic pure subgroups of direct sums of cyclics are congruent if and only if $G/H \cong G/K$, while for separable groups this is not always the case.

Another nice congruence theorem of Hill and Megibben for totally projective p–groups G appears in [**HM85**]. There they show that if H and K are isotype subgroups of G such that H and K have the same Ulm invariants and $G/H \cong G/K$ as valuated groups, then there exists an automorphism of G mapping H onto K. Moreover, any value–preserving isomorphism of G/H onto G/K can be lifted to an automorphism of G.

More recently Hill and West [**HW98**] have proved generalisations of this congruence theorem giving conditions necessary and sufficient both for extending isomorphisms between isotype subgroups H and K of G, and for lifting isomorphisms between the factor groups G/H and G/K.

In [**HJ76**], Hausen and Johnson show that if all isomorphic subgroups of G are congruent, then they are all equivalent, and in that case either G is a torsion group in which all primary components are isomorphic cocyclic groups or G is divisible with finite torsion–free rank and finite p–rank for all p.

A torsion–free group is called strongly homogeneous if its rank 1 pure subgroups are congruent. Arnold [**Ar76**] found the structure of strongly homogeneous groups. Later, Reid [**Re83**] characterized strongly homogeneous groups as modules over their endomorphism rings. Two Russian mathematicians, Bezverkhnyaya [**Be75**] and Firsov [**Fi71**] found conditions for subgroups of free abelian groups to be congruent.

More recently, Goeters and Wickless [**GW00**] have classified the torsion–free groups G of finite rank for which every isomorphism between **pure** subgroups extends to an automorphism of G, and similarly, those for which some isomorphism between pure subgroups extends. They also deal with the case in which only isomorphisms between pure rank 1 subgroups are required to extend.

For the case of acd groups, Kozhukhov [**Ko81**] described the acd groups for which every automorphism of any complete quasi–decomposition extends to an automorphism of the group.

5. Transitivity Theorems

While extension theorems deal with the problem of extending isomorphisms of similar subgroups to automorphisms of the containing group, transitivity is concerned with the problem of when similarity of elements a and b implies the existence an automorphism of G mapping a onto b. For example, Grinshpon [**Gr75**] showed that if G is torsion, then $\mathrm{Aut}(G)$ is transitive on elements of the same order if and only if each primary component of G is homocyclic or divisible. In general, if a is an element of a p–group G, define the **indicator** $I(a)$ of a to be the sequence $(h_i(a)) : i \in \mathbb{N})$ where $h_i(a)$ is the height of $p^i a$ if $p^i a \neq 0$ and $h_i(a) = \infty$ otherwise. If an automorphism of G maps a to b, then clearly $I(a) = I(b)$. Kaplansky [**K69**] called G transitive if this condition is also sufficient, i.e. G is transitive if whenever $I(a) = I(b)$ there exists $\alpha \in \mathrm{Aut}(G)$ with $a\alpha = b$, and showed that separable groups are transitive. Later, Hill [**Hi69a**] showed that totally projective p–groups are transitive.

Megibben [**Me66**] showed that not all p–groups are transitive, by exhibiting a group G of length $\omega + 1$ such that G has elements a and b each with indicator $(\omega, \infty, \ldots)$ but no automorphism mapping a onto b. Griffith [**G68**] showed that if β is a countable ordinal with $p^\beta G$ transitive and $G/p^\beta G$ a direct sum of countable groups, then G must be transitive.

Corner [**Co76**] showed that G is transitive if and only if whenever two elements of $p^\omega G$ have equal indicators, there is an automorphism of G mapping one onto the other. Carroll [**Ca94**] extended most of these results to multiply transitive p–groups. Files and Goldsmith [**FilGo98**] studied p–groups G for which $G \oplus G$ is transitive. Hennecke [**He99**] considered a stronger form of transitivity in which only equality of indicators of x and y in a distinguished subgroup of G is required to guarantee an automorphism of G mapping x onto y.

The analogous problem for transitivity in a torsion–free group G would state that if elements a and b have the same height matrices ([**Fu73**, Section 103]), then there is an automorphism of G mapping a onto b. Dobrushin [**Do81**] described several classes of transitive torsion–free groups. However, this notion of transitivity is not strong enough to deal with torsion–free abelian groups of current interest to researchers.

A more far-reaching kind of transitivity is dealt with by Hill, Megibben and Ullery. In their work, the group G may be mixed but must be local, i.e. a p–adic module for some prime p. Let α be a height sequence and let $a \in G(\alpha)$. Then a is **primitive of type** α if a has height sequence α but $h(p^n a) > \alpha_n$ for at most finitely many n. They then show that if a is not primitive then some multiple $p^n a$ is a sum of finitely many incomparable primitive elements. The **type vector** of a is the sequence of types of these primitive elements. It turns out that this type vector is an invariant of a. Their main result in [**HU98**, Theorem 8] is that if G is a local Warfield group containing elements a and b, then there is an automorphism of G mapping a onto b if and only if a and b have the same height sequence and the same type vector. In [**HMU00**], this result is extended to Σ–isotype subgroups of local Warfield groups.

More recently, Files [**Fil96, Fil97**] and Henneke and Strüngmann [**HS00**] have considered transitivity in the context of mixed groups, with emphasis on the p–local case.

6. Uniqueness Theorems

A uniqueness theorem for a class $\mathcal{A}$ of abelian groups is one which states that if G and H are in $\mathcal{A}$ and $\mathrm{Aut}(G) \cong \mathrm{Aut}(H)$, then $G \cong H$. Various generalisations also fall under the rubric of uniqueness theorems. Such generalisations include replacing isomorphism of G and H by a weaker equivalence, or recovering invariants of G from $\mathrm{Aut}(G)$. A stronger result than a uniqueness theorem is the multiplicative analogue of the Baer–Kaplansky Theorem, that is showing that every isomorphism from $\mathrm{Aut}(G)$ to $\mathrm{Aut}(H)$ is induced by an isomorphism of G onto H.

In [**Fr62**], Freedman proved that if G and H are countable reduced p–groups with $p \geq 5$ such that $p^\omega G \cong p^\omega H$ and $\mathrm{Aut}(G) \cong \mathrm{Aut}(H)$, then $G \cong H$. In [**L60a**] Leptin reached the same conclusion without the restriction on $p^\omega G$ and proved for $p = 3$ the weaker conclusion that the basic subgroups of G and H are isomorphic.

Liebert [**Li87**], based on a geometric construction due to Faltings [**Fa72**] proved the uniqueness theorem for all p–groups with $p \geq 3$, and moreover in [**Li89**] he described all isomorphisms of $\mathrm{Aut}(G)$ onto $\mathrm{Aut}(H)$. Finally in [**Sch98**], with the proof based on joint work with Hausen [**HS98**], I proved the uniqueness theorem for the class of p–groups for arbitrary primes p.

[There are some errors in [**Sch98**] which I correct in the Appendix below.]

Corner and Goldsmith [**CG94**] used similar methods to prove that if $p \neq 2$ and G and H are reduced torsion–free modules over the p–adic integers then $\mathrm{Aut}(G) \cong \mathrm{Aut}(H)$ implies that $G \cong H$.

Hausen [**Ha77**] showed that if G is an abelian p–group with $p \geq 5$, then the Ulm invariants of G can be recovered from group theoretic invariants of $\mathrm{Aut}(G)$. Later her student Abraham [**Ab98**] extended the result to $p = 3$ by proving that if $p \geq 3$, then for all positive integers n, $\mathrm{Aut}(G)$ contains a unique subgroup maximal with respect to having exponent $\leq p^n$. He also showed that this result is false for $p = 2$.

7. Appendix: Corrections to [Sch98]

In [**Sch98**] I defined the p–Loewy sequence of the torsion radical **t** of the endomorphism ring of an abelian p–group G and proved that the successive factors of this sequence are isomorphic to the corresponding successive factors of the lower p–central series of the maximal normal p–subgroup Δ of the automorphism group of G.

In the proof of Theorem A on page 378, I claimed that the inverse limit of a certain diagram of rings was in fact the torsion radical **t**. The proof depended on the unproved assertion that $\mathbf{t}_\omega = 0$. In fact, this assertion is incorrect. What is true is that $\mathbf{t}_{2\omega} = 0$ and hence the statements in the main theorem of the paper remain correct.

The paper should be modified as follows:

1. In Section 2 the definitions of the p–Loewy sequence $\{\mathbf{t}_n\}$ and the lower p–central series $\{\Delta_n\}$ should contain the additional clauses that $\mathbf{t}_\lambda = \cap_{\nu<\lambda}\mathbf{t}_\nu$ and $\Delta_\lambda = \cap_{\nu<\lambda}\Delta_\nu$ for limit ordinals λ.

2. In the second paragraph following the definition of the p–Loewy sequence, the statement that any $a \in \mathbf{t}$ is determined by its action on B should be followed by the following sentence:
   'It follows that $G\mathbf{t}_\omega \subseteq p^\omega G$. Hence for all $n < \omega$, $\mathbf{t}_{\omega+n}$ annihilates $G[p^n]$ and consequently $\mathbf{t}_{2\omega} = 0$.'

3. In the final paragraph of proof of Lemma 4 (a), replace the sentence 'It was shown in [**3**, Lemma 2.5] that for all $i \in \mathbb{N}$, $\mathbf{t}_{2i+3}$ annihilates $G[p^i]$' by 'It was shown in Praeger and Schultz, *The Loewy length of tha Jacobson radical of a bounded endomorphism ring*, Contemporary Mathematics, 130, (1992), 349–360, that for all $i \in \mathbb{N}$, $\mathbf{t}^{2i-1}$ annihilates $G[p^i]$'.

4. Following the proof of Lemma 4, insert the following sentence:
   'It follows that $\Delta_{2\omega} = 1$'.

5. In the proof of Theorem A, the phrase 'Since $\cap \mathbf{t}_n = 0$' should be replaced by 'Since $\mathbf{t}_{2\omega} = 0$'.

8. Literature

The following Bibliography is intended to be exhaustive. For the Russian language entries I have relied extensively on the masterly periodic surveys of Mishina, [**Mi67**], [**Mi72**], [**Mi79**], [**Mi85**] and [**Mi94**].

References

[Ab98] R. Abraham, *Normal p–subgroups of the automorphism group of an abelian p–group*, J. Algebra, 199, (1998) 116–123.

[Ar76] D. M. Arnold, *Strongly homogeneous torsion–free abelian groups of finite rank*, Proc. Amer. Math. Soc., 56, (1976) 67–72.

[Ba35] R. Baer, *Types of elements and characteristic subgroups of abelian groups*, Proc. London Math. Soc., Second Series, 39, (1935) 481–514.

[Ba37] R. Baer, *Primary Abelian groups and their automorphisms*, Amer. J. Math., 59 (1937) 99–117.

[Ba55] R. Baer, *Finite extensions of Abelian groups with minimum condition*, Trans. Amer. Math. Soc., 79 (1955) 521–540.

[Ba64] R. Baer, *Irreducible groups of automorphisms of abelian groups*, Pac. J. Math. , 14, (1964) 385–406.

[BK88] I. Kh. Bekker and S. F. Kozhukov, *Automorphisms of abelian torsion–free groups*, [Russian] Tomsk University press, Tomsk, (1988).

[Be86] I. Kh. Bekker, *On torsion–free abelian groups with torsion automorphism groups*, [Russian], Izv. Vyssh. Uchebn. Zaved., Mat., 2, (1986) 3–12.

[Be75] I. S. Bezverkhnyaya, *On the automorphisms of a subgroup of a free abelian group*, [Russian], Progress in Applied Mathematics, Tula, 2, (1975) 132–135.

[Br90] R. Brandl, *Abelian torsion groups with soluble automorphism groups*, Rend. Circ. Mat. Palermo, Ser. 2, 39, Suppl. 23, (1990) 43–44.

[Ca94] D. Carroll, *Multiple transitivity in abelian groups*, Arch. Math., 63 (1994) 9–16.

[Co63] A. L. S. Corner, *Every countable reduced torsion–free ring is an endomorphism ring*, Proc. London Math. Soc., (3), 13, (1963) 687–710.

[Co67] A. L. S. Corner, *Endomorphism rings of torsion–free abelian groups*, in Proc. Inter. Conf. on Theory of Groups, A.N.U., Canberra, Gordon and Breach, (1967) 59–69.

[Co76] A. L. S. Corner, *The independence of Kaplansky's notions of transitivity and full transitivity*, Quart. J. Math. Oxford, 27 (1976) 15–20.

[Co88] A. L. S. Corner, *Groups of units of orders in $\mathbb{Q}$–algebras*, Unpublished lecture notes, University of Padua, 1988.

[CG85] A. L. S. Corner and R. Göbel, *Prescribing endomorphism algebras, a unified treatment*, Proc. London Math. Soc. (3), 50, (1985) 447–479.

[CG94] A. L. S. Corner and B. Goldsmith, *Isomorphic automorphism groups of torsion–free p-adic modules*, in Abelian groups, module theory and topology, Eds. D. Dikranjan and L. Salce, Marcel Dekker Lecture Notes in Pure and Applied Mathematics, 201, (1994) 125–130.

[DE90] M. R. Dixon and M. J. Evans, *Divisible automorphism groups*, Quart. J. Math. Oxford Second Ser., 41, (1990) 179–188.

[Do81] Yu. B. Dobrusin, *On transitive and completely transitive torsion–free abelian groups*, [Russian] Proc. Algebra Section, 7th Regional Conf. on Mathematics and Mechanics, Tomsk, (1981) 16–18.

[E64] E. Enochs, *Extending isomorphisms between basic subgroups*, Arch. Math., 15, (1964) 175–178.

[Fa72] K. Faltings, *On the automorphism group of a reduced primary abelian group*, Trans. Amer. Math. Soc., 165 , (1972) 1–25.

[Fil96] S. Files, *On transitive mixed abelian groups*, in Abelian Group Theory, Lecture Notes in Pure and Appl. Math. 182, Marcel Dekker, (1996) 243–251.

[Fil97] S. Files, *Transitivity and full transitivity for nontorsion modules*, J. Algebra 197, (1997) 468–478.

[FilGo98] S. Files and B. Goldsmith, *Transitive and fully transitive groups*, Proc. Amer. Math. Soc., 126, (1998) 1605–1610.

[Fi71] Yu. M. Firsov, *On endomorphisms of the subgroups of a free abelian group*, [Russian] Mosk. Gos. Pedagog. Inst.im. V. I. Lenina, 375 (1971) 90–99

[Fr62] H. Freedman, *The automorphism groups of countable primary reduced Abelian groups*, Proc. London Math. Soc., 12, (1962) 77–99.

[Fu60] L. Fuchs, *Abelian groups*, Pergamon Press, 1960.

[Fu60a] L. Fuchs, *On the automorphism groups of abelian p–groups*, Pub. Math. Debrecen, 7, (1960) 122–129.

[Fu73] L. Fuchs, *Infinite Abelian Groups, Vol. 2*, Academic Press, 1973.

[G68] P. Griffith, *Transitive and fully transitive primary abelian groups*, Pacific J. Math., 25 (1968) 249–254.

[GW00] H. P. Goeters and W. J. Wickless, *Relative injectivity and equivalence theorems*, Houston J. of Math., 26 (2000) 223–239.

[Gr75] S. Ya. Grinshpon, *Almost holomorphically isomorphic abelian groups*, [Russian] Tr. Tomsk Univ. 220, (1975) 78–84.

[HH65] J. T. Hallett and K. A. Hirsch, *Torsion–free groups having finite automorphism groups, I*, J. Algebra, 2, (1965) 287–298.

[HH70] J. T. Hallett and K. A. Hirsch, *Die Konstruktion von Gruppen mit vorgeschriebenen Automorphismengruppen*, J. Reine Angew. Math. 241, (1970) 32–46.

[HH70a] J. T. Hallett and K. A. Hirsch, *Groups of exponent 4 as automorphism groups*, Math. Z., 117, (1970) 183–188.

[HH73] J. T. Hallett and K. A. Hirsch, *Groups of exponent 4 as automorphism groups, II*, Math. Z., 131, (1973) 1–10.

[HH77] J. T. Hallett and K. A. Hirsch, *Finite groups of exponent 12 as automorphism groups*, Math. Z., 155, (1977) 43–53.

[HS00] G. Hennecke and L. Strüngmann, *Transitivity and full transitivity for p-local modules*, Arch. der Math, 74, (2000) 28–36.

[Ha68] J. Hausen, *Automorphismengesättigte Klassen abzälbarer abelscher Gruppen*, Études sur les Groupes Abéliens, Dunod and Springer–Verlag, (1968) 147–181.

[Ha70] J. Hausen, *The hypo residuum of the automorphism group of an Abelian p-group*, Pacific J. Math., 35, (1970) 127–139.

[Ha71] J. Hausen, *Abelian torsion groups with artinian primary components and their automorphisms*, Fund. Math., 71, (1971) 273–283.

[Ha71a] J. Hausen, *Automorphisms of abelian p-groups and hypo residual finiteness*, Rend. Sem. Mat. Univ. Padova, 45, (1971) 145–156.

[Ha71b] J. Hausen, *Automorphisms of abelian torsion groups with finite p-ranks*, Arch. der Math., 22, (1971) 128–135.

[Ha72] J. Hausen, *Near central automorphisms of abelian torsion groups*, Trans. Amer. Math. Soc., 174 (1972) 199–215.

[Ha72a] J. Hausen, *On the normal structure of automorphism groups of Abelian p-groups*, J. London Math. Soc., 5 (1972) 409–413.

[Ha73] J. Hausen, *The automorphism group of an abelian p-group and its normal p-subgroups*, Trans. Amer. Math. Soc., 182, (1973) 159–164.

[Ha74] J. Hausen, *Structural relations between general linear groups and automorphism groups of abelian p-groups*, Proc. London Math. Soc., 28 (1974) 614–630.

[Ha74a] J. Hausen, *The automorphism group of an abelian p-group and its non-central normal subgroups*, J. Algebra, 30 (1974) 459–472.

[Ha75] J. Hausen, *On automorphism groups and endomorphism rings of abelian p-groups*, Trans. Amer. Math. Soc., 210 (1975) 123–128.

[Ha77] J. Hausen, *How automorphism groups reveal Ulm invariants*, J. Algebra, 44 (1977) 9–28.

[He99] G. Hennecke, *Transitivity and full transitivity over subgroups of abelian p-groups*, in Abelian Groups and Modules, Birkhäuser Verlag, (1999) 43–53.

[HJ76] J. Hausen and J. A. Johnson, *Abelian groups with many automorphisms*, Rend. Sem. Mat. Univ. Padova, 55, (1976) 1–5.

[HS98] J. Hausen and P. Schultz, *The maximal normal p-subgroup of the automorphism group of an Abelian p-group*, Proc. Amer. Math. Soc., 126, (1998) 2525–2533.

[Hi68] P. Hill, *Extending automorphisms on primary groups*, Bull. Amer. Math. Soc., 74, (1968) 1123–1124.

[Hi69] P. Hill, *Endomorphism rings generated by units*, Trans. Amer. Math. Soc., 141 (1969) 99–105.

[Hi69a] P. Hill, *On transitive and fully transitive primary groups*, Proc. Amer. Math. Soc., 22 (1969) 414–417.

[Hi70] P. Hill, *Automorphisms of countable primary Abelian groups*, Proc. Amer. Math. Soc., 25, (1970) 135–140.

[Hi71] P. Hill, *The automorphisms of primary Abelian groups*, Proc. London Math. Soc., 22, (1971) 24–38.

[HM83] P. Hill and C. E. Megibben, *On the congruence of subgroups of totally projectives*, in Abelian Group Theory, Springer–Verlag Lecture Notes in Mathematics, 1006, (1983) 513–518.

[HM85] P. Hill and C. E. Megibben, *On the theory and classification of abelian p-groups*, Math. Zeit., 190, (1985) 17–38.

[HU98] P. Hill and W. Ullery, *The transitivity of local Warfield groups*, J. of Algebra, 208, (1998) 643–661.

[HMU00] P. Hill, C. Megibben and W. Ullery, *Σ-isotype subgroups of local k-groups*, this Volume.

[HW98] P. Hill and J. K. West, *Subgroup transitivity in abelian groups*, Proc. Amer. Math. Soc., 126, (1998) 1293–1303.

[Hi61] K. A. Hirsch, *The automorphism groups of abelian p-groups*, Unpublished lecture notes, Washington University, 1961.

[HZ66] K. A. Hirsch and H. Zassenhaus, *Finite automorphism groups of torsion-free groups*, J. London math. Soc., 41, (1966) 545–549.

[K69] I. Kaplansky, *Infinite Abelian Groups*, The University of Michigan Press, 1969.

[Ko81] F. S. Kozhukov, *Extension of automorphisms in almost completely decomposable torsion-free abelian groups*, [Russian] in Abelian Groups and Modules, Tomsk, (1981) 117–127.

[Ko86] F. S. Kozhukov, *On a class of abelian groups with regular automorphisms of prime order*, [Russian] in Abelian Groups and Modules, Tomsk, 6 (1986) 50–56.

[Ko88] F. S. Kozhukov, *Finite automorphism groups of finite rank torsion-free abelian groups*, Izv. Akad. Nauk SSSR, Ser. Mat. , 52, (1988) 501–521.

[KN89] S. F. Kozhukov and V. A. Nikiforov, *Torsion-free abelian groups with finite automorphism groups*, Izv. Vyssc. Uchebn. Zaved. Mat., 9, (1989) 30–37.

[Kr95] J. Krempa, *Rings with periodic unit groups*, in Abelian Groups and Modules, Eds. A. Facchini and C. Menini, Mathematics and its Applications, Kluwer, (1995) 313–321.

[L60] H. Leptin, *Zur Theorie der überzahlbaren abelschen p–Gruppen*, Abhandl. Math. Sem. Univ. Hamburg, 24, (1960) 79–90.

[L60a] H. Leptin, *Abelsche p–Gruppen und ihre Automorphismengruppen*, Math. Z. 73, (1960) 235–253.

[L64] H. Leptin, *Einige Bemerkungen über die Automorphismen abelscher p–Gruppen*, Proc. Colloq. Abelian groups, (Tihany 1963) (1964) 99–104.

[Li87] W. Liebert, *Isomorphic automorphism groups of primary abelian groups*, in Abelian Group Theory: Proc. 3rd Conf. on Abelian Group Theory, Oberwolfach, 1985, Gordon and Breach, (1987) 9–31

[Li89] W. Liebert, *Isomorphic automorphism groups of primary abelian groups II*, Cont. Math., 87 (1989) 51–59.

[May99] W. May, *Abelian automorphism groups of countable rank*, in Abelian Groups and Modules, Birkhäuser Verlag, (1999) 23–42.

[M00] A. Mader, *Almost Completely Decomposable Groups*, Algebra, Logic and Applications Series, Vol. 13, Gordon and Breach, 2000.

[MS00] A. Mader and P. Schultz, *Endomorphism rings and automorphism groups of almost completely decomposable groups*, Comm. in Algebra, 28, (2000) 51–68

[Me66] C. K. Megibben, *Large subgroups and small homomorphisms*, Mich. Math. J. 13, (1966) 153–160.

[Mi67] A. P. Mishina, *Abelian Groups*, in Algebra, Topology, Geometry, Itogi Nauki i Tekh. VINITI Akad. Nauk SSSR (1967) 9–44 (Translated in Progress in Mathematics, Vol. 5, Plenum press, 1969, 1–37).

[Mi72] A. P. Mishina, *Abelian Groups*, in Algebra, Topology, Geometry, Itogi Nauki i Tekh. VINITI Akad. Nauk SSSR, (1972) 5–45 (Translated in Journal of Soviet Mathematics, 2, (1974) 239–263).

[Mi79] A. P. Mishina, *Abelian Groups*, in Algebra, Topology, Geometry, Itogi Nauki i Tekh. VINITI Akad. Nauk SSSR , (1979) 3–63 (Translated in Journal of Soviet Mathematics, 10, (1982) 631–668).

[Mi85] A. P. Mishina, *Abelian Groups*, in Algebra, Topology, Geometry, Itogi Nauki i Tekh. VINITI Akad. Nauk SSSR , (1985) 51–118 (Translated in Journal of Soviet Mathematics, 16, (1988) 288–331).

[Mi94] A. P. Mishina, *Abelian Groups*, in Algebra, Topology, Geometry, Itogi Nauki i Tekh. Seriya Sovremennaya Matematika i Ee Prilozheniya, 10, (1994) (Translated in Journal of Mathematical Sciences, 76, (1995) 2721–2792).

[Ni86] V. A. Nikiforov, *Torsion-free abelian groups with finite automorphism groups*, Mat. Zametki, 39, (1986) 641–646

[PW68] L. D. Parker and E. A. Walker, *An extension of the Ulm–Kolettis Theorem*, Études sur les Groupes Abéliens, Dunod and Springer–Verlag, 1968, 309–325.

[P64] R. S. Pierce, *Endomorphism rings of primary abelian groups*, Proc. Colloq. Abelian Groups (Tibany, 1963), (1964), 125–137.

[Re83] J. D. Reid, *Abelian groups cyclic over their endomorphism rings*, Abelian group Theory, Springer–Verlag Lecture Notes in Mathematics, 1006, (1983) 190–203.

[Sch94] P. Schultz, *When is an Abelian p–group determined by the Jacobson radical of its endomorphism ring?*, Contemporary Mathematics, 171, Amer. Math. Soc., (1994) 385–396 .

[Sch98] P. Schultz, *Automorphisms which determine an abelian p–group*, Abelian groups, module theory and topology, Marcel Dekker Lecture notes in Pure and Applied Mathematics, 201, Eds. D. Dikranjan and L. Salce, (1998) 373–379.

[Sch99] P. Schultz, *Groups acting on modules*, in Abelian Groups and Modules, Eds. P. Eklof and R. Goebel, Birkhäuser Trends in Mathematics (1999) , 75–85.

[Sh74] S. Shelah, *Infinite abelian groups, Whitehead Problem and some constructions*, Israel J. Maths., 18, (1974) 245–256.

[Shl85] A. Z. Shlayfer, *Finite Hamiltonian groups of automorphisms of abelian groups*, [Russian] in Abelian groups and Modules, 5, Tomsk (1985) 128–135.

[Shl88] A. Z. Shlayfer, *On solvability of the automorphism groups of abelian groups*, [Russian] in Abelian groups and Modules, 7, Tomsk (1988) 148–155.

[Sh28] K. Shoda, *Über die Automorphismen einer endlichen abelschen Gruppe*, Math. Ann., 100, (1928), 674–686.

[Sh30] K. Shoda, *Über den Automorphismenring bzw, die Automorphismengruppe einer endlichen abelschen Gruppe*, Proc. Imp. Acad. Tokyo, 6, (1930) 9–11.

[Sh30a] K. Shoda, *Über die charakteristischen Untergruppen einer endlichen abelschen Gruppe*, Math. Zeit., 31, (1930) 611–624.

[T67] J. D. Tarwater, *Galois theory of abelian groups*, Math. Zeitschr., 95, (1967) 50–59.

[T68] J. D. Tarwater, *Galois cohomology of abelian groups*, Pacific J. Math., 24, (1968) 177–179.

[T70] J. D. Tarwater, *Homogeneous primary Abelian groups*, Proc. Amer. Math. Soc., 24, (1970) 154–155.

[V61] V. G. Vilyatser, *Some examples of automorphism groups*, [Russian] Dokl. Akad. Nauk SSSR, 139, (1961) 1283–1286.

DEPARTMENT OF MATHEMATICS AND STATISTICS, THE UNIVERSITY OF WESTERN AUSTRALIA, NEDLANDS, 6907, AUSTRALIA

E-mail address: `schultz@maths.uwa.edu.au.edu`

Part III

Contributed Papers

Direct Sum Decompositions of Torsion–free Abelian Groups of Finite Rank

David M. Arnold

Dedicated to Professor Laszlo Fuchs in honor of his 75th birthday.

ABSTRACT. A torsion-free abelian group A of finite rank has the unique decomposition of subgroups (UDS) property if finite direct sums of groups quasi-isomorphic to A have unique direct sum decompositions into indecomposable summands. Classes of strongly indecomposable groups with the UDS property are characterized in terms of properties of their endomorphism rings. Strong connections between the UDS property and recent results of Goeters and Olberding on the unique decompositions into ideals (UDI) property for integral domains, and of Levy and Odenthal on the Krull-Schmidt property (TFKS) for finitely generated torsion-free modules over orders are given. Applications include an examination of minimal rank p-local torsion-free abelian groups with non-unique decompositions.

Direct sums of indecomposable groups in TF, the category of torsion-free abelian groups of finite rank, need not be unique. However, some uniqueness results are known. For example, if $A \oplus B = C_1 \oplus \cdots \oplus C_n$ and A and B have no quasi–summands in common, then for each i there is a summand C_i' of C_i with $B \cong C_1' \oplus \cdots \oplus C_n'$ [**Arnold 82**] or [**Faticoni 99**].

This paper is devoted to uniqueness of direct sums in TF for the case that A and B have all quasi–summands in common. A group $A \in$ TF has the *one-sided UDS property* if whenever $A \oplus B \cong B_1 \oplus \cdots \oplus B_n \in$ TF with each B_i quasi–isomorphic to A, then $A \cong B_i$ for some i. An $A \in$ TF has the *UDS property* if whenever $A_1 \oplus \cdots \oplus A_n \cong B_1 \oplus \cdots \oplus B_m$ with each A_i and B_j quasi-isomorphic to A, then $m = n$ and there is a permutation σ of $\{1, 2, \ldots, n\}$ with $A_i \cong B_{\sigma(i)}$ for each i.

Examples of groups with both UDS properties include *Murley groups*, those $A \in TF$ with p-rank $A \leq 1$ for each prime p, where p-rank $A = \dim_{Z/pZ}(A/pA)$. This is because there is a Krull-Schmidt theorem for the quasi–isomorphism category of TF [**Walker 64**] and quasi-isomorphic Murley groups are isomorphic [**Murley 72**].

For $A \in TF$, $N \operatorname{End} A = JQ \operatorname{End} A \cap \operatorname{End} A$ is a nilpotent ideal of $\operatorname{End} A$, where $JQ \operatorname{End} A$ is the Jacobson radical of the quasi-endomorphism ring $Q \operatorname{End} A$ of A. If A is strongly indecomposable, then $Q \operatorname{End} A / JQ \operatorname{End} A$ is a division algebra

---

1991 *Mathematics Subject Classification.* Primary 20K15.
Research supported, in part, by the Baylor Univ. Summer Sabbatical Program.

© 2001 American Mathematical Society

and there is a faithful strongly indecomposable group $G_A \in$ TF quasi–isomorphic to A with $S_A = \operatorname{End} G_A / N \operatorname{End} G_A$ a maximal order in $Q \operatorname{End} G_A / JQ \operatorname{End} G_A$ containing $\operatorname{End} A / N \operatorname{End} A$ (Lemma 1.4 (a)).

A maximal right ideal M of a ring R has the *unique maximal condition* if whenever I is a non-zero right ideal of R and M is a unique maximal right ideal containing I, then $I = M$.

THEOREM 1.5. *The following statements are equivalent for a strongly indecomposable* $A \in$ *TF:*

(a) *A has the UDS property;*
(b) *Each group in TF quasi–isomorphic to A has the one–sided UDS property;*
(c) *There is at most one prime p with p–rank $A > 1$ and either S_A / pS_A is a local ring or else S_A has exactly two maximal right ideals M_1 and M_2 containing p such that M_1 is principal, $G_A / M_1 G_A$ is a cyclic group, and M_1 / pS_A has the unique maximal condition in S_A / pS_A.*

Strongly indecomposable faithful groups in TF with the one-sided UDS property are characterized in terms of their endomorphism rings in Theorem 1.3.

The results of Section 1 are applied in Section 3 to easily construct examples of non-unique direct sum decompositions in TF, including some classical examples. M.C.R. Butler has provided an unpublished example of a p–local group in TF of rank 16 with non-unique direct sum decompositions and asked for the smallest such rank. An example of rank 10 is given in Example 3.3. Lower bounds for this rank are given in a forthcoming paper by the author. Related questions for modules over valuation domains are considered in [**Goldsmith, May 99**].

The UDS property for a strongly indecomposable group $A \in$ TF with $JQ \operatorname{End} A = 0$ is intimately related to properties of finitely generated modules over $\operatorname{End} A$. A ring R has the *UDI property* if whenever $I_1 \oplus \cdots \oplus I_n \cong J_1 \oplus \cdots \oplus J_m$ with each I_i and J_j an ideal of R, then $m = n$ and there is a permutation σ of $\{1, 2, \ldots, n\}$ with I_i isomorphic to $J_{\sigma(i)}$ for each i. One dimensional Noetherian (commutative) domains with the UDI property are characterized in [**Goeters, Olberding 99**].

If $A \in$ TF is strongly indecomposable with $Q \operatorname{End} A$ a field, then $\operatorname{End} A$ is a subring of an algebraic number field, hence a one-dimensional Noetherian domain. In this case, if A has the UDS property, then $\operatorname{End} A$ has the UDI property (Theorem 2.2). Groups $A \in$ TF with $Q \operatorname{End} A$ a field and either the one-sided UDS property or the UDS property are characterized in Corollary 2.1.

A semi-prime order Λ over a one-dimensional Noetherian domain has the TFKS property if whenever M_i and N_j are finitely generated torsion-free Λ–modules with $M_1 \oplus \cdots \oplus M_m \cong N_1 \oplus \cdots \oplus N_n$, then $m = n$ and M_i is isomorphic to $N_{\sigma(i)}$ for some permutation σ of the indices. Orders Λ with the TFKS property are characterized in [**Levy, Odenthal 96**]. An alternate characterization of the TFKS property for the special case of one-dimensional Noetherian domains is given in [**Goeters, Olberding 00**].

If $A \in$ TF is strongly indecomposable with $JQ \operatorname{End} A = 0$, then $\Lambda = \operatorname{End} A$ is a semi-prime order over a subring of an algebraic number field. It can be readily seen from the aforementioned results that if $Q \operatorname{End} A$ is a field, then the conditions on $\operatorname{End} A$ for A to have the UDS property lie strictly between the conditions for $\operatorname{End} A$ to have the TFKS property and the conditions for $\operatorname{End} A$ to have the UDI property.

This paper is a significantly revised and updated version of an unpublished manuscript by the author referred to in a review by R.B. Warfield, Jr. [Math Reviews **51**, 3147] and [**Salce, Zanardo 89**].

1. UDS Property

Define TF_Q to be the category with objects those of TF but with morphism sets $Q\operatorname{End}(A,B) = Q \otimes_Z \operatorname{End}(A,B)$ for objects A and B of TF. Isomorphism in TF_Q is called *quasi–isomorphism*. Groups A and B in TF are quasi–isomorphic if and only if there is a monomorphism $f : A \to B$ with $B/f(A)$ bounded, hence finite. Indecomposable objects in TF_Q are called *strongly indecomposable* groups and endomorphism rings in TF_Q are called *quasi–endomorphism* rings. The quasi–endomorphism ring of a strongly indecomposable group in TF is a local ring. Hence, TF_Q is a *Krull-Schmidt* category, i.e. each group in TF is quasi–isomorphic to a direct sum of strongly indecomposable groups unique up to quasi–isomorphism and order of quasi–summands.

Two right ideals I_1 and I_2 of $\operatorname{End} A$ are *co-maximal* if $I_1 + I_2 = \operatorname{End} A$. A group $A \in$ TF is *faithful* if $MA \neq A$ for each maximal right ideal M of $\operatorname{End} A$. Faithful groups include those $A \in$ TF with $\operatorname{End} A$ commutative or right hereditary [**Arnold 82**, Theorem 5.9].

LEMMA 1.1. *If $A \in$ TF has the one-sided UDS property, I_1 and I_2 are co-maximal right ideals of bounded index in $\operatorname{End} A$, and A_i is a subgroup of A containing $I_i A$, then A is isomorphic to A_1 or A_2. If, in addition, A is faithful, then I_1 or I_2 is principal.*

PROOF. Inclusion induces an epimorphism $\alpha : A_1 \oplus A_2 \to A$, observing that $A = \operatorname{End} A)A = (I_1+I_2)A = I_1A+I_2A = A_1+A_2$. Choose $f_i \in I_i$ with $1 = f_1+f_2$ and define $\beta : A \to A_1 \oplus A_2$ by $\beta(a) = (f_1(a), f_2(a))$. Then $\alpha\beta = 1_A$ and so $A_1 \oplus A_2$ is isomorphic to $A \oplus B$ for some B in TF. Now A/I_iA is bounded, since $(\operatorname{End} A)/I_i$ is bounded, and I_iA is contained in A_i. Hence, A is quasi–isomorphic to A_i for each i. Since A has the one-sided UDS property, A is isomorphic to A_1 or A_2. In particular, A is isomorphic to I_iA for some i, say $f \in \operatorname{End} A$ is an isomorphism from A to I_iA. Then $f(A) = I_iA$ and $A = f^{-1}I_i(A)$. If A is faithful, then $f^{-1}I_i = \operatorname{End} A$ and $I_i = f\operatorname{End} A$ is principal. □

LEMMA 1.2. *Let $A \in$ TF be a faithful group and p a prime of Z.*

(a) *If I is a right ideal of $\operatorname{End} A$ with $p \in I$, I is contained in a unique maximal right ideal M of $\operatorname{End} A$, and A/IA is not cyclic, then there is a subgroup of A properly containing IA that is not isomorphic to A.*

(b) *If M_1 and M_2 are distinct maximal right ideals of $\operatorname{End} A$ containing p, then A has a subgroup properly containing $(M_1 \cap M_2)A$ that is not isomorphic to A.*

PROOF. (a) Since A is faithful, there is $x \in A \setminus MA$. Define $B = IA + Zx$, a subgroup of A properly containing IA with B/IA cyclic. If $A \cong B$, say $f \in \operatorname{End} A$ with $f(A) = B$, then $f \notin M$ since $x \in B \setminus MA$. In fact, $I + f\operatorname{End} A = \operatorname{End} A$ since M is the only maximal right ideal containing I. Hence $A = IA + f(A) = B$ and so $A/IA = B/IA$ is cyclic, a contradiction.

(b) Choose $x \in A\setminus(M_1A\cup M_2A)$. To see that this is possible, pick $a \in A\setminus M_1A$, recalling that A is faithful. Now $A = M_1A + M_2A$, since $\operatorname{End} A = M_1 + M_2$, and so

$a = a_1 + a_2$ with $a_i \in M_i A$. But $a_2 \in M_2 A \setminus M_1 A$, otherwise $a \in M_1 A$. Similarly, there is $b_1 \in M_1 A \setminus M_2 A$, whence $x = b_1 + a_2 \in A \setminus (M_1 A \cup M_2 A)$.

Define $B = (M_1 \cap M_2)A + Zx$, a subgroup of A properly containing $(M_1 \cap M_2)A$. If $A \cong B$, say $f \in \operatorname{End} A$ with $f(A) = B$, then $f \notin M_1 \cup M_2$ since $x \in B \setminus (M_1 A \cup M_2 A)$. Hence $M_1 \cap M_2 + f \operatorname{End} A = \operatorname{End} A$ and so $A = (M_1 \cap M_2)A + f(A) = B$. Thus, $A/(M_1 \cap M_2)A = B/(M_1 \cap M_2)A$ is cyclic. This is a contradiction, observing that $p(A/(M_1 \cap M_2)A) = 0$, $A/(M_1 \cap M_2)A \cong A/M_1 A \oplus A/M_2 A$, and A is faithful. $\square$

Given $A \in \mathrm{TF}$ and a prime p of Z, define $m_p(A)$ to be the number of distinct maximal right ideals of $\operatorname{End} A$ containing $p \operatorname{End} A$. Then $m_p(A)$ is a nonnegative integer since $\operatorname{End} A / p \operatorname{End} A$ is finite. Notice that if $m_p(A) = 0$, then $\operatorname{End} A / p \operatorname{End} A = 0 = A/pA$.

THEOREM 1.3. *Suppose $A \in \mathrm{TF}$ is strongly indecomposable.*

(a) *If there is at most one prime p with p-rank $A > 1$ and either*
  (i) $\operatorname{End} A / p \operatorname{End} A$ *is a local ring or else*
  (ii) $\operatorname{End} A$ *has exactly two maximal right ideals M_1 and M_2 containing p such that M_1 is principal, $A/M_1 A$ is a cyclic group, and $M_1/p \operatorname{End} A$ has the unique maximal condition in $\operatorname{End} A/p \operatorname{End} A$,*
    then A has the one-sided UDS property.
(b) *The converse to (a) is true if A is faithful.*

PROOF. (a) Let $G = A \oplus B = B_1 \oplus \cdots \oplus B_n$ with each B_i quasi–isomorphic to A, π the projection of G onto A with kernel B, and π_i a projection of G onto B_i for each i with $1 = \pi_1 + \ldots + \pi_n$. Then $1_A = \beta_1 + \cdots + \beta_n$, where $\beta_i \in \operatorname{End} A$ is the restriction of $\pi \pi_i$ to A and $\beta_i(A)$ is contained in a subgroup $\pi(B_i)$ of A.

Since A is strongly indecomposable, $Q \operatorname{End} A$ is a local ring and so each β_i is either a unit in $Q \operatorname{End} A$ or is nilpotent. It suffices to assume that each β_i is a unit in $Q \operatorname{End} A$. To see this, suppose, for example, that β_1 is nilpotent. Then $g = 1_A - \beta_1 = \beta_2 + \cdots + \beta_n$ is a unit of $\operatorname{End} A$. Define $\nu : \pi(B_2) \oplus \cdots \oplus \pi(B_n) \to A$ by $\nu(\pi(b_2), \ldots, \pi(b_n)) = \pi(b_2) + \cdots + \pi(b_n)$ and $\rho : A \longrightarrow \pi(B_2) \oplus \cdots \oplus \pi(B_n)$ by $\rho(a) = (\beta_2 g^{-1}(a), \ldots, \beta_n g^{-1}(a))$.

Then $\nu\rho(a) = \beta_2 g^{-1}(a) + \ldots + \beta_n g^{-1}(a) = a$ for each $a \in A$. Hence $A \oplus \ker \nu$ is isomorphic to $\pi(B_2) \oplus \cdots \oplus \pi(B_n)$. Now induct on n, noticing that if A is isomorphic to $\pi(B_i)$, then A is isomorphic to B_i.

For each $1 \leq i \leq n$, let $I_i = \beta_i \operatorname{End} A$, a right ideal of $\operatorname{End} A$. Then $\operatorname{End} A = I_1 + \cdots + I_n$ and each $(\operatorname{End} A)/I_i$ is bounded since each β_i is a unit in $Q \operatorname{End} A$. Moreover, $I_i A$ is contained in $A_i = \pi(B_i)$ so that $[A : A_i]$ is finite. It now suffices to prove that $A \cong A_i$ for some i, in which case $A \cong B_i$.

If some $[A : A_i] = 0$, the proof is complete. The next step is to assume that each $[A : A_i] \neq 0$ and reduce to the case that each $[A : A_i]$ is prime. Suppose, by way of induction, that $[A : A_i]$ is not prime. Choose $x \in A \setminus A_i$ and a prime p such that $px \in A_i$. Then A_i is properly contained in $A_i + Zx$. If A and $A_i + Zx$ are not isomorphic, then replace A_i by $A_i' = A_i + Zx$. If A and $A_i + Zx$ are isomorphic, say $f \in \operatorname{End} A$ with $f(A) = A_i + Zx$, then replace A_i by $A_i' = f^{-1}(A_i)$. In either case, $[A : A_i']$ is a proper divisor of $[A : A_i]$ since $[A : A_i] \neq p$.

The substitution of A_i' for A_i doesn't change the hypothesis that $\operatorname{End} A = I_1 + \cdots + I_n$ for right ideals I_i of bounded index in $\operatorname{End} A$ with $I_i A$ contained in A_i. In particular, $I_i' = f^{-1} I_i$ is an ideal of $\operatorname{End} A$ (since $I_i A$ is a subgroup of

$f(A))$, $I'_i A$ is contained in a subgroup A'_i of A, and $I'_i + \Sigma\{I_j : j \neq i\} = \operatorname{End} A$ (since $f \operatorname{End} A = \Sigma f I_i$ is contained in $I_i + \Sigma\{f I_j : j \neq i\}$). If A is isomorphic to A'_i, then, by the construction of A'_i, A must also be isomorphic to A_i. By induction, and the fact that $[A : A'_i]$ is a proper divisor of $[A : A_i]$, the A_i's can be chosen with each $[A : A_i]$ prime.

By assumption, there is a prime p with q–rank $A \leq 1$ for each prime $q \neq p$. Hence, if $q = [A : A_i] \neq p$, then it follows that $A_i = qA$ is isomorphic to A.

The remaining case is that $\operatorname{End} A = I_1 + \cdots + I_n$ for right ideals I_i of finite index in $\operatorname{End} A$ with $I_i A$ contained in a subgroup A_i of A, and $[A : A_i] = p$ for each i. Replace I_i by $I_i + p\operatorname{End} A$, if necessary, to guarantee that $p\operatorname{End} A$ is contained in I_i for each i.

If $\operatorname{End} A / p\operatorname{End} A$ is a local ring, i.e. $m_p(A) = 1$, then there is a unique maximal right ideal M of $\operatorname{End} A$ containing p. Hence, each I_i is contained in M, and, since $\operatorname{End} A = I_1 + \cdots + I_n$, $I_i = \operatorname{End} A$ for some i. In this case, $I_i A = A_i = A$, a contradiction to $[A : A_i] \neq 0$.

Now assume that $m_p(A) = 2$, M_1 and M_2 are distinct maximal right ideals containing p, M_1 is a principal right ideal with $A/M_1 A$ cyclic, and $M_1/p\operatorname{End} A$ has the unique maximal condition. Since $I_1 + \cdots + I_n = \operatorname{End} A$ and each I_i is an ideal of $\operatorname{End} A$ containing p, it is sufficient to assume that I_1 is contained in M_1 but not contained in M_2. Because $M_1/p\operatorname{End} A$ has the unique maximal condition, $I_1 = M_1$. But $[A : A_1] = p$, $A/M_1 A \cong Z/pZ$, and $I_1 A = M_1 A$ is contained in A_1, whence $A_1 = M_1 A$. Since M_1 is principal, A_1 is isomorphic to A, as desired.

(b) Assume that A is faithful with the one-sided UDS property. There is at most one prime p with $m_p(A) > 1$. To see this, suppose $m_p(A) > 1$, $m_q(A) > 1$, M_1 and M_2 are distinct maximal ideals containing p, and M_3 and M_4 are distinct maximal ideals containing q. Now $\operatorname{End} A = p\operatorname{End} A + q\operatorname{End} A = M_1 \cap M_2 + M_3 \cap M_4$. By Lemma 1.2(b), there are subgroups of A containing $(M_1 \cap M_2)A$ and $(M_3 \cap M_4)A$, respectively, that are not isomorphic to A. In view of Lemma 1.1, this is a contradiction.

Next consider the case that $m_p(A) \leq 1$, i.e. $m_q(A) \leq 1$ for each prime q. If $q \neq q'$ are primes with $m_q(A) = 1 = m_{q'}(A)$, then either A/qA or $A/q'A$ is cyclic since A has the one-sided UDS property (Lemmas 1.2(a) and 1.1). In particular, either q–rank $A = 1$ or q'–rank $A = 1$. Hence, in this case, there is at most one prime p with p–rank $A > 1$ and $\operatorname{End} A/p\operatorname{End} A$ is a local ring since $m_p(A) = 1$.

Now suppose $m_p(A) > 1$ say M_1 and M_2 are distinct maximal right ideals of $\operatorname{End} A$ containing p. Since $A/(M_1 \cap M_2)A \cong A/M_1 A \oplus A/M_2 A$ and A is faithful, p–rank $A > 1$. By Lemma 1.2(b), there is a subgroup B of A containing $(M_1 \cap M_2)A$ that is not isomorphic to A. As noted above, $m_q(A) \leq 1$ for each prime $q \neq p$. If $m_q(A) = 1$, then A/qA is cyclic, again by Lemmas 1.2(a) and 1.1. Hence, in this case, p is the only prime with p–rank $A > 1$.

If $m_p(A) \geq 3$, say M_1, M_2 and M_3 are maximal right ideals of $\operatorname{End} A$ with $p \in M_i$, then $\operatorname{End} A = M_1 \cap M_2 + M_1 \cap M_3 + M_2 \cap M_3$. There is a subgroup B_{ij} of A containing $(M_i \cap M_j)A$ that is not isomorphic to A if $i \neq j$ by Lemma 1.2(b). Just as in Lemma 1.1, there is a split exact sequence $B_{12} \oplus B_{13} \oplus B_{23} \to A \to 0$. Since A has the one-sided UDS property, A is isomorphic to B_{ij} for some $i \neq j$, a contradiction.

Finally, assume that $m_p(A) = 2$ and that M_1 and M_2 are the distinct maximal right ideals of $\operatorname{End} A$ containing p. Let I_1 and I_2 be right ideals containing p such that M_i is a unique maximal right ideal containing I_i. Then $I_1 + I_2 = \operatorname{End} A$. Since

A has the one-sided UDS property for some i, say $i = 1$, there is an isomorphism $f : A \to I_1 A$ and $A/I_1 A$ is cyclic (Lemmas 1.1 and 1.2 (a)). Hence, $I_1 A = M_1 A = f(A)$. Because A is faithful, $M_1 = f \operatorname{End} A = I_1$ is principal, $A/M_1 A$ is cyclic, and $M_1/p \operatorname{End} A$ has the unique maximal condition. □

LEMMA 1.4. *Let $A \in \mathrm{TF}$ be a strongly indecomposable group.*

(a) *There is a faithful strongly indecomposable $G_A \in \mathrm{TF}$ quasi–isomorphic to A with $S_A = \operatorname{End} G_A / N \operatorname{End} G_A$ a maximal order in the division algebra $Q \operatorname{End} A / JQ \operatorname{End} A$ containing $\operatorname{End} A / N \operatorname{End} A$.*

(b) *If G_A has the one-sided UDS property and $B \in \mathrm{TF}$ is quasi-isomorphic to A, then B has the one–sided UDS property.*

PROOF. (a) Since A is strongly indecomposable, $Q \operatorname{End} A / JQ \operatorname{End} A$ is a division algebra. There is a maximal order S_A in $Q \operatorname{End} A / JQ \operatorname{End} A$ containing $\operatorname{End} A / N \operatorname{End} A$ with $S_A / (\operatorname{End} A / N \operatorname{End} A)$ bounded [**Arnold 82**, Cor. 10.14]. Moreover, there is a subring T of $\operatorname{End} A$ containing 1_A with $T \cong \operatorname{End} A / N \operatorname{End} A$ and $\operatorname{End} A$ quasi–isomorphic to $T \oplus N \operatorname{End} A$ [**Beaumont, Pierce 61**].

Define $G_A = S_A A$, viewing S_A as a subring of $Q \operatorname{End} A$ containing T with S_A / T bounded and $S_A \cap N \operatorname{End} A = 0$. Then A is a subgroup of G_A with G_A / A bounded, hence G_A is strongly indecomposable. Now S_A is a subring of finite index in $\operatorname{End} G_A / N \operatorname{End} G_A$ and $\operatorname{End} G_A / N \operatorname{End} G_A$ is an order in $Q \operatorname{End} A / JQ \operatorname{End} A$. Since S_A is a maximal order, $S_A = \operatorname{End} G_A / N \operatorname{End} G_A$, as desired.

Finally, G_A is faithful because $N \operatorname{End} G_A$ is contained in $J \operatorname{End} G_A$, S_A is a maximal order, and maximal orders are hereditary [**Arnold 82**, Theorem 11.3].

(b) Assume that G_A has the one-sided UDS property and $B \in \mathrm{TF}$ is quasi–isomorphic to A. Since G_A is faithful, G_A satisfies the hypotheses of Theorem 1.3(a) as a consequence of Theorem 1.3(b).

It is sufficient to show that B also satisfies these hypotheses, in which case B has the one-sided UDS property by Theorem 1.3(a). Now q–rank $B = q$–rank G_A since B and G_A are quasi–isomorphic [**Arnold 82**, Theorem 0.2]. Hence by Theorem 1.3(b) applied to G_A, there is at most one prime p with p-rank $B > 1$. Since $S_A = \operatorname{End} G_A / N \operatorname{End} G_A$ it follows that $m_p(G_A) = m_p(S_A)$, the number of maximal right ideals of S_A containing p. Moreover, $S_A \cong S_B$ since A and B are quasi-isomorphic, $\operatorname{End} A / N \operatorname{End} A$ and $\operatorname{End} B / N \operatorname{End} B$ are quasi–isomorphic, and S_A is a maximal order with $S_A / (\operatorname{End} A / N \operatorname{End} A)$ bounded.

Now, $m_p(S_A) = m_p(S_B) \geq m_p(\operatorname{End} B / N \operatorname{End} B) = m_p(B)$. Hence, $2 \geq m_p(G_A) \geq m_p(B)$.

The remaining case is that $m_p(B) = m_p(G_A) = 2$. Since $S_B \cong S_A$, it suffices to assume that $\operatorname{End} B$ is contained as a subring of bounded index in $\operatorname{End} G_A$. Let N_1 and N_2 be two maximal right ideals of $\operatorname{End} B$ containing p and M_1 and M_2 two maximal right ideals of $\operatorname{End} G_A$ containing p with M_1 principal, $G_A / M_1 G_A$ cyclic, and $M_1 / p \operatorname{End} G_A$ having the unique maximal condition. Then $(\operatorname{End} G_A) N_1 = M_1$, since $M_1 / p \operatorname{End} G_A$ has the unique maximal condition. It follows that $N_1 = M_1 \cap \operatorname{End} B$ is principal, $B / N_1 B$ is cyclic, and $N_1 / p \operatorname{End} B$ has the unique maximal condition. This completes the conditions of Theorem 1.3(a) for B. □

Proof of Theorem 1.5 (a) $\Rightarrow$(b) Assume that A is strongly indecomposable with the UDS property. It is sufficient to show that $G = G_A$ has the one–sided UDS property, in which case each group quasi-isomorphic to A has the one–sided UDS property by Lemma 1.4(b).

Let $G \oplus B = B_1 \oplus \cdots \oplus B_n$ with each B_i quasi–isomorphic to G. Since A has the UDS property and G is faithful, there is at most one prime p with p–rank $G > 1$, and either $m_p(G) = 1$ and $\operatorname{End} G/p \operatorname{End} G$ is a local ring; $m_p(G) = 2$ and $\operatorname{End} G$ has exactly two maximal right ideals M_1 and M_2 containing p such that M_1 is principal, $G/M_1 G$ is cyclic, and $M_1/p \operatorname{End} G$ has the unique maximal condition; or $m_p(G) \geq 3$. This is true since, except for the condition that $m_p(G) \geq 3$, the complementary summand B constructed in the proof of Theorem 1.3(b), via Lemma 1.1, is quasi–isomorphic to G, hence A.

It remains to eliminate the possibility that $m_p(G) \geq 3$, for in that case, G has the one–sided UDS condition by Theorem 1.3(a). Let M_1, M_2 and M_3 be the maximal right ideals of $\operatorname{End} G$ containing p. Since M_1 and $M_2 \cap M_3$ are co-maximal right ideals, it follows that M_1 is a principal right ideal with $G/M_1 G \cong Z/pZ$, as an application of Lemmas 1.1 and 1.2. Similarly, M_i is principal with $G/M_i G \cong Z/pZ$ for $i = 2, 3$.

If $i \neq j$, there is a subgroup G_{ij} of G containing $(M_i \cap M_j)G$ that is quasi-isomorphic, but not isomorphic, to G (Lemma 1.2(b)). Moreover, $[G : (M_i \cap M_j)G] = p^2$, since each $[G : M_i G] = p$. Hence, these G_{ij} can be chosen with $[G : G_{ij}] = p$ and $G_{12} = G_{13} \cap G_{12} + G_{23} \cap G_{12}$.

Now $G \oplus (G_{13} \cap G_{12} \oplus G_{23} \cap G_{12}) \cong G_{12} \oplus G_{13} \oplus G_{23}$. To see this, define $\sigma : G' = G_{12} \oplus G_{13} \oplus G_{23} \to G$ by $\sigma(g_{12}, g_{13}, g_{23}) = g_{12} - g_{13} - g_{23}$. Then σ is onto since $\operatorname{End} G = M_1 \cap M_2 + M_1 \cap M_3 + M_2 \cap M_3$ and each $M_i \cap M_j$ is contained in G_{ij}. Write $1_G = m_{12} + m_{13} + m_{23}$ with $m_{ij} \in M_i \cap M_j$. Define $\rho : G \to G'$ by $\rho(g) = (m_{12}g, m_{13}g, m_{23}g)$. Then $\sigma\rho = 1_G$ and so $G' = \rho(G) \oplus \ker \sigma$. But $\ker \sigma = \{(g_{13} + g_{23}, g_{13}, g_{23}) : g_{13} + g_{23} \in G_{12}\}$. Since $G_{12} = G_{13} \cap G_{12} + G_{23} \cap G_{12}$, $\ker \sigma \cong G_{13} \cap G_{12} \oplus G_{23} \cap G_{12}$, as desired.

But G is not isomorphic to any of the G_{ij}, a contradiction to the assumption that G is quasi-isomorphic to A and A has the UDS property. Thus, $m_p(G) \leq 2$, as desired.

(b) $\implies$ (c) follows from Theorem 1.3(b) and the fact that G_A is a faithful strongly indecomposable group in TF quasi–isomorphic to A with $\operatorname{End} G_A / N \operatorname{End} G_A = S_A$.

(c) $\implies$ (a) Assume that $A_1 \oplus \cdots \oplus A_n \cong B_1 \oplus \cdots \oplus B_m$ with each A_i and B_j quasi–isomorphic to A. In view of Theorem 1.3(a), G_A has the one-sided UDS property. By Lemma 1.4(b), B has the one-sided UDS property for each $B = A_i$. Hence, A_i is isomorphic to some B_j. A similar argument gives each B_j isomorphic to some A_r. It now follows by an induction argument that $m = n$ and that, after relabeling, A_i is isomorphic to B_i for each i. □

2. UDI Property

Given $A \in \mathrm{TF}$, $\operatorname{End} A$ is commutative with $N \operatorname{End} A = 0$ if and only if $Q \operatorname{End} A$ is an algebraic number field. In this case, A is faithful, as noted above. Moreover, $M_1/p \operatorname{End} A$ has the unique maximal condition if and only if $p \operatorname{End} A = M_1 J$ for some ideal J not contained in M_1. Hence, as a special case of Theorems 1.3 and 1.5:

COROLLARY 2.1. *Suppose $A \in \mathrm{TF}$ is strongly indecomposable with $Q \operatorname{End} A$ a field, S_A the integral closure of $\operatorname{End} A$ in $Q \operatorname{End} A$, and G_A in TF quasi-isomorphic to A with $\operatorname{End} G_A = S_A$. Then*

(a) *A has the one-sided UDS property if and only if there is at most one prime p with p-rank $A > 1$ and either $\mathrm{End}\, A/p\,\mathrm{End}\, A$ is a local ring or else $p\,\mathrm{End}\, A = M_1 I_2$, where M_1 is a principal maximal ideal, I_2 is a primary ideal with rad $I_2 = M_2 \neq M_1$, and $A/M_1 A \cong Z/pZ$.*

(b) *A has the UDS property if any only if there is at most one prime p with p-rank $A > 1$ and either S_A/pS_A is a local ring or else $pS_A = M_1 I_2$, where M_1 is a principal maximal ideal, I_2 is a primary ideal with rad $I_2 = M_2 \neq M_1$, and $G_A/M_1 G_A \cong Z/pZ$.*

□

THEOREM 2.2. *Assume that $A \in TF$ and $Q\,\mathrm{End}\, A$ is a field. If A has the UDS property, then $\mathrm{End}\, A$ has the UDI property.*

PROOF. Let S be the integral closure of $\mathrm{End}\, A$ in its quotient field and define $B = SA$, a subgroup of QA. Then A is quasi–isomorphic to B and B has the UDS property since $S/\mathrm{End}\, A$ is finite. However, $S = \mathrm{End}\, B$ is a Dedekind domain, since S is contained in $\mathrm{End}\, B$, S is integrally closed, and $\mathrm{End}\, B$ is quasi–isomorphic to S.

Since $\mathrm{End}\, B$ is a Dedekind domain, $I = \mathrm{End}(B, IB)$ for each ideal I of $\mathrm{End}\, B$, [**Arnold 82**, Exercise 5.5]. To see that $\mathrm{End}\, B$ has the UDI property, suppose $I_1 \oplus \cdots \oplus I_m \cong J_1 \oplus \cdots \oplus J_n$ for ideals I_i and J_i of $\mathrm{End}\, B$. Then $I_1 B \oplus \cdots \oplus I_m B \cong J_1 B \oplus \cdots \oplus J_n B$. If I is a non-zero ideal of $\mathrm{End}\, B$, then $(\mathrm{End}\, B)/I$ is finite since I contains a non-zero integer n and $\mathrm{End}\, B/n\,\mathrm{End}\, B$ is finite. Hence, each $I_i B$ and $J_j B$ is quasi-isomorphic to B. Because B has the UDS property, $m = n$ and, after relabeling, $I_i B \cong J_i B$. Hence, $I_i = \mathrm{End}(B, I_i B) \cong \mathrm{End}(B, J_i B) = J_i$ for each i, as desired. It now follows from [**Goeters, Olberding 99**] that $\mathrm{End}\, A$ has the UDI property since $\mathrm{End}\, A$ is a one–dimensional Noetherian domain with intgral closure S and $S/\mathrm{End}\, A$ is finite.

□

A variety of examples of subrings of algebraic number fields with the UDI property are given in [**Goeters, Olberding 99**].

3. Examples

The results of Section 1 can be used to find explicit examples of groups with non-unique direct sum decompositions. A group in TF is a *Krull-Schmidt group* if direct sum decompositions into indecomposable groups are unique up to isomorphism and order. The following example is illustrative of classical examples given in [**Jonsson 57**], [**Jonsson 59**] and [**Fuchs 73**]. An *almost completely decomposable group* is a group in TF quasi–isomorphic to a finite direct sum of rank–1 groups.

EXAMPLE 3.1. *Let $S = \{p_1, p_2, \ldots, p_k\}$ be a set of $k \geq 3$ distinct primes of Z. There is a rank-4 almost completely decomposable group G with $p_i G \neq G$ for each i that is not a Krull-Schmidt group.*

PROOF. Let $A_i = Z[1/p_i]$, the subring of Q generated by $1/p_i$, and define $A = A_1 \oplus A_2 + Z(1,1)/p_3$, a subgroup of $Q \oplus Q$. Since each A_i is fully invariant in A, A is indecomposable and $\mathrm{End}\, A = \{(r_1, r_2) \in A_1 \times A_2 : r_1 \equiv r_2 \pmod{p_3}\}$ is commutative. Because $A/pA = A_{(p)}/pA_{(p)}$, with $A_{(p)} = Z_{(p)} \otimes A$ the localization of A at a prime p, p_i-rank $A = 1$ for $i = 1, 2$ and p-rank $A = 2$ for $p \neq p_1, p_2$.

In view of Lemmas 1.1 and 1.2(a), there are subgroups A_1 and A_2 of A quasi-isomorphic to A with $G = A \oplus B \cong A_1 \oplus A_2$ and A not isomorphic to A_1 or A_2. Thus, G is a rank-4 almost completely decomposable group that is not a Krull-Schmidt group. □

A procedure for determining the number of non-equivalent direct sum decompositions of an almost completely decomposable group from its endomorphism ring is given in [Reid 99]. Examples of direct sums of strongly indecomposable groups with non-unique direct sum decompositions can also be easily constructed:

EXAMPLE 3.2. *Let $\{p_1, \ldots, p_k\}$ be a set of $k \geq 2$ odd primes of Z. There is a direct sum G of two strongly indecomposable groups of rank 3 with $p_i G \neq G$ for each i that is not a Krull-Schmidt group.*

PROOF. Write $n = p_1 p_2 \ldots p_k$. Then $f(x) = x(x+1)(x+2) + n(x^2+2) = x^3 + (3+n)x^2 + 2x + 2n$ is an irreducible polynomial in $Q[x]$ by Eisenstein's criterion. Let $S = Z[x]/\langle f(x)\rangle$, a subring of the algebraic number field $F = Q[x]/\langle f(x)\rangle$.

Define $R = S[1/(x+2)]$, a subring of F containing S. Then R is a torsion-free abelian group of rank 3 with p_i–rank 2 for each i. There is a subring A of F that is strongly indecomposable as a group with $A \cong \operatorname{End} A$ and R a group direct sum of m copies of A [**Reid 81**]. Moreover, A is faithful since $\operatorname{End} A$ is commutative.

For $p = p_i$, p–rank $R = 2 = m(p\text{–rank } A)$ and rank $R = 3 = m(\operatorname{rank} A)$. Hence $m = 1$ and $R = A$ is a faithful strongly indecomposable group with p_i–rank 2 and rank 3 for each $i \leq k$. Since $k \geq 2$, via Lemma 1.1, there are subgroups A_1 and A_2 of A quasi–isomorphic to A with $G = A \oplus B \cong A_1 \oplus A_2$ and A not isomorphic to either A_i. □

A group A in TF is called p–local if $qA = A$ for each prime $q \neq p$.

EXAMPLE 3.3. *If p is an odd prime, then there is a p-local group A in TF of rank 10 that is not a Krull-Schmidt group.*

PROOF. Let $f(x) = x(x+1)(x+2)(x+3)(x+4) + p(x^3+2) = x^5 + 10x^4 + (35+p)x^3 + 50x^2 + 24x + 2p$, an irreducible polynomial in $Q[x]$ by Eisenstein's criterion. Let $S = Z_p[x]/\langle f(x) \rangle$, a subring of the algebraic number field $Q[x]/\langle f(x) \rangle$ with S/pS the product of five copies of Z/pZ. Define $A = S[1/x+4]$. As above, A is a p–local group of rank 5 and p-rank 4 and $A \cong \operatorname{End} A$.

Now $pA = p\operatorname{End} A = M_1 \cap M_2 \cap M_3 \cap M_4$ with each M_i a maximal ideal of $\operatorname{End} A$. Since $M_1 \cap M_2$ and $M_3 \cap M_4$ are co–maximal right ideals, there are subgroups A_1 and A_2 of bounded index in A that are not isomorphic to A with $(M_1 \cap M_2)A$ contained in A_1 and $(M_3 \cap M_4)A$ contained in A_2. As in Lemma 1.1, there is a p-local group B with $G = A \oplus B = A_1 \oplus A_2$. Consequently, G is a rank 10 p-local group that is not a Krull-Schmidt group. □

REMARK 3.4. A similar construction could be used to find a p-local group $A \in$ TF with p-rank 3 and rank 4 such that A is isomorphic to $\operatorname{End} A$ and $p\operatorname{End} A$ is the product of three distinct maximal ideals. Hence, A does not have the one–sided UDS property. However, the proof of Theorem 1.3 results in a rank 12 group $G = A \oplus B \cong A_1 \oplus A_2 \oplus A_3$ with each A_i quasi–isomorphic, but not isomorphic, to A.

References

[Arnold 72] Arnold D. *A duality for torsion-free modules of finite rank over a discrete valuation ring*, Proc. Lond. Math. Soc. (3) 24 (1972), 204-216.

[Arnold 82] Arnold, D. *Finite Rank Torsion-Free Abelian Groups and Rings*, Lect. Notes in Math. 931, Springer-Verlag, New York, 1982.

[Arnold 00] Arnold, D. *Abelian Groups and Representations of Finite Partially Ordered Sets*, CMS Books in Mathematics, Springer-Verlag, New York, 2000.

[Arnold, Dugas 99] Arnold, D. and Dugas, M. *Co-purely indecomposable modules over a discrete valuation ring*, preprint.

[Beaumont, Pierce 61] Beaumont, R. and Pierce, R. *Torsion-free rings*, Ill. J. Math. 5 (1961), 61-98.

[Faticoni 99] Faticoni, T. *Direct Sums and refinements*, Comm. in Alg. 27 (1999), 451-464.

[Fuchs 73] Fuchs, L. *Infinite Abelian Groups*, Vol. II, Academic Press, New York, 1973.

[Goeters, Olberding 99] Goeters, P. and Olberding, B. *Unique decompositions into ideals for noetherian domains*, preprint.

[Goeters, Olberding 00] Goeters, P. and Olberding, B. *Krull-Schmidt for ideals and modules over integral domains*, preprint.

[Goldsmith, May 99] Goldsmith, B. and May, W. *The Krull-Schmidt problem for modules over valuation domains*, J. Pure and Applied Alg. 140 (1999), 57-63.

[Jonsson 57] Jonsson, B. *On direct decompositions of torsion-free abelian groups*, Math. Scand. 5 (1957), 230-235.

[Jonsson 59] Jonsson, B. *On direct decompositions of torsion-free abelian groups*, Math. Scand. 7 (1959), 361-371.

[Levy, Odenthal 96] Levy, L. and Odenthal, C. *Krull-Schmidt theorems in dimension 1*, Trans. A.M.S. 348 (1996), 3391-3455.

[Murley 72] Murley, C. E. *The classification of certain classes of torsion-free abelian groups of finite rank*, Pac. J. Math. 40 (1972), 647-665.

[Reid 63] Reid, J. *On the ring of quasi-endomorphisms of a torsion-free group*, Topics in Abelian Groups, Chicaco, 1963.

[Reid 81] Reid, J. *Abelian groups finitely generated over their endomorphism rings*, Abelian Groups, Lect. Notes in Math. 874, Springer-Verlag, New York, 41-52.

[Reid 99] Reid, J. *Some maxtrix rings associated with acd groups*, Abelian Groups and Modules, Birkhaeuser, New York, 1999.

[Salce, Zanardo 89] Salce, L. and Zanardo, P. *Finitely generated modules over valuation domains*, Abelian Groups and Modules , Contemporary Mathematics, A.M.S. 87 (1989), 249-261.

[Walker 64] Walker, E.A. *Quotient categories and quasi-isomorphisms of abelian groups*, Proc. Colloq. Abelian Groups, Budapest, 1964, 147-162.

[Warfield 72] Warfield, R.B. Jr. *Exchange rings and decompositions of modules*, Math. Ann. 199 (1972), 31-36.

[Warfield 80] Warfield, R.B. Jr. *Cancellation of modules and groups and stable range of endomorphism rings*, Pac. J. Math. 91 (1980), 457-485.

DEPARTMENT OF MATHEMATICS, BAYLOR UNIVERSITY, WACO, TEXAS, 76798-7328
E-mail address: David_Arnold@baylor.edu

Contemporary Mathematics
Volume **273**, 2001

The endomorphism ring of a bounded abelian p–group

Maria Alicia Aviñó and Phill Schultz

Dedicated to Professor Laszlo Fuchs in honour of his 75th birthday.

ABSTRACT. We classify the ideals of the endomorphism ring of a bounded abelian p–group by numerical invariants. As an application, we find the upper annihilator sequence and the Loewy sequence of the Jacobson radical.

1. Introduction and Notation

Let $G = \oplus_{i \in [1,s]} G_i$ be a bounded abelian p–group, where s is a positive integer and for all $i \in [1, s]$, G_i is a direct sum of r_i copies of a cyclic group of order p^{n_i}, where $0 < n_1 < \cdots < n_s$ are integers and the r_i are non–zero cardinal numbers. We say that G_i is homocyclic of exponent n_i and rank r_i.

We write $\mathcal{E} = \mathcal{E}(G)$ for the endomorphism ring of G. With respect to some fixed decomposition of G, denote the embedding of G_i in G by ι_i and the projection of G onto G_i by π_i.

The following facts are well–known, see for example [**F73**]:

A $\mathcal{E}$ can be represented as an $s \times s$ matrix ring $(\mathcal{E}_{ij})$, where for all $(i,j) \in [1,s] \times [1,s]$, $\mathcal{E}_{ij} = \mathrm{Hom}(G_i, G_j)$, considered as an $\mathcal{E}_{ii} - \mathcal{E}_{jj}$–bimodule.

B   The mapping $\mathcal{E} \longrightarrow \mathcal{E}_{ij}$, $f \mapsto \iota_i f \pi_j$ is an additive epimorphism which is also a surjective ring map if $i = j$.

C   If I is an ideal of $\mathcal{E}$ then $I = (I_{ij})$ where each $I_{ij} = \iota_i I \pi_j$ is an $\mathcal{E}_{ii} - \mathcal{E}_{jj}$–sub-bimodule of $\mathcal{E}_{ij}$.

The object of this paper is to classify the ideals of $\mathcal{E}$ by numerical invariants. For finite G, the ideals of $\mathcal{E}$ have been described by Shoda in [**Sh28**]. In particular, each ideal of $\mathcal{E}$ is characterized by an $s \times s$ matrix of powers of p. However, when G is infinite new phenomena arise which have been dealt with in the homocyclic case by Hausen, [**H82**]. She showed that if G is an infinite homocyclic p–group, then each ideal of $\mathcal{E}$ is characterized by a finite sequence of pairs $(k(i), \mu(i))$, $i \in [0, t]$, where t is a positive integer and for all $i \in [0, t]$, $k(i)$ is a non–negative integer and

---

1991 *Mathematics Subject Classification.* Primary: 20K10, 20K30, 20F28; Secondary: 16S50.
Key words and phrases. bounded abelian p–groups, endomorphism rings.
The first author would like to thank the Mathematics Department of New Mexico State University for its support during 1999 when this work was done.

$\mu(i)$ an ordinal. We shall show that these results may be combined to characterize ideals of $\mathcal{E}$ by $s \times s$ matrices of finite sequences of pairs $(k(i), \mu(i))$.

In Section 2 we present new proofs of the results of Shoda and and Aviñó and Bautista cited above which characterise the ideals of the endomorphism ring of a finite abelian p–group. In Section 3 we present Hausen's description of the ideals of the endomorphism ring of a homocyclic abelian p–group. Section 4 contains the classification of the ideals of the endomorphism ring $\mathcal{E}$ of an arbitrary bounded abelian p–group. Finally Section 5 contains the applications of the foregoing results to describe the upper annihilating sequence and the Loewy sequence of the Jacobson radical of $\mathcal{E}$.

Notation

1. If x is a group element of order p^n, then the exponent of x, $\exp(x) = n$. If G is a non–zero bounded abelian p–group and n the minimal positive integer such that $p^n G = 0$, then the exponent of G, $\exp(G) = n$.
2. If n and m are cardinal numbers, we denote their maximum by $n \vee m$ and their minimum by $n \wedge m$.
3. If $f \in \mathcal{E}$, then rank f is the cardinality of a maximal independent set in the abelian group Gf.

2. Ideals in the endomorphism ring of a finite abelian p–group

Henceforth we use the notation of Section 1 without comment.

LEMMA 2.1. *Let $G = \oplus_{i \in [1,s]} G_i$ be a finite abelian p-group and let $I = (I_{ij})$ be an ideal of $\mathcal{E}$. Let $(i,j) \in [1,s] \times [1,s]$ and let $G_i = \oplus_{k \in [1,r_i]} \langle a_k \rangle$ and $G_j = \oplus_{\ell \in [1,r_j]} \langle b_\ell \rangle$ be decompositions into cyclic summands. Let π_k be the projection of G_i onto $\langle a_k \rangle$ and let π_ℓ be the projection of G_j onto $\langle b_\ell \rangle$ with respect to these decompositions. For each $f \in \mathcal{E}_{ij}$ denote $\pi_k f \pi_\ell$ by $f_{k\ell}$.*

If for some $f \in I_{ij}$ and some $(s,t) \in [1,r_i] \times [1,r_j]$, f_{st} has height n, then $p^n \mathcal{E}_{ij} \subseteq I_{ij}$.

PROOF. Let $f_{st} = p^n g_{st}$ for some $g \in \mathcal{E}_{ij}$ such that g_{st} has height 0.

Let $h \in p^n \mathcal{E}_{ij}$, so for each $(k, \ell) \in [1, r_i] \times [1, r_j]$, $h_{k\ell}$ maps a_k into $p^n \langle b_\ell \rangle$. Let $d_k \in \mathcal{E}_{ii}$ map a_k onto a_s and annihilate the complement $\oplus_{k' \neq k} \langle a_{k'} \rangle$. Similarly, let $e_\ell \in \mathcal{E}_{jj}$ map b_t onto b_ℓ and annihilate the complement $\oplus_{\ell' \neq \ell} \langle b_{\ell'} \rangle$. Then $d_k f_{st} e_\ell = p^n d_k g_{st} e_\ell \in I_{ij}$ maps a_k onto $p^n b_\ell$. Hence $h_{k\ell}$ is an integral multiple of $p^n d_k g_{st} e_\ell$, so $h_{k\ell} \in I_{ij}$. Thus each $h_{k\ell} \in I_{ij}$ and hence their sum $h \in I_{ij}$. □

PROPOSITION 2.2. *Let $G = \oplus_{i \in [1,s]} G_i$ be a finite abelian p-group and let $I = (I_{ij})$ be an ideal of $\mathcal{E}$. Define a function α on $[1,s] \times [1,s]$ by $\alpha(i,j) = \min\{m : I_{ij}$ contains an element f with some component f_{st} of height $m\}$.*
Then $I_{ij} = p^{\alpha(i,j)} \mathcal{E}_{ij}$.

PROOF. Suppose $f \in I_{ij}$. Then height $f \geq \alpha(i,j)$ so $f \in p^{\alpha(i,j)} \mathcal{E}_{ij}$.

Conversely, let $g \in p^{\alpha(i,j)} \mathcal{E}_{ij}$. Now I_{ij} contains an element f with some component f_{st} of height $\alpha(i,j)$, so by Lemma 2.1, $g \in I_{ij}$. □

Proposition 2.2 describes the bimodules which can form the matrix components of an ideal. The next Theorem shows how the components corresponding to different (i,j) are linked. For $i \in [1, s-1]$, we define $\text{gap}(i) = n_{i+1} - n_i$.

THEOREM 2.3. *Let* $G = \oplus_{i \in [1,s]} G_i$ *be a finite abelian p–group and let* $I = (I_{ij})$ *be an ideal of* $\mathcal{E}$. *Define the function* α *as in Proposition 2.2. Then* α *satisfies the following conditions for all* (i,j) *for which the inequalities are defined:*

IF 1: $0 \leq \alpha(i,j) \leq n_{i \wedge j}$;
IF 2: $0 \leq (n_{i \wedge (j+1)} - \alpha(i, j+1)) - (n_{i \wedge j} - \alpha(i,j)) \leq \text{gap}(j)$
IF 3: $0 \leq (n_{(i+1) \wedge j} - \alpha(i+1, j)) - (n_{i \wedge j} - \alpha(i,j)) \leq \text{gap}(i)$.

Furthermore, any function α *defined on* $[1,s] \times [1,s]$ *satisfying conditions* **IF 1**, **IF 2** *and* **IF 3** *determines a unique ideal* I *for which* $I_{ij} = p^{\alpha(i,j)} \mathcal{E}_{ij}$.

PROOF. First note that the definition of α as well as the conditions **IF 1**–**IF 3** are independent of the ranks r_i of the homocyclic summands of G. Hence without loss of generality we may assume that each $r_i = 1$ and that $G_i = \langle a_i \rangle$.

IF 1: Since $p^{n_{i \wedge j}} \mathcal{E}_{ij} = 0$, from the definition of α it follows that $0 \leq \alpha(i,j) \leq n_{i \wedge j}$.

IF 2: Let $i \in [1,s]$ and $j \in [1, s-1]$ and suppose $\alpha(i,j) = m$ and $\alpha(i, j+1) = n$. Let $e \in \mathcal{E}_{j,j+1}$ map a_j onto $p^{\text{gap}(j)} a_{j+1}$ and let $f \in \mathcal{E}_{j+1,j}$ map a_{j+1} onto a_j.

First suppose $i \leq j$ so that **IF 2** is the statement that $n \leq m \leq n + \text{gap}(j)$. If $g \in p^m \mathcal{E}_{ij}$ then $ge \in p^m \mathcal{E}_{i,j+1}$, so $n \leq m$. Furthermore, if $h \in p^n \mathcal{E}_{i,j+1}$, then $hf \in p^{m+\text{gap}(j)} \mathcal{E}_{ij}$, so $n \leq m + \text{gap}(j)$.

Now suppose $i \geq j+1$ so that **IF 2** is the statement that $n \leq m + \text{gap}(j) \leq n + \text{gap}(j)$. If $g \in p^m \mathcal{E}_{ij}$ then $ge \in p^{m+\text{gap}(j)} \mathcal{E}_{i,j+1}$, so $n \leq m + \text{gap}(j)$. Finally, if $h \in p^n \mathcal{E}_{i,j+1}$, then $hf \in p^m \mathcal{E}_{ij}$, so $m \leq n$.

IF 3: The proof is similar to the proof of **IF 2**, using canonical maps $a_i \mapsto p^{\text{gap}(i)} a_{i+1}$ and $a_{i+1} \mapsto a_i$ in place of e and f.

Now suppose $\alpha : [1,s] \times [1,s]$ satisfies **IF 1–3**. Condition **IF 1** ensures that the sub–bimodules I_{ij} are uniquely defined. Condition **IF 2** ensures that each I_{ij} is closed under right multiplications from $\mathcal{E}_{j,j+1}$. Assuming that I_{ij} is closed under right multiplications from $\mathcal{E}_{j,j+k}$, where $j+k < s$, condition **IF 2** applied with j replaced by $j+k$ ensures that I_{ij} is closed under multiplications from $\mathcal{E}_{j,j+k+1}$. By induction, **IF 2** ensures that I is a right ideal in $\mathcal{E}$. Similarly condition **IF 3** and induction ensure that I is a left ideal in $\mathcal{E}$. □

Functions α satisfying **IF 1–IF 3** are called **ideal functions**. Note that if α and β are ideal functions and we denote $(p^{\alpha(i,j)} \mathcal{E}_{ij})$ by $\mathcal{E}(\alpha)$ and $(p^{\beta(i,j)} \mathcal{E}_{ij})$ by $\mathcal{E}(\beta)$, then $\mathcal{E}(\alpha) \subseteq \mathcal{E}(\beta)$ if and only if $\alpha(i,j) \geq \beta(i,j)$ for all $(i,j) \in [1,s] \times [1,s]$. Thus the following Corollary describes the ideal lattice of $\mathcal{E}$.

COROLLARY 2.4. *The ideal lattice of* $\mathcal{E}$ *is isomorphic to the lattice of ideal functions* α *under the reverse pointwise ordering.* □

3. Ideals in the endomorphism ring of a homocyclic abelian p–group

For G homocyclic of exponent n and infinite rank $r = \aleph_\tau$, Hausen [H82] proved the following classification of the ideals of $\mathcal{E}$. For any ordinal μ, let $\mathcal{E}^{(\mu)} = \{f \in \mathcal{E} : \text{rank } f < \aleph_\mu\}$

HAUSEN'S THEOREM. *Let* G *be homocyclic of exponent* n *and infinite rank* $r = \aleph_\tau$. I *is an ideal of* $\mathcal{E}$ *if and only if there exists a non–negative integer* t, *integers* $0 \leq k(0) < \cdots < k(t) \leq n$ *and ordinals* $0 \leq \mu(0) < \cdots < \mu(t) \leq \tau + 1$ *such that*

$$I = p^{k(0)} \mathcal{E}^{(\mu(0))} + \cdots + p^{k(t)} \mathcal{E}^{(\mu(t))}.$$

These numbers are uniquely determined by I.

We shall call a finite sequence $(\mathbf{k}, \boldsymbol{\mu}) = (\mathbf{k}(i), \boldsymbol{\mu}(i))$ of pairs satisfying the condition of Hausen's Theorem a **Hausen sequence of length** $t + 1$.

If G is homocyclic of finite rank r, then it is well–known that $\mathcal{E}$ is isomorphic to the ring $M(r, \mathbb{Z}(p^e))$ of $r \times r$ matrices over $\mathbb{Z}(p^e)$. In this case, the ideals form a descending chain $\mathcal{E} = I_0 < I_1 < \cdots < I_e = 0$ where $I_j = p^j \mathcal{E} \cong \mathcal{E}(G[p]^{e-j})$, so $\mathcal{E}$ is a local ring with radical $p\mathcal{E}$. For all $j \in [0, e]$ the natural epimorphism $\mathbb{Z}(p^e) \to \mathbb{Z}(p^j)$ induces a ring epimorphism $\mathcal{E} \to \mathcal{E}(G[p^j])$ with kernel I_j.

4. Ideals in the endomorphism ring of a bounded group

If the abelian p–group G is bounded but not finite, Hausen's Theorem shows that not every ideal of $\mathcal{E}$ has the form $\mathcal{E}(\alpha)$ for some ideal function α. We shall show that nevertheless, the lattice of ideals of $\mathcal{E}$ can be parametrised by $s \times s$ matrices of Hausen sequences. This lattice is finite if and only if $|G| < \aleph_\omega$.

Let $G = \oplus_{i \in [1,s]} G_i$, where G_i is homocyclic of exponent n_i and rank r_i. If r_i is infinite, let $r_i = \aleph_{\tau_i}$. Let $\mathcal{E} = \mathcal{E}(G) = (\mathcal{E}_{ij})$ and let $I = (I_{ij})$ be an ideal of $\mathcal{E}$. To economise on notation, we write $\mathcal{E}_i$ for $\mathcal{E}_{ii}$ and I_i for I_{ii}. By Hausen's Theorem, for each $i \in [1, s]$, if r_i is infinite then $I_i = \sum_{u=0}^{t_i} p^{k_i(u)} \mathcal{E}_i^{(\mu_i(u))}$ for some Hausen sequence $(\mathbf{k}_i(u), \boldsymbol{\mu}_i(u))$ of length $t_i + 1$. Denote this ideal by $\mathcal{E}_i(\mathbf{k}_i, \boldsymbol{\mu}_i)$. It is convenient to extend the definition to the case in which r_i is finite. In this case, the Hausen sequence for I_i has length 1, $\mu_i(0) = 0$ and $I_i = p^{k_i(0)} \mathcal{E}_i^{(0)} = p^{k_i(0)} \mathcal{E}_i$.

Thus each $I_i = \mathcal{E}_i(\mathbf{k}_i, \boldsymbol{\mu}_i)$ is uniquely determined by the Hausen sequence $(\mathbf{k}_i, \boldsymbol{\mu}_i)$. We also need to characterize the off–diagonal I_{ij} by Hausen sequences. With this in mind, we say that $(\mathbf{k}_{ij}, \boldsymbol{\mu}_{ij})$ is a Hausen sequence of length $t_{ij} + 1$ for $\mathcal{E}_{ij}$ if $0 \leq k_{ij}(0) < \cdots < k_{ij}(t_{ij}) \leq n_i \wedge n_j$ and $0 \leq \mu_{ij}(0) < \cdots \mu_{ij}(t) \leq (\tau_i \wedge \tau_j) + 1$, if both r_i and r_j are infinite; if at least one of them is finite, a Hausen sequence for $\mathcal{E}_{ij}$ has length 1 and $\mu_{ij}(0) = 0$.

Let $\mathcal{E}_{ij}^{(\mu)} = \{f \in \mathrm{Hom}(G_i, G_j) : \mathrm{rank}\ f < \aleph_\mu\}$ and let

$$\mathcal{E}_{ij}(\mathbf{k}_{ij}, \boldsymbol{\mu}_{ij}) = \sum_{u=0}^{t} p^{k_{ij}(u)} \mathcal{E}_{ij}^{(\mu_{ij}(u))}.$$

It is clear that $\mathcal{E}_{ij}(\mathbf{k}_{ij}, \boldsymbol{\mu}_{ij})$ is an $\mathcal{E}_i - \mathcal{E}_j$ sub–bimodule of $\mathcal{E}_{ij}$. To show that every sub–bimodule has the form $\mathcal{E}_{ij}(\mathbf{k}_{ij}, \boldsymbol{\mu}_{ij})$, we need the concept of an independent set of generators for G_i which is maximal with respect to having as an image an independent set in G_j.

DEFINITION 4.1. Let $i, j \in [1, s]$. A **canonical** (i, j)–**basis** for G is a pair (X_i, X_j) such that
1. for $k = i, j$, X_k is independent in G_k and each element has exponent n_k,
2. $|X_i| = |X_j| = r_i \wedge r_j$.

It is clear that canonical (i, j)–bases exist and that if $r_i \leq r_j$ then X_i is a basis for G_i, while if $r_j \leq r_i$ then X_j is a basis for G_j. Also, for $k = i$ or j, $G_k = A_k \oplus B_k$ where $A_k = \langle X_k \rangle$ and at least one of B_i, B_j is zero.

Given a canonical (i, j)–basis (X_i, X_j), define an **elementary** (i, j)–**connection** to be an endomorphism Φ of G which is zero on $\oplus_{k \neq i} G_k$, maps G_i into G_j, is zero on a complement of A_i and, if $i > j$ then Φ is a 1-1 correspondence between X_i and X_j while if $i < j$, then Φ is a 1-1 correspondence between X_i and $p^{n_j - n_i} X_j$.

LEMMA 4.2. *Let I be an ideal of $\mathcal{E}$ and let $i, j \in [1, s]$. Then $I_{ij} = \mathcal{E}_{ij}(\mathbf{k}_{ij}, \boldsymbol{\mu}_{ij})$ for some Hausen sequence $(\mathbf{k}_{ij}, \boldsymbol{\mu}_{ij})$ for $\mathcal{E}_{ij}$.*

PROOF. We first show that for any positive integer k and any ordinal μ, if $I_{ij} \cap p^k \mathcal{E}_{ij}^{(\mu)} \neq \emptyset$ then $p^k \mathcal{E}_{ij}^{(\mu)} \subseteq I_{ij}$. Let (X_i, X_j) be a canonical (i, j)–basis and let Ψ be an elementary (i, j)–connection with respect to (X_i, X_j) and Φ an elementary (j, i)–connection with respect to (X_j, X_i).

Let $f \in I_{ij} \cap p^k \mathcal{E}_{ij}^{(\mu)}$, say $f = p^k g$ for some $g \in \mathcal{E}_{ij}^{(\mu)}$. Then $g\Phi \in p^{n_j - n_i} \mathcal{E}_i^{(\mu)}$ so $f\Phi \in p^{n_j - n_i + k} I_i^{(\mu)}$. By Hausen's Theorem, $p^{n_j - n_i + k} \mathcal{E}_i^{(\mu)} \subseteq I_i$.

Now let $h \in \mathcal{E}_{ij}$ have height k and rank $\aleph_\mu$. Then $G_i h$ is contained in a free $\mathbb{Z}(p^{n_j})$–module summand of G_j having a basis Y of cardinality $\aleph_\mu$. Let F be the endomorphism of G which is zero on a complement of Y and is a 1–1 correspondence of Y onto X_j. For any $x_i \in X_i$, $x_i h F = \sum_{j \in J} s_{ij} x_j$ for some finite subset J of X_j, with s_{ij} positive integers, each divisible by $p^{n_j - n_i + k}$. Define $\eta \in p^k \mathcal{E}_i^{(\mu)}$ by $\eta = 0$ on a complement of X_i and for $x_i \in X_i$, $x_i \eta = \sum_{j \in J} t_{ij}(x_j \Phi)$, where $p^{n_j - n_i} t_{ij} = s_{ij}$. Since $\eta \in I_i$, $\eta \Psi = h \in I_{ij}$ as required.

Let $(\nu(u) : u \in [1, s])$ be the sequence of second terms in the Hausen sequence for I_i. and let $\boldsymbol{\mu} = (\mu(u) : u \in [1, t])$ be the ordered sequence of distinct elements of this sequence. For each $u \in [1, t]$, suppose that $k(u)$ is minimal such that I_{ij} contains an element of height $k(u)$ and rank $\aleph_{\mu(u)}$. Let $\mathbf{k} = (k(u) : u \in [1, t])$. Then each $p^{k(u)} \mathcal{E}_{ij}^{(\mu(u))} \subseteq I_{ij} \subset \sum_{j \in [1, t]} p^{k(u)} \mathcal{E}_{ij}^{(\mu(u))}$ and hence $I_{ij} = \mathcal{E}_{ij}(\mathbf{k}, \boldsymbol{\mu})$. □

We shall now define an order $\prec$ on Hausen sequences in such a way that if $(\mathbf{k}, \boldsymbol{\mu})$ and $(\boldsymbol{\ell}, \boldsymbol{\nu})$ are Hausen sequences (which may have different lengths) for some bimodule H (which may be $\mathcal{E}_i$ or $\mathcal{E}_{ij}$), then $(\mathbf{k}, \boldsymbol{\mu}) \prec (\boldsymbol{\ell}, \boldsymbol{\nu})$ if and only if $H(\mathbf{k}, \boldsymbol{\mu}) \leq H(\boldsymbol{\ell}, \boldsymbol{\nu})$.

Suppose $(\mathbf{k}, \boldsymbol{\mu})$ has length $t+1$ and $(\boldsymbol{\ell}, \boldsymbol{\nu})$ has length $s+1$. Define $(\mathbf{k}, \boldsymbol{\mu}) \prec (\boldsymbol{\ell}, \boldsymbol{\nu})$ if for all $u \in [0, t]$ there exists $v \in [0, s]$ such that $k(u) \geq \ell(v)$ and $\mu(u) \leq \nu(v)$.

LEMMA 4.3. *With the notation above, $(\mathbf{k}, \boldsymbol{\mu}) \prec (\boldsymbol{\ell}, \boldsymbol{\nu})$ if and only if $H(\mathbf{k}, \boldsymbol{\mu}) \leq H(\boldsymbol{\ell}, \boldsymbol{\nu})$.*

PROOF. Suppose $(\mathbf{k}, \boldsymbol{\mu}) \prec (\boldsymbol{\ell}, \boldsymbol{\nu})$. It suffices to prove that for each $u \in [0, t]$, $p^{k(u)} H^{(\mu(u))} \leq H(\boldsymbol{\ell}, \boldsymbol{\nu})$. Given u, choose $v \in [1, s]$ such that $k(u) \geq \ell(v)$ and $\mu(u) \leq \nu(v)$. Then $p^{k(u)} H^{(\mu(u))} \leq p^{\ell(v)} H^{(\nu(v))} \leq H(\boldsymbol{\ell}, \boldsymbol{\nu})$, as required.

Conversely, suppose $H(\mathbf{k}, \boldsymbol{\mu}) \leq H(\boldsymbol{\ell}, \boldsymbol{\nu})$. Then for all $u \in [0, t]$, $p^{k(u)} H^{(\mu(u))} \leq H(\boldsymbol{\ell}, \boldsymbol{\nu})$. Let $f \in p^{k(u)} H^{(\mu(u))}$, say $f = g_1 + g_2 + \cdots + g_m$ with $g_v \in p^{\ell(v)} H^{(\nu(v))}$ for all $v \in [1, m]$. Then $k(u)$ is at least the height of each g_v and $\aleph_{\mu(u)}$ is at most the rank of some $g(v)$, so there exists $v \in [0, s]$ such that $k(u) \geq \ell(v)$ and $\mu(u) \leq \nu(v)$. Since this holds for every $u \in [0, t]$, $(\mathbf{k}, \boldsymbol{\mu}) \prec (\boldsymbol{\ell}, \boldsymbol{\nu})$. □

It is not evident from the definition that $\prec$ is a partial order, but Lemma 4.3 ensures that it is, and furthermore, the set of Hausen sequences for a given bimodule H is a complete lattice with minimum $\mathbf{0} = (\mathbf{k}, \boldsymbol{\mu})$ of length 1 and maximum $\mathbf{1} = (\boldsymbol{\ell}, \boldsymbol{\nu})$ also of length 1, where $k(0) = n_s$, $\mu(0) = 0$, $\ell(0) = 0$, $\nu(0) = \tau + 1$, and $\aleph_\tau$ is the maximum rank of an element of H. Thus $H(\mathbf{0}) = 0$ and $H(\mathbf{1}) = H$. This lattice is finite if and only if there is a finite bound on the $\mu(t)$ which may occur, that is, if and only if $\tau < \omega$.

We can now develop necessary and sufficient conditions, analogous to the axioms for an ideal function, for a matrix of Hausen sequences to determine an ideal of the endomorphism ring of a bounded abelian p–group G.

Suppose that $I = (I_{ij})$ where

$$I_{ij} = \mathcal{E}_{ij}(\mathbf{k}_{ij}, \boldsymbol{\mu}_{ij}), \ \mathbf{k}_{ij} = (k_{ij}(0), \ldots, k_{ij}(t_{ij})), \ \boldsymbol{\mu}_{ij} = (\mu_{ij}(0), \ldots, \mu_{ij}(t_{ij})),$$

the t_{ij} are positive integers, the $k_{ij}(u)$ are an increasing sequence of integers and the $\mu_{ij}(u)$ are an increasing sequence of ordinals.

LEMMA 4.4. *The Hausen sequences* $(\mathbf{k}_{ij}, \boldsymbol{\mu}_{ij})$ *satisfy for all* i, j *and* u *for which the terms are defined:*

HS 1 $k_{ij}(t_{ij}) \in [0, n_{i \wedge j}]$ and $\aleph_{\mu_{ij}(t_{ij})} \leq r_i \wedge r_j + 1$.
HS 2 $(\mathbf{k}_{i,j+1} + \text{gap}(j), \boldsymbol{\mu}_{i,j+1}) \prec (\mathbf{k}_{i,j} + n_{i \wedge j+1} - n_{i \wedge j}, \boldsymbol{\mu}_{i,j}) \prec (\mathbf{k}_{i,j+1}, \boldsymbol{\mu}_{i,j+1})$
HS 3 $(\mathbf{k}_{i+1,j} + \text{gap}(i), \boldsymbol{\mu}_{i+1,j}) \prec (\mathbf{k}_{i,j} + n_{i+1 \wedge j} - n_{i \wedge j}, \boldsymbol{\mu}_{i,j}) \prec (\mathbf{k}_{i+1,j}, \boldsymbol{\mu}_{i+1,j})$.

PROOF. **HS 1** is obvious from the definition of $\mathcal{E}_{ij}(\mathbf{k}_{ij}, \boldsymbol{\mu}_{ij})$.

HS 2 Suppose $(\mathbf{k}_{i,j+1}, \boldsymbol{\mu}_{i,j+1})$ has length $t+1$ and $(\mathbf{k}_{i,j}, \boldsymbol{\mu}_{i,j})$ has length $s+1$. For the first inequality, let $u \in [0,t]$, let $f \in p^{k_{i,j+1}(u)+\text{gap}(j)} \mathcal{E}_{i,j+1}^{(\mu_{i,j+1}(u))}$ and let Φ be a canonical $(j+1, j)$–connection with respect to a $(j+1, j)$–basis (X_{j+1}, X_j). Then $f\Phi$ has height $k \geq k_{i,j+1}(u) + \text{gap}(j)$ and rank $\aleph_\nu$, where $\nu \leq \mu_{i,j+1}(u)$.

Case 1: $i \leq j$. Since $f\Phi \in p^k \mathcal{E}_{i,j}^{(\nu)}$, there exists $v \in [0, s]$ such that $k_{ij}(v) \geq k_{i,j+1}(u) + \text{gap}(j)$ and $\mu_{ij}(v) \leq \mu_{i,j+1}(u)$, as required.

Case 2: $i \geq j+1$. Since $f\Phi \in p^{k+\text{gap}(j)} \mathcal{E}_{i,j}^{(\nu)}$, there exists $v \in [0, s]$ such that $k_{ij}(v) \geq k_{i,j+1}(u) + \text{gap}(j)$ and $\mu_{ij}(v) \leq \mu_{i,j+1}(u)$, as required.

For the second inequality, suppose $f \in p^{k_{i,j}(v)} \mathcal{E}_{i,j}^{(\mu_{i,j}(v))}$ and let Ψ be a canonical $(j, j+1)$–connection with respect to the $(j, j+1)$–basis (X_j, X_{j+1}).

Case 1: $i \leq j$ Then $f\Psi$ has height $k \geq k_{i,j}(v) + \text{gap}(j)$ and rank $\aleph_\nu$ where $\nu \leq \mu_{i,j}(v)$. Since $f\Psi \in p^k \mathcal{E}_{i,j+1}^{(\nu)}$, there exists $u \in [0, t]$ such that $k_{i,j+1}(u) \geq k_{i,j}(v)$ and $\mu_{i,j+1}(u) \leq \mu_{i,j}(v)$, as required.

Case 2: $i \geq j+1$ Then $f\Psi$ has height $k \geq k_{i,j}(v)$ and rank $\aleph_\nu$ where $\nu \leq \mu_{i,j}(v)$. Since $f\Psi \in p^k \mathcal{E}_{i,j+1}^{(\nu)}$, there exists $u \in [0, t]$ such that $k_{i,j+1}(u) \geq k_{i,j}(v)$ and $\mu_{i,j+1}(u) \leq \mu_{i,j}(v)$, as required.

HS 3 The proof is similar to the proof of **HS 2**. □

The following Theorem is the infinite rank analogue of Theorem 2.3.

THEOREM 4.5. $I = (I_{ij})$ *is an ideal of* $\mathcal{E}$ *if and only if for all* $i, j \in [1, s]$, $I_{ij} = \mathcal{E}_{ij}(\mathbf{k}_{ij}, \boldsymbol{\mu}_{ij})$ *where the matrix of Hausen sequences* $(\mathbf{k}_{ij}, \boldsymbol{\mu}_{ij})$ *satisfies the axiom system* **HS 1–HS 3** *of Lemma 4.4.*

PROOF. Lemma 4.4 shows that the conditions formulated in the axiom system are necessary. To show sufficiency, suppose the Hausen matrix $(\mathbf{k}_{ij}, \boldsymbol{\mu}_{ij})$ satisfies the axioms **HS 1–HS 3**. I is clearly closed under addition, so it remains to prove that $I\mathcal{E} \subseteq I$ and $\mathcal{E}I \subseteq I$. Let $F = (F_{ij}) \in I$ and $\alpha = (\alpha_{ij}) \in \mathcal{E}$. The (i,j) component of $F\alpha$ is $\sum_v F_{iv} \alpha_{vj}$. Let $F_{iv} = \sum_{u \in [0,t]} f_u$, where $f_u \in p^{k_{iv}(u)} \mathcal{E}_{iv}^{\mu_{iv}(u)}$ according to the definition of I_{ij}. Now $f_u \alpha_{vj} \in p^{k_{iv}(u)} \mathcal{E}_{iv}^{(\mu_{iv}(u))} \mathcal{E}_{vj}$. We remark that if $v < j$ then $p^{n_j - n_v} | \mathcal{E}_{vj}$ and if $i < v$ then $p^{n_v - n_i} | \mathcal{E}_{iv}^{(\mu_{iv}(u))}$. Using **HS 1–HS 3**, it is enough to prove the result for $v = j - 1$ and $v = j + 1$.

If $v = j - 1$, then $p^{n_j - n_{j-1}} | \alpha_{j-1,j}$. First suppose $i < j$; then $p^{n_{j-1} - n_j} | \mathcal{E}_{i,j-1}^{(\mu_{i,j-1}(u))}$, so we need to prove that

$$k_{i,j-1}(u) + n_{j-1} - n_i + n_j - n_{j-1} \geq k_{ij}(u) + n_j - n_i$$

that is, $k_{i,j-1}(u) \geq k_{ij}(u)$. But this is true by **HS 2**. Now suppose $i \geq j$; we need to prove that $k_{i,j-1}(u) + n_j - n_{j-1} \geq k_{ij}(u)$. But this is a consequence of **HS 2**.

If $v = j + 1$, first suppose $i \leq j$; we need to prove that $k_{i,j+1}(u) + n_{j+1} - n_i \geq k_{ij}(u) + n_j - n_i$, or $k_{i,j+1}(u) + \text{gap}(j) \geq k_{ij}$ which is a consequence of **HS 2**. Now suppose $i > j$. Then by **HS 2** again, $k_{i,j+1}(u) \geq k_{ij}(u)$.

This concludes the proof that $I\mathcal{E} \subseteq I$, and the proof that $\mathcal{E}I \subseteq I$ is similar, using **HS 3**.

□

5. The upper annihilating sequence and the Loewy sequence

We now apply the results of Section 4 to characterize two special sequences of ideals of $\mathcal{E}$ in terms of their Hausen matrices. It is well-known ([**PS92**]) that the Jacobson radical and nilpotent radical of $\mathcal{E}(G)$ coincide if G is a bounded abelian p-group. Let $\mathcal{J}$ be this radical. The index of nilpotency λ of $\mathcal{J}$ is the least n such that $\mathcal{J}^n = 0$. λ depends on the distribution of gaps in the sequence $\{n_i : i \in [1, s]\}$, ([**PS92**]). The Loewy sequence of $\mathcal{J}$ is the descending sequence $(\mathcal{J}^n : n \in [1, \lambda])$. It was described in graph theoretic terms in [**PS92**] and used in [**S94**] to show that a bounded abelian group G is determined by the radical of $\mathcal{E}$.

The upper annihilating sequence of $\mathcal{J}$ is defined as follows: $I_0 = 0$ and for all t, $I_t / I_{t-1} = \text{Ann}(\mathcal{J}/I_{t-1})$, so

$$I_t = \{f \in \mathcal{J} : f\mathcal{J} \leq I_{t-1} \text{ and } \mathcal{J}f \leq I_{t-1}\}.$$

Thus the upper annihilating sequence (I_t) is an ascending sequence whose length, i.e. the least n such that $I_n = \mathcal{J}$, is $\lambda - 1$. The upper annihilating sequence can be used to characterise the upper central series of the automorphism group of the maximal normal p-subgroup of G.

Note that the definitions only imply that the terms of the sequences are ideals of $\mathcal{J}$. Our characterization will show that they are actually ideals of $\mathcal{E}$.

The definitions of both sequences pose no restrictions on rank and so in the infinite case, the corresponding Hausen sequences all have length 1 and the maximum possible ranks $\boldsymbol{\mu}_{ij}(0) = (\tau_i \wedge \tau_j) + 1$. In other words, the ideals in these sequences depend only on the s-tuple $(n_1, n_2, \ldots, n_s)$. Hence we may replace the Hausen matrices with the corresponding ideal functions. Furthermore, there is no loss in generality but a considerable gain in simplicity in assuming that all $r_i = 1$, as in Section 2. Thus we can take $G_i = \langle a_i \rangle$ for $i \in [1, s]$.

We characterize the sequences by means of the **digraph** $D(G)$ of G. This is the directed graph whose vertices are the elements $p^k a_i$ of G for $k \in [0, n_i]$ and whose arrows fall into four categories:

(a) horizontal arrows: for each $i \in [1, s-1]$ and for each $k \in [0, n_i - 1]$ there is an arrow from $p^k a_i$ to $p^{k + \text{gap}(i)} a_{i+1}$;

(b) diagonal arrows: for each $i \in [2, s]$ and for each $k \in [0, n_i - \text{gap}(i-1) + 1]$ there is an arrow from $p^k a_i$ to $p^k a_{i-1}$;

(c) vertical arrows: for each $i \in [1, s]$ and for each $k \in [0, n_i - 2]$ there is an arrow from $p^k a_i$ to $p^{k+1} a_i$;

(d) zero arrows: for each $i \in [1,s]$ there is an arrow from $p^{n_i-1}a_i$ to 0.

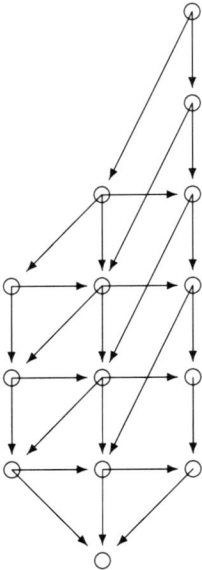

Figure 1

For example, the digraph of a group G whose sequence of exponents is $(n_1, n_2, n_3) = (3, 4, 6)$ has the diagram shown in Figure 1.

It was shown in [**PS92**, Theorem 2.9] that there exists a longest path of length t from vertex a_i to vertex $p^k a_j$ in $D(G)$ if and only if there exists an endomorphism of G in $\mathcal{J}^t$ which maps a_i onto $p^k a_j$. This property was used in Theorem 2.15 of [**PS92**] to determine the nilpotency index of $\mathcal{J}$.

If there is a directed path from a vertex u to a vertex v of $D(G)$, the distance $d(u, v)$ from u to v is defined to be the length of a longest such path. It was shown in [**PS92**] that the unique longest path in $D(G)$ goes from a_s to 0 and has length λ. Also for every vertex u there is a longest path from a_s to u and a longest path from u to 0.

For $t \in [1, \lambda]$ let $D_t = \{u \in D(G) : d(u,0) \leq t\}$. For any subset V of $D(G)$, denote by $\overline{V}$ the subgroup of G generated by V.

PROPOSITION 5.1. *With the notation introduced above:*
  (a) *The sequence $\{\overline{D_t} : t \in [1, \lambda]\}$ is a strictly increasing sequence of fully invariant subgroups of G;*
  (b) *For all $t \in [1, \lambda-1]$, $I_t = \{f \in \mathcal{J} : Gf \leq \overline{D_t}\}$;*
  (c) *For all $t \in [1, \lambda]$, $\mathcal{J}^t = \{f \in \mathcal{J} : \overline{D_t}f = 0\}$.*

PROOF. (a) It follows from the definition that $\{\overline{D_t}\}$ is a strictly increasing sequence of subgroups which are $\mathcal{J}$–invariant. To show that they are in fact fully invariant, it suffices to show that if $p^k a_i$ is a vertex in D_t and $f \in \mathcal{E}$ then every component of $p^k a_i f \in \overline{D_t}$. Now $p^k a_i f$ has exponent $\leq n_i - k$ and height $\geq k$. Let

y be the j-th component of $p^k a_i f$ and suppose y has height ℓ. Then $y \in \langle p^\ell x_j \rangle$, where $\ell \geq k$ and $n_j - \ell \leq n_i - k$. It follows that $d(p^\ell x_j, 0) \leq d(p^k a_i, 0)$, so $y \in \overline{D_t}$.

(b) The proof is by induction on t. If $t = 1$, then I_t is the annihilator of $\mathcal{J}$, so if $x \in G$, $xI_1 \in D_1$, otherwise there would be an endomorphism g in $\mathcal{J}$ such that $xI_i g \neq 0$. Hence $I_1 \leq \{f \in \mathcal{J} : Gf \in \overline{D_1}\}$.

Conversely, if $g \in \{f \in \mathcal{J} : Gf \in \overline{D_1}\}$, then $g\mathcal{J} = 0$ because every endomorphism in $\mathcal{J}$ maps D_1 to zero. Now let $h \in \mathcal{J}$, and let a_i be one of the generators of G. By [**PS92**, Theorem 2.9], each component of $a_i h$ lies in a vertex of $D(G)$ whose distance from 0 is less than that of a_i. Hence $a_i h g = 0$ for all $i \in [1, s]$ and for all $h \in \mathcal{J}$. Hence $\mathcal{J} g = 0$.

Assume the statement is true for some $t \in [1, \lambda - 1]$. Since $I_{t+1} = \{f \in \mathcal{J} : f\mathcal{J} \leq I_t$ and $\mathcal{J}f \leq I_t\}$, if $x \in G$, $xI_{t+1} \in \overline{D_{t+1}}$, otherwise there would be an endomorphism g in $\mathcal{J}$ such that $xI_{t+1}g \notin \overline{D_t}$, contradicting the inductive assumption. Hence $I_{t+1} \leq \{f \in \mathcal{J} : Gf \in \overline{D_{t+1}}\}$.

Conversely, if $g \in \{f \in \mathcal{J} : Gf \in \overline{D_{t+1}}\}$, then $g\mathcal{J} \leq I_t$, because every element of $\mathcal{J}$ maps D_{t+1} closer to 0. Now let $h \in \mathcal{J}$ and let a_i be one of the generators of G. Just as in the proof for the case $t = 1$, each component of $a_i h$ lies in a vertex of $D(G)$ whose distance from D_t is less than that of a_i. Hence $a_i h g \in \overline{D_t}$ for all $i \in [1, s]$ and for all $h \in \mathcal{J}$. Hence $\mathcal{J} g \leq I_t$.

(c) Once again the proof is by induction on t. If $t = 1$, then $\mathcal{J}$ is the annihilator of D_1, as required. Assume the statement is true for some $t \in [1, \lambda - 1]$. Since $\mathcal{J}^{t+1} = \mathcal{J}\mathcal{J}^t$, it annihilates every vertex which is distance 1 from a vertex annihilated by $\mathcal{J}^t$. Hence $\mathcal{J}^{t+1}$ annihilates $\overline{D_{t+1}}$. Conversely, if f annihilates $\overline{D_{t+1}}$, then $f = \sum_k g_k h_k$, where $g_k \in \mathcal{J}$ and h_k annihilates $\overline{D_t}$. Hence $f \in \mathcal{J}^{t+1}$. □

Proposition 5.1 provides the tool for calculating the ideal functions of I_t and $\mathcal{J}^t$ from the digraph $D(G)$. For $t \in [1, \lambda - 1]$, $(i,j) \in [1, s] \times [1, s]$, let

$$a_t(i,j) = \begin{cases} \min\{k : d(p^{k+n_{i \vee j} - n_i} a_j, 0) \leq t\} & \text{if } i \neq j \\ 1 \vee \min\{k : d(p^k a_j, 0) \leq t\} & \text{if } i = j, \end{cases}$$

and let

$$b_t(i,j) = \begin{cases} \min\{k : d(a_i, p^{k+n_{i \vee j} - n_i} a_j) \leq t\} & \text{if } d(a_i, p^{n_j - 1} a_j) \geq t \\ n_{i \wedge j} & \text{otherwise.} \end{cases}$$

THEOREM 5.2. *Let $G = \oplus_{i \in [1, s]} \langle a_i \rangle$. Let (α_t) and (β_t) be the sequences of ideal functions of the upper annihilating sequence (I_t) and the Loewy sequence $(\mathcal{J}^t)$ of $\mathcal{J}$. Then for all relevant t, i and j,*

(a) $\alpha_t(i,j) = a_t(i,j)$;

(b) $\beta_t(i,j) = b_t(i,j)$.

PROOF. (a) Suppose $i \neq j$ and let $k = a_t(i,j)$. Then $d(p^{k+n_{i \vee j} - n_i} a_j, 0) \leq t$, so $p^k a_i \mathcal{E}_{ij} \in \overline{D_t}$, and hence $p^k \mathcal{E}_{ij} \leq I_t$. Thus $a_t(i,j) \leq \alpha_t(i,j)$.

If $i = j$, then the same argument ensures $p^k a_i \mathcal{E}_i \in \overline{D_t}$ and $a_t(i,j) \geq 1$ ensures that $p^k a_i \mathcal{E}_i \in \mathcal{J}$.

Conversely, if k is minimal such that $p^k a_i \mathcal{E}_{ij} \in \overline{D_t}$, then $d(p^{k+n_{i \vee j} - n_i} a_j, 0) \leq t$ so $k \leq a_t(i,j)$. Once again, if $i = j$, $k \geq 1$ ensures $p^k a_i \mathcal{E}_i \in \mathcal{J}$. Thus $\alpha_t(i,j) \leq a_t(i,j)$.

(b) If there is a path of length t in $D(G)$ from a_i to $p^{k+n_{i \vee j} - n_i} a_j$ then each element f of $\mathcal{J}^t$ maps a_i onto an element whose j-component is in $\langle p^{k+n_{i \vee j} - n_i} a_j \rangle$.

Hence $\mathcal{J}_{ij}^t \subseteq p^k \mathcal{E}_{ij}$. If there is no such path, then $\mathcal{J}_{ij}^t = 0$; so $b_t(i,j) \leq \beta_t(i,j)$. Conversely, if $f \in p^k \mathcal{E}_{ij}$ maps a_i into $\langle p^{k+n_{i \vee j}-n_i} a_j \rangle$ then there is a path of length t in $D(G)$ from a_i to $p^{k+n_{i \vee j}-n_i} a_j$ so $f \in \mathcal{J}^t$. Thus $\beta_t(i,j) \leq b_t(i,j)$. □

EXAMPLE 5.3. Applying the algorithm of Theorem 5.2 to the endomorphism ring of the group of Figure 1, we have:

(a) the ideal functions (α_t) of the upper annihilating sequence are:

$$\alpha_0 = \begin{bmatrix} 3 & 3 & 3 \\ 3 & 4 & 4 \\ 3 & 4 & 6 \end{bmatrix} = 0 \quad \alpha_1 = \begin{bmatrix} 3 & 3 & 2 \\ 3 & 4 & 3 \\ 3 & 4 & 5 \end{bmatrix} \quad \alpha_2 = \begin{bmatrix} 3 & 2 & 1 \\ 3 & 3 & 2 \\ 3 & 3 & 4 \end{bmatrix}$$

$$\alpha_3 = \begin{bmatrix} 2 & 2 & 0 \\ 2 & 3 & 1 \\ 2 & 3 & 3 \end{bmatrix} \quad \alpha_4 = \begin{bmatrix} 2 & 1 & 0 \\ 2 & 2 & 1 \\ 2 & 2 & 3 \end{bmatrix} \quad \alpha_5 = \begin{bmatrix} 1 & 1 & 0 \\ 1 & 2 & 0 \\ 1 & 2 & 2 \end{bmatrix}$$

$$\alpha_6 = \begin{bmatrix} 1 & 0 & 0 \\ 1 & 1 & 0 \\ 1 & 1 & 2 \end{bmatrix} \quad \alpha_7 = \begin{bmatrix} 1 & 0 & 0 \\ 0 & 1 & 0 \\ 0 & 1 & 1 \end{bmatrix} \quad \alpha_8 = \begin{bmatrix} 1 & 0 & 0 \\ 0 & 1 & 0 \\ 0 & 0 & 1 \end{bmatrix} = \mathcal{J}$$

(b) the ideal functions (β_t) of the Loewy sequence are:

$$\beta_1 = \begin{bmatrix} 1 & 0 & 0 \\ 0 & 1 & 0 \\ 0 & 0 & 1 \end{bmatrix} = \mathcal{J} \quad \beta_2 = \begin{bmatrix} 1 & 0 & 0 \\ 0 & 1 & 0 \\ 0 & 0 & 2 \end{bmatrix} \quad \beta_3 = \begin{bmatrix} 1 & 1 & 0 \\ 1 & 1 & 1 \\ 0 & 1 & 2 \end{bmatrix}$$

$$\beta_4 = \begin{bmatrix} 2 & 1 & 1 \\ 1 & 2 & 1 \\ 1 & 1 & 3 \end{bmatrix} \quad \beta_5 = \begin{bmatrix} 3 & 2 & 2 \\ 2 & 2 & 2 \\ 2 & 2 & 3 \end{bmatrix} \quad \beta_6 = \begin{bmatrix} 3 & 3 & 2 \\ 3 & 3 & 2 \\ 2 & 2 & 4 \end{bmatrix}$$

$$\beta_7 = \begin{bmatrix} 3 & 3 & 3 \\ 3 & 4 & 3 \\ 3 & 3 & 4 \end{bmatrix} \quad \beta_8 = \begin{bmatrix} 3 & 3 & 3 \\ 3 & 4 & 4 \\ 3 & 4 & 5 \end{bmatrix} \quad \beta_9 = \begin{bmatrix} 3 & 3 & 3 \\ 3 & 4 & 4 \\ 3 & 4 & 6 \end{bmatrix} = 0$$

References

[F73]   L. Fuchs, **Infinite Abelian Groups, Vol. II**, Academic Press, 1973.
[H82]   J. Hausen, *Infinite general linear groups over rings*, Arch. der Math., Vol.39, (1982) 510–524.
[PS92] C. E. Praeger and P. Schultz, *The Loewy length of the Jacobson radical of a bounded endomorphism ring*, in **Abelian groups and noncommutative rings**, AMS Contemporary Mathematics 130, 1992, 349–360.
[Sh28] K. Shoda, *Über die Automorphismen einer endlichen Abelschen Gruppe*, Math. Ann., 100 (1928) 674–686.
[S94]   P. Schultz, *When is an Abelian group determined by the Jacobson radical of its endomorphism ring?*, **Abelian group theory and related topics**, Amer. Math. Soc. Contemporary Mathematics, Vol. 171, 1994, 385–396.

(Aviñó) DEPARTMENT OF MATHEMATICAL SCIENCES, NEW MEXICO STATE UNIVERSITY, LAS CRUCES, NEW MEXICO 88003
E-mail address: `mavino@nmsu.edu`

(Schultz) DEPARTMENT OF MATHEMATICS AND STATISTICS, THE UNIVERSITY OF WESTERN AUSTRALIA, NEDLANDS, AUSTRALIA 6907
E-mail address: `schultz@maths.uwa.edu.au`

The Baer–Kaplansky Theorem for Almost Completely Decomposable Groups

E. Blagoveshchenskaya, G. Ivanov, and P. Schultz

This paper is dedicated to Laszlo Fuchs on his 75th birthday.

ABSTRACT. We prove that block rigid crq groups of ring type satisfy the Baer-Kaplansky Theorem up to near isomorphism.

1. Introduction

If $\mathcal{A}$ is a class of modules, the **Baer-Kaplansky Theorem for** $\mathcal{A}$ is the statement that if $M, N \in \mathcal{A}$ satisfy $\mathrm{End}(M) \cong \mathrm{End}(N)$ as unital rings, then $M \cong N$. Since the class of finite rank torsion–free abelian groups is known to have intractable structure problems, such as failure of unique decomposition into indecomposable summands, there is some interest in finding classes of finite rank torsion–free abelian groups which satisfy the Baer-Kaplansky Theorem.

In Section 2, we prove that the class of completely decomposable groups of ring type satisfies a strong form of the Baer–Kaplansky Theorem.

Within classes of finite rank torsion–free abelian groups, there are several notions of equivalence weaker than isomorphism which have proved useful in classification. For almost completely decomposable (acd) groups, the appropriate equivalence is near isomorphism, [**OV99**], which we denote by $M \cong_n N$. The **near Baer-Kaplansky Theorem** for $\mathcal{A}$ is the statement that for all $M, N \in \mathcal{A}$, $\mathrm{End}(M) \cong \mathrm{End}(N)$ implies $M \cong_n N$. In Section 4 we prove that the class of block rigid crq groups of ring type satisfies the near Baer–Kaplansky Theorem.

The result is first established for rigid crq groups of ring type using results of Blagoveshchenskaya, Mader and Schultz ([**BM94**] and [**MS00**]). It is then extended to block rigid crq groups by techniques created by Ivanov [**I98**]. Since it is known [**L92**] that block rigid near isomorphic acd groups may have non–isomorphic endomorphism rings, the converse in this case is false. However, we show that the

---

1991 *Mathematics Subject Classification.* Primary: 20K15, 20K30.
Key words and phrases. acd groups, regulator.
Much of this work was done while the first author was visiting Macquarie University in late 1999. The first author would like to thank Macquarie University for the research grant which made her visit possible and its Department of Mathematics for its hospitality during her visit.

converse of the near Baer–Kaplansky Theorem holds for the class of rigid crq groups of ring type.

A stronger form of the theorem, which we call the **strong Baer–Kaplansky Theorem**, holds for vector spaces and abelian p–groups. This result states that for every ring isomorphism Θ of $\text{End}(M)$ onto $\text{End}(N)$ there is a module isomorphism ϕ of M onto N such that for all $f \in \text{End}(M)$, $f\Theta = \phi^{-1}f\phi$. If such a stronger form held for near isomorphism, it would state that for every ring isomorphism $\Theta : \text{End}(M) \to \text{End}(N)$ there is a monomorphism $\phi : M \to N$ and a monomorphism $\psi : N \to M$ such that for every $f \in \text{End}(M)$, $f\Theta = \psi f\phi$. We show that this statement is false in general, since it would imply that $N \cong M$.

We are deeply thankful to Adolf Mader for his interest in this work and very useful consultations.

2. Notation and the completely decomposable case

We adopt the standard notation of Fuchs [**F73**] and Mader [**M00**]. In particular, $A \leq B$ means that A is a subgroup of B, and a **type** is a subgroup τ of $\mathbb{Q}$ containing $\mathbb{Z}$. An **acd group** is a torsion–free group of finite rank containing a completely decomposable subgroup of finite index. If X is an acd group, a **regulating subgroup** of X is a completely decomposable subgroup of minimal index. The intersection of the regulating subgroups of X, denoted $R(X)$, is called the **regulator** of X. It is a fully invariant completely decomposable subgroup of X. The factor group $X/R(X)$ is the **regulator quotient** of X, its size is the **regulator index** and its exponent the **regulator exponent**.

We shall denote a generic acd group by X, its regulator by A, its regulator quotient by C and its regulator exponent by e. Thus $eA \leq eX \leq A \leq X$ so C is isomorphic to a subgroup of $e^{-1}A/A$ (where $e^{-1}A$ is a subgroup of a divisible hull of A) and hence to a subgroup of the finite group $\overline{A} = A/eA$.

The **critical typeset** of an acd group X is the finite poset T of types such that $A = \oplus_{\tau \in T} A_\tau$, where A_τ is a homogeneous completely decomposable group of type τ and finite rank r_τ, called the τ**–rank** of X. It follows that X and A have rank $r = \sum_{\tau \in T} r_\tau$. The isomorphism classes of A and C, the set T and the positive integers e and r_τ for all $\tau \in T$ are invariants of X.

We say that an acd group has **ring type** if each critical type, and hence each element type, is idempotent, that is, the type of a rank 1 ring. Since the additive group of integers and any rank 1 group whose type contains no infinities have isomorphic endomorphism rings even though they are not nearly isomorphic, it is clear that in considering Baer–Kaplansky classes of acd groups we should restrict our attention to groups whose critical types are of ring type.

An acd group X is **block rigid** if T is an antichain; if in addition each τ–rank = 1 then X is **rigid**. X is a **crq group** if its regulator quotient is cyclic. In this case of course the exponent e is the order of C. If X is a block rigid crq group, then the regulator A is the unique completely decomposable subgroup of minimal index in X ([**BM94**, p.22]).

We first show that completely decomposable groups of ring type, not necessarily of finite rank, satisfy the strong Baer–Kaplansky Theorem.

THEOREM 2.1. *Let X and Y be completely decomposable groups of ring type. If $\Theta : \text{End}(X) \to \text{End}(Y)$ is a ring isomorphism, then there exists a group isomorphism $\theta : X \to Y$ such that for all $f \in \text{End}(X)$, $f\Theta = \theta^{-1}f\theta$.*

PROOF. Let X have critical typeset T and let $\oplus_{\tau \in T} X_\tau$ be a decomposition of X into a direct sum of homogeneous summands. Let $\{\pi_\tau : \tau \in T\}$ be a complete system of orthogonal idempotents in $\text{End}(X)$ with respect to this decomposition. As an element of the torsion–free additive group $\text{End}(X)^+$, π_τ has infinite p–height if $p\tau = \tau$, since for all $n \in \mathbb{N}$, the equation $p^n f = \pi_\tau$ is solvable in $\text{End}(X)$. On the other hand, if $p\tau \neq \tau$, then X_τ contains an element x of p–height 0 for which $x = x\pi_\tau$, so π_τ has p–height 0.

For all $\tau \in T$, let $\pi'_\tau = \pi_\tau \Theta$. Then $\{\pi'_\tau : \tau \in T\}$ is a system of orthogonal idempotents in $\text{End}(Y)$ such that height $\pi'_\tau =$ height π_τ. Hence the critical typeset of Y contains T, and since the whole argument can be repeated starting with Y instead of X, T is the critical typeset of Y. Thus we have a decomposition $Y = \oplus_{\tau \in T} Y_\tau$ with $Y_\tau = Y\pi'_\tau$ homogeneous of type τ. Whenever we speak of the complement of X_τ or Y_τ in the rest of this proof, we mean with respect to these decompositions.

For each $\tau \in T$, define an isomorphism $\Theta_\tau : \text{End}(X_\tau) \to \text{End}(Y_\tau)$ as follows: let $f \in \text{End}(X_\tau)$, and consider f as an endomorphism of X which is 0 on the complement of X_τ. Then $f = \pi_\tau f \pi_\tau$ so $f\Theta = \pi'_\tau f\Theta \pi'_\tau$. Let $f\Theta_\tau = f\Theta|_{Y_\tau}$. Since $f\Theta$ maps Y_τ into Y_τ, we can consider $f\Theta_\tau$ as an element of $\text{End}(Y_\tau)$ and Θ_τ is a homomorphism. If $f\Theta_\tau = 0$, then $f = 0$, while if $g \in \text{End}(Y_\tau)$ then considering g as an endomorphism of Y which is zero on a complement of Y_τ, $g = \pi'_\tau g \pi'_\tau = (\pi_\tau f \pi_\tau)\Theta$ for some $\pi_\tau f \pi_\tau \in \text{End}(X_\tau)$. Hence Θ_τ is an isomorphism.

For each $\tau \in T$, let a_τ be an element of X_τ whose p–height is 0 if $p\tau \neq \tau$ and ∞ if $p\tau = \tau$, and let π_{a_τ} be the projection of X_τ onto the rank one direct summand which is the pure closure of a_τ. Let $\pi'_{a_\tau} = \pi_{a_\tau} \Theta_\tau$. Then π'_{a_τ} is a projection of Y_τ onto a rank one summand containing an element a'_τ whose height is the same as that of a_τ.

Now every element of X_τ is $a_\tau f$ for some $f \in \text{End}(X_\tau)$ and similarly, every element of Y_τ is $a'_\tau f'$ for some $f' \in \text{End}(Y_\tau)$. Define $\theta_\tau : X_\tau \to Y_\tau$ by $x = a_\tau f \mapsto a'_\tau(f\Theta_\tau)$. Note that θ_τ is a well–defined function, for if $a_\tau f = a_\tau g$ for some $f, g \in \text{End}(X_\tau)$, then $\pi_{a_\tau} f = \pi_{a_\tau} g$ so $\pi'_{a_\tau} f\Theta_\tau = \pi'_{a_\tau} g\Theta_\tau$, and hence $a'_\tau(f\Theta_\tau) = a'_\tau(g\Theta_\tau)$. Clearly θ_τ is an additive homomorphism which is monic for if $a'_\tau(f\Theta_\tau) = 0$ then $\pi'_{a_\tau}(f\Theta_\tau) = 0$ and hence $\pi_{a_\tau} f = 0$. Since every $y \in Y_\tau$ is in the image, θ_τ is an isomorphism.

Now for all $f \in \text{End}(X_\tau)$, $a'_\tau(\theta_\tau^{-1} f \theta_\tau) = a_\tau f \theta_\tau = a'_\tau(f\Theta_\tau)$. Hence $\theta_\tau^{-1} f \theta_\tau$ and $f\Theta_\tau$ agree at a'_τ. For any $y \in Y_\tau$, $y = a'_\tau(g\Theta_\tau)$ for some $g \in \text{End}(X_\tau)$ so $y\theta^{-1}f\theta = a'_\tau(g\Theta_\tau)\theta^{-1}f\theta = a'_\tau\theta^{-1}g\theta\theta^{-1}f\theta = a'_\tau(\theta^{-1}gf\theta) = a'_\tau(g\Theta_\tau)(f\Theta_\tau) = y(f\Theta_\tau)$. Hence $\theta_\tau^{-1} f \theta_\tau = f\Theta_\tau$.

Now define $\theta : X \to Y$ by $\theta|_{X_\tau} = \theta_\tau$. Then θ is an isomorphism for which $f\Theta = \theta^{-1} f \theta$ for all $f \in \text{End}(X)$. $\square$

3. The near Baer–Kaplansky Theorem for rigid crq groups of ring type

The nice result of Section 2 for completely decomposable groups does not hold for arbitrary acd groups. However, we can show that two acd groups with isomorphic endomorphism rings have several invariants in common.

PROPOSITION 3.1. *Let X be an acd group with regulator A, regulator exponent e and critical typeset T. Then*

1. There is a chain of rings $e\operatorname{End}(A) \subseteq \operatorname{End}(X) \subseteq \operatorname{End}(A)$.
2. $\operatorname{End}(X)^+$ is an acd group with critical typeset T, $R(\operatorname{End}(X)^+) \cong \operatorname{End}(A)^+$ and for all $\tau \in T$, $R(\operatorname{End}(X))$ has τ-rank $\sum_{\sigma \leq \tau} r_\sigma r_\tau$.

PROOF. 1. ([**MS00**, Lemma 3.1]) Since A is fully invariant in X, the restriction homomorphism maps $\operatorname{End}(X)$ monomorphically onto $\{f \in \operatorname{End}(A) : (eX)f \leq eX\}$, a subring of $\operatorname{End}(A)$. If $f \in \operatorname{End}(A)$ then $(eX)ef \leq eAf \leq eA \leq eX$ so $ef \in \operatorname{End}(X)$.

2. There is a chain of additive subgroups $e\operatorname{End}(A)^+ \leq \operatorname{End}(X)^+ \leq \operatorname{End}(A)^+$ in which $e\operatorname{End}(A)^+$ and $\operatorname{End}(A)^+$ are completely decomposable isomorphic groups. Hence $\operatorname{End}(X)^+$ contains a completely decomposable subgroup of finite index so is acd.

Furthermore, since all completely decomposable subgroups of $\operatorname{End}(X)^+$ of finite index are isomorphic, $R(\operatorname{End}(X)^+) \cong \operatorname{End}(A)^+$. The invariants of $\operatorname{End}(A)^+$ are well–known (see for example [**F73**, Chapter 13].) In particular, the critical types are of the form $\operatorname{Hom}(\sigma, \tau)$ where σ and τ are critical types of X with $\sigma \leq \tau$. Since all types in T are of ring type, $\operatorname{Hom}(\sigma, \tau) = \tau$ if $\sigma \leq \tau$, and hence the critical typeset of $\operatorname{End}(X)^+$ is T and for all $\tau \in T$, the τ-component of $\operatorname{End}(A)^+$ is isomorphic to $\oplus_{\sigma \leq \tau} \oplus_{r_\sigma r_\tau} \tau$. □

From now on we concentrate on rigid or block rigid crq groups. Let X be a rigid crq group of ring type with regulator

$$(1) \qquad R(X) = A = \bigoplus_{\tau \in T} \tau a_\tau,$$

where $\{a_\tau\} \in A$ has p–height 0 if $p\tau \neq \tau$ and p–height ∞ if $p\tau = \tau$.

It was proved in [**M00**, Chapter 6] that a crq group X with regulator A and regulator index e has **standard representation** $X = \langle A, b \rangle$, where b comes from a fixed divisible hull of A and $eb = a$ for some $a \in A$.

By suitable choice of the a_τ, by [**BM94**, Theorem 3.2, A.1] we can write

$$(2) \qquad eb = \sum_{\tau \in T} \frac{e}{m_\tau} s_\tau a_\tau$$

with e, m_τ and $s_\tau \in \mathbb{N}$ and:

1. $e = \operatorname{lcm}_{\tau \in T} m_\tau$;
2. $\gcd(s_\tau, m_\tau) = 1$ for all $\tau \in T$;
3. $\gcd(p, s_\tau) = 1$ and $\gcd(p, m_\tau) = 1$ for any prime p with $p\tau = \tau$.

The positive integers m_τ are the near-isomorphism invariants of [**BM94**, Definition 2.1] which we shall call the B–M invariants.

If we take positive integers s'_τ satisfying the conditions above instead of s_τ, we obtain a rigid crq group X', which is equal to X if and only if $s_\tau \equiv s'_\tau \pmod{m_\tau}$ for all $\tau \in T$.

If $s_\tau \not\equiv s'_\tau \pmod{m_\tau}$ for some $\tau \in T$, then X' and X are nearly isomorphic and every rigid crq group with regulator A which is nearly isomorphic to X is obtained in this way. Fix the special group $X_0 = \langle A, b \rangle$ from this near-isomorphism class which is determined by (1) and (2) with each $s_\tau = 1$, that is

$$(3) \qquad eb = \sum_{\tau \in T} \frac{e}{m_\tau} a_\tau.$$

We now show that if X is a rigid crq group with standard representation (2), then $\operatorname{End}(X)^+$ is a rigid crq group isomorphic to X_0. We use the following results from [**MS00**].

LEMMA 3.2. [**MS00**, Proposition 3.3] *If X is an acd group, then $\operatorname{End}(X)$ is commutative if and only if X is rigid.* □

LEMMA 3.3. [**MS00**, Lemma 6.2] *Let $X = \langle A, b \rangle$ be a rigid crq group of ring type with critical typeset T, regulator A and regulator quotient C of exponent e. Let X have B–M invariants m_τ. Then*

1. *there is an exact sequence of rings and natural ring homomorphisms*

$$0 \rightarrowtail \operatorname{Hom}(X, A) \rightarrowtail \operatorname{End}(X) \twoheadrightarrow \operatorname{End}(C) \twoheadrightarrow 0$$

such that
   (a): $\operatorname{Hom}(X,A)^+ \cong \operatorname{End}(A)^+ \cong A$ *is completely decomposable;*
   (b): $\operatorname{End}(C)$ *is a cyclic ring of the same order as C.*
2. $\operatorname{End}(X)^+$ *is a rigid crq group with critical typeset T and regulator $R(\operatorname{End}(X))^+ = \operatorname{Hom}(X, A)$;*
3. $\operatorname{End}(A) = \oplus_{\tau \in T} \tau 1_\tau$ *and* $\operatorname{Hom}(X, A) = \oplus_{\tau \in T} m_\tau \tau 1_\tau$*, where 1_τ is the identity map on A_τ and m_τ means multiplication by m_τ.*
4. $\operatorname{End}(X)$ *has standard representation $\operatorname{End}(X) = \langle \operatorname{Hom}(X,A), 1_X \rangle$ where 1_X is the identity map on X and $e1_X : X \to A$ is multiplication by e.*
5. *The natural map $\operatorname{End}(X) \to \operatorname{End}(C)$ such that $f + k1_X \mapsto k \pmod{e}$ for all $f \in \operatorname{Hom}(X, A)$ and $k \in \mathbb{Z}$ is an epimorphism.*

PROOF. All is proved in [**MS00**, Lemma 6.2] except the characterization $R(\operatorname{End}(X))^+ = \operatorname{Hom}(X, A)$. But this follows because $\operatorname{End}(X)^+$ is a rigid crq group and $\operatorname{Hom}(X, A)$ is a completely decomposable subgroup of minimum index e. □

We can now prove our main characterisation theorem for endomorphism rings of rigid crq groups.

THEOREM 3.4. *Let $X = \langle A, b \rangle$ be a rigid crq group of ring type. Then $X \cong_n \operatorname{End}(X)^+$. Moreover, there is exactly one group X_0 in the near isomorphism class of X such that $X_0 \cong \operatorname{End}(X_0)^+ \cong \operatorname{End}(X)^+$.*

PROOF. It follows from Proposition 3.1 that $\operatorname{End}(X)^+$ is a rigid crq group with the same typeset T as X and from Lemma 3.3 that $\operatorname{End}(X)^+$ has the same B–M invariants as X. Hence by [**M00**, Theorem 12.6.5], $X \cong_n \operatorname{End}(X)^+$. Thus $X_0 \cong \operatorname{End}(X)^+$ and the argument may be repeated with X_0 in place of X.

Now choose X_0 as in the remarks preceding Lemma 3.2. The natural mapping $f \mapsto 1f : \operatorname{End}(X) \to X_0$ is a monomorphism since the relations of the standard representations are preserved, and is surjective since $\operatorname{Hom}(X, A)$ maps onto A and the identity maps to b. □

COROLLARY 3.5. 1. *If X and Y are rigid nearly isomorphic crq groups of ring type, then $\operatorname{End}(X)^+$ and $\operatorname{End}(Y)^+$ are isomorphic crq groups from the same near-isomorphism class as X and Y.*

2. *If X is a rigid crq group of ring type then $\operatorname{End}(X)$ is an E–ring.*

PROOF. 1. Choose X_0 as in Theorem 3.4. Then $\operatorname{End}(X)^+ \cong \operatorname{End}(X_0)^+ \cong \operatorname{End}(Y)^+$.

2. $\text{End}(X_0)^+ \cong X_0$ has a commutative endomorphism ring. Hence by [**S73**, Corollary 3] $\text{End}(X) \cong \text{End}(X_0)$ is an E-ring. □

We can now complete the proof of the near Baer–Kaplansky Theorem for rigid crq groups of ring type; in fact we present a slightly stronger result.

THEOREM 3.6. *Let X be a rigid crq group of ring type and Y any crq group. Then Y is nearly isomorphic to X if and only if $\text{End}(X) \cong \text{End}(Y)$.*

PROOF. 1. Suppose $\text{End}(X) \cong \text{End}(Y)$. By Lemma 3.2, $\text{End}(Y)$ is commutative and hence Y is rigid. From Theorem 3.4 we immediately conclude that $Y \cong_n \text{End}(Y)^+$ and $X \cong_n \text{End}(X)^+$. Hence $X \cong_n Y$.

2. For the converse, let $X \cong_n Y$. By Theorem 3.4, there is a a group X_0 nearly isomorphic to X and to Y such that $X_0 \cong \text{End}(Z)^+$ for any Z in the near isomorphism class of X_0, so in particular, there is an additive isomorphism of $\text{End}(X)^+$ onto $\text{End}(Y)^+$. To see that $\text{End}(X) \cong \text{End}(Y)$ as rings, note that by Corollary 3.5, 2, $\text{End}(X)$ and $\text{End}(Y)$ are E-rings and hence by [**S73**, Corollary 4], have unital ring structures on their additive groups which are unique up to isomorphism, that is, $\text{End}(X) \cong \text{End}(Y)$ as rings. □

To see that a strong near Baer–Kaplansky cannot hold for rigid crq groups, we show that for any class of torsion–free abelian groups, strong near isomorphism implies isomorphism.

PROPOSITION 3.7. *Let M and N be nearly isomorphic but not isomorphic torsion–free groups and let $\Theta : \text{End}(M) \to \text{End}(N)$ be a unital ring isomorphism. Then there do not exist monomorphisms $\phi : M \to N$ and $\psi : N \to M$ such that for all $f \in \text{End}(M)$, $f\Theta = \psi f \phi$.*

PROOF. Assume that such monomorphisms exist Then in particular $1_N = 1_M \Theta = \psi\phi$. Hence ϕ is a bijection, so an isomorphism of M onto N, a contradiction. □

To see that it is necessary for both X and Y to be crq groups in the hypotheses of Theorem 3.6, consider the following example of a rigid acd group which is not a crq group, but whose endomorphism ring is isomorphic to that of a rigid crq group.

EXAMPLE 3.8. Let $A = s\sigma \oplus t\tau \oplus r\rho$ where σ, τ and ρ are incomparable ring types, none divisible by a prime p. Let $X = \langle A, a \rangle$ and $Y = \langle A, b, c \rangle$ where $pa = s + t + r$, $pb = s + t$ and $pc = t + r$.

Then X and Y have regulator A, X is crq with regulator index p, and Y is not nearly isomorphic to X, having regulator quotient isomorphic to $\mathbb{Z}(p) \oplus \mathbb{Z}(p)$.

Now by [**M00**, Lemma 15.3.4], $\phi \in \text{End}(A)$ is an endomorphism of X if and only if $(pa)\phi = p(ka + x)$ for some $k \in \mathbb{Z}$ and some $x \in A$.

Let $B = \langle s\sigma + t\tau, b \rangle$ and $C = \langle t\tau + r\rho, c \rangle$ so B and C are rigid crq groups which are fully invariant in Y. Then for any endomorphism ψ of Y, $pb\psi = p(\ell b + y)$ and $pc\psi = p(mc + z)$ for some $\ell, m \in \mathbb{Z}$ and some $y, z \in A$. Thus $pt\psi = p(\ell t + y) = p(mt + z)$, so $\ell = m$. Hence $\text{End}(Y) = \{\psi \in \text{End}(A) : pb\psi = p(\ell b + y) \text{ and } pc\psi = p(\ell t + z)\} = \text{End}(X)$.

4. The near Baer–Kaplansky Theorem for block rigid crq groups of ring type

To extend the results of Section 3 to all block rigid crq groups, we use the techniques developed by Ivanov [**I98**] for modules over unital rings which are generated by indecomposable summands and have the so-called **finite embedding property**. This means that if $M = \oplus_{i \in I} M_i$ is a decomposition of a module M into a direct sum of indecomposables, then every indecomposable summand of M is contained in the direct sum of finitely many M_i. Clearly all torsion–free abelian groups of finite rank have the finite embedding property.

Let $\mathcal{A}$ be the class of acd groups, considered as $\mathbb{Z}$–modules.

DEFINITION 4.1. [**I98**, p.108] Let X and $Y \in \mathcal{A}$. A ring isomorphism $\Phi : \mathrm{End}(X) \to \mathrm{End}(Y)$ is an **IP–isomorphism** if for every primitive idempotent $e \in \mathrm{End}(X)$, $Y(e\Phi) \cong Xe$.

In other words, IP–isomorphism preserves indecomposables. The next Proposition is the main ingredient of our proof of the Baer–Kaplansky Theorem for block rigid crq groups of ring type.

PROPOSITION 4.2. [**I98**, Proposition 1] *Let X and $Y \in \mathcal{A}$. Then $X \cong Y$ if and only if there is an IP-isomorphism $\mathrm{End}(X) \to \mathrm{End}(Y)$.* $\square$

We also need the following Lemma from [**F73**, Section 106].

LEMMA 4.3. *Let $A = B \oplus C$ and A' be abelian groups and let $\Phi : \mathrm{End}(A) \to \mathrm{End}(A')$ be an isomorphism between their endomorphism rings. Then $A' = B' \oplus C'$ such that Φ induces isomorphisms $\mathrm{End}(B) \to \mathrm{End}(B')$ and $\mathrm{End}(C) \to \mathrm{End}(C')$.* $\square$

To prove the near Baer–Kaplansky Theorem for block rigid crq groups, we need a slight modification of Definition 4.1 and Proposition 4.2 above.

DEFINITION 4.4. Let X and $Y \in \mathcal{A}$. A ring isomorphism $\Phi : \mathrm{End}(X) \to \mathrm{End}(Y)$ is a **near–IP–isomorphism** if for every primitive idempotent $e \in \mathrm{End}(X)$, $Y(e\Phi) \cong_n Xe$.

PROPOSITION 4.5. *Let X and Y be $\mathcal{A}$-groups of finite rank. Then $X \cong_n Y$ if and only if there is a near–IP–isomorphism $\mathrm{End}(X) \to \mathrm{End}(Y)$.*

The proof can be repeated word for word from [**I98**, Proposition 1], using the fact that if $X_i \cong_n Y_i$ for all $i \leq s$, then $\oplus_{i \leq s} X_i \cong_n \oplus_{i \leq s} Y_i$.

We now have enough machinery for our main Theorem.

THEOREM 4.6. *Let X and Y be block rigid crq groups of ring type. If $\mathrm{End}(X) \cong \mathrm{End}(Y)$ then $X \cong_n Y$.*

PROOF. First of all we show that if $\Phi : \mathrm{End}(X) \to \mathrm{End}(Y)$ is an isomorphism, then it is a near–IP–isomorphism. Let $\sigma \in \mathrm{End}(X)$ be a primitive idempotent, so $X = X\sigma \oplus X(1-\sigma)$. Let $X_1 = X\sigma$.

It follows from Lemma 4.3 that there exists the corresponding decomposition $Y = Y_1 \oplus Y_2$ with $\mathrm{End}(Y_1) = \mathrm{End}(X_1)\Phi$ and $\sigma\Phi$ is a primitive idempotent in $\mathrm{End}(Y)$ mapping Y onto Y_1.

Since X_1 is indecomposable, then according to [**BM94**, Theorems 3.3, 3.7], X_1 is either a rigid crq group or a rank one group. In both cases, $\mathrm{End}(X_1) \cong_n X_1$ by Theorem 3.4.

Since $\operatorname{End}(Y_1) \cong \operatorname{End}(X_1)$, by Theorem 3.6 Y_1 and X_1 are nearly isomorphic (or even isomorphic in case rank $X_1 = 1$). Thus Φ is a near–IP–isomorphism.

Applying Proposition 4.5 we obtain $X \cong_n Y$. □

COROLLARY 4.7. *The class of block rigid crq groups of ring type satisfies the near Baer–Kaplansky Theorem.* □

There are important classes of block rigid crq groups for which near isomorphism implies isomorphism. For such classes, the ordinary Baer–Kaplansky Theorem holds. For example, Blagoveshchenskaya and Reid [**BR00**] prove the following Isomorphism Criterion, based on matrix methods investigated by Reid in [**R99**].

DEFINITION 4.8. A block rigid crq group is called **proper** if for all $\tau \in T$, the τ–rank $r_\tau > 1$ whenever $m_\tau > 1$.

PROPOSITION 4.9. *([**BR00**, Theorem 6]) Let X and Y be proper block rigid crq groups. Then $X \cong Y$ if and only if $R(X) \cong R(Y)$ and for all critical types τ, $m_\tau(X) = m_\tau(Y)$.* □

This result implies the Baer–Kaplansky Theorem for the class of proper block rigid crq groups:

THEOREM 4.10. *Let X be a proper block rigid crq group of ring type and let Y be any crq group. If $\operatorname{End}(X) \cong \operatorname{End}(Y)$, then $X \cong Y$.*

PROOF. By Theorem 4.6, $X \cong_n Y$ so by Proposition 4.9, $X \cong Y$. □

Finally, to show that the converse of Theorem 4.6 is false, we sketch the example of Lewis of a pair of nearly isomorphic block rigid groups of ring type which have non–isomorphic endomorphism rings. Of course they cannot be crq groups. We first note that numerous examples in [**M00**, Example 8.2.8] show that in order to prove that nearly isomorphic block rigid groups are not actually isomorphic, it is necessary to consider the number theoretic relationships among the primes which divide the types.

EXAMPLE 4.11. [**L92**, Example 5.2.1] Let $\tau = \mathbb{Z}[1/11]$, $\sigma = \mathbb{Z}[1/31]$ and let $A = t\tau \oplus s\sigma$.

Let $G_1 = \langle A, a \rangle$ where $5a = t + s$, $G_2 = H_2 = \langle A, b \rangle$ where $25b = t + s$ and $H_1 = \langle A, c \rangle$ where $5c = t + 2s$. It follows that the four groups are rigid crq of ring type, and H_1 is nearly isomorphic, but not isomorphic to G_1.

Now let $X = G_1 \oplus G_2$ and $Y = H_1 \oplus H_2$, so that $X \cong_n Y$. By identifying $\operatorname{End}(X)$ and $\operatorname{End}(Y)$ with their matrix representations, for example

$$\operatorname{End}(X) = \begin{bmatrix} \operatorname{End}(G_1) & \operatorname{Hom}(G_1, G_2) \\ \operatorname{Hom}(G_2, G_1) & \operatorname{End}(G_2) \end{bmatrix},$$

Lewis then shows that $\operatorname{End}(X)^+ \cong_n \operatorname{End}(Y)^+$ but that no ring isomorphism of $\operatorname{End}(X)/R(\operatorname{End}(X))$ onto $\operatorname{End}(Y)/R(\operatorname{End}(Y))$ induces a monomorphism from $\operatorname{End}(X)$ into $\operatorname{End}(Y)$.

It follows that X and Y are nearly isomorphic block rigid groups of ring type whose endomorphism rings are not isomorphic.

On the other hand, we have not been able to construct nearly isomorphic (non block rigid) crq groups X and Y of ring type whose endomorphism rings are not isomorphic. We conjecture that the near Baer–Kaplansky Theorem holds for the class of all crq groups of ring type.

References

[BM94] E. Blagoveshchenskaya and A. Mader, *Decompositions of almost completely decomposable Abelian groups*, in **Abelian Group Theory and Related Topics**, Eds. R. Göbel, P. Hill and W. Liebert, Contemporary Mathematics, Vol. 171, Amer. Math. Soc. , 1994, 21–36

[BR00] E. Blagoveshchenskaya and J. Reid, *Classification and direct decompositions of block rigid crq groups without homogeneous subgroups of rank 1*, (preprint)

[F73] L. Fuchs, **Infinite Abelian Groups**, Vol. 2, Academic press, 1973.

[I98] G. Ivanov, *Generalizing the Baer–Kaplansky Theorem*, Journal of Pure and Applied Algebra, Vol. 133, 1998, 107–115.

[L92] W. S. Lewis, *Almost completely decomposable groups with two critical types and their endomorphism rings*, Ph.D. Dissertation, University of Hawaii, 1995, 343–366.

[M95] A. Mader, *Almost completely decomposable torsion–free abelian groups*, in **Abelian Groups and Modules**, Eds. A. Facchini and C. Menini, Mathematics and its Applications Vol. 343, Kluwer Academic Publishers, 1995, 343–366.

[M00] A. Mader. **Almost Completely Decomposable Groups**, Algebra, Logic and Applications, Vol. 13, Gordon and Breach Science Publishers, 2000.

[MS00] A. Mader and P. Schultz, *Endomorphism rings and automorphism groups of almost completely decomposable groups*, Comm. in Algebra, 28, 2000, 51–68.

[OV99] K. C. O'Meara and C. Vinsonhaler, *Separative cancellation and multiple isomorphism in torsion–free Abelian groups*, Journal of Algebra, Vol. 221, 1999, 536–550.

[R99] J. Reid, *Some matrix rings associated with ACD groups*, in **Abelian Groups and Modules**, Birkhäuser, 1999, 191–198.

[S73] P. Schultz, *The endomorphism ring of the additive group of a ring*, J. Aust. Math. Soc., 15, 1973, 60–69 .

(Blagoveshchenskaya) Department of Mathematics, St. Petersburg State Technical University, St Petersburg, 195251, Russia
E-mail address: kate@robotek.ru

(Ivanov) Department of Mathematics, Macquarie University, Sydney, NSW 2109, Australia
E-mail address: ivanov@ics.mq.edu.au

(Schultz) Department of Mathematics and Statistics, The University of Western Australia, Nedlands, Australia 6907
E-mail address: schultz@math.uwa.edu.au

Maximal Pure Independent Sets

Andreas Blass and John Irwin

Dedicated to László Fuchs on his 75th birthday.

1. Introduction

Let G be a torsion-free abelian group. A subset S of G is *pure independent* if it is linearly independent and the subgroup $\langle S \rangle$ it generates is pure in G. We shall be concerned with properties of pure independent subsets of G that are maximal with respect to inclusion. These maximal pure independent sets (or, more precisely, the subgroups they generate) generalize the basic subgroups studied in [**2, 3**]. We shall show that some of the important properties of basic subgroups hold, with minor modifications, also for maximal pure independent sets. The main advantage of working with maximal pure independent sets is that they exist in all torsion-free abelian groups. The main disadvantage is that the "minor modifications" mentioned above are, at least in one case, definitely needed; the uniqueness up to isomorphism of basic subgroups does not always carry over to maximal pure independent sets.

To place our results in perspective, we begin with a brief review of basic subgroups and their properties. The notion of basic subgroup arose first in the theory of p-groups, where it was defined in [**6**] as follows.

DEFINITION 1.1. A subgroup B of an abelian p-group G is *basic* if it is a direct sum of cyclic groups, it is pure in G, and the quotient G/B is divisible.

This concept was imported unchanged into the theory of torsion-free groups [**2, 3**]. Of course, in this new context, "direct sum of cyclic groups" amounts to "free," so the definition reads as follows.

DEFINITION 1.2. A subgroup B of a torsion-free abelian group G is *basic* if it is free, it is pure in G, and the quotient G/B is divisible.

In contrast to the p-group situation, a torsion-free abelian group need not have any basic subgroups [**3**, Theorem 14], not even if it is separable (see definition below or in [**5**]). But when they exist, basic subgroups have several pleasant properties, of which we shall need two, due to Dugas and Irwin [**2, 3**].

1991 *Mathematics Subject Classification.* Primary 20K20.

CONVENTION 1.3. From now on, when we refer to groups we shall always mean torsion-free, abelian groups.

THEOREM 1.4 (Dugas and Irwin). *Any two basic subgroups of the same group are isomorphic.*

This is Theorem 2 of [3]. Although we shall not repeat the proof here, we mention that it explicitly computes the rank of a basic subgroup of G as the dimension of G/pG as a vector space over $\mathbb{Z}/p\mathbb{Z}$, for an arbitrary prime number p. This sheds some light on why basic subgroups need not exist — this dimension may be different for different primes. Chapter VI of [5] describes a notion of p-basic that works with one prime at a time. These p-basic groups always exist, but they are not in general pure.

DEFINITION 1.5. A group is *torsionless* if it is isomorphic to a subgroup of $\mathbb{Z}^\kappa$ for some cardinal number κ. It is *separable* if it is isomorphic to a pure subgroup of $\mathbb{Z}^\kappa$ for some cardinal number κ.

THEOREM 1.6 (Dugas and Irwin). *If a separable group G has a basic subgroup of rank κ, then it is isomorphic to a pure subgroup of $\mathbb{Z}^\kappa$ for this same κ.*

This is Theorem 1 of [2].

COROLLARY 1.7. *A separable group G with a basic subgroup of infinite rank κ has cardinality at most 2^κ.*

Our primary purpose in this paper is to extend these two theorems and the corollary to deal with maximal pure independent sets rather than basic subgroups. In Section 3 we establish the analog of Theorem 1.4 except when finite ranks are involved. In Section 4, we show that this exception is necessary by constructing groups with maximal pure independent sets of many different finite cardinalities. Finally, in Section 5, we discuss analogs of Theorem 1.6. The analogy is not perfect, but it is good enough to get the exact analog of Corollary 1.7, and even a bit better for some values of κ.

2. Maximal Pure Independent Sets

This brief section is devoted to background material, namely the definition of maximal pure independent sets, their elementary properties, and their connection with the basic subgroups that they are intended to generalize. Recall our global assumption that "group" means "torsion-free abelian group."

DEFINITION 2.1. A subset S of a group G is *pure independent* if it is a set of free generators of a pure subgroup of G. It is *maximal pure independent* in G if it is pure independent and is not properly included in any other pure independent subset of G.

We record for reference an immediate consequence of Zorn's Lemma.

PROPOSITION 2.2. *Every group G has a maximal pure independent subset.*

The next proposition gives some equivalent descriptions of maximal pure independent sets.

PROPOSITION 2.3. *Let S be a subset of a group G. The following three statements are equivalent.*

1. S is a maximal pure independent set in G.
2. S freely generates a pure subgroup $\langle S \rangle$ of G which is not a proper direct summand of any other free, pure subgroup of G.
3. S freely generates a pure subgroup $\langle S \rangle$ of G such that the quotient $G/\langle S \rangle$ has no pure subgroup isomorphic to $\mathbb{Z}$.

Proof Each of the three conditions requires S to be a set of free generators for a pure subgroup $\langle S \rangle$ of G. Assuming S has this property, we show that failure of any one of the three conditions implies failure of the next in cyclic order.

If condition 1 fails, i.e., if S isn't maximal, then there is a proper superset $T \supsetneq S$ that is also a set of free generators for a pure subgroup $\langle T \rangle$ of G. Then $\langle S \rangle$ is a proper direct summand of $\langle T \rangle$, the other summand being $\langle T - S \rangle$. So condition 2 also fails.

Suppose next that condition 2 fails; say $\langle S \rangle$ is a proper direct summand of a pure, free subgroup F of G. Then $F/\langle S \rangle$ is a non-trivial, free group and is pure in $G/\langle S \rangle$. Thus condition 3 also fails.

Finally, suppose condition 3 fails. Consider any pure subgroup P of $G/\langle S \rangle$ that is isomorphic to $\mathbb{Z}$, and let $x \in G$ be an element whose coset $[x] \in G/\langle S \rangle$ corresponds to $1 \in \mathbb{Z}$ under the isomorphism. Then $S \cup \{x\}$ is an independent set generating a pure subgroup of G, namely the preimage of P under the projection $G \to G/\langle S \rangle$. In other words, $S \cup \{x\}$ is a pure independent set and, since $x \notin S$, condition 1 also fails. □

REMARK 2.4. In connection with condition 2 in Proposition 2.3, we emphasize that $\langle S \rangle$ need not be a maximal, pure, free subgroup of G. It may well be a proper subgroup — just not a summand — of another pure, free subgroup of G.

COROLLARY 2.5. *If B is a basic subgroup of G, then every set S of free generators of B is a maximal pure independent subset of G.*

Proof Condition 3 of the theorem is trivial to verify, since a divisible group $G/B = G/\langle S \rangle$ cannot have a pure subgroup isomorphic to $\mathbb{Z}$. □

REMARK 2.6. It is well known that there are torsion-free groups with no basic subgroups. Thus, Corollary 2.5 and Proposition 2.2 show that the notion of (subgroup generated by a) maximal pure independent set is strictly more general that that of basic subgroup. In fact, even when basic subgroups exist, maximal pure independent sets are more general. For example, let F be a free group of countably infinite rank, and let $h : F \to \mathbb{Q}$ be a homomorphism whose range is a proper, non-free subgroup of $\mathbb{Q}$, such as the group of dyadic rational numbers. Then the kernel of h is not basic in F (for the quotient, the image of h, isn't divisible), but any free generating set for this kernel is a maximal pure independent set in F (for the quotient contains no pure copy of $\mathbb{Z}$). Of course F has basic subgroups, for example all of F or the kernel of a homomorphism from F onto $\mathbb{Q}$.

We close this section with an observation that will exclude some exceptional cases when we deal with torsionless groups.

PROPOSITION 2.7. *Suppose G is a torsionless group with a finite maximal pure independent set S. Then S generates the whole group G, and therefore G is a free group of finite rank.*

Proof Suppose, toward a contradiction, that $\langle S \rangle$ is a proper subgroup of G. Fix some element $x \in G - \langle S \rangle$, and consider the purification H of $\langle S, x \rangle$ in G. This H is a torsionless group of rank $n+1$, where n is the number of elements of S, so H is free. But then, since $\langle S \rangle$ is a pure subgroup of rank n, the quotient $H/\langle S \rangle$ is isomorphic to $\mathbb{Z}$. This quotient is pure in $G/\langle S \rangle$, so, by Proposition 2.3, S cannot be a maximal pure independent set in G. □

3. Invariance of Cardinality

Our main goal in this section is to establish an analog of Theorem 1.4 for maximal pure independent sets. A perfect analog would say that any two such sets in the same group have the same cardinality. This is not generally true — counterexamples are given in Section 4 — but it is true if the maximal pure independent sets are both infinite. That will be a corollary of the following more precise result.

THEOREM 3.1. *Let G be a group with a pure, free subgroup F of rank κ. Suppose that G also has a subgroup M of cardinality μ such that G/M has no pure subgroup isomorphic to $\mathbb{Z}$. Then $\kappa \leq \mu$.*

Proof We first eliminate the trivial case where μ is finite. Since G is torsion-free, this case would make M the trivial group, so G would have no pure subgroup isomorphic to $\mathbb{Z}$. Then F would also have to be trivial. So we would have $\kappa = 0$ and $\mu = 1$, which verifies the conclusion of the theorem.

From now on, we assume that μ is infinite. Thus $\mu \cdot \aleph_0 = \mu$.

Suppose, toward a contradiction, that $\mu < \kappa$. Define H to be the purification in G of the subgroup $M + F$, and observe that H/M, being a pure subgroup of G/M, has no pure subgroup isomorphic to $\mathbb{Z}$.

For each element $m \in M$ and each non-negative integer n, if there is some $f \in F$ such that $m + f$ is divisible by n, then choose one such f and call it $f_{m,n}$. We adopt the natural convention that "divisible by 0" means equal to 0, so $f_{m,0}$ is $-m$ if $m \in F$ and undefined otherwise. Since there are only μ values for m and only $\aleph_0$ values for n, there are at most μ elements of the form $f_{m,n}$. These elements lie in the free group F, so they lie in a direct summand F_1 of F of rank at most μ. Write F_2 for a complementary summand of F. Notice that F_2 is nontrivial, because F has rank κ and F_1 has strictly smaller rank, at most μ.

Let H_1 be the purification in G of $M + F_1$. We intend to show that H is the direct sum $H_1 \oplus F_2$. If we can show this, then the proof will be complete. Indeed, since $M \subseteq H_1$, we shall have
$$\frac{H}{M} = \frac{H_1}{M} \oplus F_2,$$
which contradicts, since F_2 is a nontrivial free group, the fact that H/M contains no pure copy of $\mathbb{Z}$.

So it remains only to verify that $H = H_1 \oplus F_2$. Consider any $h \in H$. By definition of H, we have $nh = m + f$ for some positive integer n, some $m \in M$, and some $f \in F$. So $f_{m,n}$ exists, and n divides both $m + f$ and $m + f_{m,n}$ and therefore also their difference $f - f_{m,n}$. As F is pure in G, we have $f - f_{m,n} = nf'$ for some $f' \in F$. Write $f' = f'_1 + f'_2$ with $f'_1 \in F_1$ and $f'_2 \in F_2$. Then

(1) $\quad n(h - f'_2) = nh - nf'_2 = m + f - nf'_2 = m + f_{m,n} + f - f_{m,n} - nf'_2$
$$= m + f_{m,n} + nf' - nf'_2 = m + f_{m,n} + nf'_1 \in M + F_1.$$

By definition of H_1, it follows that $h - f_2' \in H_1$, and therefore $h \in H_1 + F_2$.

This establishes that $H = H_1 + F_2$. It remains to prove that this sum is direct. Suppose $x \in H_1 \cap F_2$. By definition of H_1, we can write $nx = m + f$ with n a positive integer, $m \in M$, and $f \in F_1$. Since $x \in F_2 \subseteq F$, we have $m = nx - f \in F$. Therefore, $f_{m,0}$ exists and equals $-m$, which gives us that $m \in F_1$. Now nx is in F_1 (because it equals $m + f$ and both m and f are in F_1) and also in F_2 (because $x \in F_2$). Since $F = F_1 \oplus F_2$ is a direct sum, this implies $nx = 0$ and thus $x = 0$. $\square$

COROLLARY 3.2. *If a group has maximal pure independent sets of cardinalities κ and λ, then $\kappa \cdot \aleph_0 = \lambda \cdot \aleph_0$.*

Proof It is easy to check (as at the beginning of the proof of the theorem), that if one of κ and λ is zero then so is the other. We assume from now on that they are not zero.

Apply the theorem with F being the subgroup generated by a maximal pure independent set of cardinality κ and M being the subgroup generated by a maximal pure independent set of cardinality λ. The cardinality μ of M is thus $\lambda \cdot \aleph_0$. By the theorem, $\kappa \leq \lambda \cdot \aleph_0$, and it follows, since $\aleph_0^2 = \aleph_0$, that $\kappa \cdot \aleph_0 \leq \lambda \cdot \aleph_0$.

Since the hypotheses of the theorem are symmetrical between κ and λ, the reverse inequality also holds. $\square$

COROLLARY 3.3. *If a group has maximal pure independent sets of infinite cardinalities κ and λ, then $\kappa = \lambda$.*

Proof For infinite cardinals κ, we have $\kappa \cdot \aleph_0 = \kappa$. $\square$

COROLLARY 3.4. *In a torsionless group, all maximal pure independent sets have the same cardinality.*

Proof If the group has infinite rank, then, by Proposition 2.7, its maximal pure independent sets are infinite, so the preceding corollary applies. If the group has finite rank, then, again by Proposition 2.7, the cardinality of each maximal pure independent set equals the rank of the group. $\square$

Thus, we have the analog of Theorem 1.4 except that we might have, in one and the same (non-torsionless) group G, two maximal pure independent sets of different (non-zero) finite cardinalities, or one of finite cardinality and one of cardinality $\aleph_0$. We shall see in Section 4 that this exception really occurs.

4. Non-Invariance of Cardinality

In this section, we shall construct examples showing that a group may have maximal pure independent sets of different finite cardinalities or of a finite cardinality and $\aleph_0$. In fact, the following theorem says that "anything is possible", i.e., we can arbitrarily prescribe the cardinalities of maximal pure independent sets, subject only to Corollary 3.2 and the trivial observation that, if the empty set is maximal pure independent in G, then no other set is.

THEOREM 4.1. *Let R be a nonempty subset of $\{1, 2, \ldots, \aleph_0\}$. There exists a group G with maximal pure independent sets of all cardinalities in R and of no other cardinalities.*

This theorem will be an easy consequence of the following more explicit description of the pure, free subgroups of the G's that we construct.

THEOREM 4.2. *Let R be a nonempty subset of $\{1, 2, \ldots, \aleph_0\}$. There exists a group G with pure independent sets S_r, for $r \in R$, such that:*
1. *S_r has cardinality r.*
2. *The different S_r's are linearly independent.*
3. *Every pure, free subgroup of G is included in one of the groups $\langle S_r \rangle$.*

Proof We begin by observing that requirement 3 in the theorem will be satisfied if every pure copy of $\mathbb{Z}$ in G is included in some $\langle S_r \rangle$. Indeed, if this weakened form of 3 holds and if F is any pure, free subgroup of G, then each element of F is in some $\langle S_r \rangle$ (for it lies in a pure copy of $\mathbb{Z}$). And we cannot have two non-zero elements of F lying in different $\langle S_r \rangle$'s, for then their sum could not be in any $\langle S_r \rangle$, because of requirement 2. Thus, all of F lies in a single $\langle S_r \rangle$.

It is convenient to isolate part of the construction of G in the following technical lemma.

LEMMA 4.3. *Suppose we are given, in the rational vector space $\mathbb{Q}^n$,*
- *finitely many rational subspaces H_i,*
- *a finitely generated additive subgroup L of $\mathbb{Q}^n$, and*
- *a vector $v \in L$ lying in none of the H_i.*

Then there is an integer $p > 1$ such that the subgroup $\langle L, v/p \rangle$ generated by L and v/p contains no elements in any H_i except those already in L, i.e.,

$$\langle L, v/p \rangle \cap H_i \subseteq L \quad \text{for each } i.$$

Proof We may assume without loss of generality that each H_i is a hyperplane not containing v, because each of the given H_i is included in such a hyperplane.

Temporarily fix i. Since $\mathbb{Q}^n = H_i \oplus \mathbb{Q}v$, let $\pi_i : \mathbb{Q}^n \to \mathbb{Q}v$ be the projection with kernel H_i. Then $\pi_i(L)$ is a finitely generated subgroup of $\mathbb{Q}v$.

Now un-fix i. The subgroup of $\mathbb{Q}v$ generated by all the $\pi_i(L)$ together is again finitely generated, so it has the form $\mathbb{Z}\alpha v$ for some rational number α. Since $v \in L$, this subgroup contains v, so $\alpha = 1/k$ for some $k \in \mathbb{Z} - \{0\}$.

Let p be a prime number not dividing k. We shall show that this p satisfies the conclusion of the lemma. That is, $\langle L, v/p \rangle \cap H_i \subseteq L$ for each i.

So consider any vector in $\langle L, v/p \rangle$, say $l + (mv/p)$ where $l \in L$ and $m \in \mathbb{Z}$, and assume that $l + \frac{mv}{p} \in H_i$ for a certain i. We must show $l + (mv/p) \in L$. Recall that π_i has kernel H_i, so we have $\pi_i(l + (mv/p)) = 0$. That is, since π_i fixes v,

$$\frac{mv}{p} = \pi_i\left(\frac{mv}{p}\right) = -\pi_i(l) \in \pi_i(L) \subseteq \mathbb{Z}\frac{1}{k}v.$$

So $\frac{m}{p} = \frac{j}{k}$ for some $j \in \mathbb{Z}$. Then $km = pj$ and, since p was chosen as a prime not dividing k, we infer that p divides m. Thus, m/p is an integer and $l + \frac{mv}{p} \in \langle L, v \rangle = L$ because $v \in L$. □

Returning to the proof of Theorem 4.2, we construct the desired G as an additive subgroup of the rational vector space

$$\bigoplus_{r \in R} \bigoplus_{0 \leq i < r} \mathbb{Q}e_{r,i},$$

i.e., the rational vector space having as basis a doubly indexed family of vectors $e_{r,i}$ where the first index r ranges over R and, for each fixed r, the second index i takes

r values. For each $r \in R$, let $S_r = \{e_{r,i} : 0 \leq i < r\}$. We begin the construction of G by putting into G all members of

$$G_0 = \bigoplus_{r \in R} \bigoplus_{0 \leq i < r} \mathbb{Z} e_{r,i},$$

i.e., the group freely generated by the union of all the S_r's.

The rest of the construction of G is an induction, putting into G at each stage one new element along with whatever that element and previous members of G generate. We write G_k for the group obtained after k steps. So $G_{k+1} = \langle G_k, x_k \rangle$ for some x_k, and the final result is

$$G = \bigcup_{k=0}^{\infty} G_k.$$

Notice that each of our S_r's is an independent set and they are independent of each other. In view of the remarks at the beginning of the proof (and the obvious fact that S_r has cardinality r), the proof of the theorem will be complete if we can arrange the construction so that each $\langle S_r \rangle$ is pure in G and so that, for every $v \in G$ that is not in any $\langle S_r \rangle$, the cyclic subgroup $\langle v \rangle$ is not pure in G.

To make each $\langle S_r \rangle$ pure in G, it suffices to make it pure in each G_k. It is obviously pure in G_0 (being a direct summand of G_0); we shall arrange each step, from G_k to G_{k+1}, so as to preserve this purity. That is, this step should not add to G any new elements of the rational vector subspace

$$H_r = \bigoplus_{0 \leq i < r} \mathbb{Q} e_{r,i}.$$

By the *support* of a vector $v \in \bigoplus_{r \in R} \bigoplus_{0 \leq i < r} \mathbb{Q} e_{r,i}$, we mean the set of $r \in R$ such that, for some i, the component of v in $\mathbb{Q} e_{r,i}$ is not zero. Thus, to maintain purity of $\langle S_r \rangle$, we must avoid putting into G any new elements whose support is $\{r\}$.

List, in a sequence (v_k), all the vectors in $\bigoplus_{r \in R} \bigoplus_{0 \leq i < r} \mathbb{Q} e_{r,i}$ whose support contains at least two elements. This is possible because this vector space is countable. For technical reasons, choose the sequence so that each such vector occurs infinitely often in it.

At stage k of our inductive construction of G, when we have G_k and want to enlarge it to $G_{k+1} = \langle G_k, x_k \rangle$, we shall choose x_k so as to satisfy the following two requirements.

Purity: $G_{k+1} \cap H_r \subseteq G_k$ for all $r \in R$.
Division: If $v_k \in G_k$ then $v_k/p \in G_{k+1}$ for some integer $p > 1$.

As mentioned earlier (and as its name suggests), the purity requirement ensures that each $\langle S_r \rangle$ is pure in G.

The division requirement ensures that no $v \in G - \bigcup_{r \in R} \langle S_r \rangle$ generates a pure copy of $\mathbb{Z}$ in G. Indeed, any such v must have at least two elements in its support, because of the purity requirement, and v is in some G_m. But then $v = v_k$ for infinitely many k, including some $k \geq m$. Then the division requirement puts some v/p into G_{k+1} and therefore into G. So $\mathbb{Z}v$ is not pure in G.

So all that remains is to carry out the step from G_k to G_{k+1} in such a way as to satisfy the two requirements. If $v_k \notin G_k$, then we simply set $G_{k+1} = G_k$, and the requirements are trivially satisfied.

From now on, assume $v_k \in G_k$. We intend to choose an integer $p > 1$ and set $G_{k+1} = \langle G_k, v_k/p \rangle$. This will certainly satisfy the division requirement; we show that p can be chosen so as to also satisfy the purity requirement.

Let R_0 be a finite subset of R so large that
$$V = \bigoplus_{r \in R_0} \bigoplus_{0 \leq i < r} \mathbb{Q} e_{r,i}$$
contains v_k and all the generators x_j ($j < k$) from earlier stages of the inductive construction. Thus,
$$G_k = (G_k \cap V) \oplus \bigoplus_{r \in R - R_0} \bigoplus_{0 \leq i < r} \mathbb{Z} e_{r,i}.$$
No matter what p we choose, v_k/p will be in V, so G_{k+1} will have the form
$$(G_{k+1} \cap V) \oplus \bigoplus_{r \in R - R_0} \bigoplus_{0 \leq i < r} \mathbb{Z} e_{r,i}$$
and will therefore satisfy the purity requirement for all $r \in R - R_0$. To satisfy this requirement also for $r \in R_0$, we apply Lemma 4.3. For notational simplicity, suppose first that $\aleph_0 \notin R_0$.

Start with the finite-dimensional rational vector space V (which can be identified with a $\mathbb{Q}^n$ as in the lemma), the finitely many subspaces H_r for $r \in R_0$, the subgroup $L = G_k \cap V$, and the vector $v = v_k$. Note that L is finitely generated, the generators being the $e_{r,i}$ for $r \in R_0$ and $i < r$ along with the x_j for $j < k$. So the hypotheses of the lemma are satisfied, and we obtain an integer $p > 1$ such that, for each $r \in R_0$,
$$\langle G_k \cap V, v_k/p \rangle \cap H_r \subseteq G_k \cap V, \quad \text{i.e.,} \quad (G_{k+1} \cap V) \cap H_r \subseteq G_k.$$
But for $r \in R_0$, we have $H_r \subseteq V$, so the previous formula simplifies to $G_{k+1} \cap H_r \subseteq G_k$.

If $\aleph_0 \in R_0$ the argument is similar, but V should have $\mathbb{Q} e_{\aleph_0, i}$ as a summand for only finitely many i, enough so that v_k and all earlier x_j are in V. This completes the proof of the purity requirement and thus of the theorem. $\square$

An immediate consequence of the theorem, by taking $R = \mathbb{N}$, is the following answer to a question posed (in private communication) by J. Reid.

COROLLARY 4.4. *There is a group G with pure free subgroups of all finite ranks but no pure free subgroup of infinite rank.*

Finally, we infer Theorem 4.1 from Theorem 4.2.

Proof of Theorem 4.1 Given R, let G and the sets S_r be as in Theorem 4.2. This G has maximal pure independent sets of all cardinalities in R, namely the S_r, which are maximal by condition 3.

Now consider any maximal pure independent set S in G, say of cardinality s; we must show that $s \in R$. By condition 3 of Theorem 4.2, S is included in $\langle S_r \rangle$ for some $r \in R$; so clearly $s \leq r$. If s is infinite, then so is r and we have $s = r = \aleph_0$, so the proof is complete. If s is finite then, as a finite-rank, pure subgroup of the free group $\langle S_r \rangle$, $\langle S \rangle$ must be a direct summand of $\langle S_r \rangle$. But by Proposition 2.3, it cannot be a proper direct summand. So $\langle S \rangle = \langle S_r \rangle$ and therefore $s = r \in R$. $\square$

5. Embeddings and Cardinalities

In this section, we connect the size of a maximal pure independent set in a group G with the number κ of factors needed in a product $\mathbb{Z}^\kappa$ in order that G be isomorphic to a subgroup (or to a pure subgroup) of this product. We also consider related bounds on the cardinality of G.

We begin by stating for reference a well-known lemma describing embeddings of the sort we seek.

LEMMA 5.1. *A group G is embeddable as a subgroup of $\mathbb{Z}^\kappa$ if and only if there is a family $\mathcal{F}$ of at most κ homomorphisms $G \to \mathbb{Z}$ such that, for any non-zero $x \in G$, there is at least one $f \in \mathcal{F}$ such that $f(x) \neq 0$.*

A reduced group G is embeddable as a pure subgroup of $\mathbb{Z}^\kappa$ if and only if there is a family $\mathcal{F}$ of at most κ homomorphisms $G \to \mathbb{Z}$ such that, for any non-zero $x \in G$, the only integers that divide $f(x)$ for every $f \in \mathcal{F}$ are those that divide x. Furthermore, if no non-zero element of G is divisible by infinitely many primes, then $\mathcal{F}$ will satisfy this condition provided, for each non-zero $x \in G$, the only primes that divide $f(x)$ for every $f \in \mathcal{F}$ are those that divide x.

Proof For both directions of both parts, just take the homomorphisms $f \in \mathcal{F}$ to be the coordinate projections $\mathbb{Z}^\kappa \to \mathbb{Z}$ restricted to a copy of G. □

The hypotheses "reduced" in the second sentence of the lemma and "no non-zero element of G is divisible by infinitely many primes" in the third serve only to ensure that $\mathcal{F}$ satisfies the condition in the first sentence. Notice that these hypotheses are satisfied by all torsionless groups; these are the only groups to which we shall apply the lemma.

COROLLARY 5.2. *If G is embeddable as a pure subgroup of $\mathbb{Z}^\kappa$, then there is a family $\mathcal{F}$ of at most κ homomorphisms $G \to \mathbb{Z}$ such that, for any $x \in G$, there is $f_x \in \mathcal{F}$ with $f_x(x)$ a divisor of x in G.*

Proof Beginning with a family $\mathcal{F}$ as given by the second part of the lemma, close it under the group operations in $\text{Hom}(G, \mathbb{Z})$. The resulting $\mathcal{F}'$ will satisfy the requirement of the corollary. Indeed, for any non-zero $x \in G$ (the case of $x = 0$ being trivial), $V = \{f(x) : f \in \mathcal{F}\}$ is a set of integers with no common divisor except for divisors of x. The greatest common divisor d of V is thus a divisor of x, and it is a linear combination, with integer coefficients, of members of V. It is therefore of the form $f(x)$ for some $f \in \mathcal{F}'$, and this f can serve as f_x. □

COROLLARY 5.3. *If G is a separable group of cardinality κ, then G is embeddable as a pure subgroup of $\mathbb{Z}^\kappa$.*

Proof As G is separable, there is a family $\mathcal{F}$ of homomorphisms as in the preceding corollary, except that it may have cardinality greater than κ. But if we choose, for each $x \in G$, one $f_x \in \mathcal{F}$ as in that corollary, then the resulting subfamily of $\mathcal{F}$ still satifies the conditions of the second part of the lemma. □

After these preliminaries, we are ready to present our results relating maximal pure independent sets to embeddings in powers of $\mathbb{Z}$.

THEOREM 5.4. *Suppose G is a separable group with a maximal pure independent set of cardinality κ. Then G can be embedded as a subgroup of $\mathbb{Z}^\kappa$. In particular, the cardinality of G is at most 2^κ.*

Proof We assume that κ is infinite, for otherwise the result is trivial by Proposition 2.7.

Fix a maximal pure independent set S of size κ. Since κ is infinite, the subgroup $\langle S \rangle$ of G generated by S also has cardinality κ. For each $m \in \langle S \rangle$, apply Corollary 5.2 (with some possibly larger cardinal in place of κ) to choose a homomorphism $f_m : G \to \mathbb{Z}$ such that $f_m(m)$ divides m in G. We shall show that the κ (or fewer) homomorphisms so chosen satisfy the conditions in the first part of Lemma 5.1 and thus establish the desired embeddability of G in $\mathbb{Z}^\kappa$.

So consider any non-zero element $x \in G$. We must find $m \in \langle S \rangle$ with $f_m(x) \neq 0$. Since G is separable and $x \neq 0$, the purification in G of the cyclic group generated by x is also cyclic; let y be a generator of it. So $x = ny$ for some non-zero integer n, and y is not divisible in G by any primes.

Recall from Proposition 2.3 that $G/\langle S \rangle$ has no pure subgroup isomorphic to $\mathbb{Z}$. This applies in particular to the subgroup generated by the coset $[y]$ of y. Thus, in $G/\langle S \rangle$, this coset is divisible by some prime p, say $[y] = p[z]$. Back in G, this means that $y = pz + m$ for some $m \in \langle S \rangle$. Notice that p doesn't divide m, because it doesn't divide y. Therefore $f_m(m)$ is not divisible by p either. But then $f_m(y) = pf_m(z) + f_m(m)$ isn't divisible by p either. In particular, $f_m(y) \neq 0$. Therefore $f_m(x) = nf_m(y) \neq 0$. $\square$

REMARK 5.5. Unlike Theorem 1.6, the preceding theorem does not guarantee that the image of the embedding can be taken to be pure in $\mathbb{Z}^\kappa$. We do not know whether the theorem can be improved to provide this guarantee. Of course the cardinality bound in the theorem implies, via Corollary 5.3, that G is embeddable as a pure subgroup of $\mathbb{Z}^{2^\kappa}$. We shall obtain a smaller exponent, for some values of κ, in Corollary 5.9 below.

REMARK 5.6. Leaving the conclusion of the theorem as it is, without purity, one might hope to reduce the hypothesis from "separable" to "torsionless." But this too is an open problem.

The following result is the analog, for maximal independent sets in torsion-free groups, of a result for basic subgroups in p-groups given in [**5**, Theorem 34.3]. An analog for basic subgroups of torsion-free groups is mentioned as background in [**1**, Proposition 1].

THEOREM 5.7. *Suppose G is a torsionless group with a maximal pure independent set of infinite cardinality κ. Then G has cardinality at most $\kappa^{\aleph_0}$.*

Proof Fix a maximal pure independent set S in G of cardinality κ. We shall define a one-to-one function from G into the set Φ of all functions whose domains are infinite subsets of the positive integers and whose values are in $\langle S \rangle$, the subgroup of G generated by S. Since $\langle S \rangle$, like S, has cardinality κ, Φ has cardinality $\kappa^{\aleph_0}$, so the existence of j will suffice to establish the theorem.

To define $j(x)$ for an arbitrary $x \in G$, proceed as follows. The domain of $j(x)$ is the set of those positive integers that divide $[x]$, the coset of x in $G/\langle S \rangle$. Since $G/\langle S \rangle$ has no pure subgroup isomorphic to $\mathbb{Z}$ (by Proposition 2.3), each of its elements is divisible by infinitely many integers, and so the domain of $j(x)$ is infinite. To complete the definition of $j(x)$, consider any n in its domain. Then since n divides $[x]$, we have $x = ny + h$ for some $y \in G$ and $h \in \langle S \rangle$. For given x and n, there will be many choices for y and h; pick one such pair arbitrarily. Then set $j(x)(n) = h$.

This completes the definition of our function $j : G \to \Phi$. It remains to check that j is one-to-one. So suppose $j(x) = j(x')$. Thus, for infinitely many positive integers n (namely all n in the domain of $j(x)$) we can write $x = ny+h$ and $x' = ny'+h$, with the same h. Then $x-x'$ is divisible by these infinitely many positive integers n. But G is torsionless, so $x-x'$ must be 0, i.e., $x = x'$, and the proof is complete. $\square$

REMARK 5.8. In this theorem, the hypothesis that G is torsionless can be weakened. All we needed in the proof is that the only element divisible by infinitely many integers is 0. This is equivalent to requiring all subgroups of rank 1 to be isomorphic to $\mathbb{Z}$.

COROLLARY 5.9. *If a separable group G has a maximal pure independent set of infinite cardinality κ, then G can be embedded as a pure subgroup in $\mathbb{Z}^{\kappa^{\aleph_0}}$.*

Proof   Combine the theorem with Corollary 5.3. $\square$

REMARK 5.10. To clarify the relationship between the cardinality estimates in Theorems 5.4 and 5.7, we mention some of the relevant facts of cardinal arithmetic. For all infinite cardinals, $\kappa^{\aleph_0} \leq 2^\kappa$, so Theorem 5.7 is never worse than the estimate in Theorem 5.4. Whether it is actually better depends on κ. There are cardinals κ for which $\kappa^{\aleph_0} = 2^\kappa$; an obvious example is $\kappa = \aleph_0$. But there are also cardinals κ for which $\kappa^{\aleph_0} < 2^\kappa$. Examples include the cardinality $\mathfrak{c} = 2^{\aleph_0}$ of the continuum as well as cardinals finitely many steps beyond it, i.e., the next cardinal $\mathfrak{c}^+$, and $\mathfrak{c}^{++}$, and so forth for a simple (not transfinite) sequence. In fact, for these particular examples, $\kappa^{\aleph_0} = \kappa$, so we recover in Theorem 5.7 the cardinality estimate of Theorem 1.6 for these values of κ.

The following theorem provides improvements of some of the preceding work if we consider groups satisfying stronger hypotheses than separability. We refer to [4, Sections IV.2 and VII.4] for the definitions of the hypotheses "Whitehead group" and "coseparable" used here.

THEOREM 5.11. *Suppose G is a Whitehead group (or only a coseparable group) of cardinality at most 2^κ, where κ is an infinite cardinal. Then G can be embedded as a pure subgroup in $\mathbb{Z}^\kappa$.*

Proof   Given G as in the hypothesis, we intend to produce a family $\mathcal{F}$ of at most κ homomorphisms $G \to \mathbb{Z}$ satisfying the conditions of the second part of Lemma 5.1. Thus, for every $g \in G$ and every prime number p not dividing g, there should be an $f \in \mathcal{F}$ with p not dividing $f(g)$. Equivalently, the induced homomorphisms $\bar{f} : G/pG \to \mathbb{Z}/p\mathbb{Z}$, for $f \in \mathcal{F}$, should not all simultaneously annihilate any non-zero element of G/pG.

Thus, to complete the proof, it suffices to find, for each prime p, a collection of at most κ homomorphisms $G/pG \to \mathbb{Z}/p\mathbb{Z}$ that do not all annihilate any non-zero element of G/pG and to lift these to homomorphisms $G \to Z$. Then, by taking all these lifted homomorphisms, for all primes p (for a total of at most $\kappa \cdot \aleph_0 = \kappa$ homomorphisms), we obtain the desired $\mathcal{F}$.

Consider, therefore, any fixed prime p. Since G has cardinality at most 2^κ, the dimension of G/pG as a vector space over $\mathbb{Z}/p\mathbb{Z}$ is at most 2^κ. On the other hand, $(\mathbb{Z}/p\mathbb{Z})^\kappa$ has dimension exactly 2^κ. So there is an embedding of vector spaces, $G/pG \to (\mathbb{Z}/p\mathbb{Z})^\kappa$. The coordinates of this embedding are κ homomorphisms $G/pG \to \mathbb{Z}/p\mathbb{Z}$ that do not all annihilate any non-zero element of G/pG.

All that remains is to lift these homomorphisms $G/pG \to \mathbb{Z}/p\mathbb{Z}$ to homomorphisms $G \to \mathbb{Z}$. Of course we can compose our homomorphisms with the projection $G \to G/pG$; we must then lift the resulting homomorphisms $G \to \mathbb{Z}/p\mathbb{Z}$ to homomorphisms $G \to \mathbb{Z}$. In fact, we claim that *every* homomorphism $G \to \mathbb{Z}/p\mathbb{Z}$ can be so lifted.

To see this, consider the short exact sequence
$$0 \to \mathbb{Z} \to \mathbb{Z} \to \frac{\mathbb{Z}}{p\mathbb{Z}} \to 0$$
where the map $\mathbb{Z} \to \mathbb{Z}$ is multiplication by p. Applying $\mathrm{Hom}(G, -)$, we get a long exact sequence, part of which is
$$\mathrm{Hom}(G, \mathbb{Z}) \to \mathrm{Hom}(G, \frac{\mathbb{Z}}{p\mathbb{Z}}) \to \mathrm{Ext}(G, \mathbb{Z}) \to \mathrm{Ext}(G, \mathbb{Z}),$$
where the last map is multiplication by p. Our claim about lifting arbitrary homomorphisms amounts to saying that the first map in this (segment of the) exact sequence is surjective, or equivalently (by exactness) that the second map is 0, or equivalently (by exactness) that the third map is one-to-one. For Whitehead groups, this last reformulation is obviously true, since the Ext terms are zero. For coseparable G, we need to show that multiplication by the prime p is one-to-one on $\mathrm{Ext}(G, \mathbb{Z})$. But one of the equivalent characterizations of coseparability in [**4**, Section IV.2] is that $\mathrm{Ext}(G, \mathbb{Z})$ is torsion-free. So the proof is complete. □

COROLLARY 5.12. *Let G be a Whitehead (or coseparable) group with a maximal pure independent set of infinite cardinality κ. Then G can be embedded as a pure subgroup of $\mathbb{Z}^\kappa$.*

Proof Combine Theorems 5.4 and 5.11. □

Thus, for Whitehead groups, a maximal pure independent set gives a pure embedding in as small a product of $\mathbb{Z}$'s as a basic subgroup would give via Theorem 1.6.

References

[1] A. Blass, "On the divisible parts of quotient groups," in *Abelian Group Theory and Related Topics*, edited by R. Göbel, P. Hill, and W. Liebert, Contemp. Math. 171, Amer. Math. Soc. (1994) 37–50.

[2] M. Dugas and J. Irwin, "On pure subgroups of Cartesian products of integers," *Resultate Math.* 15 (1989), 35–52.

[3] M. Dugas and J. Irwin, "On basic subgroups of $\prod \mathbf{Z}$," *Comm. Algebra* 19 (1991), 2907–2921.

[4] P. Eklof and A. Mekler, *Almost Free Modules: Set-Theoretic Methods*, North-Holland, 1990

[5] L. Fuchs, *Infinite Abelian Groups, vol. I*, Academic Press, 1970.

[6] L. Ya. Kulikov, "On the theory of abelian groups of arbitrary cardinality" [Russian], *Mat. Sbornik* 16 (1945) 129–162.

MATHEMATICS DEPARTMENT, UNIVERSITY OF MICHIGAN, ANN ARBOR, MI 48109–1109, U.S.A.
E-mail address: `ablass@umich.edu`

9001 STARMOUNT, LAS VEGAS, NV 89134, U.S.A.
E-mail address: `drjmirwin@lvcm.com`

Characterization of the Tori via density of the Solution Set of Linear Equations

Dikran Dikranjan and Michael Tkachenko

Dedicated to Professor Laszlo Fuchs in honour of his 75th birthday

ABSTRACT. We say that a compact metrizable Abelian group G with an invariant metric d satisfies $\mathcal{E}$ if for every $a \in G$ and for every natural n there exists $\delta_n > 0$ such that $\delta_n \to 0$ and for the solution set $S_n(a)$ of the equation $nx = a$, we have $d(y, S_n(a)) \leq \delta_n$ for each $y \in G$. We characterize the finite-dimensional tori among the class of all compact Abelian groups G by proving that such a group G is a torus if and only if every closed connected one-dimensional subgroup of G satisfies $\mathcal{E}$. We prove then that the finite-dimensional tori satisfy a much stronger density condition for the solution sets of linear equations with several variables.

1. Introduction

An Abelian group G is *divisible* when every linear equation
$$n \cdot x = a \tag{1}$$
has a solution in G, where $a \in G$, $n \in \mathbb{Z}$ and $n \neq 0$. While divisibility guarantees the existence of a solution of (1), such solutions need not form a big set, e.g. when G is torsion-free then there exists precisely one solution. In order to better measure the size of the solution set $S_n(a) = \{x \in G : nx = a\}$ we need to consider a *topological* group G. Here we shall be mostly interested in compact Abelian groups, where divisibility is equivalent to connectedness [**HR**, Section 24].

The following example explains the title of the article:

EXAMPLE 1.1. The torus $\mathbb{T}^m$ is divisible, so that every linear equation (1) has a solution and the solutions are relatively dense in the sense that one can find a solution of (1) in every open ball of diameter $> 1/n$. Indeed, for every $\varepsilon > 1/n$ the set $S_n(a)$ (coinciding with a coset of the subgroup $\mathbb{T}^m[n] = \{x \in \mathbb{T}^m : n \cdot x = 0\}$) is ε-dense in $\mathbb{T}^m$ in the sense that the open balls of diameter ε and centers placed at the points of $\mathbb{T}^m[n]$ cover $\mathbb{T}^m$ (cf. Definition 2.1).

2000 *Mathematics Subject Classification.* Primary 22A05, 22B05, 54D25, 54H11; Secondary 54A35, 54B30, 54D30, 54H13.

The second author was partially supported by the Mexican National Council of Sciences and Technology (CONACyT), grant no. 400200-5-3012PE.

The aim of this note is:
1) to show that the property described in Example 1.1 determines the tori within the class of compact Abelian groups;
2) to consider a more general form of this phenomenon for linear equations (3) in Section 4 (necessarily valid *only* on the tori).

The relevance of the stronger form (3) of this property on the tori is used essentially in [**DT1, DT2, DT3**] for the construction of countably compact group topologies on the non-torsion Abelian groups of cardinality $\mathfrak{c} = 2^\omega$ (see Section 5 for more details).

1.1. Notation and terminology.
We denote by $\mathbb{N}$ and $\mathbb{P}$ the sets of positive naturals and primes, respectively, by $\mathbb{Z}$ the integers, by $\mathbb{Q}$ the rationals, by $\mathbb{R}$ the reals, by $\mathbb{T}$ the unit circle group in $\mathbb{C}$, by $\mathbb{Z}_p$ the p-adic integers ($p \in \mathbb{P}$), by $\mathbb{Z}(n)$ the cyclic group of order n ($n \in \mathbb{N}$). For $k, n \in \mathbb{N}$, the fact that k divides n abbreviates to $k|n$. The cardinality of continuum 2^ω will also be denoted by $\mathfrak{c}$.

All groups in the paper are Abelian and will be written additively. Let G be a group and A be a subset of G. We denote by 0 the neutral element of G and by $\langle A \rangle$ the subgroup of G generated by A. We denote by $r(G)$ the free-rank of G and we set $G[n] = \{x \in G : nx = 0\}$, $n \in \mathbb{N}$. For a non-empty set π of prime numbers define the π-*socle* of G as $soc_\pi(G) = \bigoplus_{p \in \pi} G[p]$. For $\pi = \{p\}$ or $\pi = \mathbb{P}$ we briefly say p-socle and socle, respectively. The torsion subgroup of G is denoted by $t(G)$, and the p-torsion subgroup — by $t_p(G)$.

For a prime number p and a topological Abelian group G, an element $x \in G$ is *quasi-p-torsion* if the subgroup $\langle x \rangle$ is either p-torsion or isomorphic to $\mathbb{Z}$ equipped with the p-adic topology. By $td_p(G)$ we denote the subgroup of all quasi-p-torsion elements of G.

For a compact or discrete group G, we denote by $\widehat{G}$ the Pontryagin dual of G. If G is a compact Abelian group and Y is a subgroup of $X = \widehat{G}$, we denote by $A(Y)$ the annihilator $\{x \in G : (\forall \chi \in Y)\chi(x) = 0\}$ of Y. Clearly, $A(Y)$ is always a closed subgroup of G. It is known that $\widehat{A(Y)} \cong X/Y$ and $\widehat{G/A(Y)} \cong Y$ [**HR**, Section 24].

2. Relative density of the solutions of equation (1)

We need some notation regarding a metrizable topological group G with invariant metric ρ. For a positive number ε, we denote by $B_\varepsilon^d(x)$ (or simply by $B_\varepsilon(x)$) the ε-ball with center x in (G, d).

DEFINITION 2.1. Let G be a topological group and let X be a subset of G.
(a) If V is a neighborhood of 0 in G, then X is V-*dense in* G if $X + V = G$;
(b) if G is metrizable and d is an invariant compatible metric, then for $\varepsilon > 0$ the set X is ε-dense in G if it is $B_\varepsilon(0)$-dense in G.

Clearly, $X \subseteq G$ is dense in G iff it is V-dense in G for every neighborhood V of 0 (in particular, if G is metrizable, X is dense iff X is ε-dense in G for every $\varepsilon > 0$). We show below that the above property of the tori remains true also for linear equation with many variables. Let us introduce some more notions in order to be able to formulate it rigorously.

DEFINITION 2.2. We say that a metrizable topological group G satisfies the condition $\mathcal{E}$ if for every $n \in \mathbb{N}$ and for every $a \in G$ there exists $\varepsilon_n > 0$ such that the solution set $S_n(a)$ is ε_n-dense in G and $\varepsilon_n \to 0$ when $n \to \infty$.

Clearly, every torus $\mathbb{T}^m$ satisfies $\mathcal{E}$ (see Example 1.1).

One may be left with the impression that other versions of $\mathcal{E}$ are possible: one with a fixed $a \in G$ and the other when $a_n \in G$ varies when n varies. In order to simplify $\mathcal{E}$ we note now that for a compact metrizable Abelian group G the choice of a_n is absolutely irrelevant and can therefore be replaced by the neutral element 0 of G.

LEMMA 2.3. *Let G be a metrizable Abelian group and let $n \in \mathbb{N}$. Then the following are equivalent for all $a \in G$ and $\varepsilon > 0$:*
  (i) *the set $S_n(a) = \{x \in G : n \cdot x = a\}$ is ε-dense in G;*
  (ii) *there exists $t_0 \in G$ such that $n \cdot t_0 = a$ and $G[n]$ is ε-dense in G.*

PROOF. Suppose that $S_n(a) \neq \emptyset$, i.e., $n \cdot t_0 = a$ for some $t_0 \in G$. Then $n \cdot t = a$ iff $t \in t_0 + G[n]$. Therefore, $S_n(a) = G[n] + t_0$. □

This lemma suggests to consider instead of $\mathcal{E}$ the following condition for a topological Abelian group G which is technically easier to deal with:

$$\forall n \in \mathbb{N}\ \exists \varepsilon_n > 0 \text{ such that } G[n] \text{ is } \varepsilon_n\text{-dense in } G \text{ and } \varepsilon_n \to 0 \text{ when } n \to \infty. \quad (*)$$

Seemingly one has to pay the price of loosing divisibility by passing from $\mathcal{E}$ to $(*)$. We see in the next lemma that for compact metrizable Abelian groups these two properties coincide, i.e., $(*)$ implies divisibility. Another important point of the lemma is that the apparent dependence of $\mathcal{E}$ on the metric of G disappears, since we offer a third equivalent property $(\mathcal{D})$ that depends only on the topology of G.

For an Abelian group G and $k \in \mathbb{Z}$, we denote by φ_k the endomorphism $G \to G$ defined by $\varphi_k(x) = kx$.

LEMMA 2.4. *For every compact metrizable Abelian group, the properties $\mathcal{E}$ and $(*)$ coincide and are equivalent to the following one:*

  $(\mathcal{D})$ *$\forall p \in \mathbb{P}$ and for every infinite $\pi \subseteq \mathbb{P}$, the subgroups $t_p(G)$ and $soc_\pi(G)$ of G are dense in G.*

PROOF. Let G be a compact Abelian group. By Lemma 2.3, $G \in \mathcal{E}$ is equivalent to the fact that G is divisible and $(*)$ holds. Our plan is to show that $(*)$ implies $(\mathcal{D})$, while the latter implies $(*)$ and divisibility of G.

$(*) \Rightarrow (\mathcal{D})$. Assume that $(*)$ holds. Let us show first that $soc_\pi(G)$ is dense in G for every infinite $\pi \subseteq \mathbb{P}$ and the subgroup $t_p(G)$ is dense in G for every $p \in \mathbb{P}$. Indeed, for the subgroup $t_p(G)$ consider for every $n \in \mathbb{N}$ the homomorphism $g_n = \varphi_{p^n} : G \to G$ and note that $G[p^n] = \ker g_n$ must be ε_n-dense in G for an appropriate $\varepsilon_n > 0$ with $\lim_n \varepsilon_n = 0$. This obviously yields that $t_p(G) = \bigcup_{n=1}^\infty G[p^n]$ is ε-dense in G for every $\varepsilon > 0$. Hence $t_p(G)$ is dense in G.

For $soc_\pi(G)$, it suffices to order $\pi = \{p_1, p_2, \ldots, p_n, \ldots\}$ and argue similarly with the homomorphism $f_n = \varphi_{p_1 \ldots p_n} : G \to G$, noting that $\ker \varphi_{f_n} = \oplus_{i=1}^n soc_{p_i}(G)$ and $soc_\pi(G) = \bigcup_{n=1}^\infty \ker \varphi_{f_n}$.

Let us see now that $(\mathcal{D})$ implies G is divisible. We start the proof of this implication by noting that the property $(\mathcal{D})$ is preserved by taking quotients. Now suppose that G satisfies $(\mathcal{D})$. Then also $G/c(G)$ satisfies $(\mathcal{D})$, where $c(G)$ is the connected component of G. Since the compact group $G/c(G)$ is totally disconnected, every subgroup $td_p(G/c(G))$ is closed [**DPS**]. By $(\mathcal{D})$, $t_p(G)$ is dense in G, so $t_p(G/c(G))$ is dense in $G/c(G)$. This yields that $td_p(G/c(G))$ must coincide with $G/c(G)$ as it contains the dense subgroup $t_p(G/c(G))$. Now pick any prime

$q \neq p$. Then $td_q(G/c(G)) = \{0\}$. On the other hand, ($\mathcal{D}$) yields that $t_q(G/c(G))$ is dense in $G/c(G)$. This is possible only if $G/c(G)$ is trivial. This proves that G is connected, hence divisible.

($\mathcal{D}$) $\Rightarrow$ ($*$). For every $k \in \mathbb{N}$, let δ_k be the infimum of all positive $\delta > 0$ such that $G = G[k] + B_\delta(0)$. Since G is compact, such an infimum exists. Obviously, the subgroup $G[k]$ is ε-dense for every $\varepsilon > \delta_k$. Therefore, to prove that ($\mathcal{D}$) $\Rightarrow$ ($*$) it suffices to take $\varepsilon_k = 2\delta_k$ and see that $\delta_k \to 0$ when $k \to \infty$ (then obviously $\varepsilon_k \to 0$). We shall repeatedly use the fact that the sequence δ_k is non-increasing with respect to divisibility, i.e., if $k|l$, then $\delta_k \geq \delta_l$.

Assume for contradiction that δ_n fails to converge to 0. Then there exist a real number $r > 0$ and an infinite sequence of naturals $n_1 < \ldots < n_s < \ldots$ such that $\delta_{n_s} \geq r$ for each $s \in \mathbb{N}$. Then either

1) there exists a prime p such that for every $k \in \mathbb{N}$ there exists $m_k \in \mathbb{N}$ with $p^k | n_m$ for every $m > m_k$, or
2) there exist infinitely many primes $p_1 < \ldots < p_s < \ldots$ that divide the numbers $n_{i_1} < \ldots < n_{i_s} < \ldots$.

Assume that 1) occurs. Then by ($\mathcal{D}$) the subgroup $t_p(G)$ is dense in G. Since δ_k is non-increasing in the sense described above, for every $k \in \mathbb{N}$ there exists $m_k \in \mathbb{N}$ with $r \leq \delta_{n_m} \leq \delta_{p^k}$ for every $m > m_k$. This leads to $G[p^k] + B_s(0) \neq G$ for $s = r/2$ and every $k \in \mathbb{N}$. Since G is compact, and $V_k = G[p^k] + B_s(0) \neq G$ ($k \in \mathbb{N}$) form an ascending chain of proper open subsets of G, there exists an element $z \in G \setminus \bigcup_{k=1}^\infty (G[p^k] + B_s(0))$. Then the ball $B_s(z)$ contains no points of the dense set $t_p(G) = \bigcup_{k=1}^\infty G[p^k]$, a contradiction.

A similar proof shows that if case 2) occurs, then our hypothesis and the fact that the subgroup $soc_\pi(G)$ of G is dense for $\pi = \{p_s : s \in \mathbb{N}\}$ according to ($\mathcal{D}$), lead to the same kind of contradiction. Therefore, $\delta_n \to 0$. $\square$

This lemma suggests that along with the class $\mathcal{E}$ also the larger class $\mathcal{T}$ of compact Abelian groups K in which $t(K)$ is dense should be studied (see [**Di2**]). Its hereditary hull her($\mathcal{T}$), namely, the biggest subclass of $\mathcal{T}$ that consists of compact groups K such that every closed subgroup N of K belongs to $\mathcal{T}$, was thoroughly studied in [**DPr**], where groups of this class were named *exotic tori* (see [**Di2, DS**] for the counterpart of exotic tori in the non-Abelian case).

2.1. A different approach: the mesh of a subgroup. Let G be a metric compact group and let d be an invariant metric on G. For a subgroup H of G, we set $\mathbf{m}(H) = \inf\{\delta > 0 : H \cdot B_\delta(0) = G\}$. Clearly, $\mathbf{m}(H) = 0$ iff H is dense in G. Therefore, if H is closed, then $\mathbf{m}(H) = 0$ iff $H = G$, so that the *mesh* $\mathbf{m}(H)$ of H measures how "large", in a sense, is the coset space G/H. So, another way to get this number is to consider the continuous function $d(x, H)$ with $x \in G$. Since G is compact, its maximum exists and it is easy to see that it coincides with $\mathbf{m}(H)$.

We introduce the parameter $\mathbf{m}(H)$ related to H in order to measure the set $S_n(a)$ of solutions of the linear equation $n \cdot x = a$ in an Abelian group G. In case this equation has a solution x_0, one has $S_n(a) = x_0 + G[n]$. Hence it suffices to measure the size of the subgroup $G[n]$. Our measure will be the mesh $\mathbf{m}(G[n])$. Since our objective is $\mathcal{E}$, we will be interested merely in the asymptotic behavior of the numbers $\mathbf{m}(G[n])$ for $n \to \infty$.

Here are some easy to prove general property of the number $\mathbf{m}(H)$.

1) $\mathbf{m}(-)$ is monotone: $H \leq K \leq G$ implies $\mathbf{m}(H) \leq \mathbf{m}(K)$;

2) m(-) is (sequentially) continuous in the sense that if H is an increasing union of the subgroups H_n of G, then $m(H) = \inf_n m(H_n)$;

3) with H as in 2), H is dense in G iff $m(H_n)$ converges to 0.

In these terms, G satisfies (∗) (and consequently, $\mathcal{E}$) iff $m(G[n]) \to 0$. Formally, the latter property of G depends on the choice of the invariant metric d on G. However, every two metrics on a compact space are equivalent, so the asymptotic behaviour of m on the subgroups $G[n]$ does not depend on the metric d.

3. Compact Abelian groups hereditarily satisfying $\mathcal{E}$

Since closed connected subgroups of a torus are tori, the implication (b) ⇒ (c) of the next theorem follows from Example 1.1, where we established that all tori possess the property $\mathcal{E}$. More surprising is the fact that the still weaker "hereditary" version (a) of $\mathcal{E}$ implies that such a group must necessarily be a torus. Clearly, we have to consider only connected subgroups in (a) and (c), since $\mathcal{E}$ implies connectedness by Lemma 2.4. As far as connectedness of G is concerned, it may be obviously omitted, if we ask G to be covered by closed subgroups satisfying $\mathcal{E}$.

THEOREM 3.1. *The following are equivalent for a connected compact Abelian group G:*

(a) *every closed connected one-dimensional subgroup of G satisfies $\mathcal{E}$;*
(b) *G is a usual torus;*
(c) *every closed connected subgroup of G satisfies $\mathcal{E}$.*

PROOF. (c) ⇒ (a) is obvious.

(b) ⇒ (c). As mentioned above, this is a direct consequence of Example 1.1 (see also Lemma 4.1 below).

(a) ⇒ (b). We denote by $\mathbb{K}$ the Pontryagin dual of the group $\mathbb{Q}$. Since $\mathbb{Q}$ is divisible and torsion-free, $\mathbb{K}$ is torsion-free and connected [**HR**, Th. 24.25]. In addition, $\dim \mathbb{K} = r(\mathbb{Q}) = 1$ by [**HR**, Th. 24.28], so $\mathbb{K}$ is a one-dimensional compact connected Abelian group.

We first prove (a) yields that the compact group G is finite-dimensional. Indeed, assume not. Then such a group G will contain a closed subgroup N isomorphic to $\mathbb{K}$. In fact, the discrete Pontryagin dual X of G satisfies $r(X) = \infty$, so there exists a surjective homomorphism $h\colon X \to \mathbb{Q}$. Taking the dual of h we obtain an injective continuous homomorphism $\mathbb{K} \to G$ as desired. Since $\mathbb{K}$ is a one-dimensional connected subgroup of G, (a) implies that $\mathbb{K}$ satisfies $\mathcal{E}$. However, $\mathbb{K}$ is torsion-free, a contradiction.

Let us prove the implication (a) ⇒ (b) for one-dimensional groups G. Our assumption, $\dim G = 1$ implies that the discrete Pontryagin dual X of G satisfies $r(X) = 1$. Since G is connected, X is torsion-free. Therefore, there exists an exact sequence

$$0 \longrightarrow X \xrightarrow{\psi} \mathbb{Q} \longrightarrow T \longrightarrow 0,$$

where ψ is an embedding of X in $\mathbb{Q}$. Here $T = \bigoplus_{p \in \mathbb{P}} T_p$ is a torsion divisible group isomorphic to a quotient of $\mathbb{Q}$. Hence, for every $p \in \mathbb{P}$ either $T_p = \{0\}$ or $T_p \cong \mathbb{Z}(p^\infty)$. Clearly, $T_p = \{0\}$ iff X is p-divisible, i.e., $X = pX$. But in such a case $t_p(G) = \{0\}$ as pX is the annihilator of $G[p]$ [**DPS, HR**]. By Lemma 2.4, the latter contradicts (a). Hence, $T_p \cong \mathbb{Z}(p^\infty)$ for every prime p. Let N be the compact subgroup of $\mathbb{K}$ corresponding to X under the annihilator correspondence, i.e., $N = A(X)$. Then the Pontryagin dual of N is isomorphic to T, so $N \cong \widehat{T}$ is a totally disconnected subgroup of $\mathbb{K}$ of the form $N = \prod_{p \in \mathbb{P}} N_p$, where $N_p \cong \widehat{T_p} \cong \mathbb{Z}_p$ for every $p \in \mathbb{P}$ (see [**HR**, Th. 23.22]), and there exists an exact sequence

$$0 \longrightarrow N \longrightarrow \mathbb{K} \xrightarrow{\varphi} G \longrightarrow 0,$$

where φ is the adjoint of ψ. Since ψ is an embedding, φ is surjective. Let $f \colon G \to \mathbb{T}$ be a non-trivial continuous homomorphism and $\xi = f \circ \varphi$. Since G is connected and since the only proper closed subgroups of $\mathbb{T}$ are finite, f must be surjective. So $\xi \colon \mathbb{K} \to \mathbb{T}$ is surjective as well. Let $\mathbb{H} = \ker \xi$. This gives rise to the exact sequence

$$0 \longrightarrow \mathbb{H} \longrightarrow \mathbb{K} \xrightarrow{\xi} \mathbb{T} \longrightarrow 0.$$

Since the dual of $\mathbb{T}$ is $\mathbb{Z}$, we can conclude that the dual of the compact group $\mathbb{H}$ is isomorphic to $\widehat{\mathbb{Q}/\mathbb{Z}} \cong \oplus_{p \in \mathbb{P}} \mathbb{Z}(p^\infty)$. Hence $\mathbb{H} = \prod_{p \in \mathbb{P}} \mathbb{H}_p$, with $\mathbb{H}_p \cong \mathbb{Z}_p$. Clearly, $N = \ker \varphi \leq \mathbb{H}$, so that $N_p \leq \mathbb{H}_p$ for all p. Since $N_p \cong \mathbb{Z}_p$, for every $p \in \mathbb{P}$ there exists $k_p \in \mathbb{N} \cup \{0\}$ such that $N_p = p^{k_p}\mathbb{H}_p$ [**HR**, 10.16 (a)]. Suppose that the set $\pi = \{p \in \mathbb{P} : k_p \neq 0\}$ is infinite. For $p \in \pi$, the quotient $F_p = \mathbb{H}_p/N_p$ is a non-trivial cyclic p-group. Let $F = \mathbb{H}/N$, so that $F = \prod_{p \in \pi} F_p$. Note that $\mathbb{H}/N \leq \mathbb{K}/N \cong G$, so we can identify $\mathbb{H}/N$ with the subgroup $\varphi(\mathbb{H})$ of G. Let us see now that for every prime $p \in \pi$ the group $F_p = \mathbb{H}_p/N_p$ contains $soc_p(G)$. Indeed, let $g \in G$ satisfy $pg = 0$, where $p \in \pi$. Then $g = \varphi(x)$ for some $x \in \mathbb{K}$, and $px \in N$. Now decompose $px = y + z$, with $y \in N_p$ and $z \in \prod_{q \neq p} N_q$. Since $k_p \geq 1$, we have $N_p \leq p\mathbb{H}_p$. Hence $y = pu$, with $u \in \mathbb{H}_p$. Note that the subgroup $\prod_{q \neq p} N_q$ of $\mathbb{K}$ is p-divisible, so there exists $z' \in \prod_{q \neq p} N_q$ such that $z = pz'$. Thus $px = pu + pz'$, and hence $x = u + z'$ since $\mathbb{K}$ is torsion-free. Now $g = \varphi(x) = \varphi(u) \in F_p$. This proves that $soc_p(G) \subseteq soc_p(F)$ for each $p \in \pi$, whence $soc_\pi(G) \subseteq soc_\pi(F)$. Since F is a proper closed subgroup of G, $soc_\pi(G)$ cannot be dense in G contrary to Lemma 2.4. This proves that the set π and the group F are finite. Hence there exists $m \in \mathbb{N}$ such that $N = m\mathbb{H}$. Thus $G \cong \mathbb{K}/N = \mathbb{K}/m\mathbb{H} \cong \mathbb{K}/\mathbb{H} \cong \mathbb{T}$.

In the general case, we know that $n = \dim G$ is finite. The discrete Pontryagin dual X of G satisfies $r(X) = n$. Let F be a free subgroup of X of rank n and let F_i, $i = 1, 2, \ldots, n$ be rank $n-1$ subgroups of F such that

$$\bigcap_{i=1}^{n} F_i = \{0\}. \tag{2}$$

Let $Y_i = \{x \in X : mx \in F_i \text{ for some } m > 0\}$. Then $r(Y_i) = r(F_i) = n - 1$, hence X/Y_i is a rank-one torsion-free group. Therefore the closed subgroup $L_i = A(Y_i)$ of G is a connected one-dimensional subgroup of G. Moreover, (2) yields $\bigcap_{i=1}^{n} Y_i = \{0\}$, and hence $G = L_1 + \cdots + L_n$. By our assumption, each $L_i \cong \mathbb{T}$. Therefore, G is an n-dimensional quotient of $\mathbb{T}^n$, hence $G \cong \mathbb{T}^n$. $\square$

REMARK 3.2. (a) It should be noted that the other consequence of $\mathcal{E}$ in Lemma 2.4 may strongly fail for exotic tori (see [**DPS**, Ch.4] for examples and the description of the one-dimensional exotic tori with dense socle).

(b) Our choice to consider the property $\mathcal{E}$ in "hereditary version" is due to the fact that $\mathcal{E}$ alone does not suffice to prove such a characterization as the following examples show.

  (b$_1$) Obviously, the property ($\mathcal{D}$) is productive, so also $\mathcal{E}$ is productive in the class of of compact Abelian groups by Lemma 2.4. Therefore, $\mathbb{T}^\omega \in \mathcal{E}$.

(b$_2$) Here is an example of a two-dimensional compact connected Abelian group $G \not\cong \mathbb{T}^2$ in $\mathcal{E}$. It was proved in [**Di1**] (following ideas of an old constructions of Kurosch [**Ku**]), that there exists a rank-two subgroup X of the group $\mathbb{Z}_p$ such that for a rank-two free subgroup F of X one has $X/F \cong \mathbb{Z}(p^\infty)$. Then X has no infinitely p-divisible elements, as a subgroup of $\mathbb{Z}_p$. This means that $\bigcap_n p^n X = \{0\}$. For any prime $q \neq p$, the subgroup F of X is q-pure, i.e., $F \cap q^n X = q^n F$. Therefore, the subgroup $X_q = \bigcap_{n=1}^\infty q^n X$ of X satisfies $X_q \cap F = \bigcap_n q^n F = \{0\}$. Since F non-trivially meets every non-zero subgroup of X, this gives $X_q = \{0\}$. Then for the compact Pontryagin dual G of X, the subgroup $t_q(G)$ is dense in G. Analogously, $t_p(G)$ is dense in G as $\bigcap_n p^n X = \{0\}$. Now let π be an infinite set of primes with $p \notin \pi$ and $Y = \bigcap_{q \in \pi} qX$. From q-purity of F it follows as before that $F \cap Y = \bigcap_{q \in \pi} qF = \{0\}$. Hence $Y = \{0\}$. Since $A(Y)$ must coincides with the closure of $soc_\pi(G)$, we conclude that $soc_\pi(G)$ is dense in G. Hence $G \in \mathcal{E}$. On the other hand, X is not finitely generated, so that G cannot be isomorphic to any finite-dimensional torus.

(c) We do not know whether one can get rid of the hereditary version of $\mathcal{E}$ in the smaller class of exotic tori, i.e., whether every exotic torus satisfying $\mathcal{E}$ is a torus in the usual sense. This is obviously true in dimension one, since our proof shows that every one-dimensional group with $\mathcal{E}$ is isomorphic to $\mathbb{T}$. In higher dimensions one should use the fact that $\mathcal{E}$ is stable under taking quotients and produce a lower-dimensional quotient of an exotic torus $G \in \mathcal{E}$ that is not a torus in order to apply induction. Note that the two-dimensional group in (b$_2$) is not an exotic torus.

4. The tori satisfy a stronger version of $\mathcal{E}$

By a linear equation in an Abelian group G we mean an equation

$$g(x) = \sum_{i=1}^{m} k_i x_i = a, \; x = (x_1, \ldots, x_m) \in G^m, \tag{3}$$

where $a \in G$, $m \in \mathbb{N}$ and $k_i \in \mathbb{Z}$, with at least one of k_i's distinct from zero. The class of divisible groups is defined by the requirement that every equation (1) (i.e., (3) with $m = 1$) has a solution. In such a case also every equation (3) has obviously a solution. For $m \in \mathbb{N}$ and a homomorphism $g \colon G^m \to G$ defined by (3), set $\|g\| = \sum_{i=1}^{m} |k_i|$.

If $m \in \mathbb{N}$, let ρ be the metric in $\mathbb{R}^m$ defined by $\rho(x, y) = \max\{|x_i - y_i| : i = 1, \ldots, m\}$. This distance defines an invariant metric d on $\mathbb{T}^m = \mathbb{R}^m/\mathbb{Z}^m$ in the standard way. Note that the exponential map $\varphi \colon \mathbb{R}^m \to \mathbb{T}^m$ is contracting with respect to these distances.

Note that every continuous homomorphism $g \colon \mathbb{T}^m \to \mathbb{T}$ has the form $g(t) = \sum_{i=1}^{m} k_i t_i$ for each $t = (t_1, \ldots t_m) \in \mathbb{T}^m$, where $k_1, \ldots, k_m \in \mathbb{Z}$.

Let us say that a metrizable topological group G satisfies the condition $\mathcal{E}_m$ if for every equation $g(x) = a$ as in (3) there exists $\varepsilon_g > 0$ depending on g such that solution set $g^{-1}(a)$ is ε_g-dense in G and $\varepsilon_g \to 0$ whenever $\|g\| \to \infty$. Clearly, $\mathcal{E}_1$ coincides with $\mathcal{E}$ and $\mathcal{E}_{m+1}$ implies $\mathcal{E}_m$ for each $m \in \mathbb{N}$. It follows from Lemma 4.1 below that $\mathbb{T}$ satisfies $\mathcal{E}_m$ for every $m \in \mathbb{N}$.

LEMMA 4.1. *Let $g\colon \mathbb{T}^m \to \mathbb{T}$ be a homomorphism defined by $g(y) = k_1 \cdot y_1 + \cdots + k_m \cdot y_m$, where $y = (y_1, \ldots, y_m) \in \mathbb{T}^m$ and $k_1, \ldots, k_m \in \mathbb{Z}$. Suppose that $\varepsilon > 0$ and $||g|| > 1/2\varepsilon$. Then for every $a \in \mathbb{T}$ and $c = (c_1, \ldots, c_m) \in \mathbb{T}^m$, there exists $y = (y_1, \ldots, y_m) \in \mathbb{T}^m$ such that $g(y) = a$ and $d(y, c) < \varepsilon$.*

PROOF. Consider the linear function $f\colon \mathbb{R}^m \to \mathbb{R}$ defined by the same formula $f(x_1, \ldots, x_m) = k_1 \cdot x_1 + \cdots + k_m \cdot x_m$ for every $(x_1, \ldots, x_m) \in \mathbb{R}^m$. Let $\pi\colon \mathbb{R} \to \mathbb{T}$ be the natural quotient map, $\pi(x) = \mathbb{Z} + x$. Choose a point $b = (b_1, \ldots, b_m) \in \mathbb{R}^m$ such that $\pi(b_i) = c_i$ for each $i = 1, \ldots, m$. Take $\mu > 0$ satisfying $1/2||g|| < \mu < \varepsilon$ and let $\Pi_\mu = \prod_{i=1}^m [b_i - \mu, b_i + \mu]$ be the closed m-dimensional cube with the center at b. One easily verifies that $f(\Pi_\mu) = [f(b) - \mu \cdot ||g||, f(b) + \mu \cdot ||g||]$. It is clear that $2\mu \cdot ||g|| > 1$, so $\pi(f(\Pi_\mu)) = \mathbb{T}$. Therefore, we can find a point $x = (x_1, \ldots, x_m) \in \Pi_\mu$ such that $\pi(f(x)) = a$. The point $y = (y_1, \ldots, y_m) \in \mathbb{T}^m$ with $y_i = \pi(x_i)$, $i = 1, \ldots, m$, is as required. □

5. An application

The problem of characterization of the class of abelian groups that admit compact group topologies was set by Halmos in 1944 [**Hal**] and resolved in the next decade by the common efforts of Kaplansky, Harrison and Hulanicki. The counterpart of Halmos' problem for pseudocompact groups was faced by various authors ([**CvM, CR1, CR2, DS1, DS2**]) after the pioneering work of van Douwen [**vD**] where the first restraint was given on the cardinality of an infinite pseudocompact group (a topological group G is said to be pseudocompact if every continuous function $f : G \to \mathbb{R}$ is bounded).

A topological group G is said to be *countably compact* if every countable open cover of G has a finite subcover. Countably compact groups are pseudocompact, while compact groups are obviously countably compact. We will briefly discuss here the recent progress in resolving the counterpart of Halmos' problem for small countably compact groups ([**DT1, DT2, DT3, TY**]). More precisely, under the assumption of Martin's Axiom (briefly, MA), a complete description of the class of all abelian groups of size at most continuum that admit countably compact group topologies can be given.

Let us briefly recall some of the main results and the impact of the property $\mathcal{E}_m$ of the tori for the construction of countably compact group topologies on the non-torsion Abelian groups of cardinality $\mathfrak{c}$. The property $\mathcal{E}_m$ in the strongest form given in 4.1 is substantially used in [**DT1**] to prove the following seemingly technical property of infinite Abelian groups (see also [**TY**] for torsion-free groups).

LEMMA 5.1. [**DT1**, Lemma 4.2] *Let X be an infinite subset of an Abelian group G such that $X \cap S_n(a)$ is finite for every $n \in \mathbb{N}$ and $a \in G$. Then the set*

$$H_X = \{h \in \widehat{G} : h(X) \text{ is dense in } \mathbb{T}\}$$

is an intersection of countably many open dense subsets of $\widehat{G}$, hence dense in $\widehat{G}$.

This lemma is used further in [**DT1**], combined with ideas from [**Tk**] to prove the following theorem.

THEOREM 5.2. [MA] *For an Abelian non-torsion group G of cardinality $\mathfrak{c}$, the following conditions are equivalent:*

1) G admits a countably compact group topology;

2) $r(G) = \mathfrak{c}$ and for all $d, n \in \mathbb{N}$ with $d|n$, the group $dG[n] \cong G[n]/G[d]$ is either finite or has cardinality $\mathfrak{c}$.

In the case of Abelian torsion groups of cardinality $\mathfrak{c}$, a similar construction proves the following theorem in [**DT1**]:

THEOREM 5.3. [MA] *For a torsion Abelian group G of cardinality $\mathfrak{c}$, the following conditions are equivalent:*
  (a) *G admits a countably compact group topology;*
  (b) *G has finite exponent n and dG is either finite or has cardinality $\mathfrak{c}$ for every proper divisor d of n.*

It was known from [**DS1, DS2**] that (b) is equivalent to the property "G admits a pseudocompact group topology".

This technique can be used further to prove (again under MA) that the every precompact Abelian group of non-measurable size belongs to the smallest class of groups that contains all Abelian countably compact groups and is closed under direct products, taking closed subgroups and continuous isomorphic images ([**DT2, DT3**]). Proofs and further details can be found in [**DT1**]–[**DT3**].

References

[CvM] W. COMFORT and J. VAN MILL, *Concerning connected, pseudocompact Abelian groups*, Topology Appl. **33** (1989), 21–45.

[CR1] W. COMFORT and D. REMUS, *Imposing pseudocompact group topologies on Abelian groups*, Fund. Math. **142** (1993), 221–240.

[CR2] W. COMFORT and D. REMUS, *Abelian torsion groups with a pseudocompact group topology*, Forum Math. **6** (1994), 323–337.

[Di1] D. DIKRANJAN, *On a conjecture of Prodanov*, C. R. Acad. Bulgare Sci. **38** (1985), no. 9, 1117–1120.

[Di2] D. DIKRANJAN, *Density and total density of the torsion part of a compact group*, Rend. Accad. Naz. dei XL, Memorie di Mat. 108, Vol. XIV, fasc. 13 (1990) 235–252.

[DPr] D. DIKRANJAN and IV. PRODANOV, *A class of compact Abelian groups*, Annuaire Univ. Sofia, Fac. Math. Méc. **70**, (1975/76) 191–206.

[DPS] D. DIKRANJAN, I. PRODANOV and L. STOYANOV, *Topological groups (Characters, Dualities and Minimal group topologies)*. Marcel Dekker, Inc., New York–Basel, 1990.

[DS1] D. DIKRANJAN and D. SHAKHMATOV, *Pseudocompact topologizations of groups*, Zb. radova Filozof. fakulteta u Nišu, Ser. Mat. **4** (1990), 83–93.

[DS2] D. DIKRANJAN and D. SHAKHMATOV, *Algebraic structure of the pseudocompact groups*, Memoirs Amer. Math. Soc., 133/633, April 1998, pp. viii+83.

[DS] DIKRANJAN and L. STOYANOV, *Compact groups with totally dense torsion part*, Baku Internat. Topological Conf. (Russian) (Baku, 1987), 231–238, "Èlm", Baku, 1989.

[DT1] D. DIKRANJAN and M. G. TKACHENKO, *Algebraic structure of small countably compact Abelian groups*, submitted.

[DT2] D. DIKRANJAN and M. G. TKACHENKO, *Varieties generated by countably compact Abelian groups*, submitted.

[DT3] D. DIKRANJAN and M. G. TKACHENKO, *Varieties generated by countably compact Abelian groups II*, work in progress.

[vD] E. K. VAN DOUWEN, *The weight of a pseudocompact (homogeneous) space whose cardinality has countable cofinality*, Proc. Amer. Math. Soc. **80** (1980), 678–682.

[Hal] P. HALMOS, *Comments on the real line*, Bull. Amer. Math. Soc. **50** (1944), 877–878.

[HR] E. HEWITT and K. A. ROSS, *Abstract Harmonic Analysis*, Springer Verlag, Berlin-Heidelberg-New York, 1970.

[Ku] A. KUROSCH, *Primitive torsionfreie abelsche Gruppen vom endlichen Range*, Ann. of Math. **38** (1937) 175–203.

[Sto] L. STOYANOV, *Weak periodicity and minimality of topological groups*, Annuaire Univ. Sofia Fac. Math. Méc. **73** (1978/79) 155–167.

[Tk] M. G. TKACHENKO, *On countably compact and pseudocompact topologies on free Abelian groups,* Soviet Math. (Izv. VUZ) **34** (5) (1990), 79–86. Russian original in: Izv. Vyssh. Uchebn. Zaved. Ser. Matem. **5** (1990), 68–75.

[TY] M. G. TKACHENKO and I. YASCHENKO, *Independent group topologies on Abelian groups,* Topology Appl., to appear.

DIPARTIMENTO DI MATEMATICA E INFORMATICA, UNIVERSITÀ DI UDINE VIA DELLE SCIENZE 206, 33100 UDINE, ITALY
E-mail address: `dikranja@dimi.uniud.it`

DEPARTAMENTO DE MATEMÁTICAS, UNIVERSIDAD AUTÓNOMA METROPOLITANA, C.P. 09340 IZTAPALAPA, MÉXICO D.F.
E-mail address: `mich@xanum.uam.mx`

Quotient divisible mixed groups

Alexander A. Fomin

Dedicated to Professor Laszlo Fuchs in honour of his 75th birthday.

ABSTRACT. A characterization of quotient divisible mixed abelian groups is given with a help of finitely generated modules over the ring of pseudo-rational numbers. Some applications to torsion-free groups of finite rank are considered.

1. Introduction

This paper is a continuation of [13]. Throughout the paper "group" will mean abelian group. Z, Q and $\widehat{Z}_p$ are rings of integers, rationals and p-adic integers respectively. If S is a subset of an abelian group M, $\langle S \rangle$ is a subgroup generated by the set S, $\langle S \rangle_*$ is the *pure hull* of S in M, which consists of all the elements $x \in M$ such that $nx \in \langle S \rangle$ for an integer $0 \neq n \in Z$. In particular, $\langle S \rangle_*$ contains all torsion elements of M. Whenever a notion or notation is not defined, it is adopted from [17]. The following definition is a little different from the original one in [16]. A new condition is added according to which the torsion part has to be reduced.

DEFINITION 1.1. ([16]) A group A is called quotient divisible, if it does not contain torsion divisible subgroups but contains a free finite-rank subgroup F such that the quotient A/F is torsion divisible.

The torsion-free rank of A, that is the rank of F, will be called the rank of a quotient divisible group A for short. Quotient divisible torsion-free groups are exactly the classical quotient divisible groups introduced and researched by R. Beaumont and R. Pierce, [6]. Groups of the class $\mathcal{G}$, introduced by S.Glaz and W.Wickless in [18], are examples of properly mixed quotient divisible groups.

DEFINITION 1.2. The class $\mathcal{G}$ consists of all groups of the form $V \oplus G$, where V is finite-rank torsion-free divisible and G is a reduced group with torsion subgroup $T = \underset{p}{\oplus} T_p$ satisfying:

- each T_p is finite and
- G is (can be embedded as) a pure subgroup of $\prod_p T_p$ and
- G/T is finite-rank (necessarily divisible) and

1991 *Mathematics Subject Classification.* Primary 20K15, 20K21.

- (the projection condition) there exists a full free subgroup F of G such that F projects onto T_p for all primes p.

The class $\mathcal{G}$ is a focus of research. A number of papers have dealt with various properties of groups in $\mathcal{G}$ (see [1]–[4], [7]–[8], [14]–[16], [18], [20], [22]).

We consider a subring of the ring $\prod_p \widehat{Z_p}$ generated by the ideal $\oplus_p \widehat{Z_p}$ and by the unit of $\prod_p \widehat{Z_p}$. The pure hull of this subring is a subring R of the ring $\prod_p \widehat{Z_p}$, too.

DEFINITION 1.3. ([13]) The ring $R = \left\langle 1, \oplus_p \widehat{Z_p} \right\rangle_*$ is called the ring of pseudo-rational numbers.

Basic properties of the ring R and modules over R are considered in [13]. In particular, it is proved that the category of groups $\mathcal{G}$ is a full subcategory of the category of finitely generated modules over R. We are interested in the category $\mathcal{QD}$. Objects of $\mathcal{QD}$ are quotient divisible groups, morphisms are quasi–homomorphisms, that is, elements of the group $Q \otimes Hom(A, B)$ for two quotient divisible groups A and B. The main reason of this interest is that the category $\mathcal{QD}$ is dual to the category of torsion-free finite-rank groups with quasi–homomorphisms as morphisms, [16].

In the present paper, we introduce a new category $\mathcal{F}$. An object of this category is a pair consisting of a free finite–rank abelian group F and an R–module M, such that $F \subset M$ and $M = \langle F \rangle_R$. The main result of the paper is Theorem 3.3 which says that the categories $\mathcal{F}$ and $\mathcal{QD}$ are equivalent.

Some applications to torsion–free groups of finite rank are considered in the last section of the paper.

2. Quotient divisible groups and modules over the ring of pseudo-rational numbers

Let A be a group. Accoding to [17], the Z-adic completion $\widehat{A}$ of the group A is the inverse limit of the inverse system of natural homomorphisms

$$A/mA \to A/nA$$

over all pairs of natural numbers such that n divides m. The p-adic completion $\widehat{A_p}$ of the group A is the inverse limit of the system

$$\ldots \to A/p^n A \to \ldots \to A/p^2 A \to A/pA \to 0.$$

There are natural homomorphisms $\mu : A \to \widehat{A}$ and $\mu_p : A \to \widehat{A_p}$ which are called also the Z–adic and p–adic completions respectively. $T(A)$ and $T_p(A)$ are the torsion part of a group A and its p-primary component. The number of elements of a p-basis of the group A is called the p-rank of A, see details in [17, Section 34].

THEOREM 2.1. *Let A be a quotient divisible group of rank n. Then:*

1. $r_p(A) \leq n$ *for every prime number p.*
2. $T_p(A)$ *is a finite group with* $r_p(T_p(A)) \leq r_p(A)$ *and it is a direct summand of the group A,* $A = T_p(A) \bigoplus A'_p$, *for every prime number p.*
3. *The group $T(A)$ maps isomorphically onto the torsion part of the group $\hat{A}$ by the Z-adic completion* $\mu : A \to \hat{A}, T(A) = T(\hat{A}).$

4. *The factor-group $A/T(A)$ is a torsion free quotient divisible group of rank n.*

PROOF. Let $F = \langle x_1, \ldots, x_n \rangle$ be a free subgroup of the group A such that A/F is torsion divisible. If p is a prime number and B is a p-basic subgroup of A, then $A/pA \cong B/pB$ ([**17**, Section 34]). On the other hand, $A = F + pA$ and the elements $x_1 + pA, \ldots, x_n + pA$ generate the group A/pA. Therefore, the number of elements of a p-basis of A can not be greater than n. It follows immediately that $T_p(A)$ is a finite group with $r_p(T_p(A)) \leq r_p(A)$ and $T_p(A)$ is a direct summand of the group A ([**17**, Section 32]). Since $A = T_p(A) \oplus A'_p$, then $\hat{A}_p = \widehat{T_p(A)}_p \oplus \widehat{A'}_p$. The group $\widehat{A'}_p$ is isomorphic to a direct sum of k copies of $\widehat{Z}_p$, where $k = r_p(A'_p)$, because A'_p does not have p-torsion, and $\widehat{T_p(A)}_p = T_p(A)$, because it is finite. Hence $T_p(A) = T_p(\hat{A}_p)$ and $T(A) = \bigoplus_p T_p(A)$ is the torsion part of the group $\hat{A} = \prod_p \hat{A}_p$. By the natural homomorphism $A \to A/T(A)$, the group F is embedded isomorphically in the group $A/T(A)$, that is $(F + T(A))/T(A) \subset A/T(A)$. Moreover, the factor-group of the group $A/T(A)$ with respect to the image of F, that is the group $A/(F + T(A))$, is a divisible torsion group as a homomorphic image of the divisible torsion group A/F. □

Every element r of the ring of pseudo-rational numbers R can be presented in the form
$$r = (\alpha_p) \in \prod_p \widehat{Z}_p,$$
where almost all the p-components α_p are equal to one rational number which is denoted as $|r|$.

An R-module M is called *divisible* ([**13**]) if its additive group is torsion free divisible and the multiplication is defined by $rm = |r|m$, $r \in R$, $m \in M$. Note that torsion-free divisible groups of finite rank can be considered as finitely generated R-modules. The maximal divisible submodule of an R-module M is denoted by $divM$.

THEOREM 2.2. ([**13**]) *Every divisible submodule of an R-module M is a direct summand. In particular, $M = divM \oplus A$, where A is a reduced R-module.* □

Pseudo-rational numbers of the form $\varepsilon_p = (0, \ldots, 0, \underset{p}{1}, 0, \ldots)$ are idempotents of the ring R. Moreover, every idempotent of R is of the form $\varepsilon = \varepsilon_{p_1} + \varepsilon_{p_2} + \ldots + \varepsilon_{p_n}$ or of the form $1 - \varepsilon = 1 - (\varepsilon_{p_1} + \varepsilon_{p_2} + \ldots + \varepsilon_{p_n})$, where $p_1, p_2, \ldots, p_n$ are distinct prime numbers. Every pseudo-rational number $r \in R$ can be presented in the form $r = \varepsilon r + (1 - \varepsilon)|r|$.

An R-module M is a *pure* submodule of an R-module N if the solvability of every equation $rx = m$, $r \in R$, $m \in M$ in the module N implies its solvability in the module M.

Submodules $M_p = \varepsilon_p M \subset M$ and $\bigoplus_p M_p \subset M$ are pure. They serve for R-modules like the p-primary component and the torsion part of abelian groups respectively.

THEOREM 2.3. *([13])* Every module over the ring of pseudo-rational numbers is of the form $D \bigoplus M$, where D is a divisible R-module and M is reduced. Moreover, M is (can be embedded as) a pure submodule of $\prod_p M_p$. Thus, $\bigoplus_p M_p \subset M \subset \prod_p M_p$ and both embeddings are pure. □

Note that every submodule M_p may be considered as a module over the ring of p–adic integers $\widehat{Z_p} = \varepsilon_p R$. The structure of finitely generated R–modules is particularly transparent.

COROLLARY 2.4. *A module over the ring of pseudo-rational numbers is finitely generated if and only if it is of the form $D \bigoplus M$, where D is a finitely generated divisible R-module and M is reduced satisfying:*

- *each M_p is a finitely generated $\widehat{Z_p}$-module and*
- *M is (can be embedded as) a pure submodule of $\prod_p M_p$ and*
- *$M/\bigoplus_p M_p$ is a finitely generated divisible R-module and*
- *there exists a free finite-rank subgroup F of the additive group of M such that F_p generates M_p over $\widehat{Z_p}$ for all prime numbers p.*

□

The fourth condition of the Corollary 2.4 implies the first one. The second condition is redundant, because every reduced R-module satisfies it. But we use this formulation to compare Corollary 2.4 with Definition 1.2. It is immediately seen that if all the M_p are torsion, then M becomes a group of the class $\mathcal{G}$. To be exact, the additive group of M belongs to the class $\mathcal{G}$ if the set of generators of M is linearly independent over Z. Otherwise, it is quasi–equal to a group of the class $\mathcal{G}$ (see [**13**, Theorem 5.2]).

Quasi-homomorphisms of modules $M_1 \to M_2$ over the ring of pseudo-rational numbers R are elements of the group $Q \otimes Hom_R(M_1, M_2)$. Analogously, the notions of *quasi-isomorphism* and *quasi-equality* make sense for modules over the ring of pseudo-rational numbers R.

LEMMA 2.5. *Every finitely generated R-module is quasi-equal to an R-module having a finite set of generators linearly independent over Z.*

PROOF. A finite set of generators of an R-module M generates a subgroup $A \oplus F$ of the additive group of M, where A is finite and F is free with a system of free generators $x_1, \ldots, x_n$. Let $p_1, \ldots, p_k$ be all the prime divisors of the order of A. Then $M = T_{p_1}(M) \oplus \ldots \oplus T_{p_k}(M) \oplus M_1$ and projections of the elements $x_1, \ldots, x_n$ on the direct summand M_1 are linearly independent over Z and generate M_1 as an R-module. Since the module $T_{p_1}(M) \oplus \ldots \oplus T_{p_k}(M)$ is finite the modules M and M_1 are quasi-equal. □

THEOREM 2.6. *Let $x_1, \ldots, x_n$ be a set of generators of an R-module M linearly independent over Z. Then the pure hull $A = \langle x_1, \ldots, x_n \rangle_*$ of these elements in the R-module $M = \langle x_1, \ldots, x_n \rangle_R$ is a quotient divisible group.*

PROOF. $F = \langle x_1, \ldots, x_n \rangle_Z$ is a free subgroup of A. The group A does not contain divisible torsion subgroups, because M does not contain them either. We prove that the factor-group A/F is torsion divisible. Let $a \in A$. For a sufficiently large integer m, the element ma maps to an element $s_1\bar{x}_1 + \ldots + s_n\bar{x}_n \in M/T(M)$ by the homomorphism $M \to M/T(M)$, $\bar{x}_i = x_i + T(M)$, $s_i \in Z$, $i = 1, \ldots, n$. Then $ma - (s_1 x_1 + \ldots + s_n x_n) \in T(M)$ and $m_1 ma = m_1 s_1 x_1 + \ldots + m_1 s_n x_n$ for a sufficiently large integer m_1. Therefore, A/F is torsion.

Let $a = r_1 x_1 + \ldots + r_n x_n \in A$, where $r_1 = \left(\alpha_p^{(1)}\right), \ldots, r_n = \left(\alpha_p^{(n)}\right) \in R$, $\alpha_p^{(i)} \in \widehat{Z}_p$. For any prime number p, let
$$a_p^{(i)} \equiv \alpha_p^{(i)} \pmod{p},\ a_p^{(i)} \in Z,\ i = 1, \ldots, n.$$
Then the element $a_p^{(1)} x_1 + \ldots + a_p^{(n)} x_n$ belongs to $F \subset A$ and
$$b = a - \left(a_p^{(1)} x_1 + \ldots + a_p^{(n)} x_n\right) = \sum_{i=1}^{n} \left(r_i - a_p^{(i)}\right) x_i$$
is divisible by p in M, because $r_i - a_p^{(i)}$ is divisible by p in R for each i. Moreover, $b \in A$ and $p^{-1} b \in A$ due to the purity as well. It follows that A/F is p-divisible for every p and A is quotient divisible. $\square$

COROLLARY 2.7. *Groups of the class $\mathcal{G}$ are quotient divisible.*

PROOF. Let $A \in \mathcal{G}$ satisfy the projection condition with respect to a linearly independent set $x_1, \ldots, x_n$. Then $A = M = \langle x_1, \ldots, x_n \rangle_R$ is an R-module ([**13**, Theorem 5.2]) and $A = \langle x_1, \ldots, x_n \rangle_*$ is quotient divisible by Theorem 2.6. $\square$

The next theorem shows that every quotient divisible group can be obtained in the same way as in Theorem 2.6.

THEOREM 2.8. *Let A be a quotient divisible group of rank n with a free subgroup $F = \langle x_1, \ldots, x_n \rangle$ such that A/F is torsion divisible. Then there exists an R-module M such that $A \subset M$ and $M = \langle A \rangle_R = \langle F \rangle_R$ and A coincides with the pure hull of the set of elements $x_1, \ldots, x_n$ in the module M, $A = \langle x_1, \ldots, x_n \rangle_*$.*

PROOF. Consider a homomorphism of the Z-adic completion $\mu_A : A \to \widehat{A}$ ([**17**, Section 39]). It is shown in [**16**] that the first Ulm subgroup D of the group A is a torsion-free divisible group of finite rank. Hence $A = D \oplus A'$, $\widehat{A'} = \widehat{A}$. The homomorphism $\mu_{A'} : A' \to \widehat{A'}$ is a restriction of the homomorphism $\mu_A : A \to \widehat{A}$ on the subgroup A', that is $\mu_A(d + a) = \mu_A(a) = \mu_{A'}(a)$ for every $d \in D$ and $a \in A'$. The homomorphism $\lambda_A = id \oplus \mu_{A'} : D \oplus A' \to D \oplus \widehat{A}$ is an embedding. Here, D is a divisible R-module and $\widehat{A}$ is a module over the ring $\prod_p \widehat{Z}_p$ and, therefore, over the ring of pseudo–rational numbers R as well. We may identify elements by the monomorphism λ_A and assume that A is a subgroup of the additive group of the R-module $D \oplus \widehat{A}$, $A = \Im \lambda_A \subset D \oplus \widehat{A}$.

The set A generates over R a submodule $M = \langle A \rangle_R \subset D \oplus \widehat{A}$. We shall prove that M is generated by the elements $x_1, \ldots, x_n$ as well. It is sufficient to show that every element $a \in A$ may be presented in the form $a = r_1 x_1 + \ldots + r_n x_n$ with $r_1, \ldots, r_n \in R$. Since A/F is torsion it follows that $ma = m_1 x_1 + \ldots + m_n x_n$ for integer coefficients with $m \neq 0$. Further, $ma = (1 - \varepsilon)(m_1 x_1 + \ldots + m_n x_n) + \varepsilon(m_1 x_1 + \ldots + m_n x_n)$, where $\varepsilon = \varepsilon_{p_1} + \ldots + \varepsilon_{p_s} \in R$, $p_1, p_2, \ldots, p_s$ are all the prime divisors of m and $1 - \varepsilon$ is an idempotent divisible by m in R. Then $m\left(a - \frac{1-\varepsilon}{m}(m_1 x_1 + \ldots + m_n x_n)\right) \in \varepsilon M = \varepsilon_{p_1} M \oplus \ldots \oplus \varepsilon_{p_s} M$. A $\widehat{Z}_p$-module $\varepsilon_p M$ coincides with the p–adic completion $\widehat{A}_p$. It follows that $a - \frac{1-\varepsilon}{m}(m_1 x_1 + \ldots +$

$m_n x_n) \in \widehat{A}_{p_1} \oplus \ldots \oplus \widehat{A}_{p_s}$. But this submodule of the R–module M is generated by elements $\varepsilon_{p_i} x_j$, $i = 1, \ldots, s$; $j = 1, \ldots, n$, as shown in [16] (the projection condition). Eventually we obtain $a - \frac{1-\varepsilon}{m}(m_1 x_1 + \ldots + m_n x_n) = \varepsilon r'_1 x_1 + \ldots + \varepsilon r'_n x_n$, where $r'_1, \ldots, r'_n \in R$. This implies the required representation of the element a. Therefore, $M = \langle A \rangle_R = \langle F \rangle_R$.

Since the group $\mu_A(A)$ is pure in $\widehat{A}$, the group A is pure in M, too. By Theorem 2.1 (3), $T(A) = T(M)$. The group $A/T(M)$ is pure in the group $M/T(M)$. Therefore, $A/T(M)$ is the pure hull of the set $x_1 + T(M), \ldots, x_n + T(M)$ in the group $M/T(M)$ and $A = \langle x_1, \ldots, x_n \rangle_*$. □

DEFINITION 2.9. The R–module $\langle \Im \lambda_A \rangle_R$ defined in the proof of Theorem 2.8 is called the **pseudo–rational type** of a quotient divisible group A.

3. The category $\mathcal{F}$

Each group of the class $\mathcal{G}$ admits uniquely the structure of an R–module. Besides, homomorphisms of groups coincide with the homomorphisms of R–modules for this class of groups as shown in the following theorem.

THEOREM 3.1. ([13]) Let $A_1, A_2 \in \mathcal{G}$. Then $Hom_R(A_1, A_2) = Hom_Z(A_1, A_2)$. □

While it is not true for all R–modules, but the next theorem complements Theorem 3.1.

THEOREM 3.2. Let N and M be finitely generated R-modules. If M is reduced or N is divisible, then every additive homomorphism $f : N \to M$ is an R-module homomorphism.

PROOF. Let M be reduced. For any prime number p the group $(1 - \varepsilon_p)N$ is p-divisible and the group $\varepsilon_p N$ is q-divisible for every prime number $q \neq p$. Hence $f((1 - \varepsilon_p)N) \subset (1 - \varepsilon_p)M$ and $f(\varepsilon_p N) \subset \varepsilon_p M \oplus div(M)$. Since M is reduced, $div(M) = 0$ and $f(\varepsilon_p N) \subset \varepsilon_p M$. Therefore $f(\varepsilon_p x) = \varepsilon_p f(\varepsilon_p x)$ and $0 = \varepsilon_p f((1 - \varepsilon_p)x) = \varepsilon_p f(x) - \varepsilon_p f(\varepsilon_p x) = \varepsilon_p f(x) - f(\varepsilon_p x)$, that is $f(\varepsilon_p x) = \varepsilon_p f(x)$ for every $x \in N$. Since $\varepsilon_p N$ and $\varepsilon_p M$ are complete in the p–adic topology, the homomorphism $f : \varepsilon_p N \to \varepsilon_p M$ is a $\widehat{Z}_p$-module homomorphism. Therefore $f(r\varepsilon_p x) = r\varepsilon_p f(x)$ for every $x \in N$ and for every $r \in R$. Further, it follows that $f : N \to M$ is an R-module homomorphism as in the proof of Theorem 3.1 (see [13, Theorem 4.5], for details).

Let N be divisible. Then $f(N) \subset div(M)$ and $f(rx) = f(|r|x) = |r|f(x) = rf(x)$ for every $r \in R, x \in N$. □

REMARK 3.3. If the module N in the last Theorem is reduced, for example $N = \widehat{Z}_p$, and the module M is divisible, for example $M = Q$, then the group $Hom_Z(\widehat{Z}_p, Q)$ is obviously not equal to zero. But the module $Hom_R(\widehat{Z}_p, Q)$ is equal to zero. In fact, $f(x) = f(\varepsilon_p x) = \varepsilon_p f(x) = |\varepsilon_p| f(x) = 0$ for every $x \in N$ and for every $f \in Hom_R(\widehat{Z}_p, Q)$.

DEFINITION 3.4. Objects of the category $\mathcal{F}$ are free groups of finite rank generating R-modules. That is if $F \subset M$ is a free finite–rank subgroup of the additive group of an R-module M, then the embedding $F \to \langle F \rangle_R$ is an object of the category $\mathcal{F}$. To avoid troubles with divisible modules, discussed above, we define morphisms of the category $\mathcal{F}$ in the following way. A morphism from one object $F \to \langle F \rangle_R$ to another one $F_1 \to \langle F_1 \rangle_R$ is an arbitrary pair (f, φ) consisting of a group quasi–homomorphism $f : F \to F_1$ and an R–module quasi–homomorphism $\varphi : K \to K_1$, where $K = \langle F \rangle_R / div(\langle F \rangle_R)$ and $K_1 = \langle F_1 \rangle_R / div(\langle F_1 \rangle_R)$, such that the following diagram is commutative

$$\begin{array}{ccc} F & \xrightarrow{f} & F_1 \\ \downarrow & & \downarrow \\ \langle F \rangle_R & & \langle F_1 \rangle_R \\ \downarrow & & \downarrow \\ K & \xrightarrow{\varphi} & K_1 \end{array}$$

The lower vertical arrows mean the natural homomorphisms of the R–modules $\langle F \rangle_R \to \langle F \rangle_R / div(\langle F \rangle_R)$ and $F_1 \to \langle F_1 \rangle_R / div(\langle F_1 \rangle_R)$.

Two categories $\mathcal{A}$ and $\mathcal{B}$ are *equivalent*, if there exist two covariant functors $f : \mathcal{A} \to \mathcal{B}$ and $f' : \mathcal{B} \to \mathcal{A}$ such that $ff' \sim id_{\mathcal{B}}$ and $f'f \sim id_{\mathcal{A}}$. The notation $f'f \sim id_{\mathcal{A}}$ means that for every object A of the category $\mathcal{A}$ there exists an isomorphism $g_A : A \to f'(f(A))$ such that for every morphism $h : A \to B$ of the category $\mathcal{A}$ the following diagram is commutative

$$\begin{array}{ccc} A & \xrightarrow{h} & B \\ g_A \downarrow & & \downarrow g_B \\ f'(f(A)) & \xrightarrow{f'(f(h))} & f'(f(B)) \end{array}$$

We remind the reader that $\mathcal{QD}$ is the category of quotient divisible groups with quasi–homomorphisms as morphisms. Now we are able to prove the main result of this paper.

THEOREM 3.5. *The categories $\mathcal{F}$ and $\mathcal{QD}$ are equivalent.*

PROOF. We extend the category $\mathcal{QD}$ to the following category $\mathcal{QDE}$. Objects of $\mathcal{QDE}$ are pairs $F \subset A$, where A is a quotient divisible group and F is its free subgroup such that A/F is torsion divisible. Morphisms from an object $F \subset A$ to an object $F_1 \subset B$ are quasi-homomorphisms from A to B. Note that every quasi–homomorphism $f : A \to B$ induces a quasi–homomorphism of their full free subgroups $f : F \to F_1$. The categories $\mathcal{QD}$ and $\mathcal{QDE}$ are obviously equivalent. Hence it is sufficient to prove the equivalence of the categories $\mathcal{QDE}$ and $\mathcal{F}$, that is, to define two covariant functors $c : \mathcal{QDE} \to \mathcal{F}$ and $c' : \mathcal{F} \to \mathcal{QDE}$ such that $cc' \sim id_{\mathcal{F}}$ and $c'c \sim id_{\mathcal{QDE}}$.

Let $F \subset A$ be an object of the category $\mathcal{QDE}$. According to Theorem 2.8, we have an embedding $\lambda_A : A \to M$, where $M = \langle F \rangle_R$ is the pseudo–rational type of A. The restriction $\lambda_A : F \to M$ is an object of the category $\mathcal{F}$ which is denoted as $c(F \subset A)$. Consider a morphism f from the object $F \subset A$ to an object $F_1 \subset B$ of the category $\mathcal{QDE}$, that is a quasi–homomorphism of quotient divisible groups $f : A \to B$. The object $c(F_1 \subset B)$ of the category $\mathcal{F}$ is a map $\lambda_B : F_1 \to M_1$. Then the restriction $f : F \to F_1$ is a quasi–homomorphism of free groups and $mf : A \to B$ is a homomorphism for an integer $m \neq 0$. Since $M/divM$ is an R–submodule of the R–module $\widehat{A}$, $M/divM = \langle \mu_A(A) \rangle_R$, where $\mu_A : A \to \widehat{A}$ is the homomorphism of the Z-adic completion as above, the $\prod_p \widehat{Z_p}$-module homomorphism $\widehat{mf} : \widehat{A} \to \widehat{B}$ is an R-module homomorphism as well. Its restriction on $\langle \mu_A(A) \rangle_R$ is an R-module homomorphism $\widehat{mf} : M/divM \to M_1/divM_1$. We define $c(f) = (f, \varphi)$, where $\varphi = \frac{1}{m} \otimes \widehat{mf}$. The diagram

$$\begin{array}{ccc} F & \xrightarrow{f} & F_1 \\ \downarrow & & \downarrow \\ M & & M_1 \\ \downarrow & & \downarrow \\ M/divM & \xrightarrow{\varphi} & M_1/divM_1 \end{array}$$

is commutative and c is a covariant functor $c : \mathcal{QDE} \to \mathcal{F}$ (see [**17**, Section 39] for details).

Let now F be a free group with free generators $x_1, \ldots, x_n$ and $F \to M$ an object of the category $\mathcal{F}$, that is $M = \langle x_1, \ldots, x_n \rangle_R$. By Theorem 2.6, the pure hull $A = \langle x_1, \ldots, x_n \rangle_*$ of the elements $x_1, \ldots, x_n$ in the additive group of M is a quotient divisible group. We define $c'(F \to M)$ is $F \subset A$, as an object of the category $\mathcal{QDE}$, because A/F is torsion divisible as proved in Theorem 2.6.

Let a morphism of the category $\mathcal{F}$ be given

$$\begin{array}{ccc} F & \xrightarrow{f} & F_1 \\ \downarrow & & \downarrow \\ M & & M_1 \\ \theta \downarrow & & \downarrow \theta_1 \\ K & \xrightarrow{\varphi} & K_1 \end{array}$$

where $\theta : M \to M/divM$ and $\theta_1 : M_1 \to M_1/divM_1$ are natural homomorphisms of R-modules. For a sufficiently large integer k, the map kf is a homomorphism of groups and $k\varphi$ is a homomorphism of R-modules, and the following diagram is commutative as well

$$\begin{array}{ccc} F & \xrightarrow{kf} & F_1 \\ \downarrow & & \downarrow \\ M & & M_1 \\ \theta \downarrow & & \downarrow \theta_1 \\ K & \xrightarrow{k\varphi} & K_1 \end{array}$$

Let $F \subset A$ and $F_1 \subset A_1$, where $A = \langle F \rangle_*$, $A_1 = \langle F_1 \rangle_*$, be objects $c'(F \to M)$ and $c'(F_1 \to M_1)$ respectively. I prove that there exists a unique homomorphism of groups $k\Phi : A \to A_1$ making the diagram

$$\begin{array}{ccc} & kf & \\ F & \longrightarrow & F_1 \\ \downarrow & k\Phi \downarrow & \\ A & \longrightarrow & A_1 \\ \theta \downarrow & & \downarrow \theta_1 \\ K & \longrightarrow & K_1 \\ & k\varphi & \end{array}$$

commutative.

Let $a \in A$. We have $ma = m_1 x_1 + \ldots + m_n x_n \in F$ for a sufficiently large integer m and integer coefficients m_i, so $z = (kf)(m_1 x_1 + \ldots + m_n x_n) \in F_1 \subset A_1$. Due to the commutativity of the diagram, the element $\theta_1(z)$ is divisible by m in $K_1 = M_1/div M_1$, because $\theta_1(z) = m(k\varphi\theta(a))$. It follows that z is divisible by m in M_1. The equation $my = z$ is solvable in the module M_1. Due to the definition of A_1, this equation is solvable in A_1 and, moreover, all the solutions lie in A_1. Since the difference of any two solutions of the equation is torsion, the homomorphism θ_1 maps different solutions to different elements of K_1. Then only one of them, say y_1, maps to $(k\varphi)(\theta(a))$. We therefore define $k\Phi(a) = y_1$. Thus, the map $k\Phi$ is unique such that the diagram becomes commutative and it is easy to check that $k\Phi$ is a homomorphism of groups. We define $c'((f,\varphi)) = \frac{1}{k} \otimes k\Phi$. The uniqueness of $k\Phi$ implies that $c' : \mathcal{F} \to \mathcal{QD}$ is a functor. All other checks are standard and the Theorem is proved. $\square$

4. Torsion-free groups of finite rank

In this section we consider the category $\mathcal{QTF}$: the objects are torsion-free abelian groups of finite rank, and the morphisms are quasi–homomorphisms.

THEOREM 4.1. *The category $\mathcal{QTF}$ is dual to the category $\mathcal{F}$.*

PROOF. A duality $d : \mathcal{QTF} \to \mathcal{QD}$ is introduced in [16]. The composition $cd : \mathcal{QTF} \to \mathcal{F}$ is a duality as well. $\square$

Let M be a finitely generated R–module with a system of generators $x_1, \ldots, x_n$. Then obviously the $\widehat{Z}_p$–module $M_p = \varepsilon_p M$ is generated by the elements $\varepsilon_p x_1, \ldots, \varepsilon_p x_n$. The finitely generated p–adic module M_p may be presented as a direct sum of cyclic $\widehat{Z}_p$–modules.

(4.1) $$M_p = \langle a_1 \rangle_{\widehat{Z}_p} \oplus \ldots \oplus \langle a_n \rangle_{\widehat{Z}_p}.$$

Some direct summands may be equal to 0. Every cyclic $\widehat{Z}_p$–module is isomorphic either to $Z/p^{k_{pi}}Z$, where $0 \leq k_{pi} \in Z$, or to $\widehat{Z}_p$. Hence the isomorphism

(4.2) $$M_p \cong Z(p^{k_{p1}}) \oplus \ldots \oplus Z(p^{k_{pl}}) \oplus \left(\underset{s}{\oplus} \widehat{Z}_p \right), \ l + s = n,$$

determines the following ordered sequence of non-negative integers and symbols ∞

(4.3) $$0 \leq k_{p1} \leq \ldots \leq k_{pn} \leq \infty,$$

where the last s terms are symbols $\infty, 0 \leq s \leq n$. The sequences 4.3 over all the prime numbers p determine a sequence of types (of rank 1 groups) $\sigma_1 \leq \ldots \leq \sigma_n$.

Some initial subsequence of types can be equal to zero. Suppressing these, we obtain finally a sequence of non-zero types.

(4.4) $$\tau_1 \leq \ldots \leq \tau_k$$

DEFINITION 4.2. The sequence 4.4 is called the **Richman type** of the finitely generated R–module M, the number k is called **universal rank** of M. Since the factor ring $R/\oplus_p \hat{Z}_p$ is isomorphic to the field of rationals Q, the R–module $M/\oplus_p M_p$ may be considered as a vector space over Q. We call $\dim_Q \left(M/\oplus_p M_p \right)$ the **pseudo–rational rank** of the finitely generated R–module M. Obviously, both $\dim_Q M/\oplus_p M \leq n$ and $k \leq n$.

The duality cd makes correspond to each finite rank torsion-free group A an object $F \to M$ of the category $\mathcal{F}$, where $M = \langle F \rangle_R$ which will be called the *pseudo-rational type* of the group A as well. We say $\mathcal{R}(A) = M$. All the notions of the Definition 4.2 may be applied to a torsion–free group of finite rank, having in mind, for example, the pseudo–rational rank of $\mathcal{R}(A)$ instead of the pseudo–rational rank of A.

As is shown in the next theorem, the Richman type of $\mathcal{R}(A)$ is not a new invariant for a torsion-free abelian group A of finite rank.

THEOREM 4.3. *The Richman type of a torsion-free group A of finite rank coincides with the Richman type of the R–module $\mathcal{R}(A)$.*

PROOF. Let F be any full free subgroup of A, so that A/F is torsion. Then the p-primary component of A/F is a direct sum of cyclic and quasi-cyclic groups.

(4.5) $$(A/F)_p \cong Z(p^{k_{p1}}) \oplus \ldots \oplus Z(p^{k_{pl_p}}) \oplus \left(\oplus_{s_p} Z(p^\infty) \right),$$

where $k_{p1} \leq \ldots \leq k_{pl_p}$ are non-negative integers and $l_p + s_p = \text{rank} A$. It follows from the definition of the functor d in [**16**] that the p-adic completion of the quotient divisible group $d(A)$ is a direct sum of cyclic $\hat{Z}_p$-modules

(4.6) $$\widehat{d(A)}_p \cong Z(p^{k_{p1}}) \oplus \ldots \oplus Z(p^{k_{pl_p}}) \oplus \left(\oplus_{s_p} \hat{Z}_p \right)$$

On the other hand, $\mathcal{R}(A)_p = \varepsilon_p \mathcal{R}(A) = \widehat{d(A)}_p$. Comparing the definitions of Richman types of A and $\mathcal{R}(A)$, we obtain the result. □

The notation $A \cong B$ $\left(A \doteq B \right)$ means that the groups (R–modules) A and B are quasi–isomorphic (quasi–equal). A torsion-free group A of finite rank is said to be *determined by its pseudo-rational type* if $\mathcal{R}(A) \cong \mathcal{R}(B) \Rightarrow A \cong B$ for every torsion–free group B of finite rank. A torsion-free group A of finite rank is called *locally free* if it contains a free subgroup F such that the factor-group A/F is reduced and torsion. The last definition is a little bit different from the original definition of R.Warfield [**21**]. A tensor product of a locally free group in our sense by a subring of Q is locally free in the sense of R.Warfield as well.

THEOREM 4.4. *A torsion–free group of finite rank is determined by its pseudo-rational type if and only if it is locally free.*

PROOF. Let A be a locally free group. Then $s_p = 0$ in the formula 4.5 for all primes p and $\mathcal{R}(A)$ is a finitely generated R–module with finite $\mathcal{R}(A)_p$ for all p, that is, the additive group of $\mathcal{R}(A)$ is a group of the class $\mathcal{G}$ (Corollary 2.4 and the remark after it). By Corollary 2.7 and Theorem 2.8, $\mathcal{R}(A) \doteq d(A)$. If $\mathcal{R}(A) \cong \mathcal{R}(B)$, then $d(A) \doteq \mathcal{R}(A) \cong \mathcal{R}(B) \doteq d(B)$ and $d(A) \cong d(B)$ implies $A \cong B$.

If A is not locally free then $s_p \neq 0$ at least for one prime p and $\mathcal{R}(A)$ contains at least one direct summand $\widehat{Z}_p$, $\mathcal{R}(A) = \widehat{Z}_p \oplus M$. A basis $x_1, \ldots, x_n$ of the group F in the formula (5) determines a dual basis $x_1^*, \ldots, x_n^*$ of the group $d(A)$ (see [16]), which maps to a system of generators $y_1 = \lambda_{d(A)}(x_1^*), \ldots, y_n = \lambda_{d(A)}(x_n^*)$ of the R-module $\mathcal{R}(A)$ (Theorem 2.8). Since the rank of the additive group of $\widehat{Z}_p$ is $2^{\aleph_0}$, we can choose elements $y_{n+1}, \ldots, y_{n+k} \in \widehat{Z}_p$ for every natural k such that the set $y_1, \ldots, y_n, y_{n+1}, \ldots, y_{n+k}$ is linearly independent over Z. Then the pure hull $B = \langle y_1, \ldots, y_n, y_{n+1}, \ldots, y_{n+k} \rangle_*$ in $\mathcal{R}(A)$ is a quotient divisible group by Theorem 2.6. A finite rank torsion–free group C which is dual to B, has got the same pseudo-rational type, $\mathcal{R}(C) = \mathcal{R}(A)$, but can not be quasi-equal to A because $rank(C) = rank(A) + k$. □

THEOREM 4.5. *The following statements are equivalent for a torsion-free finite-rank group A:*

1. *The pseudo-rational rank of A is equal to zero,*
2. *A is an extension of a free group by a torsion divisible group of finite rank,*
3. *A is a minimax group.*

PROOF. The pseudo–rational rank of the R–module $\mathcal{R}(A)$ is zero if and only if $\mathcal{R}(A) = \varepsilon \mathcal{R}(A)$ for an idempotent $\varepsilon = \varepsilon_{p_1} + \ldots + \varepsilon_{p_k} \in R$. This means that $\widehat{d(A)}_p = 0$ in the formula 4.6 for almost all primes p and that $(A/F)_p = 0$ in the formula 4.5 for almost all primes p. But this is equivalent to the second statement. A (not necessarily abelian) group is called minimax [5] if it contains a normal subgroup which satisfies the maximum condition on subgroups, such that the factor–group satisfies the minimum condition. It is shown in [19] that the statements 2. and 3. are equivalent for torsion–free finite rank groups. □

COROLLARY 4.6. *If a torsion–free group has a countable number of subgroups then its pseudo–rational rank is equal to zero.*

PROOF. It is shown in [11] that a torsion–free group having a countable number of subgroups is an extension of a free finite-rank group by a group of the form $Z(p_1^\infty) \oplus Z(p_2^\infty) \oplus \ldots \oplus Z(p_n^\infty)$, where $p_1, p_2, \ldots, p_n$ are distinct prime numbers. In particular, such a group satisfies the second condition of Theorem 4.4. □

The universal rank of a torsion-free finite rank group coincides with the number of Baer types in its Richman type (the sequence 4.4). Groups of universal rank 1 and 2 are considered in [9] and [10] respectively. The universal rank is called there the number of τ–adic relations. Finally, we note some resemblance between the Theorem 4.1 and the main result of [12].

References

[1] U. Albrecht, *A-projective resolutions and an Azumaya theorem for a class of mixed abelian groups*, to appear.

[2] U. Albrecht, H.P. Goeters and W. Wickless, *The flat dimension of mixed abelian groups as E-modules,* Rocky Mt. J. Math. **25** (1995), 569–590.

[3] U. Albrecht and J. Hausen, *Mixed abelian groups with the summand intersection property,* Lecture Notes in Pure and Applied Math. **182,** Marcel Dekker, New York (1996), 123–132.

[4] U. Albrecht and J-W Jeong, *Homomorphisms between A-projective abelian groups and left Kasch rings,* to appear.

[5] R. Baer, *Polyminimaxgruppen,* Math. Annalen **175** (1968), 1–43.

[6] R. Beaumont and R. Pierce, *Torsion-free rings,* Ill. J. Math. **5** (1961), 61–98.

[7] S. Files and W. Wickless, *The Baer-Kaplansky Theorem for a class of global mixed abelian groups,* Rocky Mt. J. Math. **26** (1996), 593–613.

[8] S. Files and W. Wickless, *Direct sums of self-small mixed groups,* J. of Algebra, **222** (1999), 1–16

[9] A.A. Fomin, *Abelian groups with one τ-adic relation,* Algebra and Logic **28** (1989), 83–104 (Russian, English translation).

[10] A.A. Fomin, *Torsion-free abelian groups of rank 3,* Math USSR Sbornik **68** (1991), 1–17 (English translation).

[11] A.A. Fomin and S.V. Rychkov, *Abelian groups having countably many subgroups,* in **Abelian groups and modules**, Tomsk, **10** (1991), 91–105 (Russian).

[12] A.A. Fomin, *Finitely presented modules over the ring of universal numbers,* Contemp. Math. **171** (1995), 109–120.

[13] A.A. Fomin, *Some mixed abelian groups as modules over the ring of pseudo-rational numbers,* in **Abelian Groups and Modules,** Trends in Mathematics (1999), Birkhaeuser Verlag Basel/Switzerland, 87–100.

[14] A.A. Fomin and W. Wickless, *Categories of Mixed and Torsion-Free Finite Rank Abelian Groups,* in **Abelian Groups and Modules** (A. Facchini and C. Menini, eds) Kluver Acad. Publ. (1995), 185–192.

[15] A.A. Fomin and W. Wickless, *Self-small mixed abelian groups G with $G/T(G)$ finite rank divisible,* Comm. in Algebra **26** (1998), 3563–3580.

[16] A.A. Fomin and W. Wickless, *Quotient divisible abelian groups,* Proc. Amer. Math. Soc. **126** (1998), 45–52.

[17] L.Fuchs, **Infinite abelian groups**, vols I and II, Academic Press (1971, 73).

[18] S. Glaz and W. Wickless, *Regular and principal projective endomorphism rings of mixed abelian groups,* Comm. in Algebra **22** (1994), 1161–1176.

[19] O. Mutzbauer, **Fastabelsche Minimaxgruppen**, Dissertation, Erlangen (1971).

[20] C. Vinsonhaler and W. Wickless, *Realizations of finite dimensional algebras over tha rationals,* Rocky Mt. J. Math. **24** (1994), 1553–1565.

[21] R. Warfield, *Homomorphisms and duality for torsion-free groups,* Math. Z. **107** (1968), 189–200.

[22] W. Wickless, *A functor from mixed groups to torsion-free groups,* Contemp. Math. **171** (1995), 407–419.

OLIMPIADAS COLOMBIANAS EN MATEMATICAS, CRA 38 # 58 A 77, BOGOTA, COLOMBIA.
E-mail address: `mfomin@cable.net.co`

Stacked Bases over h-local Prüfer Domains

Laszlo Fuchs and Sang Bum Lee

ABSTRACT. The Stacked Basis Theorem is proved for h-local Prüfer domains R: If M is a direct sum of cyclic R-modules of projective dimension one, then in every presentation $0 \to H \to F \to M \to 0$ of M, the projective modules F and H have stacked bases.

1. Introduction

The classical Stacked Bases Theorem on abelian groups states that if H is a subgroup of a finitely generated free abelian group F, then F has a basis $\{x_1, \ldots, x_k\}$ such that $\{n_1 x_1, \ldots, n_k x_k\}$ is a basis of H, for suitable non-negative integers n_i. Moreover, here $n_1 | n_2 | \ldots | n_k$ can be assumed.

Answering a question raised by I. Kaplansky, Cohen-Gluck [2] generalized the Stacked Bases Theorem to infinitely generated free modules F and submodules H over principal ideal domains under the (necessary) hypothesis that F/H was a direct sum of cyclic modules. Another proof was given by Hill-Megibben [5] as a corollary of a more general result on the equivalence of presentations. The Stacked Bases Theorem was generalized to Dedekind domains by Generalov-Zheludev [4], and to local domains (even in a stronger form) by Salce-Zanardo [8].

It is known that the Stacked Bases Theorem holds for certain finitely generated modules over special rings, more general than P.I.D. Indeed, Levy [6] has established the existence of stacked bases for finitely presented modules M over Prüfer domains R of finite character ('finite character' means that every non-zero element is contained in but a finite number of maximal ideals): M has a presentation $M \cong F/H$ where the free R-module F and its submodule H have 'stacked' decompositions:

$$F = J_1 x_1 \oplus \ldots \oplus J_n x_n$$

with invertible ideals J_i, and

$$H = I_1 J_1 x_1 \oplus \ldots \oplus I_m J_m x_m \quad (m \leq n)$$

1991 *Mathematics Subject Classification.* Primary 13F05, 13C13; Secondary 20K25.
Key words and phrases. stacked basis, Prüfer domain, h-local .

© 2001 American Mathematical Society

for suitable finitely generated ideals I_j satisfying $I_1 \geq \cdots \geq I_m$. The module M is said to have the *stacked bases property* if each of its free presentations $0 \to H \to F \to M \to 0$ admits stacked bases.

Recall that a commutative ring R is said to be *local-global* if every polynomial (in possibly several indeterminates) that admits unit values locally admits unit values globally.

THEOREM 1. (Brewer-Klingler [1], Levy [6]) *Let R be a Prüfer domain such that every proper homomorphic image of R is a local-global ring (in particular, it is a Prüfer domain of finite character or a Prüfer domain of Krull dimension 1). Then every finitely presented R-module admits stacked bases.*

Our goal here is to generalize this theorem to the infinitely generated case. First observe that, over Prüfer domains, a finitely presented module is nothing else than a finitely generated module of projective dimension 1. Accordingly, in the generalization, we will consider direct sums of cyclic modules of projective dimension 1.

Unfortunately, we have been able to prove the infinite stacked bases theorem only for h-local Prüfer domains, but not for arbitrary local-global Prüfer domains. Recall that a domain R is called *h-local* (Matlis [7]) if it is of finite character and each non-zero prime ideal is contained in a unique maximal ideal. We intend to show that direct sums of cyclic modules of projective dimension 1 over h-local Prüfer domains admit stacked bases.

The proof of the stacked bases theorem is divided into three parts according as the module M is finitely, countably, or uncountably generated. For finitely generated M we just refer to Levy [6]. We shall concentrate on the countable case where most of the difficulty occurs. Though the basic idea of our proof is similar to the one used by Cohen-Gluck [2] for principal ideal domains, most details (where specific results on abelian groups were used) had to be modified in order to handle the general case. Finally, the uncountable case can routinely be reduced to the countably generated case.

2. The countably generated case.

In the proof a simple technical lemma will be required on finitely presented modules over valuation domains.

LEMMA 1. *Let R be a valuation domain, and M a finitely presented R-module. If A, B are submodules with $M = A + B$, then there exists a non-zero cyclic summand of M contained either in A or in B.*

PROOF. Let P denote the maximal ideal of R. Passing mod PM, we clearly have
$$M/PM = (A + PB)/PM + (B + PA)/PM .$$
The first term is an R/P-subspace, so a summand, and therefore we can write $M/PM = (A + PB)/PM \oplus B'/PM$ for some submodule $B' \leq B + PA$. Choose representatives $a_1, \ldots, a_k \in A$ and $b_1, \ldots, b_{n-k} \in B$ of bases of the two summand subspaces. Using Nakayama's lemma, one can argue that these representatives

generate M, and one of them, say a_1, generates a pure submodule of M (see e.g. [3, (II.3.6)]. Then Ra_1 must be a summand of M, since M/Ra_1 is finitely presented. Thus, $M = Ra_1 \oplus M'$ for some $M' < M$. □

Another preparatory lemma that will be needed is as follows. This is valid over arbitrary rings.

LEMMA 2. *Let $X = X_1 \oplus X_2$ be a projective R-module, and $\phi : X \to M = M_1 \oplus M_2$ an epimorphism onto a direct sum such that $M_1 \leq \phi X_1$. Then in the direct decomposition of X, the summand X_2 can be replaced by some $Y_2 \leq X$ satisfying $\phi Y_2 \leq M_2$.*

Moreover, if X' is a summand of X_2 with $\phi X' \leq M_2$, then Y_2 can be chosen so as to contain X'.

PROOF. Let $\pi : M \to M_1$ denote the projection with kernel M_2. Consider the commutative square

$$\begin{array}{ccc} X_2 & \xrightarrow{\rho} & X_1 \\ \phi \downarrow & & \pi\phi \downarrow \\ M_1 \oplus M_2 & \xrightarrow{\pi} & M_1. \end{array}$$

The existence of ρ is guaranteed by the projectivity of X_2. If we define $Y_2 = (1-\rho)X_2$, then evidently $\pi\phi Y_2 = \pi(\phi - \phi\rho)X_2 = 0$. Clearly, $X_1 \oplus X_2 = X_1 \oplus Y_2$, establishing the first claim.

To complete the proof, it suffices to observe that the diagram above shows that ρ can be chosen to act trivially on the summand X'. □

We shall make use of the well-known fact that a torsion module M over an h-local domain R is the direct sum of its P-components $M_P = M \otimes_R R_P$ with P running over the set Max R of maximal ideals of R; see Matlis [7]. In the Prüfer case, if M_P is finitely generated and of projective dimension 1, then it is finitely presented, so it is a direct sum of cyclically presented modules.

The following lemma is the backbone of the proof of the main theorem.

LEMMA 3. *Let R be an h-local Prüfer domain, and*

$$0 \to H \to F \xrightarrow{\phi} M \to 0$$

a free presentation of a torsion R-module M. Assume that M is a countable direct sum of cyclic submodules of projective dimension 1, say, $M = \oplus_{i<\omega} Ru_i$.

If $M_0 = Ru_0 \oplus \cdots \oplus Ru_n$ is a finitely generated summand of M, then there is a decomposition

$$F = F_1 \oplus F_2$$

such that

(i) *F_1 is finitely generated and $M_0 \leq \phi F_1$; and*
(ii) *$M = \phi F_1 \oplus \phi F_2$.*

PROOF. Since p.d.$M = 1$, H must be a projective submodule of F.

Without loss of generality we may assume that the cyclic modules Ru_i are already decomposed into their P-components ($P \in \text{Max}\, R$); the P-components of these cyclic modules are cyclically presented.

Starting with M_0, write $M = M_0 \oplus M_0'$ where M_0' is the complement of M_0 in the given direct decomposition of M. Pick a finitely generated free summand X_1 in F, say $F = X_1 \oplus X_2$ such that $M_0 \leq \phi X_1$. By the preceding lemma, we can replace X_2 by a summand Y_2 of X such that $\phi Y_2 \leq M_0'$. Choose a summand $M_2 \leq \phi Y_2$ of M_0', say $M_0' = M_1 \oplus M_2$ with finitely generated M_1. Again by Lemma 2, X_1 can be replaced by a summand Y_1 of X such that $\phi Y_1 \leq M_0 \oplus M_1$. In this way, we obtain decompositions $F = Y_1 \oplus Y_2$ and
$$M = M_0 \oplus M_1 \oplus M_2,$$
where $M_0 \leq \phi Y_1$ and $M_2 \leq \phi Y_2$. Manifestly, $M_1 = (M_1 \cap \phi Y_1) + (M_1 \cap \phi Y_2)$. As M_1 is a finite direct sum of its P-components, Lemma 1 shows that M_1 has a non-trivial cyclic summand C contained either in ϕY_1 or in ϕY_2. If C is contained in ϕY_1, then write $M_1 = C \oplus M_{10}$, and with the aid of Lemma 2 we change Y_2 to a summand Y_2' such that $\phi Y_2'$ has trivial projection on $M_0 \oplus C = M_{00}$, and at the same time replace M_0 by M_{00}, and M_1 by M_{10} to obtain $M = M_{00} \oplus M_{10} \oplus M_2$.

Continuing in a similar way, adjoining a summand of M_{10} to M_2, etc., it is readily seen that after a finite number of steps we arrive at a decomposition $F = F_1 \oplus F_2$, as required. Then $H = (H \cap F_1) \oplus (H \cap F_2)$. □

The proof of Theorem 2 below requires a careful consideration of the generators of F. The next lemma will take care of that.

LEMMA 4. *Assume the hypotheses of Lemma 3. If $F_0 = Rx_1 \oplus \cdots \oplus Rx_n$ is a summand of F, then there are decompositions $F = F_1 \oplus F_2$ and $H = (H \cap F_1) \oplus (H \cap F_2)$ such that*
  (a) F_1 *is finitely generated and* $F_0 \leq F_1$;
  (b) $M = \phi F_1 \oplus \phi F_2$.

PROOF. ϕF_0 is contained in a finitely generated summand M_0 of M. Apply Lemma 3 to this M_0 in order to obtain a decomposition as described there. It remains to show that F_1 can be chosen so as to satisfy (a). That this is possible is assured by the second statement in Lemma 2. □

We are now ready to settle the countably generated torsion case.

THEOREM 2. *Let R be an h-local Prüfer domain, and M a countable direct sum of cyclic torsion modules of projective dimension 1. Then M has the stacked bases property.*

PROOF. The free module F can also be assumed to be countably generated, say $F = \oplus_{i \in \omega} Rx_i$. It will be sufficient to reduce the problem to the finitely generated case, because then a reference to Theorem 1 will complete the proof.

By the preceding lemma, there is a decomposition $F = F_{11} \oplus F_{12}$ such that F_{11} is finitely generated, $x_0 \in F_{11}$, and $H = (H \cap F_{11}) \oplus (H \cap F_{12})$. Then F has a decomposition $F = F_{21} \oplus F_{22}$ where F_{21} is finitely generated, contains both x_1 and F_{11}, and H splits accordingly. Continuing in the same way, we obtain an ascending chain $F_{11} \leq F_{21} \leq \ldots$ of summands of F; the union of this chain must be all of F.

We can write $A_0 = 0$ and $F_{n1} = F_{n-1,1} \oplus A_n$ and $B_n = H \cap A_n$ $(n > 0)$ with suitable submodules A_n. Then evidently $F = \oplus_{n<\omega} A_n$ and $H = \oplus_{n<\omega} B_n$ are stacked finite decompositions of F and H. Thus the problem has been reduced to the finitely generated case. This concludes the proof. □

3. The general case.

We shall need a lemma in order to reduce the general problem to countably generated modules.

In the next two lemmas, R can be an arbitrary domain and A an arbitrary R-module. Roughly speaking, the first lemma is a kind of 'stacked bases' theorem for countable generation. It was proved by Cohen-Gluck for principal ideal domains, but their proof extends easily *mutatis mutandis* to arbitrary rings.

LEMMA 5. *Let R be any ring. If $\phi: F \to M$ is an epimorphism of a free R-module F onto an R-module M which is a direct sum of countably generated modules, then there are decompositions*

$$F = \oplus_{\sigma < \tau} F_\sigma \quad \text{and} \quad M = \oplus_{\sigma < \tau} M_\sigma \tag{1}$$

for some ordinal τ such that for each $\sigma < \tau$
  (i) *F_σ is countably generated, and*
  (ii) *$\phi F_\sigma = M_\sigma$.* □

The next lemma will be needed to reduce the proof to the torsion case.

LEMMA 6. *Let F be a free R-module and $M = A \oplus B$ an R-module, where B is projective. Given an epimorphism $\phi: F \to M$, there is a decomposition $F = F_1 \oplus F_2$ such that $\phi(F_1) = A$ and $F_2 \cong B$.*

PROOF. Let $\beta: M \to B$ denote the projection along A. Then $F = F_1 \oplus F_2$ with $F_1 = \operatorname{Ker} \beta\phi$ and $F_2 = \operatorname{Im} \beta\phi = B$. Clearly, the inclusion $A \leq \phi F_1$ cannot be proper. □

4. The main result.

We are now ready to formulate and prove the result announced above concerning the existence of stacked bases.

THEOREM 3. *Let R be an h-local Prüfer domain, and M a direct sum of cyclic R-modules of projective dimension one. Then every presentation $0 \to H \to F \to M \to 0$ of M has stacked bases.*

PROOF. The first step is to reduce the proof to the torsion case. This is done with the aid of Lemma 6. The next step is to observe that Lemma 4 shows that it suffices to deal with countably generated modules M.

Thus for the rest of the proof, we may assume that M is a countably generated torsion module. An appeal to Theorem 2 completes the proof. □

Let us point out that the proof above is considerably simpler than the proofs given by Cohen-Gluck [2] for principal ideal domains and by Generalov-Zheludev [4] for Dedekind domains.

References

1. J. Brewer and L. Klingler, *Pole assignability and the invariant factor theorem in Prüfer domains and Dedekind domains*, J. Algebra **111** (1987), 536–545.
2. J. Cohen and H. Gluck, *Stacked bases for modules over principal ideal domains*, J. Algebra **14** (1970), 493–505.
3. L. Fuchs and L. Salce, **Modules over Valuation Domains**, Lecture Notes in Pure Appl. Math. vol. 97 (Marcel Dekker, 1985).
4. A. I. Generalov and M. V. Zheludev, *The stacked bases theorem for modules over Dedekind domains*, St. Petersburg Math. J., **7** (1996), 619–661.
5. P. Hill and C. Megibben, *Generalizations of the stacked bases theorem*, Trans. Amer. Math. Soc., **312** (1989), 377–402.
6. L. S. Levy, *Invariant factor theorem for Prüfer domains of finite character*, J. Algebra **106** (1987), 259–264.
7. E. Matlis, *Cotorsion modules*, Mem. Amer. Math. Soc., **49** (1964).
8. L. Salce and P. Zanardo, *Presentation of modules over local domains*, J. Algebra **207** (1998), 182–204.

(Fuchs) Department of Mathematics, Tulane University, New Orleans, Louisiana 70118, USA
E-mail address: `fuchs@tulane.edu`

(Lee) Department of Mathematical Education, Sangmyung University, Seoul 110–743, S. Korea
E-mail address: `sblee@pine.sangmyung.ac.kr`

Groups with Locally Defined Heights and Products of $\mathfrak{R}^*$ Groups

Anthony J. Giovannitti

Dedicated to Professor Laszlo Fuchs in honour of his 75th birthday.

ABSTRACT. $\mathfrak{R}^*$ groups are torsion free abelian groups with the property that their pure rank one subgroups are precobalanced. They are a generalization of finite rank Butler Groups. In this paper we give necessary and sufficent conditions on the typesets of the factor groups of a product for the product to be an $\mathfrak{R}^*$ group. Toward this we introduce the concept of locally defined heights and a new class of groups that are closely related to the class of Baer Separable Groups.

1. Introduction

The class $\mathfrak{C}$ of finite rank torsion free completely decomposable abelian groups is a well understood class of torsion free abelian groups (henceforth referred to as *groups*). Important to the classification of these groups are the concepts of height sequences and types. For a group G and a $g \in G$, we denote the p-height of g in G for the prime p by $ht_p^G(g)$, the height sequence of g in G is denoted $ht^G(g) = \left(ht_p^G(g)\right)_{p \in P}$ where P is the set of all primes. A subgroup B of a group G is a *pure subgroup* (denoted $B \lhd G$) if $ht^B(b) = ht^G(b)$ for all $b \in B$, or equivalently if G/B is torsion free. We let $\mathbb{Z}$ be the group of integers and $\mathbb{Q}$ be the group of the rational field of $\mathbb{Z}$. A rank one group is any group isomorphic to a subgroup of $\mathbb{Q}$. The *type* of a rank one group R is the isomorphism class of that group and is denoted by $[R]$. For a group G, we let $typeset(G) = \{[R] : R \lhd G \text{ and } R \text{ is rank one}\}$. We refer the reader to the book by D. Arnold [2] for the basic properties and relationships of the height sequences and types. Butler [4] introduced the study of pure subgroups of groups in $\mathfrak{C}$. This class (which we denote by $\mathfrak{R}$) has become known as the class of Butler groups and has been a subject of intense interest for the past twenty years. There are two equivalent homological conditions on the subgroups of Butler groups that define this class. The condition that a finite rank group is in $\mathfrak{R}$ if and only if every subgroup is prebalanced was first noted by F. Richman [10] and

---

1991 *Mathematics Subject Classification.* Primary 20K20; Secondary 20K25, 20K27.
Key words and phrases. Torsion Free Abelian Groups, Vector Groups.
This paper is in final form and no version of it will be submitted for publication elsewhere.

© 2001 American Mathematical Society

independently by L. Fuchs and C. Viljoen [5]. The other condition is that a finite rank group is in $\mathfrak{R}$ if and only if every subgroup is precobalanced was shown in the paper by K. Rangaswamy and the author [8]. In[6], the author investigated the class of groups with the property that each rank one pure subgroup is precobalanced. (This class is denoted by $\mathfrak{R}^*$.) It was shown that the finite rank groups in this class are precisely the finite rank Butler groups and that the class of pure subgroups of separable (in the sense of Baer [3]) groups are contained in this class. In this paper, we investigate when a product of groups will be an $\mathfrak{R}^*$-group. Our main results are as follows.

THEOREM 1. *A vector group $\prod_{\lambda \in I} R_\lambda$ is an $\mathfrak{R}^*$-group if and only if*

1. *there is no infinite sequence of distinct labels $\lambda_1, \lambda_2, \ldots$ such that for all n*

$$\bigwedge_{k=1}^{n+1} [R_{\lambda_k}] < \bigwedge_{k=1}^{n} [R_{\lambda_k}];$$

2. *there is no infinite sequence of distinct labels $\lambda_1, \lambda_2, \ldots$ and distinct primes $p_1, p_2, \ldots$ such that for all n*

$$p_n R_{\lambda_n} \neq R_{\lambda_n} \qquad \text{and } [R] \leq [R_{\lambda_n}]$$

where $\mathbb{Z} \subseteq R \subseteq \mathbb{Q}$ with

$$ht_p^R(1) = \begin{cases} 1 & \text{if } p \in \{p_1, p_2, \ldots\}, \\ 0 & \text{otherwise.} \end{cases}$$

and

THEOREM 2. *A group $\prod_{\lambda \in I} B_\lambda$ is an $\mathfrak{R}^*$-group if and only if*

1. *each B_λ is an $\mathfrak{R}^*$-group;*
2. *there is no infinite sequence of distinct labels $\lambda_1, \lambda_2, \ldots$ and types $\tau_{\lambda_n} \in \text{typeset}(B_{\lambda_n})$ such that for all n,*

$$\bigwedge_{k=1}^{n+1} \tau_{\lambda_k} < \bigwedge_{k=1}^{n} \tau_{\lambda_k};$$

3. *there is no infinite sequence of distinct labels $\lambda_1, \lambda_2, \ldots$, types $\tau_{\lambda_n} \in \text{typeset}(B_{\lambda_n})$, and distinct primes $p_1, p_2, \ldots$ such that*

$$[R] \leq \tau_{\lambda_n}$$

where τ_{λ_n} has finite p_n-component for all n, and $\mathbb{Z} \subseteq R \subseteq \mathbb{Q}$ with

$$ht_p^R(1) = \begin{cases} 1 & \text{if } p \in \{p_1, p_2, \ldots\}, \\ 0 & \text{otherwise.} \end{cases}$$

We also introduce a new type of purity that defines the precobalanced subgroups of a vector group.

2. Locally Defined Heights

A subgroup B of a group G is said to be *precobalanced* [8] if whenever

$$0 \longrightarrow B \xrightarrow{f} G$$
$$\alpha \downarrow$$
$$R$$

where f is the inclusion map and R is a rank one group, there is an $H \in \mathfrak{C}$ and maps $\beta : R \to H$ and $g : G \to H$ such that β is a pure monomorphism and $g\alpha = \beta f$. (We say that α has a *semi-extension*.) In [6], it was shown that for rank one B, we need only show that the identity map $1_B : B \to B$ has a semi-extension to imply that B is precobalanced in G. If G is an $\mathfrak{R}^*$-group, then for each $0 \neq x \in G$ there is an $H_x \in \mathfrak{C}$ and map $g_x : G \to H_x$ such that $ht^G(x) = ht^{H_x}(g_x(x))$. Then the evaluation map $e : G \to \prod_{x \in G \setminus \{0\}} H_x = V$ is a pure monomorphism with the properties that for any nonempty finite $X \subset G$ the projection map $\pi_X : V \to \prod_{x \in X} H_x = V_X$ is such that for all $x \in X$, $ht^G(x) = ht^{V_X}(\pi_X(e(x)))$. Furthermore, if X is linearly independent in G, then $\{\pi_X(x) : x \in X\}$ is linearly independent in V_X with the same cardinality as X. But V_X is a finite rank completely decomposable summand of V. So the heights of elements of G are defined locally in V.

DEFINITION 1. *Let B be a subgroup of G. The subset $X \subseteq B$ is said to have locally defined heights in G with respect to B (shortened to X is ldh in G) if there is a decomposition $G = C \oplus D$ where $C \in \mathfrak{C}$ and for the projection map $\pi : G \to C$, $ht^B(x) = ht^C(\pi(x)) = ht^G(\iota\pi(x))$ for all $x \in X$ where $\iota : C \to G$ is the inclusion map.*

1. *B is said to be weakly sharp in G (denoted $B \leq^{(\#)} G$) if for all $b \in B$, $\{b\}$ is ldh in G.*
2. *B is said to be sharp in G (denoted $B \leq^{\#} G$) if any finite $X \subset B$ is ldh in G.*
3. *B is said to be strongly sharp in G (denoted $B \leq^{\Delta} G$) if $B \leq^{\#} G$ and whenever $X \subset B$ is a finite and linearly independent, C can be chosen so that the projection map of G onto C not only preserves heights in B but is also monic and preserves linear independence for the elements of X.*

Thus if $G \in \mathfrak{R}^*$, then there is a vector group V with $G \leq^{\Delta} V$. All three of these properties are inherited by pure subgroups. It is evident that strongly sharp implies sharp which implies weakly sharp. The reverse implications are not known except in some special cases as noted below.

PROPOSITION 1. *Let B be a rank one subgroup of G. Then the following are equivalent*
1. *$B \leq^{(\#)} G$*
2. *$B \leq^{\#} G$*
3. *$B \leq^{\Delta} G$*

Furthermore, B is precobalanced in G.

PROOF. If B is divisible, then it is a summand of G. Thus, all three purities hold.

Since the only linearly independent subsets of B are singletons and the empty set and the empty set is always ldh. We need only show that weakly sharp implies sharp when B is reduced. Chose a nonzero $b \in B$ and let $G = C \oplus D$ with $C \in \mathfrak{C}$ and for the projection map $\pi : G \to C$, $ht^B(b) = ht^C(\pi(b))$. This implies that $\pi(b) \neq 0$ and it then an easy exercise to show that π restricted to B is a pure monomorphism. Thus B is precobalanced in G and $ht^B(x) = ht^C(\pi(x))$ for all $x \in B$. Thus every subset of B is ldh in G. □

In [8], it was shown that if B is a subgroup of G and A is a subgroup of B that is precobalanced in G, then A is precobalanced in B.

PROPOSITION 2. *If $B \leq^{(\#)} G$, then $B \triangleleft G$ and $B \in \mathfrak{R}^*$.*

PROOF. $B \triangleleft G$ is evident from the definition of weak sharp. Suppose A is a pure rank one subgroup of B. Then $A \leq^{(\#)} G$. Thus A is precobalanced in G and hence in B. Thus $B \in \mathfrak{R}^*$. □

3. Vector Groups

A vector group is a direct product of rank one groups. The author in [7] investigated rank one quotients of vector groups. Using these results and the Los-Eda Theorem, we are able to classify precobalanced rank one subgroups of vector groups with respect to sharp purity. (We refer the reader to this paper for explanation of notations when applying this theorem and other results from that paper.)

For convenience, we let $\langle x \rangle_*$ denote the pure rank one subgroup of a group G generated by $0 \neq x \in G$.

LEMMA 1. *Let $V = \prod_{\lambda \in I} R_\lambda$ where each R_λ is a rank one group. Then for $x = (x_\lambda) \in V$, the following are equivalent*

1. $\langle x \rangle_*$ *is precobalanced in V*
2. $\langle x \rangle_* \leq^{(\#)} V$
3. *there are $\lambda_1, \ldots, \lambda_n \in I$ such that $ht^V(x) = \bigwedge_{k=1}^n ht^{R_{\lambda_k}}(x_{\lambda_k})$.*

PROOF. 2. implies 1. is a consequence of Proposition 1.

Assume 1. If x is divisible by all primes, then each component is also divisible by all primes. Thus for any $\lambda \in I$, $ht^V(x) = ht^{R_\lambda}(x_\lambda)$.

Otherwise, there is a reduced $C \in \mathfrak{C}$ and map $\alpha : V \to C$ such that $ht^V(x) = ht^C(\alpha(x))$. Since C is slender, the Los-Eda Theorem (cf., [7]) implies that there are ω_1-complete ultrafilters $D_1, \ldots, D_n$ of I and maps $q : V \to \bigoplus_{k=1}^n V/D_k = H$ and $\rho : H \to C$ such that $\rho q(x) = \alpha(x)$. We let $q_k(x)$ be the image of $q(x)$ in V/D_k. Then Corollary 1 in [7] implies that there is a $\lambda_k \in I$ for each $1 \leq k \leq n$, such that there is an isomorphism $\beta_k : R_{\lambda_k} \to V/D_k$ with $\beta_k(x_{\lambda_k}) = q_k(x)$. Thus

$$ht^V(x) \leq \bigwedge_{k=1}^n ht^{R_{\lambda_k}}(x_{\lambda_k}) \leq \bigwedge_{k=1}^n ht^{V/D_k}(\beta_k(x_{\lambda_k})) = \bigwedge_{k=1}^n ht^{V/D_k}(q_k(x))$$
$$= ht^H(q(x)) \leq ht^C(\rho q(x)) = ht^C(\alpha(x)) = ht^V(x)$$

and so 3. holds.

Suppose $x \in V$ and there are $\lambda_1, \ldots, \lambda_n \in I$ such that $ht^V(x) = \bigwedge_{k=1}^n ht^{R_{\lambda_k}}(x_{\lambda_k})$. Then $\bigoplus_{k=1}^n R_{\lambda_k} = C$ is a finite rank completely decomposable summand of V and for the projection map π of V onto C, $ht^V(x) = ht^C(\pi(x))$. Thus 2. holds. □

The next theorem shows that for subgroups of vector groups these three purities are equivalent.

THEOREM 3. *Let B be a subgroup of the vector group $V = \prod_{\lambda \in I} R_\lambda$. Then the following are equivalent*
1. $B \leq^\Delta V$
2. $B \leq^\# V$
3. $B \leq^{(\#)} V$.

PROOF. We have seen that 1. implies 2. which implies 3.

Assume 3. holds and $X = \{x_1, ..., x_n\} \subset B$ with $x_i = (x_{i\lambda})$. The previous lemma implies that for each $1 \leq i \leq n$, there is a finite set $J_i \subseteq I$ such that $ht^V(x_i) = \bigwedge_{\lambda \in J_i} ht^{R_\lambda}(x_{i\lambda})$. Then $J = J_1 \cup \cdots \cup J_n$ is finite and

$$ht^V(x_i) \leq \bigwedge_{\lambda \in J} ht^{R_\lambda}(x_{i\lambda}) \leq \bigwedge_{\lambda \in J_i} ht^{R_\lambda}(x_{i\lambda}) = ht^V(x_i).$$

If X is also linearly independent, we can find a finite $J^* \subset I$ such that $\{(x_{i\lambda})_{\lambda \in J^*} : 1 \leq i \leq n\}$ is linearly independent in $\prod_{\lambda \in J^*} R_\lambda$. Then $J' = J \cup J^*$ is a finite subset of I and $\{(x_{i\lambda})_{\lambda \in J'} : 1 \leq i \leq n\}$ is linearly independent in $\prod_{\lambda \in J'} R_\lambda = V'$. If π is the projection of V onto V', then for each i,

$$ht^V(x_i) \leq ht^{V'}(\pi(x_i)) = \bigwedge_{\lambda \in J'} ht^{R_\lambda}(x_{i\lambda}) \leq \bigwedge_{\lambda \in J} ht^{R_\lambda}(x_{i\lambda}) = ht^V(x_i).$$

Since $V' \in \mathfrak{C}$, 3. implies 1. □

From the above theorem and the comments follow the definition, we have as an immediate consequence.

COROLLARY 1. *$G \in \mathfrak{R}^*$ if and only if there is a vector group V with $G \leq^\Delta V$.*

By a separable group, we are referring to groups that satisfy the concept of separability introduced by Baer [3]. These are groups with the property that every finite subset is contained in a finite rank completely decomposable summand. Let S denote the class of separable groups, $S^{(\#)} = \{A : A \leq^{(\#)} A\}$, $S^\# = \{A : A \leq^\# A\}$, and $S^\Delta = \{A : A \leq^\Delta A\}$. Then $S \subset S^\Delta \subseteq S^\# \subseteq S^{(\#)}$. It is easily shown that each class is closed with respect to direct sums.

COROLLARY 2. *Let V be a vector group. Then the following are equivalent*
1. $V \in S^\Delta$
2. $V \in S^\#$
3. $V \in S^{(\#)}$
4. $V \in \mathfrak{R}^*$.

PROOF. The equivalence of 1., 2., and 3. follows from the previous theorem. Proposition 2 shows that 3. implies 4.

Suppose $V \in \mathfrak{R}^*$ and $x \in V$. Then $\langle x \rangle_*$ is precobalanced in V and Lemma 1 implies that $\{x\}$ is ldh in V. Thus 4. implies 3. □

Homogeneous groups are groups such that all their pure rank one subgroups are isomorphic. In [7], the author has shown that homogeneous groups are in $\mathfrak{R}^*$ if and only if they are separable.

COROLLARY 3. *Let A be a homogeneous group. Then the following are equivalent*

1. $A \in S$
2. $A \in S^{\triangle}$
3. $A \in S^{\#}$
4. $A \in S^{(\#)}$
5. $A \in \mathfrak{R}^*$.

4. Products of $\mathfrak{R}^*$ Groups

Separable groups are not closed with respect to arbitrary products. Mishina in [9] gave necessary and sufficient condition on the rank one factors of a vector group for that group to be separable. Albrecht and Hill [1] reaffirmed this result and gave conditions on the typesets of the factor groups of a product for that product to be separable.

Let $V = \prod_{\lambda \in I} R_\lambda$ be a vector group with each R_λ a rank one group. The set I satisfies:

M1 if there is no infinite sequence of distinct elements $\lambda_1, \lambda_2, ...$ of I such that
$$[R_{\lambda_1}] > [R_{\lambda_2}] > \cdots$$

M2 if there is no infinite sequence of distinct elements $\lambda_1, \lambda_2, ...$ of I such that the $[R_{\lambda_i}]$'s are pairwise incomparable.

M3 if for each $\lambda \in I$, (M3$_\lambda$) there is a nonzero $r \in R_\lambda$ such that for each prime p that divides r, the set $\Gamma(p, r, \lambda) = \{\mu \in I : [R_\lambda] \leq [R_\mu] \text{ and } pR_\mu \neq R_\mu\}$ is finite.

PROPOSITION 3. *If $V = \prod_{\lambda \in I} R_\lambda$, a vector group with each R_λ a rank one group, is in $\mathfrak{R}^*$, then I satisfies M3.*

PROOF. For $r \in R_\lambda$, let $P(r) = \{p : 0 < ht_p^{R_\lambda}(r) \text{ is finite.}\}$. If $P(r)$ is empty, then $ht_p^{R_\lambda}(r) = \infty$ whenever p divides r. Thus if $[R_\lambda] \leq [R_\mu]$, then $pR_\mu = R_\mu$. Thus $\Gamma(p, r, \lambda)$ is empty. If $P(r) = \{p_1, ..., p_n\}$, then $ht_{p_i}^{R_\lambda}(r) = k_i$ is finite and
$$r' = \frac{1}{p_1^{k_1} \cdots p_n^{k_n}} r \in R_\lambda$$
with $P(r')$ empty. In either case M3$_\lambda$ is satisfied.

Suppose for some λ_1, M3$_{\lambda_1}$ is not satisfied. Then for each nonzero $r \in R_{\lambda_1}$, $P(r)$ is infinite. (Note that this implies that $[R_{\lambda_1}]$ is of non-idempotent type.) Fix a nonzero $r_1 \in R_{\lambda_1}$ and choose a $p_1 \in P(r_1)$ such that $\Gamma(p_1, r_1, \lambda_1)$ is infinite. Choose $\lambda_2 \in \Gamma(p_1, r_1, \lambda_1) \setminus \{\lambda_1\}$. Then $[R_{\lambda_1}] \leq [R_{\lambda_2}]$ and so there is a nonzero $r \in R_{\lambda_2}$ such that $ht^{R_{\lambda_1}}(r_1) \leq ht^{R_{\lambda_2}}(r)$. Since $p_1 R_{\lambda_2} \neq R_{\lambda_2}$, $ht_{p_1}^{\lambda_2}(r) = k$ is finite. Let $r_2 = \frac{1}{p_1^k} r$. Then $r_2 \in R_{\lambda_2}$, $ht_p^{R_{\lambda_1}}(r_1) \leq ht_p^{R_{\lambda_2}}(r_2)$ for $p \neq p_1$, while $ht_{p_1}^{R_{\lambda_2}}(r_2) = 0 < ht_{p_1}^{R_{\lambda_1}}(r_1) = k_1$. We choose $p_2 \in P\left(\frac{1}{p_1^{k_1}} r_1\right) = P(r_1) \setminus \{p_1\}$ (which is infinite) such that $\Gamma\left(p_2, \frac{1}{p_1^{k_1}} r_1, \lambda_1\right)$ is infinite. Choose $\lambda_3 \in \Gamma\left(p_2, \frac{1}{p_1^{k_1}} r_1, \lambda_1\right) \setminus \{p_1, p_2\}$ and

$r_3 \in R_{\lambda_3}$ such that

$$ht_{p_2}^{R_{\lambda_3}}(r_3) = 0 < ht_{p_2}^{R_{\lambda_1}}(r_1) = ht_{p_2}^{R_{\lambda_1}}(r_1) \bigwedge ht_{p_2}^{R_{\lambda_2}}(r_2)$$

and

$$ht_p^{R_{\lambda_1}}(r_1) \le ht_p^{R_{\lambda_3}}(r_3)$$

for all other primes p. Thus there are distinct $\lambda_1, \lambda_2, \dots$ in I and distinct primes $p_1, p_2, \dots$ and $r_n \in R_{\lambda_n}$ for each n such that for $n > 1$

$$ht_{p_{n-1}}^{R_{\lambda_n}}(r_n) = 0 < ht_{p_{n-1}}^{R_{\lambda_1}}(r_1) \le \bigwedge_{k=1}^{n-1} ht_{p_{n-1}}^{R_{\lambda_k}}(r_k)$$

and

$$ht_p^{R_{\lambda_1}}(r_1) \le ht_p^{R_{\lambda_n}}(r_n)$$

for all other primes p. Then for $e = (e_\lambda) \in V$ where

$$e_\lambda = \begin{cases} r_n & \text{if } \lambda = \lambda_n, \\ 0 & \text{otherwise,} \end{cases}$$

$ht_{p_n}^V(e) = 0$ for all n. For any given finite subset $X \subset I$, there is a N such that for all $n \ge N$, $\lambda_n \notin X$. Thus for $n \ge N$,

$$ht_{p_n}^V(e) = 0 < ht_{p_n}^{R_{\lambda_1}}(r_1) \le \bigwedge_{\lambda \in X} ht_{p_n}^{R_\lambda}(e_\lambda).$$

Lemma 1 implies that $\langle e \rangle_*$ is not precobalanced in V. Thus $V \notin \mathfrak{R}^*$. So $V \in \mathfrak{R}^*$ implies M3 holds for I. □

Note that condition 1 of the Theorem 1 implies M1.

EXAMPLE 1. *Partition the primes by $\bigcup_{n=1}^\infty J_n$ where each J_n is a nonempty finite set. Let R_n be the subring of $\mathbb{Q}$ such that $\frac{1}{p} \in R_n$ if and only if $p \notin J_n$. Thus R_n has type with 0 in the p-components with $p \in J_n$ and with ∞ for all other p-components Thus $\bigwedge_{k=1}^n [R_k]$ has a type with 0 in the p-components for $p \in \bigcup_{k=1}^n J_k$ and ∞ for all other p-components. Thus for all n,*

$$\bigwedge_{k=1}^{n+1} [R_k] < \bigwedge_{k=1}^n [R_k].$$

Thus condition 1 of Theorem 1 does not hold for $\prod_{n=1}^\infty R_n$. But since $[R_n]$'s are pairwise incomparable M1 does hold.

PROOF OF THEOREM 1. ($\Longrightarrow$) First consider $V = \prod_{n=1}^\infty R_n$. Let us assume that $\bigwedge_{k=1}^{n+1} [R_k] < \bigwedge_{k=1}^n [R_k]$ for all n. For each n, choose a nonzero $r_n \in R_n$. Then for $r = (r_n) \in V$, $ht^V(r) < \bigwedge_{k=1}^n ht^{R_k}(r_k)$ for all n. Thus if X is a finite set of natural numbers, $ht^V(r) < \bigwedge_{k \in X} ht^{R_k}(r_k)$. As was shown above, this implies that $V \notin \mathfrak{R}^*$. Suppose there are distinct primes $p_1, p_2, \dots$ such that $p_k R_k \ne R_k$ and $\mathbb{Z} \subseteq R \subseteq \mathbb{Q}$ is such that

$$ht_p^R(1) = \begin{cases} 1 & \text{if } p \in \{p_1, p_2, \dots\}, \\ 0 & \text{otherwise.} \end{cases}$$

with $[R] \leq [R_k]$ for all k. For each k, chose $r_k \in R_k$ such that $ht_{p_k}^{R_k}(r_k) = 0$ and $ht_p^R(r) \leq ht_p^{R_k}(r_k)$ for any $p \neq p_k$. Let $r = (r_k) \in V$. Thus for each $n > 1$, $ht_{p_n}^V(r) = 0 < ht_{p_n}^R(1) \leq \bigwedge_{k=1}^{n-1} ht_{p_n}^{R_n}(r_k)$. Thus $ht^V(r) < \bigwedge_{k=1}^{n} ht^{R_k}(r_k)$ for all n and so $V \notin \mathfrak{R}^*$. Thus if such a $V \in \mathfrak{R}^*$, then conditions 1 and 2 must hold.

For $V = \prod_{\lambda \in I} R_\lambda$ where R_λ is a rank one group and distinct $\lambda_1, \lambda_2, \ldots$ in I, $V' = \prod_{n=1}^{\infty} R_{\lambda_n}$ is a summand of V of the same form as above. If $V \in \mathfrak{R}^*$, then $V' \in \mathfrak{R}^*$. Thus conditions 1 and 2 hold.

($\Longleftarrow$) Let $g = (g_\lambda) \in V = \prod_{\lambda \in I} R_\lambda$. For each prime p, there is a $\lambda_p \in I$ with $ht_p^V(g) = ht_p^{R_{\lambda_p}}(g_{\lambda_p})$ while $ht_p^V(g) \leq ht_p^{R_\lambda}(r_\lambda)$ for all other $\lambda \in I$. If $I(g) = \{\lambda_p : p \in P\}$ is finite, then $\langle g \rangle_*$ is precobalanced in V. Otherwise, we let $I_0 = \{\lambda_p : ht_p^V(g) = 0\}$ and $I_1 = \{\lambda_p : 0 < ht_p^V(g) < \infty\}$. If I_1 is infinite, then $[\langle g \rangle_*]$ is of non-idempotent type and for each $\lambda_p \in I_1$, $pR_{\lambda_p} \neq R_{\lambda_p}$. Since $[\langle g \rangle_*] \leq [R_{\lambda_p}]$. Condition 2 does not hold. Thus for condition 2 to hold when $I(g)$ is infinite, I_1 must be finite and $[\langle g \rangle_*]$ is of idempotent type.

Suppose for any finite subset $X \subset P$, there is a $q \in P \backslash X$ (depending on X) such that $\bigwedge_{p \in X \cup \{q\}} [R_{\lambda_p}] < \bigwedge_{p \in X} [R_{\lambda_p}]$. Then there are distinct primes $\{p_1, p_2, \ldots\}$ such that

$$\bigwedge_{k=1}^{n+1} [R_{\lambda_{p_k}}] < \bigwedge_{k=1}^{n} [R_{\lambda_{p_k}}].$$

Thus condition 1 does not hold. Given that condition 1 holds, there is a finite set $X \subset P$, such that for any finite set F with $X \subseteq F \subset P$

$$\bigwedge_{p \in X} [R_{\lambda_p}] = \bigwedge_{p \in F} [R_{\lambda_p}].$$

Since $[\langle g \rangle_*] \leq \bigwedge_{p \in X} [R_{\lambda_p}]$, the set $W = \{q \in P : qR_{\lambda_p} \neq R_{\lambda_p} \text{ for some } p \in X\}$ is a subset of $P^* = \{p : \lambda_p \in I_0 \cup I_1\}$. If $\bigwedge_{p \in X} [R_{\lambda_p}]$ is of non-idempotent type, then W is infinite which implies $P^* \backslash X$ is also infinite. For $q \in P^* \backslash X$, $qR_{\lambda_q} \neq R_{\lambda_q}$ and $\bigwedge_{p \in X} [R_{\lambda_p}] = \bigwedge_{p \in X \cup \{q\}} [R_{\lambda_p}] \leq [R_{\lambda_q}]$. Thus condition 2 does not hold. Hence for conditions 1 and 2 to hold when $I(g)$ is infinite, $\bigwedge_{p \in X} [R_{\lambda_p}]$ is of idempotent type. This implies

$$P_3 = \left\{ q \in P : 0 < \bigwedge_{p \in X} ht_q^{R_{\lambda_p}}(g_{\lambda_p}) \text{ and } ht_q^V(g) \neq \infty \right\}$$

is finite. For any prime q, $ht_q^V(g) \leq \bigwedge_{p \in X \cup P_3} ht_q^{R_{\lambda_p}}(g_{\lambda_p})$ and $X \cup P_3$ is finite. For primes q where $ht_q^V(g) = \infty$, $\bigwedge_{p \in X \cup P_3} ht_q^{R_{\lambda_p}}(g_{\lambda_p}) = ht_q^V(g)$. For $q \in P_3$,

$$ht_q^V(g) \leq \bigwedge_{p \in X \cup P_3} ht_q^{R_{\lambda_p}}(g_{\lambda_p}) \leq ht_q^{R_{\lambda_q}}(g_{\lambda_q}) = ht_q^V(g)$$

and for $q \in P^* \backslash P_3$,

$$\bigwedge_{p \in X \cup P_3} ht_q^{R_{\lambda_p}}(g_{\lambda_p}) \leq \bigwedge_{p \in X} ht_q^{R_{\lambda_p}}(g_{\lambda_p}) = 0.$$

Thus for all primes q, $ht_q^V(g) = \bigwedge_{p \in X \cup P_3} ht_q^{R_{\lambda_p}}(g_{\lambda_p})$ and hence $ht^V(g) = \bigwedge_{p \in X \cup P_3} ht^{R_{\lambda_p}}(g_{\lambda_p})$. Lemma 1 implies that $V \in \mathfrak{R}^*$. □

Note that $V \in \mathfrak{R}^*$ implies M1 and M3, which makes these groups close to being separable.

EXAMPLE 2. *Partition the primes by $\bigcup_{n=1}^{\infty} J_n$ where each J_n is infinite. Let R_n be the subring of $\mathbb{Q}$ such that $\frac{1}{p} \in R_n$ if and only if $p \in J_n$. Thus R_n has type with ∞ in these p-components and 0 for p-components with $p \notin J_n$ and $[R_m] \bigwedge [R_n] = [\mathbb{Z}]$ whenever $m \neq n$. Thus conditions 1 and 2 of Theorem 1 hold for $\prod_{n=1}^{\infty} R_n = V$. But since $[R_n]$'s are pairwise incomparable M2 does not hold. Thus $V \in \mathfrak{R}^*$ but V is not separable.*

PROOF OF THEOREM 2. ($\Longrightarrow$) For each λ, $B_\lambda \triangleleft B$ and is thus in $\mathfrak{R}^*$. For a sequence of distinct $\lambda_1, \lambda_2, \ldots \in I$ and $\tau_{\lambda_n} \in typeset(B_{\lambda_n})$, we let R_{λ_n} be a pure rank one subgroup of B_{λ_n} with type τ_{λ_n}. Then $\prod_{n=1}^{\infty} R_{\lambda_n} = V \triangleleft B$ which implies that it is in $\mathfrak{R}^*$. Thus conditions 1 and 2 of Theorem 1 hold which will imply that conditions 2 and 3 of this theorem hold.
($\Longleftarrow$) For a nonzero $x = (x_\lambda) \in B$, let $R_\lambda = \langle x_\lambda \rangle_* \triangleleft B_\lambda$. Then $\langle x \rangle_*$ is a pure subgroup of $\prod_{\lambda \in I} R_\lambda = V$. Theorem 1 implies that $V \in \mathfrak{R}^*$. Thus there are $\lambda_1, \ldots, \lambda_n \in I$ such that $ht^V(x) = \bigwedge_{k=1}^n ht^{R_{\lambda_k}}(x_{\lambda_k}) = \bigwedge_{k=1}^n ht^{B_{\lambda_k}}(x_{\lambda_k})$. Thus there is a pure monomorphism $\alpha : \langle x \rangle_* \to B' = \bigoplus_{k=1}^n B_{\lambda_k}$. Since each $B_{\lambda_k} \in \mathfrak{R}^*$, $B' \in \mathfrak{R}^*$ and so Im(α) is precobalanced in B'. This implies that there is an $H \in \mathfrak{C}$ and maps $\beta : \langle x \rangle_* \to H$ and $\theta : B' \to H$ such that β is a pure monomorphism and $\theta \alpha = \beta$. Let $\pi : B \to B'$ be the projection map and $f : \langle x \rangle_* \to B$ be the inclusion map. Then $\theta \pi : B \to H$ and $(\theta \pi) f = \theta (\pi f) = \theta \alpha = \beta$. Thus $\langle x \rangle_*$ is precobalanced in B and so $B \in \mathfrak{R}^*$. $\square$

Though the class of separable groups is closed under summands, it is not known whether any of $S^\triangle$, $S^\#$, or $S^{(\#)}$ are closed under summands. We are able to show that under certain set conditions that these classes are closed under products if conditions 2 and 3 of Theorem 2 hold.

THEOREM 4. *Suppose $B = \prod_{\lambda \in I} B_\lambda$ and each $B_\lambda \in S^\triangle$ (or $B_\lambda \in S^\#$ or $B_\lambda \in S^{(\#)}$). If either I or $\{[B_\lambda] : \lambda \in I\}$ have cardinality less than the first non–measurable cardinal, $B \in S^\triangle$ (or $B \in S^\#$ or $B \in S^{(\#)}$, respectively) if and only if conditions 2 and 3 of Theorem 2 hold.*

PROOF. ($\Longrightarrow$) If B is in any of these classes, then it is in $\mathfrak{R}^*$. Therefore Theorem 2 holds.
($\Longleftarrow$) Suppose conditions 2 and 3 of Theorem 2 hold and suppose each $B_\lambda \in S^\triangle$ (or $B_\lambda \in S^\#$ or $B_\lambda \in S^{(\#)}$). Then (in any of the above cases) this implies $B \in \mathfrak{R}^*$. Let X be a finite subset of B and $R(X)$ be the pure subgroup of B generated by the set X. Then $R(X)$ is a finite rank Butler group and is precobalanced in B (cf.,[6]). Let $f : R(X) \to B$ be the inclusion map. Then there is an $H \in \mathfrak{C}$ and maps $\alpha : R(X) \to H$ and $\beta : B \to H$ with α a pure monomorphism and $\beta f = \alpha$. The Łos-Eda Theorem implies that there are ω_1-complete ultrafilters $D_1, \ldots, D_n$ of I and maps $\gamma : B \to \bigoplus_{k=1}^n B/D_k = S$ with γ the natural quotient map and $\theta : S \to H$ and $\gamma \theta = \beta$. The author in [7] has shown that if either I or $\{[B_\lambda] : \lambda \in I\}$ have cardinality less than the first non–measurable cardinal there are $\lambda_1, \ldots, \lambda_n \in I$ and an isomorphism $\eta : \bigoplus_{k=1}^n B_{\lambda_k} \to S$ such that for the projection

$\pi : B \to \bigoplus_{k=1}^{n} B_{\lambda_k} = B'$, $\pi\eta = \gamma$. Let $b = (b_\lambda) \in X$. Then

$$ht^B(b) \leq ht^{B'}(\pi(b)) = \bigwedge_{k=1}^{n} ht^{B_{\lambda_k}}(b_{\lambda_k}) \leq ht^S(\eta\pi(b)) = ht^S(\gamma f(b))$$

$$\leq ht^H(\theta\gamma f(b)) = ht^H(\beta f(b)) = ht^H(\alpha(b)) = ht^B(b)$$

and $Y = \{\pi(b) : b \in X\} \subset B'$ has the same cardinality of X.

First consider when each $B_\lambda \in S^\triangle$. If $X = \{b_1, ..., b_n\}$ is linearly independent and $m_1\pi(b_1) + \cdots + m_n\pi(b_n) = 0$, then $0 = \theta\eta\pi(m_1 b_1 + \cdots + m_n b_n) = \alpha(m_1 b_1 + \cdots + m_n b_n)$. Thus $m_1 b_1 + \cdots + m_n b_n = 0$ which implies $m_1 = \cdots = m_n = 0$. Thus Y is linearly independent and so there is a decomposition $B' = C \oplus D$ with $C \in \mathfrak{C}$, $\{\rho\pi(b_1), ..., \rho\pi(b_n)\}$ is linearly independent, and $ht^{B'}(\pi(b_k)) = ht^C(\rho\pi(b_k))$ for each k where $\rho : B' \to C$ is the appropriate projection. But C is also a summand of B where $\rho\pi$ is the projection of B onto C. Thus $B \in S^\triangle$.

The proofs that if each $B_\lambda \in S^\#$ or $B_\lambda \in S^{(\#)}$ and conditions 2 and 3 of Theorem 2 hold, then $B \in S^\#$ or $B \in S^{(\#)}$, respectively, are similar. $\square$

References

[1] Albrecht, U. and P. Hill, *Separable Vector Groups*, in **Abelian Group Theory: Proceedings of the 1987 Perth Conference**, Cont. Math., Vol. 87, (1987), 155–160.

[2] Arnold, D., **Finite Rank Torsion Free Abelian Groups and Rings**, Lecture Notes in Math., Vol. 931, (1982).

[3] Baer, R., *Abelian Groups without elements of finite order*, Duke Math. J., 3, (1937) 68–122.

[4] Butler, M.C.R., *A class of torsion-free abelian groups*, Proc. London Math. Soc., 15, (1965) 680–698.

[5] Fuchs, L. and G. Viljeon, *Notes on Extensions of Butler Groups*, Bull. Aust. Math. Soc., 41, (1990) 117–122.

[6] Giovannitti, A.J., *Torsion-Free Abelian Groups with Precobalanced Finite Rank Pure Subgroups*, in **Abelian Groups: Proceedings of the 1991 Curacao Conference**, Lecture Notes in Pure and Applied Math., Vol. 146, (1993) 157–166.

[7] Giovannitti, A.J., *Rank One Quotients of Vector Groups*, Comm. in Algebra, Vol. 25 (3), (1997) 709–714.

[8] Giovannitti, A. J. and K. Rangaswamy, *Precobalanced Subgroups of Abelian Groups*, Comm. in Algebra, Vol. 19 (1), (1991) 249–269.

[9] Mishina, A.P., *Separability of Complete Direct Sums of Torsion-free Groups of Rank One*, (Russian), Mat. Sb., 57, (1962) 375–383.

[10] Richman, F., *Butler Groups, Valuated Vector Spaces, and Duality*, Rend. Sem. Mat. Univ. Padova, Vol 72, (1984) 13–19.

STATE UNIVERSITY OF WEST GEORGIA CARROLLTON, GA 30118 USA
E-mail address, A. Giovannitti: agiovann@westga.edu

Reflexive subgroups of the Baer-Specker group and Martin's axiom

Rüdiger Göbel and Saharon Shelah

Dedicated to Professor Laszlo Fuchs in honour of his 75th birthday

ABSTRACT. In two recent papers [9, 10] we answered a question raised in the book by Eklof and Mekler [7, p. 455, Problem 12] under the set theoretical hypothesis of $\diamondsuit_{\aleph_1}$ which holds in many models of set theory, respectively of the special continuum hypothesis (CH). The objects are reflexive modules over countable principal ideal domains R, which are not fields. Following H. Bass [1] an R-module G is reflexive if the evaluation map $\sigma : G \longrightarrow G^{**}$ is an isomorphism. Here $G^* = \text{Hom}(G, R)$ denotes the dual module of G. We proved the existence of reflexive R-modules G of infinite rank with $G \not\cong G \oplus R$, which provide (even essentially indecomposable) counter examples to the question [7, p. 455]. Is CH a necessary condition to find 'nasty' reflexive modules? In the last part of this paper we will show (assuming the existence of supercompact cardinals) that large reflexive modules always have large summands. So at least being essentially indecomposable needs an additional set theoretic assumption. However the assumption need not be CH as shown in the first part of this paper. We will use Martin's axiom to find reflexive modules with the above decomposition which are submodules of the Baer-Specker module R^ω.

1. Introduction

We will derive our results for abelian groups, but it is an easy exercise to replace the ground ring $\mathbb{Z}$ by any countable principal ideal domain which is not a field. Just notice that we could work with one prime only! For supercompact cardinals we refer either to Jech [13] or to Kanamori [14]. If G is any abelian group then $G^* = \text{Hom}(G, \mathbb{Z})$ denotes its dual group, and G is a *dual* if $G \cong D^*$ for some abelian group D.

1991 *Mathematics Subject Classification.* Primary: 13C05, 13C10, 13C13, 20K15, 20K25, 20K30; Secondary: 03E05, 03E35.

Key words and phrases. almost free modules, reflexive modules, duality theory, modules with particular monomorphism.

This work is supported by the project No. G-545-173.06/97 of the German-Israeli Foundation for Scientific Research & Development

GbSh 727 in Shelah's list of publications.

Particular dual groups are the reflexive groups D, see Bass [**1**, p. 476]. Recall that
$$\sigma = \sigma_D : D \longrightarrow D^{**} \quad (d \longrightarrow \sigma(d))$$
with $\sigma(d) \in D^{**}$ and
$$\sigma(d) : D^* \longrightarrow \mathbb{Z} \quad (\varphi \longrightarrow \varphi(d))$$
is the evaluation map and D is *reflexive* if the evaluation map σ_D is an isomorphism. Recent results about reflexive and dual abelian groups are discussed in [**7, 9, 10**].

In the third section we will show that dual groups, in particular reflexive groups may have large summands, hence can't be essentially indecomposable without any set-theoretic restrictions.

THEOREM 1.1. *If κ is a supercompact cardinal and H is a dual group of cardinality $\geq \kappa$, then there is a direct summand H' of H with $\chi \leq |H'| < \kappa$ for any cardinal $\chi < \kappa$.*

This theorem shows that generally we will encounter set theoretic restrictions for finding natural classes of reflexive groups. As CH implies Martin's axiom, our main result (Theorem 1.2) below gives a new proof of the existence of reflexive groups as in [**10**].

In order to prove a result in contrast to Theorem 1.1 we use scalar products on the Baer-Specker group P. Recall that
$$P = \mathbb{Z}^\omega$$
is the set of all elements
$$\mathbf{x} = \sum_{i \in \omega} x_i \mathbf{e}_i \text{ with } x_i \in \mathbb{Z}$$
where $\mathbf{e}_i \in P$ is defined by the Kronecker symbol and addition is defined componentwise. Throughout this paper we will adopt the convention in writing elements of P as displayed in the last formula. The Baer-Specker group P has the subgroup S of all elements $\mathbf{x}$ of finite support, that is $x_i = 0$ for almost all $i \in \omega$. The crucial subgroup for constructing reflexive groups is the $\mathbb{Z}$-adic closure $\mathbb{D}$ of S in P. This will be our target in Section 3. We will also show that the endomorphism ring of such a reflexive abelian group can be $\mathbb{Z}$ *modulo* the ideal of all endomorphisms of finite rank. We have the following

THEOREM 1.2. *(ZFC + MA) There are two subgroups H_i ($i = 1, 2$) of the Baer-Specker group P with the following properties:*
  (i) $S \subseteq H_i \subseteq_* \mathbb{D}$ *are pure.*
  (ii) H_i *is $\aleph_1$-free and slender.*
  (iii) *There is a natural bilinear form $\Phi : H_1 \times H_2 \longrightarrow \mathbb{Z}$ induced by $\Phi(\mathbf{e}_i, \mathbf{e}_j) = \delta_{i,j}$, ($i, j \in \omega$) which yields $H_1^* \cong H_2$ and $H_2^* \cong H_1$ such that H_1 and H_2 are reflexive.*
  (iv) $H_i \oplus \mathbb{Z} \not\cong H_i$ *for $i = 1, 2$.*
  (v) $\operatorname{End} H_i = \mathbb{Z} \oplus \operatorname{Fin} H_i$.

Note that $\varphi \in H_1^*$ is induced by Φ if there is $h \in H_2$ such that $\varphi = \Phi(\ , h)$. The set $\operatorname{Fin} H_i$ of all endomorphisms of H_i with finite rank image is an ideal of the endomorphism ring $\operatorname{End} H_i$ and the last statement of the theorem means that this ideal is a split extension in $\operatorname{End} H_i$.

Hence each H_i is separable and essentially indecomposable, which means any decomposition $H_i = C \oplus E$ must have a summand E or C of finite rank. New algebraic and combinatorial methods and some old techniques from earlier papers like [**11**] or [**4**] will be used to prove Theorem 1.2.

2. Reflexive groups of cardinality $\leq 2^{\aleph_0}$ under Martin's axiom

In this section we will now construct essentially indecomposable reflexive groups under Martin's axiom MA. This contrasts with the results in Section 3 concerning the existence of arbitrarily large summands of reflexive groups larger than a supercompact cardinal. As above let $P = \prod_{n \in \omega} \mathbf{e}_n \mathbb{Z}$ be the Baer-Specker group of all elements

$$P = \{\mathbf{x} = \sum_{i \in \omega} x_i \mathbf{e}_i : (x_i \in \mathbb{Z})\}.$$

Here $\mathbf{e}_i$ can be viewed as the element $\mathbf{x} = (x_{ij})_j$ with coefficients $x_{ij} = \delta_{ij}$ the Kronecker symbol. Hence

$$S = \langle \mathbf{e}_i : i \in \omega \rangle = \bigoplus_{i \in \omega} \mathbf{e}_i \mathbb{Z}$$

is a subgroup of P of all elements $\mathbf{x}$ of finite support

$$[\mathbf{x}] = \{i \in \omega : x_i \neq 0\}$$

and P/S is algebraically compact by an old result of Balcerzyk (see Fuchs [**8**]). Obviously P/S is torsion-free or, equivalently, S is pure in P. Pure subgroups $X \subseteq P$ are denoted by $X \subseteq_* P$. Moreover, let $\mathbb{D}$ be the $\mathbb{Z}$-adic closure of S in P, so $\mathbb{D}/S$ is the maximal divisible (torsion-free) subgroup of P/S which has size $2^{\aleph_0}$. If H is an abelian group, then $\operatorname{Fin} H$ denotes the ideal of all endomorphisms $\sigma \in \operatorname{End} H$ with $\operatorname{Im} \sigma$ of finite rank. The groups we want to construct will be sandwiched between S and $\mathbb{D}$.

We will use Martin's axiom for σ-centered sets, which is a (proper) consequence of the well-known Martin's axiom and equivalent to the combinatorial principle $P(2^{\aleph_0})$ (see below) as shown by Bell [**2**]. Recall that $D \subseteq \mathfrak{P}$ is *dense* in the poset $\mathfrak{P}$ if for any $p \in \mathfrak{P}$ there exists $d \in D$ such that $p \leq d$. Martin's axiom is based on posets $\mathfrak{P}$ with c.c.c. using that $p, q \in \mathfrak{P}$ are compatible if there is $r \in \mathfrak{P}$ with $\{p, q\} \leq r$. Recall that $F \subseteq \mathfrak{P}$ is *bounded* by r, say $F \leq r$ if $f \leq r$ for all $f \in F$. A set $X \subseteq \mathfrak{P}$ is *directed* if all finite subsets of X are bounded in X and X is called σ-*centered* (or σ-directed) if it is the countable union of directed subsets. Replacing c.c.c. by 'σ-centered' MA turns into *Martin's axiom for σ-centered sets:*

Let $\mathfrak{D}$ be a collection of dense subsets D of the poset $\mathfrak{P}$. If $|\mathfrak{D}| < 2^{\aleph_0}$ and $(\mathfrak{P}, \leq)$ is a σ-centered poset then there is a $\mathfrak{D}$-generic subset $G \subset \mathfrak{P}$. Hence G is directed and meets every $D \in \mathfrak{D}$, i.e. $G \cap D \neq \emptyset$.

See [**7**, p. 164] for MA with c.c.c. Note that the main result in Bell [**2**] is that Martin's axiom for σ-centered sets is equivalent to

The combinatorial principle $P(2^{\aleph_0})$: If $\mathfrak{D}$ is a collection of subsets of ω such that $|\mathfrak{D}| < 2^{\aleph_0}$ and $\bigcap F$ is infinite for every finite $F \subseteq \mathfrak{D}$, then there is an infinite $B \subseteq \omega$ such that $B \setminus D$ is finite for all $D \in \mathfrak{D}$.

Martin's axiom will help us to define a scalar product or bilinear form Φ on suitable pairs $\mathbb{H} = (H_1, H_2)$ of pure subgroups H_j of $\mathbb{D}$. We begin with

$$\Phi : S \times S \longrightarrow \mathbb{Z} \text{ with } \Phi(\mathbf{e}_i, \mathbf{e}_j) = \delta_{ij}.$$

Hence Φ is the unique integer valued, bilinear form on $S \times S$. By continuity it extends uniquely to the non-degenerate, symmetric bilinear form

$$\Phi : \mathbb{D} \times \mathbb{D} \longrightarrow \widehat{\mathbb{Z}} \text{ where } \widehat{\mathbb{Z}} \text{ is the } \mathbb{Z}\text{-adic completion of } \mathbb{Z}.$$

We keep this map fixed throughout this section and also denote restrictions to pairs of subgroups by Φ. Note that $\widehat{\mathbb{Z}}$ is the cartesian product of the additive groups of p-adic integers over all primes p and if

$$\mathbf{a} = \sum_{i \in \omega} a_i \mathbf{e}_i \in \mathbb{D} \text{ and } \mathbf{b} = \sum_{i \in \omega} b_i \mathbf{e}_i \in \mathbb{D}, \text{ then } \Phi(\mathbf{a}, \mathbf{b}) = \sum_{i \in \omega} a_i b_i$$

is well-defined and symmetry $\Phi(\mathbf{a}, \mathbf{b}) = \Phi(\mathbf{b}, \mathbf{a})$ is obvious. Now we consider pairs $\mathbb{H} = (H_1, H_2)$ such that $\Phi \restriction (H_1, H_2)$ takes only values in $\mathbb{Z}$. More precisely, let $\mathbb{H} \in \mathfrak{P}$ if and only if the following hold for $j = 1, 2$:

(i) $S \subseteq H_j \subseteq_* \mathbb{D}$
(ii) $|H_j| < 2^{\aleph_0}$
(iii) $\Phi : H_1 \times H_2 \longrightarrow \mathbb{Z}$.

We now define a partial order on $\mathfrak{P}$.

DEFINITION 2.1. *If* $\mathbb{H}, \mathbb{H}' \in \mathfrak{P}$ *then* $\mathbb{H} \subseteq \mathbb{H}'$ *if and only if* $H_1 \subseteq H_1'$ *and* $H_2 \subseteq H_2'$.

The next crucial lemma of this paper will show under MA that $\mathfrak{P}$ is a rich structure.

MAIN LEMMA 2.2. *(ZFC + MA) Let* $\mathbb{H} = (H_1, H_2) \in \mathfrak{P}, \mathbf{b} \in P \setminus \mathbb{D}$ *and* $\mathbf{b}^n \in H_1$ *for* $n \in \omega$. *Then there is* $\mathbf{a} = \sum_{i \in \omega} a_i \mathbf{e}_i \in \mathbb{D}$ *such that for* $H_1' = \langle H_1, \mathbf{a} \rangle_* \subseteq \mathbb{D}$ *and* $\mathbb{H}' = (H_1', H_2)$ *the following hold.*

(i) $\mathbb{H} \subseteq \mathbb{H}' \in \mathfrak{P}$ *and* $\Phi(\mathbf{a}, \mathbf{b}) \in \widehat{\mathbb{Z}} \setminus \mathbb{Z}$.
(ii) (a) *Either* $\sum_{i \in \omega} a_i \mathbf{b}^i \notin H_1'$
     (b) *or there is* $t \in \mathbb{Z}$ *such that* $\langle \mathbf{b}^j - t\mathbf{e}_j : j \in \omega \rangle$ *is a free direct summand of finite rank.*

Remark. By symmetry we obtain a dual result of the Main Lemma 2.2 with $\mathbf{a} \in H_2'$ and $\Phi(\mathbf{b}, \mathbf{a}) \in \widehat{\mathbb{Z}} \setminus \mathbb{Z}$ and *(ii)* accordingly. From $\mathbf{a} \in \mathbb{D}$, it follows that $\sum_{i \in \omega} a_i \mathbf{b}^i \in \widehat{P}$ is well-defined as a member of the $\mathbb{Z}$-adic completion of P.

Proof. Let $\mathbf{b} = \sum_{i \in \omega} b_i \mathbf{e}_i \in P \setminus \mathbb{D}$ and $\mathbb{H} = (H_1, H_2) \in \mathfrak{P}$ be given by the lemma. Moreover we assume that condition *(ii)(b)* of the lemma does not hold. This is to say that we must show *(ii)(a)* of the lemma. This implication will follow at the end of the proof from density of the sets $D_{\mathbf{d}tn_0}^5$ and density will be a consequence of the assumption just made.

We want to approximate $\mathbf{a} \in H_1'$ by a forcing notion $\mathfrak{F}$, a partially ordered set, used for application of MA. The elements $p \in \mathfrak{F}$ are triples

$$(M^p, A^p, n^p) \text{ with } A^p = \langle a_l^p : l < l^p \rangle, M^p = \{m_\mathbf{x}^p = \sum_{l < l^p} x_l a_l^p : \mathbf{x} = \sum_{i \in \omega} x_i \mathbf{e}_i \in u^p\}$$

subject to the following conditions

(i) u^p is a finite subset of H_2,
(ii) $l^p \in \omega$, $a_l^p, m_\mathbf{x}^p \in \mathbb{Z}$, and $n^p \in \mathbb{N}$.

We call l^p the length of the finite sequence of integers A^p and note that $n|m$ means n divides m in $\mathbb{Z}$. In order to turn $\mathfrak{F}$ into a partially ordered set let $p \leq q$ for some $p, q \in \mathfrak{F}$ if the following holds:

$$u^p \subseteq u^q, \ l^p \leq l^q, \ A^p = A^q \upharpoonright l^p,$$

$$n^p | n^q, \text{ and if } l^p \leq l < l^q \text{ then } n^p | a_l^q,$$

if $\mathbf{x} = \sum_{l \in \omega} x_l \mathbf{e}_l \in u^p$ then $m_\mathbf{x}^q = m_\mathbf{x}^p =: \sum_{l < l^p} x_l a_l^p$ or equivalently $\sum_{l^p \leq l < l^q} x_l a_l^p = 0$.

If $p, q \in \mathfrak{F}$, then let

$$p \sim q \ \Leftrightarrow \ (l^p = l^q, A^p = A^q, n^p = n^q)$$

and note that $\sim$ is an equivalence relation on $\mathfrak{F}$. If $p \in \mathfrak{F}$, then let

$$\mathfrak{F}_p = \{q \in \mathfrak{F} : q \sim p\}.$$

Surely $\mathfrak{F}$ decomposes into countably many such uncountable equivalence classes $\mathfrak{F}_p$. We claim that each of them is directed. If $q_1, q_2 \in \mathfrak{F}_p$ then $n^{q_i} = n^p, A^{p_i} = A^p, l^{p_i} = l^p$, hence $q_i = (M^{q_i}, A^p, n^p)$, and if $\mathbf{x} = \sum_{i \in \omega} x_i \mathbf{e}_i \in u^{q_1} \cap u^{q_2}$, then

$$m_\mathbf{x}^{q_1} = \sum_{i < l^{q_1}} x_i a_i^{q_1} = \sum_{i < l^p} x_i a_i^p = m_\mathbf{x}^{q_2}.$$

If we define $q' \in \mathfrak{F}$ by $u^{q'} = u^{q_1} \cup u^{q_2}, A^{q'} = A^p, l^{q'} = l^p, n^{q'} = n^p$, then

$$M^{q'} = \{m_\mathbf{x}^{q'} = \sum_{i < l^p} x_i a_i^p : \mathbf{x} \in u^{q'}\} = M^{q_1} \cup M^{q_2},$$

hence $q' = (M^{q'}, A^p, n^p)$ is a member of $\mathfrak{F}$ and $q_1, q_2 \leq q'$. The claim is shown and by definition

(2.1) $\qquad\qquad (\mathfrak{F}, \leq)$ is a σ-centered poset,

as required for applications of MA for σ-centered sets.

In order to apply MA effectively we must define dense subsets of $\mathfrak{F}$ which describe 'local properties' of the desired $\mathbf{a} \in \mathbb{D}$. If $\mathbf{x} = \sum_{i \in \omega} x_i \mathbf{e}_i \in H_2, m \in \mathbb{N}, l_0 \in \omega$, then let

$$D_\mathbf{x}^1 = \{p \in \mathfrak{F} : \mathbf{x} \in u^p\}, \ D_m^2 = \{p \in \mathfrak{F} : m | n^p\},$$
$$D_{l_0}^3 = \{p \in \mathfrak{F} : l_0 \leq l^p\}, \ D_m^4 = \{p \in \mathfrak{F} : \sum_{l < l^p} b_l a_l^p \not\equiv m \mod n^p\}$$

and for $\mathbf{d} \in H_1, t \in \mathbb{Z}$ and $n_0 \in \mathbb{N}$, let

$$D_{\mathbf{d}tn_0}^5 = \{p \in \mathfrak{F} : \exists m \in \mathbb{N} \ (m | n^p, \ n_0 \sum_{i < l^p} a_i^p \mathbf{b}^i - t \sum_{i < l^p} a_i^p \mathbf{e}_i - \mathbf{d} \not\equiv 0 \mod m\mathbb{D})\}.$$

First note that we defined $< 2^{\aleph_0}$ subsets of $\mathfrak{F}$ as required for MA. Next we want to show that all these sets are dense in $\mathfrak{F}$. The first three cases are easy while the remaining two cases need work. For $D_\mathbf{x}^1$ with $\mathbf{x} = \sum_{i \in \omega} x_i \mathbf{e}_i$ we take any $p \in \mathfrak{F}$ and define q like p just by enlarging $u^q = u^p \cup \{\mathbf{x}\}$, let $m_\mathbf{x}^q = \sum_{l < l^p} x_l a_l^p$ and enlarge $M^q = \{m_\mathbf{y}^p : \mathbf{y} = \sum_{i \in \omega} y_i \mathbf{e}_i \in u^p\} \cup \{m_\mathbf{x}^q\}$ as well, hence $p \leq q$ and $D_\mathbf{x}^1$ is dense in $\mathfrak{F}$. Similarly take any $p \leq q \in \mathfrak{F}$ with $m | n^q$, hence D_m^2 is dense. For $D_{l_0}^3$ replace any A^p by $A^q = (A^p)^\frown(0,\ldots,0)$ with $(0,\ldots,0)$ a vector of l_0 zeros and let $u^q = u^p$, $l^q = l^p + l_0$, $n^q = n^p$. In the fourth case we first notice that $\mathbf{b} = \sum_{i \in \omega} b_i \mathbf{e}_i \in P \setminus \mathbb{D}$ by hypothesis, hence there is $s' \in \mathbb{N}$ such that the set

$$W = \{k \in \omega : b_k \in \mathbb{Z} \setminus s'\mathbb{Z}\} \text{ is infinite.}$$

Suppose $p \in \mathfrak{F}$ contradicts the density of D_m^4 for some $m \in \mathbb{N}$, hence

(2.2) \qquad there is no $q \in D_m^4$ with $p \leq q$.

We write

$$u^p = \{\mathbf{a_1}, \ldots \mathbf{a_{k-1}}\} \subseteq H_2 \text{ and let } \mathbf{a_j} = \sum_{i \in \omega} a_{ji} \mathbf{e}_i.$$

Also consider the $k \times \omega$-matrix ($s \in \omega$)

$$(\mathbf{G}) = \begin{pmatrix} a_{11} & a_{12} & \cdots & a_{1s} & \cdots \\ a_{21} & a_{22} & \cdots & a_{2s} & \cdots \\ \vdots & & & & \\ a_{k-1,1} & a_{k-1,2} & \cdots & a_{k-1,s} & \cdots \\ b_1 & b_2 & \cdots & b_s & \cdots \end{pmatrix}$$

as well as the $(k-1) \times \omega$-matrix

$$(\mathbf{H}) = \begin{pmatrix} a_{11} & a_{12} & \cdots & a_{1s} & \cdots \\ a_{21} & a_{22} & \cdots & a_{2s} & \cdots \\ \vdots & & & & \\ a_{k-1,1} & a_{k-1,2} & \cdots & a_{k-1,s} & \cdots \end{pmatrix}$$

which is obtained by deleting the last row of b_s's of the matrix $(\mathbf{G})$. We pick finite subsets w of $[l^p, \omega)$ and consider the column vectors g_l^p ($l \in w$) of the first matrix $(\mathbf{G})$ and h_l^p ($l \in w$) of the second matrix $(\mathbf{H})$ accordingly and claim that for all finite

(2.3) \qquad $w \subseteq [l^p, \omega)$ and $d_l \in \mathbb{Q}$ $[\sum_{l \in w} d_l h_l^p = 0 \Leftrightarrow \sum_{l \in w} d_l g_l^p = 0]$.

The proof "$\Leftarrow$" is trivial. For "$\Rightarrow$", suppose for contradiction that

$$\sum_{l \in w} d_l h_l^p = 0 \text{ but } \sum_{l \in w} d_l g_l^p \neq 0$$

for some finite $w \subseteq [l^p, \omega)$ and $d_l \in \mathbb{Q}$. Hence

(2.4) \qquad $\sum_{l \in w} d_l a_{jl} = 0$ for $j < k$ and $v = \sum_{l \in w} d_l b_l \neq 0$.

Multiplying this homogeneous system of equations and the inequality by a large enough natural number we may assume that

$$d_l \in n^p \mathbb{Z} \text{ for all } l \in w.$$

We now want to define $q > p$ with $q \in D_m^4$ and distinguish two cases. If $\sum_{l < l^p} b_l a_l^p \not\equiv m$ then choose n^q large enough such that $n^p | n^q$ and $\sum_{l < l^p} b_l a^p - m \not\equiv 0 \mod n^q$ and put $u^p = u^q, M^p = M^q, A^p = A^q$. Then $p < q$ and $\sum_{l < l^q} b_l a_l^q \not\equiv m \mod n^q$

hence $q \in D_m^4$ is a contradiction, see (2.2). If $\sum_{l<l^p} b_l a_l^p = m$, then choose $l^q > \sup(w \cup \{l^p\})$ and define q such that

$$a_l^q(t) = \begin{cases} a_l^p(t) & \text{if } t \in [0, l^p) \\ d_l & \text{if } l \in w \\ 0 & \text{if } l \in [l^p, \omega) \setminus w. \end{cases}$$

Set $u^q = u^p \subseteq H_2$ and using (2.4) let n^q be large enough such that $n^p | n^q$ but $v \not\equiv 0 \mod n^q$. It follows $p < q$ and

$$\sum_{l<l^q} b_l a_l^q = \sum_{l<l^p} b_l a_l^q + \sum_{l \in w} b_l d_l = m + v.$$

Hence $q \in D_m^4$ is another contradiction, see (2.2). The linear dependence (2.3) between the h_l^p's and g_l^p's is shown. Now we want to use (2.3) to derive a final contradiction for (2.2). For each finite $w \subseteq \omega$ we have a $\mathbb{Q}$-vector space $V_w = \langle h_l^p : l \in w \rangle$ of finite dimension $\leq k$. Hence there is an $r \in \omega$ and a finite $w^* \subseteq [l^p, \omega)$ such that h_l^p ($l \in w^*$) is a maximal independent set and V_{w^*} has maximal dimension $|w^*| = r \leq k$. If $w^* \subseteq w \subseteq [l^p, \omega)$ for some finite w, then the sub-matrix $(\mathbf{H}_w) = (h_l^p, l \in w)$ of $(\mathbf{H})$ has finite column rank r, hence row rank r as well and there is a subset $z \subset \{1, \ldots, k-1\}$ of size r such that

$$\{\mathbf{a_j} \upharpoonright w : j \in z\} \text{ is maximal independent.}$$

By (2.3) $\mathbf{b} \upharpoonright w$ is a linear combination of the $\{\mathbf{a_j} \upharpoonright w : j \in z\}$ and there are *unique* elements $c_l \in \mathbb{Q}$, $l \in z$ such that $\mathbf{b} \upharpoonright w = \sum_{l \in z} c_l \mathbf{a_l} \upharpoonright w$. If we increase w we have the same coefficients by maximal independence. Hence

(2.5) $$\mathbf{b} \upharpoonright [l^p, \omega) = \sum_{l \in z} c_l \mathbf{a_l} \upharpoonright [l^p, \omega).$$

We can choose $m' \in \mathbb{N}$ large enough such that $m' c_l \in s'\mathbb{Z}$ for all $l \in z$. If $t \in W$ is large enough, then $m' | a_{lt}$ for all $l \in z$. Using (2.5) we get

$$b_t = \sum_{l \in z} c_l a_{lt} \in s'\mathbb{Z}$$

contradicting W. Hence D_m^4 is dense in $\mathfrak{F}$.

In order to show density of the last collection of subsets, suppose there are $\mathbf{d} \in H_1, t \in \mathbb{Z}$ and $n_0 \in \mathbb{N}$ such that

(2.6) $$D_{\mathbf{d}tn_0}^5 \text{ is not dense in } \mathfrak{F}.$$

Hence there is $p \in \mathfrak{F}$ such that

(2.7) $$\text{no } q \in D_{\mathbf{d}tn_0}^5 \text{ satisfies } p \leq q.$$

Let $u^p = \{\mathbf{c}^i = \sum_{j \in \omega} c_j^i \mathbf{e}_j : i < k\}$ and $l^p < l < \omega$. We want to consider extensions $p \leq q$ with $l^q = l$ and hence let

$$F_l = \{(y_{l^p}, \ldots, y_{l-1}) \in \mathbb{Z}^{l - l_p} : \sum_{j=l^p}^{l-1} c_j^i y_j = 0, i < k\}$$

which is a non-trivial subgroup of the free group $\mathbb{Z}^{l - l_p}$ for any large enough l. Also let

$$s(y_{l^p}, \ldots, y_{l-1}) = n_0 (\sum_{i<l^p} a_i^p \mathbf{b}^i + \sum_{i=l^p}^{l-1} y_i \mathbf{b}^i) - t(\sum_{i<l^p} a_i^p \mathbf{e}_i + \sum_{i=l^p}^{l-1} y_i \mathbf{e}_i) - \mathbf{d}.$$

We claim that

(2.8) $\quad (y_{l^p}, \ldots, y_{l-1}) \in F_l \Rightarrow s(y_{l^p}, \ldots, y_{l-1}) = 0$ holds in $\mathbb{D}$.

If $s(y_{l^p}, \ldots, y_{l-1}) \neq 0$ for some $(y_{l^p}, \ldots, y_{l-1}) \in F_l$, then there is some $m \in \mathbb{N}$ such that

(2.9) $\quad s(y_{l^p}, \ldots, y_{l-1}) \not\equiv 0 \bmod m\mathbb{D}$.

We now define some $q \in \mathfrak{F}$ taking

$$l^q = l, n^q = n^p \cdot m, u^q = u^p, M^q = \{m_{\mathbf{x}}^q = \sum_{i<l^q} x_i a_i^q : \mathbf{x} \in u^q\}$$

where

$$a_i^q = \begin{cases} a_i^p & \text{if } i < l^p \\ y_i & \text{if } l^p \leq i < l. \end{cases}$$

Clearly $q \in \mathfrak{F}$ and also $q \in D_{\mathbf{d}tn_0}$ from (2.9), hence $p \not\leq q$ from (2.7). On the other hand $\sum_{j=l^p}^{l-1} c_j^i a_j^q = 0$ from F_l and definition of a_i^q would imply $p \leq q$, a contradiction which proves the claim (2.8).

If we let

$$\mathbf{s}^i = \sum_{j \in \omega} s_j^i \mathbf{e}_j = n_0 \mathbf{b}^i - t\mathbf{e}_i \in \mathbb{D} \quad (l^p \leq i < \omega),$$

then the implication of (2.8) can be written as

$$\sum_{i=l^p}^{l-1} y_i \mathbf{s}^i = \mathbf{d} + t \sum_{i<l^p} a_i \mathbf{e}_i - n_0 \sum_{i<l^p} a_i^p \mathbf{b}^i.$$

From $(0, \ldots, 0) \in F_l$ follows

(2.10) $\quad n_0 \sum_{i<l^p} a_i^p \mathbf{b}^i = \mathbf{d} + t \sum_{i<l^p} a_i \mathbf{e}_i$

and from $(y_{l^p}, \ldots, y_{l-1}) \in F_l$ also follows

(2.11) $\quad \sum_{i=l^p}^{l-1} y_i \mathbf{s}^i = 0.$

If we view $\mathbf{s}^i = \sum_{j \in \omega} s_j^i \mathbf{e}_j$ as an infinite row vector ($l^p \leq i < l$), then from the matrix

$$\begin{pmatrix} s_0^{l^p} & s_1^{l^p} & \cdots & s_k^{l^p} & \cdots \\ s_0^{l^p+1} & s_1^{l^p+1} & \cdots & s_k^{l^p+1} & \cdots \\ \vdots & & & & \\ s_0^{l-1} & s_1^{l-1} & \cdots & s_k^{l-1} & \cdots \end{pmatrix}$$

we have finite column vectors $\mathbf{s}_n = (s_n^i : l^p \leq i < l)$ for any $n \in \omega$. Let $\mathbf{c}^i \restriction [l^p, l)$ be the restriction of $\mathbf{c}^i$ viewed as an infinite column vector restricted to the coordinates j such that $l^p \leq j < l$, then

$$\langle \mathbf{c}^i \restriction [l^p, l) : i < k \rangle$$

denotes the vector space over $\mathbb{Q}$ generated by these finite column vectors. We claim that

$$\mathbf{s}_n \in \langle \mathbf{c}^i \restriction [l^p, l) : i < k \rangle \quad \text{for all } n \in \omega.$$

Naturally $F_l \subseteq \mathbb{Z}^{l-l^p} \subseteq \mathbb{Q}^{l-l^p}$. If $\overline{F}_l = \langle F_l \rangle$ denotes the subspace of $\mathbb{Q}^{l-l^p}$ generated by F_l, then $\overline{F}_l = \langle \mathbf{c}^i \upharpoonright [l^p, l) : i < k \rangle^\perp$ where orthogonality is defined naturally by

$$U^\perp = \{ x \in \mathbb{Q}^{l-l^p} : \; x \cdot u = 0 \;\; \forall u \in U \}$$

for $U \subseteq \mathbb{Q}^{l-l^p}$ and the obvious scalar product $x \cdot u = \sum_{i \leq l-l^p} x_i u_i$. From (2.11) follows

$$\overline{F}_l = \langle \mathbf{s}_n : n \in \omega \rangle^\perp.$$

Using $\perp$ again, we have

$$\langle \mathbf{s}_n : n \in \omega \rangle^{\perp\perp} \subseteq \langle \mathbf{c}^i \upharpoonright [l^p, l) : i < k \rangle^{\perp\perp}$$

which is

$$\langle \mathbf{s}_n : n \in \omega \rangle \subseteq \langle \mathbf{c}^i \upharpoonright [l^p, l) : i < k \rangle$$

as $\dim \mathbb{Q}^{l-l^p}$ is finite. This shows the claim.

Now let l be large enough such that $\langle \mathbf{c}^i \upharpoonright [l^p, l) : i < k \rangle$ has maximal dimension $k' \leq k$ and let $\mathbf{c}^i \upharpoonright [l^p, l)$ ($i < k'$) be a basis of this vector space. We now can write

$$\mathbf{s}_n = \sum_{i<k'} r_i^{nl} \mathbf{c}^i \upharpoonright [l^p, l)$$

with *unique* coefficients $r_i^{nl} \in \mathbb{Q}$. By uniqueness these coefficients are independent of l for any larger l, say that $r_i^{nl} = r_i^n$. In the system of equations

$$\mathbf{s}_n = \sum_{i<k'} r_i^n \mathbf{c}^i \upharpoonright [l^p, l), \; (l^p \leq l < \omega, n \in \omega)$$

we can also eliminate l and get

$$\mathbf{s}_n = \sum_{i<k'} r_i^n \mathbf{c}^i \upharpoonright [l^p, \omega), \; n \in \omega.$$

From $\mathbf{s}^j$ and $\mathbf{b}^j = \sum_{n \in \omega} b_n^j \mathbf{e}_n$ we have that $s_n^j = n_0 b_n^j - t\delta_{jn} = \sum_{i<k'} r_i^n c_n^i$ for any $n \geq l^p$, hence $(n_0 \mathbf{b}^j - t\mathbf{e}_j) \upharpoonright [l^p, \omega) \in \langle \mathbf{c}^i \upharpoonright [l^p, \omega) : i < k' \rangle$ and

$$U = \langle n_0 \mathbf{b}^j - t\mathbf{e}_j : \; j \in \omega \rangle_* \subseteq \mathbb{D}$$

has finite rank. Hence U is a free direct summand of $\mathbb{D}$, see Fuchs [8]. If n_0 does not divide t, then *modulo* $n_0 \mathbb{D}$ the image of U is $\langle t\mathbf{e}_j + n_0 \mathbb{D} : \; j \in \omega \rangle_*$ and has infinite rank, which is impossible. Hence $n_0 | t$ and we rename tn_0^{-1} by t. Using purity, we get that $U = \langle \mathbf{b}^j - t\mathbf{e}_j : \; j \in \omega \rangle_*$ is a free direct summand of $\mathbb{D}$ which contradicts our assumption that condition $(ii)(b)$ does not hold. Hence $D_{\mathbf{d}tn_0}$ is dense in $\mathfrak{F}$ indeed, see (2.6).

We are ready to apply Martin's axiom. There is a generic set $\mathbb{G} \subseteq \mathfrak{F}$ which meets the dense subsets of $\mathfrak{F}$ just constructed. We define $\mathbf{a} = \sum_{i \in \omega} a_i \mathbf{e}_i$ such that $a_i = a_i^p$ for any $p \in \mathbb{G}$ with $i < l^p$. Here we applied $D_{l_0}^3$ and note that $\mathbb{G}$ is directed, hence $\mathbf{a}$ is well-defined. Also $\mathbf{a} \in \mathbb{D}$ by D_m^2. Let $H_1' = \langle H_1, \mathbf{a} \rangle_* \subseteq \mathbb{D}$ be the *pure* subgroup of $\mathbb{D}$ generated by $H_1'' = H_1 + \mathbb{Z}\mathbf{a}$ and $\mathbb{H}' = (H_1', H_2)$. Then clearly $\mathbb{H} \subseteq \mathbb{H}'$ and we claim that $\mathbb{H}' \in \mathfrak{P}$. It is enough to show (iii) for $\mathfrak{P}$. If $\mathbf{c} \in H_1''$ then $\mathbf{c} = k\mathbf{a} + \mathbf{e}$ for some $k \in \mathbb{N}, \mathbf{e} \in H_1$. If $\mathbf{y} \in H_2$, then consider $\Phi(\mathbf{c}, \mathbf{y}) = k\Phi(\mathbf{a}, \mathbf{y}) + \Phi(\mathbf{e}, \mathbf{y})$. From density of $D_{\mathbf{y}}^1$ and $p \in D_{\mathbf{y}}^1 \cap \mathbb{G}$ and the choice of $\mathbf{a}$ follows $\Phi(\mathbf{a}, \mathbf{y}) = m_{\mathbf{y}}^p \in \mathbb{Z}$ and therefore $\Phi(\mathbf{c}, \mathbf{y}) \in \mathbb{Z}$. The map Φ extends to $H_1'' \times H_2 \longrightarrow \mathbb{Z}$. If $\mathbf{x} \in H_1'$

then $t\mathbf{x} = \mathbf{h} \in H_1''$ for some $t \in \mathbb{N}$ and if $\mathbf{x} = \sum_{i \in \omega} x_i \mathbf{e}_i$, $\mathbf{h} = \sum_{i \in \omega} h_i \mathbf{e}_i$ then $t\mathbf{x} = \sum_{i \in \omega} tx_i \mathbf{e}_i = \sum_{i \in \omega} h_i \mathbf{e}_i$ and $h_i = tx_i$ for all $i \in \omega$. Hence

$$\Phi(\mathbf{h}, \mathbf{y}) = \Phi(t\mathbf{x}, \mathbf{y}) = \sum_{i \in \omega} tx_i y_i = t(\sum_{i \in \omega} x_i y_i) = t\Phi(\mathbf{x}, \mathbf{y}) \in t\widehat{\mathbb{Z}} \cap \mathbb{Z}$$

and by purity of $\mathbb{Z} \subseteq_* \widehat{\mathbb{Z}}$ also $t\Phi(\mathbf{x}, \mathbf{y}) \in t\mathbb{Z}$ and by torsion-freeness $\Phi(\mathbf{x}, \mathbf{y}) \in \mathbb{Z}$. We have seen that $\mathbb{H}' \in \mathfrak{P}$. Next we claim that

(2.12) by definition of $\mathbf{a}$ and $\mathbf{b}$ we have $z = \Phi(\mathbf{a}, \mathbf{b}) = \sum_{i \in \omega} b_i a_i \in \widehat{\mathbb{Z}} \setminus \mathbb{Z}$.

Note that $a_i \longrightarrow 0$ in the $\mathbb{Z}$-adic topology, hence $b_i a_i \longrightarrow 0$ and $z \in \widehat{\mathbb{Z}}$ is well-defined. If $z \in \mathbb{Z}$ and $n \in \mathbb{N}$ then $\sum_{i<k} b_i a_i \equiv z \mod n$ for any large enough k, which contradicts $D_{|z|}^4$.

Finally we show that $\sum_{i \in \omega} a_i \mathbf{b}^i \notin H_1'$. Otherwise there are $t, n \in \mathbb{N}$ and $\mathbf{d} \in H_1$ such that

(2.13) $$n \sum_{i \in \omega} a_i \mathbf{b}^i - t\mathbf{a} - \mathbf{d} = 0.$$

Let $p \in \mathbb{G} \cap D_{\mathbf{d}tn}^5$ from density of $D_{\mathbf{d}tn}^5$ and choose m from the definition of $D_{\mathbf{d}tn}^5$. Hence

$$n \sum_{i<l^p} a_i^p \mathbf{b}^i - t \sum_{i<l^p} a_i^p \mathbf{e}_i - \mathbf{d} \in \mathbb{D} \setminus m\mathbb{D}.$$

On the other hand $a_i^p = a_i$ for all $i < l^p$ from $p \in \mathbb{G}$ and $m | n^p$ by $p \in D_{\mathbf{d}tn}^5$. The set $\mathbb{G}$ is directed, hence $m | a_i$ for all $i \geq l^p$. So $n \sum_{i \geq l^p} a_i \mathbf{b}^i \in m\mathbb{D}$ as well as $t \sum_{i \geq l^p} a^i \mathbf{e}_i \in m\mathbb{D}$. The last displayed expression becomes $n \sum_{i \in \omega} a_i \mathbf{b}^i - t\mathbf{a} - \mathbf{d} \in \mathbb{D} \setminus m\mathbb{D}$ which contradicts (2.13). The Main Lemma 2.2 is shown. $\square$

From the proof of the Main Lemma 2.2 we have an immediate

COROLLARY 2.3. *If $\mathbb{H} = (H_1, H_2) \in \mathfrak{P}$, $\mathbf{a} \in \mathbb{D}$ with $\Phi(\mathbf{a}, \mathbf{y}) \in \mathbb{Z}$ for all $\mathbf{y} \in H_2$ and $\mathbb{H}_1' = \langle H_1, \mathbf{a} \rangle_* \subseteq \mathbb{D}$ then $(H_1', H_2) \in \mathfrak{P}$, in particular $\Phi : H_1' \times H_2 \longrightarrow \mathbb{Z}$.*

In order to show Theorem 1.2 we want to use an *ad hoc* and preliminary definition. Here we also use that Φ is symmetric.

DEFINITION 2.4. *A pair $\mathbb{H} = (H_1, H_2)$ of pure subgroups of $\mathbb{D}$ is a full pair if the following holds.*
  (i) *There is an increasing continuous chain $\mathbb{H}_\alpha = (H_{\alpha 1}, H_{\alpha 2}) \in \mathfrak{P}$ with $\alpha \in 2^{\aleph_0}$ whose union is (H_1, H_2).*
  (ii) *If $\mathbf{b} \in P \setminus \mathbb{D}$ and $d \in \{1, 2\}$, there is $\mathbf{a} \in H_d$ such that $\Phi(\mathbf{a}, \mathbf{b}) \in \widehat{\mathbb{Z}} \setminus \mathbb{Z}$.*
  (iii) *If $\mathbf{b} \in \mathbb{D}$, then for all $d \in \{1, 2\}$ either $\mathbf{b} \in H_d$ or for some $\mathbf{a} \in H_{3-d}$ we have $\Phi(\mathbf{a}, \mathbf{b}) \in \widehat{\mathbb{Z}} \setminus \mathbb{Z}$.*
  (iv) *If $d \in \{1, 2\}$ and $\mathbf{b}^n \in H_d$, $(n \in \omega)$, there is $\mathbf{a} = \sum_{i \in \omega} a_i \mathbf{e}_i \in H_d$ such that*
    (a) *either $\sum a_i \mathbf{b}^i \notin H_d$*
    (b) *or there is $t \in \mathbb{Z}$ such that $\langle \mathbf{b}^j - t\mathbf{e}_j : j \in \omega \rangle$ is a free direct summand of finite rank.*

Remark. As in the Main Lemma 2.2, the element $\sum a_i \mathbf{b}^i$ is a well-defined member of the $\mathbb{Z}$-adic closure $\widehat{P}$ of P.

LEMMA 2.5. *(ZFC + MA) There is a full pair $\mathbb{H} = (H_1, H_2)$.*

Proof. Enumerate $P \setminus \mathbb{D} = \{\mathbf{b}_\alpha : \alpha \in 2^{\aleph_0}\}$, $\mathbb{D} = \{\mathbf{c}_\alpha : \alpha \in 2^{\aleph_0}\}$ and $\mathbb{D}^\omega = \{(\mathbf{b}_\alpha^n)_{n \in \omega} : \alpha \in 2^{\aleph_0}\}$ with $2^{\aleph_0}$ repetitions such that any element appears $2^{\aleph_0}$ times. We want to construct the $\mathfrak{P}$-chain inductively and let $(H_{01}, H_{02}) = (S, S)$. By continuity we only have to define $\mathbb{H}_{\alpha+1}$. Alternatively we switch between 1 and 2, say we are in case $H_{\alpha 1}$ and consider $\mathbf{b}_\alpha$, $\mathbf{c}_\alpha$ and $(\mathbf{b}_\alpha^n)_{n \in \omega}$. By the Main Lemma 2.2 there is $\mathbf{a}_\alpha \in \mathbb{D}$ such that $(H'_{(\alpha+1)1}, H_{\alpha 2}) \in \mathfrak{P}$ where $H'_{(\alpha+1)1} = \langle H_{\alpha 1}, \mathbf{a}_\alpha \rangle_* \subseteq \mathbb{D}$ and $\Phi(\mathbf{a}_\alpha, \mathbf{b}_\alpha) \in \widehat{\mathbb{Z}} \setminus \mathbb{Z}$. Moreover $(\mathbf{b}_\alpha^n)_{n \in \omega}$ satisfies condition (ii) of the Main Lemma 2.2 for $\mathbf{b}^n = \mathbf{b}_\alpha^n$. If $\mathbf{c}_\alpha \in H'_{(\alpha+1)1}$, then let $H'_{(\alpha+1)2} = H_{\alpha 2}$ and if $\mathbf{c}_\alpha \notin H'_{(\alpha+1)1}$, then by Main Lemma 2.2 there is $\mathbf{d}_\alpha \in \mathbb{D}$ such that $\Phi(c_\alpha, d_\alpha) \in \widehat{\mathbb{Z}} \setminus \mathbb{Z}$. We let $H'_{(\alpha+1)2} = \langle H_{\alpha 2}, \mathbf{d}_\alpha \rangle_* \subseteq \mathbb{D}$ and treat $(H'_{(\alpha+1)1}, H'_{(\alpha+1)2})$ by a dual argument (case 2). Hence we get $\mathbb{H}_{\alpha+1} = (H_{(\alpha+1)1}, H_{(\alpha+1)2}) \in \mathfrak{P}$. This finishes the construction of $\mathbb{H}$ and Definition 2.4 is easily checked. $\square$

LEMMA 2.6. *If $\varphi \in H_1^*$ for a full pair $\mathbb{H} = (H_1, H_2)$, then there is $\mathbf{b} \in H_2$ with $\varphi = \Phi(\ , \mathbf{b})$*

Remark A similar result holds for $\varphi \in H_2^*$.

Proof. Let $b_j = \mathbf{e}_j \varphi \in \mathbb{Z}$ for all $j \in \omega$, and set $\mathbf{b} = \sum_{j \in \omega} b_j \mathbf{e}_j \in P$. If $\mathbf{a} \in H_1 \subseteq \mathbb{D}$, then write $\mathbf{a} = \sum_{j \in \omega} a_j \mathbf{e}_j$ and by continuity $\mathbf{a}\varphi = (\sum_{j \in \omega} a_j \mathbf{e}_j)\varphi = \sum_{j \in \omega} a_j(\mathbf{e}_j \varphi) = \sum_{j \in \omega} a_j b_j = \Phi(\mathbf{a}, \mathbf{b})$. Hence $\varphi = \Phi(\ , \mathbf{b})$. If $\mathbf{b} \in P \setminus \mathbb{D}$, then by Definition 2.4 there is $\mathbf{x} \in H_1$ with $\mathbf{x}\varphi = \Phi(\mathbf{b}, \mathbf{x}) \in \widehat{\mathbb{Z}} \setminus \mathbb{Z}$ contradicting $\varphi \in H_1^*$, hence $\mathbf{b} \in \mathbb{D}$. Similarly by Definition 2.4 (iii) we have $\mathbf{b} \in H_2$ and the lemma follows. $\square$

The pair $\mathbb{H} = (H_1, H_2)$ in Lemma 2.6 satisfies conditions (i) and (iii) of Theorem 1.2. Reflexivity follows easily as in [9] or [10] because the dual maps are induced by scalar multiplication. As a subgroup of P, each H_i is $\aleph_1$-free (see Fuchs [8]). Slenderness can easily be checked and is left to the reader, hence (ii) of Theorem 1.2 follows. Condition (iv) can be derived using the arguments in [9] or [10]. The final condition (v) will follow immediately from our next Lemma 2.7.

LEMMA 2.7. *If $\mathbb{H} = (H_1, H_2)$ is a full pair and $\sigma \in \operatorname{End} H_1$, then there is $s \in \mathbb{Z}$ such that $\sigma - s1 \in \operatorname{Fin} H_1$, where $\operatorname{Fin} H_1$ is the ideal of $\operatorname{End} H_1$ of all endomorphisms of finite rank.*

Proof. If $\mathbf{e}_j \sigma = \mathbf{b}^j$, $j \in \omega$, then using that $\mathbb{H}$ is a full pair, we find $\mathbf{a} = \sum_{i \in \omega} a_i \mathbf{e}_i \in H_1$ such that Definition 2.4(iv) holds. By continuity,

$$\mathbf{b} = \mathbf{a}\sigma = \left(\sum_{i \in \omega} a_i \mathbf{e}_i\right)\sigma = \sum_{i \in \omega} a_n \mathbf{b}^i \in H_1$$

which shows that we are in case (b) of Definition 2.4(iv). The subgroup $U = \langle \mathbf{b}^j - t\mathbf{e}_j : j \in \omega \rangle$ is a free direct summand of finite rank of $\mathbb{D}$. However the image of $S = \bigoplus_{i \in \omega} \mathbf{e}_i \mathbb{Z}$ under $\sigma - t\,\mathrm{id}$ is in U, hence $S(\sigma - t\,\mathrm{id})$ has finite rank, and by continuity the same holds for $H_1(\sigma - t\,\mathrm{id})$, this is to say that $\sigma - t1 \in \operatorname{Fin} H_1$. $\square$

3. Large reflexive groups

Let κ be a fixed supercompact cardinal. Then there is a κ-complete, fine ultrafilter U over κ such that the constant function

$$j : V \longrightarrow M = \mathrm{Ult}\,(V, U)\ (x \longrightarrow j(x))\ (j(x)_\alpha = x \text{ for all } \alpha \in \kappa)$$

is an elementary embedding of the universe V into the ultrapower M; for details see Kanamori [14, pp. 471, 298–306, 37–56]. If ρ is a cardinal, then

$$\mathfrak{H}(\rho) = \{x \in V : |TC(x)| < \rho\}$$

is the set of all sets in V hereditarily $< \rho$ where $TC(x)$ denotes the transitive closure of the set x.

THEOREM 3.1. *If κ is a supercompact cardinal and H is a dual group of cardinality $\geq \kappa$, then for any $\chi < \kappa$ there is a direct summand H' of H with $\chi \leq |H'| < \kappa$.*

The following corollary is immediate.

COROLLARY 3.2. *Every reflexive group of cardinality $\geq \kappa$, with κ supercompact, has arbitrarily large summands $< \kappa$.*

Proof of Theorem 3.1: Let $H = G^* = \mathrm{Hom}\,(G, \mathbb{Z})$ be as in the theorem. If $|G| = \lambda_1, |H| = \lambda_2$, then let $\lambda > 2^{\lambda_1 + \lambda_2}$ and assume $G = \lambda_1, H = \lambda_2$ as sets and $\chi < \kappa$. If $\mathfrak{P} = \mathfrak{P}_\kappa(\mathfrak{H}(\lambda))$ is the poset of all subsets of $\mathfrak{H}(\lambda)$ of cardinality $< \kappa$, then by the above there is a κ-complete (normal and fine) ultrafilter D on $\mathfrak{P}$ with elementary embedding

$$(\mathfrak{H}(\lambda), \epsilon) \prec M := \mathrm{Ult}\,(\mathfrak{P}, D).$$

From $H = G^*$ each $h \in H$ gives rise to a homomorphism

$$\Phi(h,\ ) : G \longrightarrow \mathbb{Z}$$

and $\Phi : H \oplus G \longrightarrow \mathbb{Z}$ is a bilinear form. Moreover

$$\Phi(h,\ ) = 0 \Rightarrow h = 0,$$

hence Φ is not degenerate. Let $\mathfrak{C}$ be the set of all $N \in \mathfrak{P}$ subject to the conditions
   (i) $G, H, \Phi \in N$
   (ii) $\chi + 1 \subseteq N$
   (iii) N is an elementary submodel of $(\mathfrak{H}(\lambda), \epsilon)$.
   (iv) If $\tau = \mathrm{otp}\,(N \cap \lambda)$ is the order type of $N \cap \lambda$, then (N, ϵ) is isomorphic to $(\mathfrak{H}(\tau), \epsilon))$, say by an isomorphism j_N.

By supercompactness $\mathfrak{C} \in D$, hence $\mathfrak{C} \neq \emptyset$ and we can choose $N \in \mathfrak{C}$. By Łoś's theorem ([14, p. 47, Theorem 5.2] the desired properties of $\mathfrak{H}(\lambda)$ carry over to N. Now define

$$H' = H \cap N \text{ and } G' = G \cap N.$$

From $\chi + 1 \subseteq N \in P$ and $\chi + 1 \subseteq \lambda_1 = G, \lambda + 1 \subseteq \lambda_2 = H$ follows $\chi + 1 \subseteq H'$ and $\chi + 1 \subseteq G'$, hence

$$\chi \leq |H'| < \kappa \text{ and } \chi \leq |G'| < \kappa$$

and by (*iii*)

(3.1) $\qquad\qquad H' \subseteq H,\ G' \subseteq G$ are subgroups.

Similarly, if $\Phi' = \Phi \restriction H' \oplus G'$, then
$$\Phi' : H' \oplus G' \longrightarrow \mathbb{Z}$$
and from (iii) and Φ' we have
$$H' = G'^*.$$

We are ready to use an old trick from functional analysis to show that H' is also a summand of H. Let
$$G'^\perp = \{h \in H : \Phi(h, G') = 0\} \text{ where } \Phi(h, G') = \{\Phi(h, g) : g \in G'\}.$$

Clearly $G'^\perp \subseteq H$, and consider any $h \in H' \cap G'^\perp$. We have $\Phi(h, G') = 0$ and from $h \in H'$ follows that in the submodel N the following holds
$$N \models (\forall x \in G'^N \longrightarrow \Phi(h, x) = 0).$$

By (iii) we also have
$$(\mathfrak{H}(\lambda), \epsilon) \models (\forall x \in G \longrightarrow \Phi(h, x) = 0),$$
hence $\Phi(h, \ ) = 0$ and $h = 0$ because Φ is not degenerate. We conclude
$$H' \cap G'^\perp = 0, G'^\perp \subseteq H.$$

In order to show

(3.2) $$H' + G'^\perp = H$$

we consider any $h \in H = G^*$ and let $\phi = \Phi(h, \ ) \restriction G'$ which belongs to G'^*. From (3.1) we find $h' \in H'$ such that $\Phi(h', \ ) = \phi$. If $g' \in G'$ we have
$$\Phi(h - h', g') = \Phi(h, g') - \Phi(h', g') = g'\phi - g'\phi = 0,$$
hence $h - h' \in G'^\perp$ and $h \in H' + G'^\perp$, and (3.2) follows. Altogether we see that H' is a summand of H of the right size. □

References

[1] H. Bass, Finitistic dimension and a homological generalization of semi-primary rings, Transact. Amer. Math. Soc. **95**, 466 – 488 (1960).
[2] M. G. Bell, On the combinatorial principle $P(c)$, Fundamenta math. **114** (1981) 149 – 157.
[3] K. Devlin, S. Shelah, A weak version of $\diamondsuit$ which follows from $2^{\aleph_0} < 2^{\aleph_1}$, Israel J. Math. **6**, 239 – 247 (1978).
[4] M. Dugas, J. Irvin, S. Khabbaz, Countable rings as endomorphism rings, Quart. J. Math. Oxford (2) **39** (1988), 201–211.
[5] K. Eda, On $\mathbb{Z}$-kernel groups, Archiv der Mathematik **41**, 289 – 293 (1983).
[6] K. Eda, H. Ohta, On abelian groups of integer-valued continuous functions, their $\mathbb{Z}$-dual and $\mathbb{Z}$-reflexivity, in *Abelian Group Theory*, pp. 241 – 257, Gordon and Breach, London 1986.
[7] P. Eklof, A. Mekler, Almost free modules, Set-theoretic methods, North-Holland, Amsterdam 1990.
[8] L. Fuchs, Infinite abelian groups - Volume 1,2 Academic Press, New York 1970, 1973.
[9] R. Göbel, S. Shelah, Some nasty reflexive groups, to appear in Mathematische Zeitschrift
[10] R. Göbel, S. Shelah, Decompositions of reflexive modules, to appear in Archiv der Mathematik
[11] R. Göbel, B. Wald, Martin's axiom implies the existence of certain growth types, Mathematische Zeitschrift **172** (1980), 107 – 121.
[12] R. Göbel, B. Wald, Separable torsion–free modules of small type, Houston Journal of Math. **16** (1990), 271 – 287.
[13] T. Jech, Set theory, Academic Press, New York 1978
[14] A. Kanamori, The higher infinite, Springer, Berlin 1994.

Fachbereich 6, Mathematik und Informatik, Universität Essen, 45117 Essen, Germany
E-mail address: R.Goebel@Uni-Essen.De

Department of Mathematics, Hebrew University, Jerusalem, Israel, and Rutgers University, Newbrunswick, NJ, U.S.A
E-mail address: Shelah@math.huji.ac.il

Σ-isotype subgroups of local k-groups

Paul Hill, Charles Megibben, and William Ullery

Dedicated to Professor Laszlo Fuchs in honor of his 75th birthday.

ABSTRACT. A local Warfield group G is weakly transitive in the sense that if a and b are elements of G with the same height sequence and equal type vectors, then there is an automorphism of G that maps a to b. The type vector is an invariant associated with each element in a local k-group, and hence the notion of weak transitivity is meaningful for all such groups. In this paper, we consider Σ-isotype subgroups of arbitrary local abelian groups; this is an abundant class that includes all torsion isotype subgroups, all isotype knice subgroups, and all p^σ-high subgroups. Our main result is that a Σ-isotype subgroup of a k-group inherits the property of being a k-group. As an application of this result and some new information regarding the structure of knice subgroups of k-groups, we are able to prove that a Σ-isotype subgroup of a Warfield group is weakly transitive. We also show that a Σ-isotype subgroup of a Warfield group is transitive if and only if it is fully transitive.

1. Introduction

In this paper, we continue the study of weak transitivity initiated in [**HU**]. It was established in the latter paper that local Warfield groups are weakly transitive, and here we show that the same conclusion holds for a certain class of isotype subgroups of local Warfield groups. These isotype subgroups are called Σ-isotype subgroups. In section 2, we establish the basic properties of Σ-isotype subgroups and prove that a Σ-isotype subgroup of a local k-group inherits the property of being a k-group. Techniques developed in that section then lead to new insights into the structure of knice subgroups of k-groups in section 3. In section 4, we prove a theorem on the extension of height-preserving maps between knice subgroups of a Σ-isotype subgroup of a local Warfield group. This extension theorem is then utilized in section 5 to prove that a Σ-isotype subgroup H of a local Warfield group is weakly transitive; that is, if a and b are elements of H with the same height sequence and equal type vectors $V(a)$ and $V(b)$, then there is an automorphism of H that maps a to b. The type vector of an element in a k-group is the new

---

1991 *Mathematics Subject Classification*. Primary: 20K21, 20K30; Secondary: 20K27.

Key words and phrases. k-group, Warfield group, Σ-isotype, knice subgroup, primitive element, type vector, weakly transitive .

invariant introduced in [**HU**]. Using this result on weak transitivity, we then show that a Σ-isotype subgroup of a local Warfield group is transitive if and only if it is fully transitive. By contrast, we conclude section 5 with an example of a mixed k-group that is neither weakly transitive nor fully transitive. Finally, in section 6, we make certain observations about the relation between $*$-isotype and Σ-isotype subgroups. In particular, it is shown that Σ-isotype subgroups of a Warfield group are not necessarily Warfield groups; indeed, they need not even have decomposition bases. Thus, our results regarding the transitivity and weak transitivity of Σ-isotype subgroups of Warfield groups are far more general than the corresponding results for Warfield groups that were obtained in [**HU**].

In the remainder of this introduction, we establish notation and review facts about primitive elements and knice subgroups. The reader familiar with [**HM2**] and [**HU**] will find little new here except for the characterization of knice subgroups in Proposition 1.3, and two lemmas which represent the formalization of observations implicit in [**HM2**]. Nonetheless, these lemmas are crucial in the proofs of results in sections 2, 3 and 5 below.

All groups G in this paper are p-local abelian groups for an arbitrary but fixed prime p; that is, G is a module over $\mathbb{Z}_p$, the ring of integers localized at p, and all subgroups of G are understood to be $\mathbb{Z}_p$-submodules. We use the notation $\langle S \rangle$ to denote the subgroup of G generated by a subset S.

A *height sequence* is any sequence $\bar{\alpha} = \{\alpha_n\}_{n<\omega}$ where each α_n is an ordinal or one of the symbols ∞ or ∞^+ and where the inequality $\alpha_n < \alpha_{n+1}$ holds for each n, with the convention that $\alpha < \infty < \infty < \infty^+ < \infty^+$ for all ordinals α. If k is a nonnegative integer, $p^k\bar{\alpha}$ denotes the height sequence $\{\alpha_{k+n}\}_{n<\omega}$. Also, the ordering of the class of ordinals (with the symbols ∞ and ∞^+ adjoined) induces in a pointwise manner the lattice relations $\leq$ and $\wedge$ on the class of all height sequences.

For an element $x \in G$, denote the *height* of x in G by $|x|$. Thus, if we take $p^\alpha G$ to have the usual meaning for all ordinals α, $|x| = \alpha$ provided that $x \in p^\alpha G$ but $x \notin p^{\alpha+1}G$. In the exceptional case where $x \in p^\alpha G$ for every ordinal α, we define $|x| = \infty$ whenever $x \neq 0$, but we set $|0| = \infty^+$. With every $x \in G$, we associate its *height sequence* $\|x\| = \{|p^n x|\}_{n<\omega}$. Notice that the convention $|0| = \infty^+$ allows $\|x\|$ to distinguish between elements of finite order and those that are merely finite modulo $p^\infty G = \bigcap_{\alpha \leq \infty} p^\alpha G$. If $\bar{\alpha} = \{\alpha_n\}_{n<\omega}$ is any height sequence, $G(\bar{\alpha})$ denotes the subgroup of G consisting of all those x for which $\|x\| \geq \bar{\alpha}$. Also, we define, in case that α_n is always an ordinal,

$$G(\bar{\alpha}^*) = \langle x \in G(\bar{\alpha}) : |p^n x| > \alpha_n \text{ for infinitely many values of } n \rangle.$$

On the other hand, if $\alpha_n \geq \infty$ for some n, we let $G(\bar{\alpha}^*)$ be the maximal torsion subgroup of $G(\bar{\alpha})$. Recall that $x \in G$ is *primitive* of type $\bar{\alpha}$ if $\|x\| = \bar{\alpha}$ but $x \notin G(\bar{\alpha}^*)$. Clearly, a primitive element must have infinite order. A noteworthy property of a primitive element x is that if $x \in G(\bar{\beta}^*)$ for some height sequence $\bar{\beta} = \{\beta_n\}_{u<\omega}$, then $|p^n x| > \beta_n$ for infinitely many $n < \omega$. A direct sum $A = \bigoplus_{i \in I} A_i$ of subgroups of G is a *valuated coproduct* if, for every height sequence $\bar{\alpha}$,

$$A \cap G(\bar{\alpha}) = \bigoplus_{i \in I}(A_i \cap G(\bar{\alpha})).$$

A valuated coproduct is called $*$-*valuated* if it satisfies the additional condition

$$A \cap G(\bar{\alpha}^*) = \bigoplus_{i \in I}(A_i \cap G(\bar{\alpha}^*)).$$

Recall that G is said to be *simply presented* if it can be defined in terms of generators and relations where all relations are of the form $px = y$ or $px = 0$. Torsion simply presented groups are classified by means of their Ulm-Kaplansky invariants in [**H1**]. A *local Warfield group* is defined to be a direct summand of a simply presented p-local group. In general, a local Warfield group is a nontorsion (mixed) group. Such groups have been classified [**HRW**] in terms of their Ulm-Kaplansky invariants and Warfield invariants (see the discussion in section 4 below). One characteristic property of a Warfield group G is the existence of a *decomposition basis*, a subset X of G consisting of elements of infinite order with the properties that $G/\langle X \rangle$ is torsion and $\langle X \rangle = \bigoplus_{x \in X} \langle x \rangle$ is a valuated coproduct.

In [**HM2**], knice subgroups were introduced as a vehicle for establishing an Axiom 3 characterization of local Warfield groups in the spirit of the familiar combinatorial description of simply presented torsion groups via a family of nice subgroups. Recall that the subgroup N is a *knice subgroup* of G if (1) N is a nice subgroup of G (that is, $p^\sigma(G/N) = (p^\sigma G + N)/N$ for all ordinals σ) and (2) for each finite subset S of G, there exists a finite number $r \geq 0$ of primitive elements $x_1, x_2, \ldots, x_r \in G$ such that

$$N' = N \oplus \langle x_1 \rangle \oplus \langle x_2 \rangle \oplus \cdots \oplus \langle x_r \rangle$$

is a $*$-valuated coproduct for which $p^m \langle S \rangle \subseteq N'$ for some $m < \omega$. (We do not intend to exclude the possibility that the collection of primitive elements is vacuous.) Finite extensions of knice subgroups are knice subgroups; more generally, if N is knice in G and N'/N is knice in G/N, then N' is knice in G. Indeed, from the proof of this latter fact in [**HM2**], we can extract the following useful result.

LEMMA 1.1. *Let N be a knice subgroup of the p-local group G and suppose that*

$$\langle y_1 + N \rangle \oplus \langle y_2 + N \rangle \oplus \cdots \oplus \langle y_r + N \rangle$$

is a $*$-valuated coproduct in G/N with each $y_i + N$ primitive in G/N. Then, there exist a nonnegative integer m and primitive elements $a_1, a_2, \ldots, a_r$ in G such that $p^m y_i + N = a_i + N$ for each i and

$$N' = N \oplus \langle a_1 \rangle \oplus \langle a_2 \rangle \oplus \cdots \oplus \langle a_r \rangle$$

is a $*$-valuated coproduct in G

PROOF. Since N is a knice subgroup of G, there is a subgroup A and a nonnegative integer m such that $N \oplus A$ is a $*$-valuated coproduct that contains each $p^m y_i$. For each i, write $p^m y_i = z_i + a_i$, where $z_i \in N$ and $a_i \in A$. It follows, as in the proof of Proposition 2.7 in [**HM2**], that the height sequence of a_i is the same as the height sequence of $p^m y_i + N$ in G/N and that each a_i is a primitive element of G. Furthermore, the fact that $N' = N \oplus \langle a_1 \rangle \oplus \langle a_2 \rangle \oplus \cdots \oplus \langle a_r \rangle$ is a $*$-valuated coproduct is a consequence of the argument appearing at the end of the cited proof in [**HM2**]. □

A p-local group G is said to be a *k-group* if 0 is a knice subgroup of G. Thus, k-groups may be thought of as p-local groups that have "decomposition bases" for finitely generated subgroups. Every Warfield group is a k-group since a Warfield group must satisfy Axiom 3 with respect to knice subgroups (see [**HM2**]). When G is a k-group, any $*$-valuated coproduct $\langle x_1 \rangle \oplus \langle x_2 \rangle \oplus \cdots \oplus \langle x_r \rangle$ with the x_i's primitive in G is a knice subgroup (see Proposition 2.5 in [**HM2**]). Since nonzero multiples of primitive elements are primitive, it follows that if x is an element of infinite

order in a k-group G, there exist a nonnegative integer m and primitive elements $x_1, x_2, \ldots, x_r$ such that $p^m x = x_1 + x_2 + \cdots + x_r$. Furthermore, as observed in [**HU**], this representation can be chosen in such a manner that the collection of primitive elements $\{x_1, x_2, \ldots, x_r\}$ is an *incomparable set* in the sense that $|p^n x_i| \gneq |p^n x_j|$ for infinitely many $n < \omega$ whenever $i \neq j$. Moreover, given any incomparable set of primitive elements $\{x_1, x_2, \ldots, x_r\}$ in a k-group G, there exists a nonnegative integer n such that
$$\langle p^n x_1 \rangle \oplus \langle p^n x_2 \rangle \oplus \cdots \oplus \langle p^n x_r \rangle$$
is a $*$-valuated coproduct [**HU**, Lemma 3]. We can glean from the proof of Theorem 2.10 in [**HM2**] another valuable observation.

LEMMA 1.2. *Let $\{x_1, x_2, \ldots, x_r\}$ be an incomparable set of primitive elements in the k-group G. If $a_1, a_2, \ldots, a_r$ are elements of G such that $\|x_i\| \leq \|a_i\|$ for all i and if*
$$x_1 + x_2 + \cdots + x_r = a_1 + a_2 + \cdots + a_r,$$
then each a_i is primitive and there exists a nonnegative integer m such that $\langle p^m a_1 \rangle \oplus \langle p^m a_2 \rangle \oplus \cdots \oplus \langle p^m a_r \rangle$ is a $*$-valuated coproduct and $\|p^m a_i\| = \|p^m x_i\|$ for all i.

PROOF. From what we have just observed above, there is no loss of generality in assuming that $N = \langle x_1 \rangle \oplus \langle x_2 \rangle \oplus \cdots \oplus \langle x_r \rangle$ is a $*$-valuated coproduct. In particular, N is a knice subgroup of G and thus there exist a nonnegative integer m and a subgroup B of G such that $N' = N \oplus B$ is a $*$-valuated coproduct with $p^m a_i \in N'$ for all i. Thus, we have r equations
$$p^m a_i = \sum_{j=1}^{r} c_{i,j} x_j + y_i$$
where the $c_{i,j}$'s are in $\mathbb{Z}_p$ and the y_i's are in B. The remainder of the proof is exactly as in [**HM2**]. □

We shall also require a characterization of knice subgroups that does not appear in [**HM2**], but can be found in globalized form as Proposition 1.7 in [**HM5**].

PROPOSITION 1.3. *A subgroup N of a p-local group G is a knice subgroup if and only if the following conditions are satisfied:*

(1) *N is a nice subgroup of G.*
(2) *G/N is a k-group.*
(3) *For each $g \in G$, there exist an $x \in G$ and a nonnegative integer n such that $p^n g + N = x + N$ and $\|x\| = \|p^n g + N\|$.*

COROLLARY 1.4. *If N is a knice subgroup of the p-local group G and if A is a subgroup of N, then N/A is a knice subgroup of G/A.*

PROOF. As is well known, N being nice in G implies that N/A is nice in G/A. Furthermore, if N is a knice subgroup of G, then by the preceding proposition $(G/A)/(N/A) \cong G/N$ is a k-group. Finally, this canonical isomorphism can be applied to show that N/A inherits from N the properties required by condition (3) of Proposition 1.3. Indeed, $\|(g + A) + N/A\| = \|g + N\|$ where these height sequences are computed in $(G/A)/(N/A)$ and G/N, respectively. □

2. Σ-isotype subgroups

A subgroup H of the p-local group G is said to be *isotype* in G if $H \cap p^\sigma G = p^\sigma H$ for all ordinals σ; or, equivalently, $H \cap G(\bar{\alpha}) = H(\bar{\alpha})$ for each height sequence $\bar{\alpha}$. Call an isotype subgroup H *-isotype in G if it also satisfies the condition $H \cap G(\bar{\alpha}^*) = H(\bar{\alpha}^*)$ for each height sequence $\bar{\alpha}$. We should mention that *-isotype subgroups have been considered elsewhere in the literature (for example, see [**HM2, HM3**]). We say that H is a *Σ-isotype subgroup* of G if $H \cap \sum_{i=1}^r G(\bar{\alpha}_i) = \sum_{i=1}^r H(\bar{\alpha}_i)$ for each finite collection of height sequences $\bar{\alpha}_1, \bar{\alpha}_2, \ldots, \bar{\alpha}_r$. In the torsion-free setting, Σ-isotype subgroups were introduced in [**HM3**]. It is clear that every Σ-isotype subgroup of G is *-isotype in G. As we shall see below, Σ-isotype subgroups are abundant in any p-local group. Obviously direct summands are Σ-isotype and, moreover, isotype torsion subgroups are Σ-isotype as a consequence of the following useful observation.

LEMMA 2.1. *Let H be an isotype subgroup of the p-local group G. If $h \in H$ has finite order and if $h \in \sum_{i=1}^r G(\bar{\alpha}_i)$ for some height sequences $\bar{\alpha}_1, \bar{\alpha}_2, \ldots, \bar{\alpha}_r$, then $h \in \sum_{i=1}^r H(\bar{\alpha}_i)$.*

PROOF. For each i, let $\alpha_{i,0}$ denote the leading term of $\bar{\alpha}_i$. Now select j so that $\alpha_{j,0} = \min\{\alpha_{1,0}, \alpha_{2,0}, \ldots, \alpha_{r,0}\}$ and note, by the triangle inequality, that $|h| \geq \alpha_{j,0}$. If the order of h is p, then the desired conclusion follows since $h \in H \cap G(\bar{\alpha}_j) = H(\bar{\alpha}_j)$. Proceeding by induction on the order of h, we may assume that $ph = x_1 + x_2 + \cdots + x_r$ where $x_i \in H(p\bar{\alpha}_i)$ for each i. But then, since H is isotype in G, there exist elements $h'_i \in H(\bar{\alpha}_i)$ such that $ph'_i = x_i$ for each i. Clearly then $z = h - (h'_1 + h'_2 + \cdots + h'_r) \in H$ and $|z| \geq \alpha_{j,0}$. Therefore, if $h_j = h'_j + z$ and $h_i = h'_i$ for $i \neq j$, then $h = \sum_{i=1}^r h_i \in \sum_{i=1}^r H(\bar{\alpha}_i)$. □

COROLLARY 2.2. *Let $h \in H$, where H is an isotype subgroup of the p-local group G. If $h \in \sum_{i=1}^r G(\bar{\alpha}_i)$ for some height sequences $\bar{\alpha}_1, \bar{\alpha}_2, \ldots, \bar{\alpha}_r$, and there exists a nonnegative integer m such that $p^m h \in \sum_{i=1}^r H(p^m \bar{\alpha}_i)$, then $h \in \sum_{i=1}^r H(\bar{\alpha}_i)$.*

PROOF. From the hypotheses, $p^m h = h_1 + h_2 + \cdots + h_r$, where $h_i \in H(p^m \bar{\alpha}_i)$ for all i, $1 \leq i \leq r$. Thus, we have elements $h'_i \in H(\bar{\alpha}_i)$ with $p^m h'_i = h_i$ for each i. But then $h' = h - (h'_1 + h'_2 + \cdots + h'_r) \in H$ is an element of finite order in $\sum_{i=1}^r G(\bar{\alpha}_i)$. Therefore, by Lemma 2.1, $h = h' + (h'_1 + h'_2 + \cdots + h'_r) \in \sum_{i=1}^r H(\bar{\alpha}_i)$. □

Another extensive class of Σ-isotype subgroups is given by our next proposition. If H is maximal among those subgroups of G having trivial intersection with $p^\sigma G$, H is said to be a p^σ-*high subgroup* of G.

PROPOSITION 2.3. *For each ordinal σ, every p^σ-high subgroup H of a p-local group G is Σ-isotype.*

PROOF. Suppose $h \in H \cap \sum_{i=1}^r G(\bar{\alpha}_i)$ for some height sequences $\bar{\alpha}_1, \bar{\alpha}_2, \ldots, \bar{\alpha}_r$, and write $h = g_1 + g_2 + \cdots + g_r$ with $g_i \in G(\bar{\alpha}_i)$ for each i. Since $G/(H + p^\sigma G)$ is a torsion group, there exist $h_1, h_2, \ldots, h_r \in H$ and a nonnegative integer m such that $-h_i + p^m g_i \in p^\sigma G$ for all i. For each i, set $\bar{\alpha}_i = \{\alpha_{i,n}\}_{n<\omega}$. If $\sigma \geq \sup\{\alpha_{i,n} : 1 \leq i \leq r \text{ and } n < \omega\}$, then $h_i \in G(p^m \bar{\alpha}_i)$; otherwise, we may take m large enough to ensure that $h_i \in H \cap p^\sigma G$. Thus, we may assume that $h_i \in G(p^m \bar{\alpha}_i)$ for each i. Therefore,

$$p^m h - (h_1 + h_2 + \cdots + h_r) = \sum_{i=1}^r (-h_i + p^m g_i) \in H \cap p^\sigma G = 0.$$

Since p^σ-high subgroups are well known to be isotype and $p^m h = h_1 + h_2 + \cdots + h_r \in \sum_{i=1}^r H(p^m \bar{\alpha}_i)$, Corollary 2.2 implies that $h \in \sum_{i=1}^r H(\bar{\alpha}_i)$, as desired. □

Our next proposition is crucial to the proof of Theorem 2.6 below.

PROPOSITION 2.4. *Suppose N is a knice subgroup of the p-local group G and that $N \subseteq H$, where H is an isotype subgroup of G. Then, H is a Σ-isotype subgroup of G if and only if H/N is Σ-isotype in G/N.*

PROOF. First assume that H is a Σ-isotype subgroup of G, and suppose that $h + N = (g_1 + N) + (g_2 + N) + \cdots + (g_r + N)$, where $h \in H$ and $g_i + N \in (G/N)(\bar{\alpha}_i)$ for all i. Then, by Proposition 1.3, there exist a nonnegative integer n and elements $x_1, x_2, \ldots, x_r \in G$ such that $p^n g_i + N = x_i + N$ and $\|x_i\| = \|p^n g_i + N\|$ for each i. Therefore, there is an $x \in N \subseteq H$ with

$$p^n h + x = x_1 + x_2 + \cdots + x_r \in H \cap \sum_{i=1}^r G(p^n \bar{\alpha}_i) = \sum_{i=1}^r H(p^n \bar{\alpha}_i),$$

and so

$$p^n h + N \in \sum_{i=1}^r (H/N)(p^n \bar{\alpha}_i).$$

Since N is a nice subgroup of G, H/N is isotype in G/N; hence, by an application of Corollary 2.2, $h + N \in \sum_{i=1}^r (H/N)(\bar{\alpha}_i)$.

Conversely, assume that H/N is Σ-isotype in G/N and suppose $h = g_1 + g_2 + \cdots + g_r$ where $h \in H$ and $g_i \in G(\bar{\alpha}_i)$ for each i. Then, there exist a nonnegative integer n and a $*$-valuated coproduct $N' = N \oplus A$ in G that contains each $p^n g_i$. Thus, for each i, we can write $p^n g_i = x_i + a_i$, with $x_i \in N$ and $a_i \in A$. Notice that it follows that both x_i and a_i are contained in $G(p^n \bar{\alpha}_i)$; in fact, $x_i \in H(p^n \bar{\alpha}_i)$ because $N \subseteq H$. Furthermore, $a = a_1 + a_2 + \cdots + a_r \in A \cap H$ and, since H/N is Σ-isotype in G/N, $a + N = (h_1 + N) + (h_2 + N) + \cdots + (h_r + N)$, where each $h_i + N \in (H/N)(p^n \bar{\alpha}_i)$. But $N' = N \oplus A$ may also be assumed to be a knice subgroup [**HM2**, Proposition 2.5] and so, by enlarging A if necessary, there exists an $l \geq n$ such that, for each i, $p^l h_i = y_i + b_i$, with $y_i \in N$ and $b_i \in A$. As noted in the proof of Lemma 1.1, $\|b_i\| = \|p^l h_i + N\| \geq p^m \bar{\alpha}_i$, where $m = n + l$. Now select $z \in N$ such that $a + z = h_1 + h_2 + \cdots + h_r$ and observe that $p^l(a + z) = (y_1 + y_2 + \cdots + y_r) + (b_1 + b_2 + \cdots + b_r)$. Consequently, $p^l a = b_1 + b_2 + \cdots + b_r$, where $b_i \in H \cap G(p^m \bar{\alpha}_i) = H(p^m \bar{\alpha}_i)$, and

$$p^m h = p^l(x_1 + x_2 + \cdots + x_r) + p^l a = (p^l x_1 + b_1) + (p^l x_2 + b_2) + \cdots + (p^l x_r + b_r)$$

with $p^l x_i + b_i \in H$ and $\|p^l x_i + b_i\| = \|p^l x_i\| \wedge \|b_i\| \geq p^m \bar{\alpha}_i$ for all i. Therefore, $p^m h \in \sum_{i=1}^r H(p^m \bar{\alpha}_i)$ and $h \in \sum_{i=1}^r H(\bar{\alpha}_i)$ by Corollary 2.2. □

COROLLARY 2.5. *If H is an isotype knice subgroup of the p-local group G, then H is a Σ-isotype subgroup of G.*

PROOF. Since the trivial subgroup is a Σ-isotype subgroup, the conclusion follows by taking $H = N$ in Proposition 2.4. □

We are now in position to establish the following fundamental result.

THEOREM 2.6. *Suppose that G is a k-group and that H is a $*$-isotype subgroup of G. Then, H is a Σ-isotype subgroup of G if and only if H is a k-group.*

PROOF. Since H is a $*$-isotype subgroup, an element of H is primitive in H if and only if it is primitive in G. Similarly, a direct sum in H is a $*$-valuated coproduct in H if and only if it is a $*$-valuated coproduct in G. First suppose H is also a k-group and let $h = g_1 + g_2 + \cdots + g_r$, where $h \in H$ and $g_i \in G(\bar{\alpha}_i)$ for each i. In order to show that H is Σ-isotype, Lemma 2.1 allows us to assume that h has infinite order and to write $p^m h = x_1 + x_2 + \cdots + x_s$, where $\{x_1, x_2, \ldots, x_s\}$ is an incomparable set of primitive elements in H and m is a nonnegative integer. By increasing m if necessary, we may assume that $N = \langle x_1 \rangle \oplus \langle x_2 \rangle \oplus \cdots \oplus \langle x_s \rangle$ is a $*$-valuated coproduct in H and therefore a knice subgroup of G. Then, as in the proof of Proposition 2.4, we have a $*$-valuated coproduct $N' = N \oplus A$ in G and a nonnegative integer n with $p^n g_i = y_i + a_i$, where $y_i \in N \subseteq H$ and $a_i \in A$ for each i. But since we may take $n \geq m$, $p^n h \in N$ and $\sum_{i=1}^{r} a_i \in N \cap A = 0$. Furthermore, $y_i \in H \cap G(p^n \bar{\alpha}_i) = H(p^n \bar{\alpha}_i)$ for each i and therefore

$$p^n h = y_1 + y_2 + \cdots + y_r \in \sum_{i=1}^{r} H(p^n \bar{\alpha}_i).$$

Since H is an isotype subgroup, Corollary 2.2 yields the desired conclusion that $h \in \sum_{i=1}^{r} H(\bar{\alpha}_i)$.

Conversely, assume that H is a Σ-isotype subgroup of G. To establish the fact that H is a k-group, we need to prove the following: given a finite subset S of H, there exist a $*$-valuated coproduct $N = \langle h_1 \rangle \oplus \langle h_2 \rangle \oplus \cdots \oplus \langle h_r \rangle$, where each h_i is a primitive element of H, and a nonnegative integer m such that $p^m \langle S \rangle \subseteq N$. We shall prove this by induction on $|S|$

First we consider the case $|S| = 1$. Thus, $S = \{h\}$ where, without loss of generality, h is an element of infinite order in H. Since G is a k-group, there are incomparable primitive elements $x_1, x_2, \ldots, x_r$ in G and a nonnegative integer m such that $p^m h = x_1 + x_2 + \cdots + x_r$. But then, because H is a Σ-isotype subgroup of G, we know that $p^m h = h_1 + h_2 + \cdots + h_r$ where $h_i \in H$ and $\|x_i\| \leq \|h_i\|$, for $i = 1, 2, \ldots, r$. By Lemma 1.2, each h_i is a primitive element and $N = \langle p^n h_1 \rangle \oplus \langle p^n h_2 \rangle \oplus \cdots \oplus \langle p^n h_r \rangle$ is a $*$-valuated coproduct for some nonnegative integer n. Since $p^{m+n} h \in N$, we have reached the desired conclusion when $|S| = 1$.

Now suppose $|S| > 1$ and $S = S_1 \cup \{h\}$ with $h \notin S_1$. By induction there exist a $*$-valuated coproduct $N = \langle y_1 \rangle \oplus \langle y_2 \rangle \oplus \cdots \oplus \langle y_s \rangle$, with each y_i a primitive element of H, and a nonnegative integer m such that $p^m \langle S_1 \rangle \subseteq N$. Note that Proposition 1.3 and Proposition 2.4 imply that H/N is a Σ-isotype subgroup of the k-group G/N. Hence, we may apply the special case $|S| = 1$ to the group H/N and conclude that there are primitive elements $h_1 + N, h_2 + N, \ldots, h_r + N$ in H/N and a nonnegative integer n such that $\langle h_1 + N \rangle \oplus \langle h_2 + N \rangle \oplus \cdots \oplus \langle h_r + N \rangle$ is a $*$-valuated coproduct in G/N that contains $p^n(h + N)$. But then, by Lemma 1.1, we have primitive elements $a_1, a_2, \ldots, a_r$ and a nonnegative integer l such that $p^l h_i + N = a_i + N$ for each i and $N \oplus \langle a_1 \rangle \oplus \langle a_2 \rangle \oplus \cdots \oplus \langle a_r \rangle$ is a $*$-valuated coproduct in G. But since $N \subseteq H$, each a_i is contained in H and therefore we have the desired conclusion that some p-power multiple of $\langle S \rangle$ is contained in the $*$-valuated coproduct $\langle y_1 \rangle \oplus \langle y_2 \rangle \oplus \cdots \oplus \langle y_s \rangle \oplus \langle a_1 \rangle \oplus \langle a_2 \rangle \oplus \cdots \oplus \langle a_r \rangle \subseteq H$. □

COROLLARY 2.7. *A p^σ-high subgroup of a k-group is a k-group.*

PROOF. This is an immediate consequence of Proposition 2.3 and Theorem 2.6. □

COROLLARY 2.8. *An isotype knice subgroup of a k-group is a k-group.*

PROOF. Apply Corollary 2.5 and Theorem 2.6. □

COROLLARY 2.9. *If H is a countable Σ-isotype subgroup of a k-group G, then H is a Warfield group.*

PROOF. By Theorem. 4.2 of [**HM2**] and Theorem 2.6, it is enough to observe that every countable k-group H has a decomposition basis. However, that H has a decomposition basis is obvious since any such H is the ascending union of finite subsets S_n, with some p-power multiple of $\langle S_n \rangle$ contained in a $*$-valued coproduct $\langle x_1 \rangle \oplus \langle x_2 \rangle \oplus \cdots \oplus \langle x_r \rangle$ with primitive x_i's. □

More generally, any k-group having countable torsion-free rank has a decomposition basis. Also, the following is not difficult to establish: if A is a countable subgroup of a k-group G with $A \cap p^{\omega_1} G = 0$, then G contains a countable Σ-isotype subgroup H with $A \subseteq H$.

3. Structure of Knice Subgroups

By modifying the proof of Theorem 2.6, we can gain insights into the structure of knice subgroups of k-groups.

THEOREM 3.1. *If M is a knice subgroup of the k-group G and if S is a finite subset of M, then there exists a $*$-valued coproduct $N = \langle y_1 \rangle \oplus \langle y_2 \rangle \oplus \cdots \oplus \langle y_s \rangle \subseteq M$, with each y_i a primitive element of G, for which $p^m \langle S \rangle \subseteq N$ is satisfied for some nonnegative integer m.*

PROOF. We prove the theorem by induction on $|S|$. First consider the case where $S = \{x\}$ where, without loss of generality, x is an element of infinite order in M. Since G is a k-group, there are incomparable primitive elements $x_1, x_2, \ldots, x_r$ in G and a nonnegative integer m such that $p^m x = x_1 + x_2 + \cdots + x_r$. But then, because M is a knice subgroup of G, there exist a $*$-valued coproduct $M \oplus B$ and a nonnegative integer n such that, for each i, $p^n x_i = a_i + b_i$ where $a_i \in M$ and $b_i \in B$. Note that $\|p^n x_i\| \leq \|a_i\|$ for $i = 1, 2, \ldots, r$ and that $\sum_{i=1}^r b_i \in M \cap B = 0$. Therefore, by Lemma 1.2, each a_i is a primitive element and $N = \langle p^l a_1 \rangle \oplus \langle p^l a_2 \rangle \oplus \cdots \oplus \langle p^l a_r \rangle$ is a $*$-valued coproduct for some nonnegative integer l. Since $p^k x \in N$ where $k = m + n + l$, we have reached the desired conclusion when $|S| = 1$.

Now asume that $|S| > 1$ and $S = S_1 \cup \{x\}$ with $x \notin S_1$. By induction, there exist a $*$-valued coproduct $N = \langle y_1 \rangle \oplus \langle y_2 \rangle \oplus \cdots \oplus \langle y_r \rangle$, where the primitive elements y_i are all contained in M, and a nonnegative integer m such that $p^m \langle S_1 \rangle \subseteq N$. Then since Corollary 1.4 implies that M/N is a knice subgroup of G/N, we may apply the special case $|S| = 1$ to the group M/N and conclude that there are primitive elements $z_1 + N, z_2 + N, \ldots, z_s + N$ contained in M/N and a nonnegative integer n such that $\langle z_1 + N \rangle \oplus \langle z_2 + N \rangle \oplus \cdots \oplus \langle z_s + N \rangle$ is $*$-valued coproduct in G/N that contains $p^n(x + N)$. By Lemma 1.1, we have primitive elements $a_1, a_2, \ldots, a_s$ and a nonnegative integer l such that $p^l z_i + N = a_i + N$ for each i and $N \oplus \langle a_1 \rangle \oplus \langle a_2 \rangle \oplus \cdots \oplus \langle a_s \rangle$ is a $*$-valued coproduct in G. Noting that each $a_i \in M$, we have the desired conclusion that some p-power multiple of $\langle S \rangle$ is contained in the $*$-valued coproduct $\langle y_1 \rangle \oplus \langle y_2 \rangle \oplus \cdots \oplus \langle y_r \rangle \oplus \langle a_1 \rangle \oplus \langle a_2 \rangle \oplus \cdots \oplus \langle a_s \rangle \subseteq M$. □

COROLLARY 3.2. *If M is a finitely generated knice subgroup of the k-group G, then there is a $*$-valued coproduct $N = \langle x_1 \rangle \oplus \langle x_2 \rangle \oplus \cdots \oplus \langle x_r \rangle \subseteq M$ where each x_i is primitive in G and M/N is finite.*

COROLLARY 3.3. *If $\pi : M \to M'$ is a height-preserving isomorphism between knice subgroups of the k-group G, then π preserves primitivity.*

PROOF. Let $y = \pi(x)$ where $x \in M$ is a primitive element of G with height sequence $\bar{\alpha} = \{\alpha_n\}_{n<\omega}$. Then, by Theorem 3.1, there exist a $*$-valuated coproduct $N = \langle y_1 \rangle \oplus \langle y_2 \rangle \oplus \cdots \oplus \langle y_r \rangle \subseteq M'$, where each y_i is primitive, and a nonnegative integer m such that $p^m y \in N$. Since $\bar{\alpha}$ is also the height sequence of y, it will follow that y is primitive if we can show that $p^m y \notin G(p^m \bar{\alpha}^*)$. Without loss of generality, we may assume that $p^m y = y_1 + y_2 + \cdots + y_r$. If to the contrary $p^m y \in G(p^m \bar{\alpha}^*)$, then $y_i \in G(p^m \bar{\alpha}^*)$ for each i. But then for each i, the primitivity of y_i implies that $|p^n y_i| > \alpha_{m+n}$ for infinitely many values of n. Since, however, $p^m x = x_1 + x_2 + \cdots + x_r$ where $x_i = \pi^{-1}(y_i)$ for each i, $p^m x \in G(p^m \bar{\alpha}^*)$ and the primitivity of x is contradicted. □

4. An Extension Theorem

To establish the weak transitivity of a Σ-isotype subgroup H of a local Warfield group G, we require a result that ensures that a height-preserving isomorphism $\pi : M \to M'$ between finitely generated knice subgroups of H can be extended to an automorphism of H. In the special case $H = G$, this is done directly in [**HU**] using the isomorphism theorem for Warfield groups that appears in [**HM2**]. For a general H, however, we need an *equivalence theorem*; that is, we need to know that there is an automorphism of G that extends π and at the same time maps H onto itself. Fortunately, in [**HM4**], there is a theorem delineating precisely when two isotype subgroups H and H' are equivalent. But this equivalence theorem is quite technical, and it requires the introduction of certain invariants.

Two height sequences $\bar{\alpha}$ and $\bar{\beta}$ are said to be *equivalent* provided that there exist nonnegative integers m and n such that $p^m \bar{\alpha} = p^n \bar{\beta}$. We regard a given equivalence class of height sequences $\mathbf{e}$ as a directed set by defining $\bar{\alpha} < \bar{\beta}$ to mean that $p^m \bar{\alpha} = \bar{\beta}$ for some $m < \omega$. Now suppose that H and M are subgroups of the p-local group G. If $p^m \bar{\alpha} = \bar{\beta}$, multiplication by p^m defines a homomorphism $\rho_m : H(\bar{\alpha}) \to H(\bar{\beta})$ that maps $H(\bar{\alpha}) \cap (M + G(\bar{\alpha}^*))$ to $H(\bar{\beta}) \cap (M + G(\bar{\beta}^*))$. Therefore, the quotient groups $H(\bar{\alpha})/H(\bar{\alpha}) \cap (M + G(\bar{\alpha}^*))$ together with the maps induced by the various ρ_m's form a direct system over $\mathbf{e}$. We define

$$W_G^H(\mathbf{e}, M) = \varinjlim \frac{H(\bar{\alpha})}{H(\bar{\alpha}) \cap (M + G(\bar{\alpha}^*))}.$$

In particular, we take $W_G^H(\mathbf{e}) = W_G^H(\mathbf{e}, 0)$. (In order to appreciate the need to introduce direct limits at this juncture, the reader should refer to [**HM2**, Proposition 3.3].) Note that $W_G^H(\mathbf{e}, M)$ is a vector space over $\mathbb{Z}/p\mathbb{Z}$ whenever $\bar{\alpha} \in \mathbf{e}$ does not contain ∞ or ∞^+; otherwise, it is a rational vector space. When G is a Warfield group and $H = G$, $\dim W_G^H(\mathbf{e})$ is a Warfield invariant of G and is equal to the number of primitive elements of type $\mathbf{e}$ in any decomposition basis for G.

If we specialize the hypotheses of Theorem 2.2 in [**HM4**] to the case where $G' = G$, $H' = H$, and ϕ is the identity map of G/H, and then observe how the proof of that theorem proceeds using Lemma 2.3 in the same paper, we arrive at the following result.

THEOREM 4.1. *Let H be an isotype subgroup of the Warfield group G and suppose that $\pi : M \to M'$ is a height-preserving isomorphism, where M and $M'*

are knice subgroups of G. Then, there exists an automorphism ψ of G such that $\psi(H) = H$ and ψ extends π provided that the following conditions are satisfied:
   (1) $\pi(x) + H = x + H$ for all $x \in M$.
   (2) For all ordinals σ,
$$\dim \frac{p^\sigma H[p]}{p^\sigma H[p] \cap (M + p^{\sigma+1}G)} = \dim \frac{p^\sigma H[p]}{p^\sigma H[p] \cap (M' + p^{\sigma+1}G)}.$$
   (3) For each $\mathbf{e}$, there is an automorphism $\eta_\mathbf{e}$ of $W_G^H(\mathbf{e})$ that induces an isomorphism of $W_G^H(\mathbf{e}, M)$ onto $W_G^H(\mathbf{e}, M')$.

We now apply the preceding theorem to the situation where H is Σ-isotype in G and the knice subgroups M and M' are contained in H.

COROLLARY 4.2. *Let H be a Σ-isotype subgroup of the Warfield group G and suppose that $\pi : M \to M'$ is a height-preserving isomorphism, where M and M' are finitely generated knice subgroups of H. Then, there exists an automorphism of H that extends π.*

PROOF. Since H is Σ-isotype in G, it follows from Theorem 2.6 and Corollary 3.2 that M and M' are also knice in G. Moreover, condition (1) of Theorem 4.1 is satisfied trivially when M and M' are contained in H. To see why conditions (2) and (3) are also satisfied in the present context, it helps to make the following general observations about vector spaces of possibly transfinite dimension. Suppose A and A' are finite-dimensional subspaces of a vector space V with $\dim A = \dim A'$. Under these hypotheses, $\dim V/A = \dim V/A'$ and there exists an automorphism η of V with $\eta(A) = A'$. Hence η induces an isomorphism between the quotient spaces V/A and V/A'.

Now, to see that condition (2) is satisfied, a standard argument (see, for example, the discussion in [**HM2**]) shows that the height-preserving isomorphism $\pi : M \to M'$ induces an isomorphism between the subspaces $p^\sigma H[p] \cap (M + p^{\sigma+1}G)/p^{\sigma+1}H[p]$ and $p^\sigma H[p] \cap (M' + p^{\sigma+1}G)/p^{\sigma+1}H[p]$ of the vector space $p^\sigma H[p]/p^{\sigma+1}H[p]$. Moreover, both M and M' being finitely generated implies that these subspaces are finite-dimensional. Therefore, according to the above discussion, the respective quotient spaces have the same dimension; that is,
$$\dim \frac{p^\sigma H[p]}{p^\sigma H[p] \cap (M + p^{\sigma+1}G)} = \dim \frac{p^\sigma H[p]}{p^\sigma H[p] \cap (M' + p^{\sigma+1}G)}.$$

To verify condition (3), note that since H is $*$-isotype and contains the subgroups M and M', $H(\bar{\alpha}) \cap (M + G(\bar{\alpha}^*)) = H(\bar{\alpha}) \cap (M + H(\bar{\alpha}^*))$ and $H(\bar{\alpha}) \cap (M' + G(\bar{\alpha}^*)) = H(\bar{\alpha}) \cap (M' + H(\bar{\alpha}^*))$. It readily follows from Corollary 3.3 that π induces an isomorphism from $H(\bar{\alpha}) \cap (M + H(\bar{\alpha}^*))/H(\bar{\alpha}^*)$ to $H(\bar{\alpha}) \cap (M' + H(\bar{\alpha}^*))/H(\bar{\alpha}^*)$. Moreover, M and M' being finitely generated implies the latter are finite-dimensional subspaces of $H(\bar{\alpha})/H(\bar{\alpha}^*)$. In fact, there is a nonnegative integer n such that the dimension of these subspaces does not exceed n for all $\bar{\alpha} \in \mathbf{e}$. Consequently,
$$\varinjlim \frac{H(\bar{\alpha}) \cap (M + H(\bar{\alpha}^*))}{H(\bar{\alpha}^*)} \cong \varinjlim \frac{H(\bar{\alpha}) \cap (M' + H(\bar{\alpha}^*))}{H(\bar{\alpha}^*)}$$
and each of these is a finite-dimensional subspace of $W_G^H(\mathbf{e})$. From the above observation, there is an automorphism $\eta_\mathbf{e}$ of $W_G^H(\mathbf{e})$ that induces an isomorphism between the respective quotient spaces $W_G^H(\mathbf{e}, M)$ and $W_G^H(\mathbf{e}, M')$. □

5. Transitivity and Weak Transitivity

Given two elements a and b in a p-local group G, when does there exist an automorphism of G that maps a to b? Obviously, a necessary condition is that a and b have the same height sequence. But even when G is a Warfield group, this condition is not sufficient since there can exist primitive and nonprimitive elements with the same height sequence. In [**HU**], the notion of the type vector $V(x)$ associated with each element x in an arbitrary k-group is introduced; and it is shown for a Warfield group G that $\|a\| = \|b\|$ and $V(a) = V(b)$ are necessary and sufficient conditions for there to exist an automorphism of G that maps a to b. Here we shall prove that the same holds if G is replaced by an arbitrary Σ-isotype subgroup of a Warfield group.

For the convenience of the reader, we now include the definition of a type vector as presented in [**HU**]. If $\bar{\alpha}_1, \bar{\alpha}_2, \ldots, \bar{\alpha}_r$ is a finite collection of height sequences with $r \geq 0$, we call the r-tuple

$$\boldsymbol{\alpha} = (\bar{\alpha}_1, \bar{\alpha}_2, \ldots, \bar{\alpha}_r)$$

a *height-sequence vector*. Observe that the notation does not exclude the possibility that $\boldsymbol{\alpha}$ is the empty vector. If $\boldsymbol{\beta} = (\bar{\beta}_1, \bar{\beta}_2, \ldots, \bar{\beta}_s)$ is another height-sequence vector, we say that $\boldsymbol{\alpha}$ and $\boldsymbol{\beta}$ are *equivalent* if $r = s$ and, after a suitable reindexing, there are nonnegative integers k and l such that $p^k \bar{\alpha}_i = p^l \bar{\beta}_i$ for all i, $1 \leq i \leq r$. Call the equivalence class containing the height-sequence vector $\boldsymbol{\alpha}$ the *type vector* of $\boldsymbol{\alpha}$.

With each element x in a k-group G we associate a type vector $V(x)$. First, if x has finite order, $V(x)$ is defined to be the singleton class of the empty vector Φ. If x is a primitive element, $V(x)$ is the type vector of $(\|x\|)$, the height-sequence vector with the single entry $\|x\|$. Finally, if x is a nonprimitive element of infinite order, there exist a nonnegative integer n and an incomparable set of primitive elements $\{x_1, x_2, \ldots, x_r\}$ with $r \geq 2$ such that

$$p^n x = x_1 + x_2 + \cdots + x_r.$$

In this case, define $V(x)$ to be the type vector of $(\|x_1\|, \|x_2\|, \ldots, \|x_r\|)$. In the case where x is a nonprimitive element of infinite order, it is established in [**HU**, Lemma 5] that $V(x)$ is independent of n and the representation of $p^n x$ as a sum of incomparable primitive elements. Therefore, $V(x)$ is well defined for each element x of a k-group.

Recall that a p-local group G is said to be *transitive* if, for any pair of elements $a, b \in G$, $\|a\| = \|b\|$ implies that there exists an automorphism of G that maps a to b. Although local Warfield groups are not in general transitive, torsion simply presented groups are transitive, as are all their isotype subgroups [**HM1**, Theorem 4.2]. Thus, in the presence of elements of infinite order, a more general notion of transitivity is called for. If G is a p-local k-group, we follow [**HU**] and say that G is *weakly transitive* if there exists an automorphism of G that maps a to b whenever a and b are elements of G with $\|a\| = \|b\|$ and $V(a) = V(b)$. (Of course the latter two conditions are necessary in order for an automorphism of G to map a to b). As a final preface to our next result, observe that one consequence of Theorem 2.6 is that any Σ-isotype subgroup of a local Warfield group is a k-group.

THEOREM 5.1. *A Σ-isotype subgroup of a local Warfield group is weakly transitive.*

PROOF. Let H be a Σ-isotype subgroup of a local Warfield group and suppose that a and b are elements of H with $\|a\| = \|b\|$ and $V(a) = V(b)$.

If either a or b has finite order, then they both must have finite order since they have the same type vector. Likewise, if either a or b is primitive, then they are both primitive. In each case, $M = \langle a \rangle$ and $M' = \langle b \rangle$ are finitely generated knice subgroups of H. Moreover, since $\|a\| = \|b\|$, there is a height-preserving isomorphism $\pi : M \to M'$ obtained by mapping a to b. Therefore, by Corollary 4.2, π extends to an automorphism of H that maps a to b.

It remains to consider the case in which both a and b are nonprimitive elements of infinite order. The hypotheses $\|a\| = \|b\|$ and $V(a) = V(b)$ imply that there exist a nonnegative integer n and incomparable sets of primitive elements $\{x_1, x_2, \ldots, x_r\}$ and $\{y_1, y_2, \ldots, y_r\}$ contained in H such that $p^n a = x_1 + x_2 + \cdots + x_r$ and $p^n b = y_1 + y_2 + \cdots + y_r$ with $\|x_i\| = \|y_i\|$ for all $i \le r$. (For a detailed justification, see [**HU**, Proposition 7].) Therefore, by increasing n if necessary, $N = \langle x_1 \rangle \oplus \langle x_2 \rangle \oplus \cdots \oplus \langle x_r \rangle$ and $N' = \langle y_1 \rangle \oplus \langle y_2 \rangle \oplus \cdots \oplus \langle y_r \rangle$ are $*$-valued coproducts and hence knice subgroups of H. Moreover, the correspondence $x_i \mapsto y_i$ induces a height-preserving isomorphism between these subgroups. Since finite extensions of knice subgroups are knice, $M = \langle N, a \rangle$ and $M' = \langle N', b \rangle$ are also finitely generated knice subgroups of H. Recalling that $N = \langle x_1 \rangle \oplus \langle x_2 \rangle \oplus \cdots \oplus \langle x_r \rangle$ is a $*$-valued coproduct and that $p^n a = x_1 + x_2 + \cdots + x_r$, one sees that it is impossible for an element of the coset $p^m a + N$ to have height strictly greater than $|p^m a|$ whenever $m < n$. Consequently, $p^m a$ is proper with respect to N for $m = 0, 1, \ldots, n-1$. But the same holds for b and N', and therefore a well-known argument shows that the indicated map between N and N' extends to a height-preserving isomorphism $\pi : M \to M'$ with $\pi(a) = b$. Another application of Corollary 4.2 completes the proof. □

COROLLARY 5.2. *A p^σ-high subgroup of a local Warfield group is weakly transitive.*

PROOF. This is an immediate consequence of Proposition 2.3 and Theorem 5.1. □

COROLLARY 5.3. *An isotype knice subgroup of a local Warfield group is weakly transitive.*

PROOF. By Corollary 2.5, an isotype knice subgroup is Σ-isotype. □

COROLLARY 5.4. *If H is a Σ-isotype subgroup of a transitive local Warfield group G, then H is also transitive.*

PROOF. Suppose a and b are elements of H such that $\|a\| = \|b\|$. Since G is transitive, it must be the case that $V(a) = V(b)$. Because type vectors of elements of H are the same whether computed in H or in G, the corollary is a consequence of Theorem 5.1. □

From the preceding, it is clear that a weakly transitive k-group G is transitive if and only if it satisfies the transitivity condition formulated in the following definition.

DEFINITION 5.5. A k-group G is said to satisfy the *transitivity condition* provided that for any pair of incomparable sets of primitive elements $\{x_1, x_2, \ldots, x_r\}$ and $\{y_1, y_2, \ldots, y_s\}$, $\bigwedge_{i=1}^r \|x_i\| = \bigwedge_{j=1}^s \|y_j\|$ implies that the height-sequence vectors $(\|x_1\|, \|x_2\|, \ldots, \|x_r\|)$ and $(\|y_1\|, \|y_2\|, \ldots, \|y_s\|)$ are equivalent.

In other words, a k-group G satisfies the transitivity condition if and only if whenever two elements of G have the same height sequence, they also have the same type vector.

Recall that a p-local group G is *fully transitive* if, whenever a and b are elements of G with $\|a\| \leq \|b\|$, there exists an endomorphism of G that maps a to b. In [**F**] and in [**HU**, Theorem 12] it is proved that a local Warfield group is transitive if and only if it is fully transitive. The same condition holds for Σ-isotype subgroups of local Warfield groups. To establish this fact, we require the next result.

PROPOSITION 5.6. *The following hold for any k-group G.*
(1) *If G satisfies the transitivity condition, then so does $G \oplus G$.*
(2) *If G is fully transitive, then G satisfies the transitivity condition.*

PROOF. For (1), assume that G satisfies the transitivity condition and let $K = G \oplus G$. Then the canonical isomorphism
$$K(\bar{\alpha})/K(\bar{\alpha}^*) \cong (G(\bar{\alpha})/G(\bar{\alpha}^*)) \oplus (G(\bar{\alpha})/G(\bar{\alpha}^*))$$
implies that if (x, y) is a primitive element of K with $\|(x, y)\| = \bar{\alpha}$, then both x and y belong to $G(\bar{\alpha})$ and at least one of x and y is not in $G(\bar{\alpha}^*)$. Accordingly, there exists a nonnegative integer n such that at least one of $p^n x$ or $p^n y$ is a primitive element of G with height sequence $p^n \bar{\alpha}$. Therefore, if $\{(x_1, y_1), (x_2, y_2), \ldots, (x_r, y_r)\}$ is an incomparable set of primitive elements in K with $\|(x_i, y_i)\| = \bar{\alpha}_i$ for each i, then there corresponds a nonnegative integer n and an incomparable set of primitive elements $\{z_1, z_2, \ldots, z_r\}$ in G with $\|z_i\| = p^n \bar{\alpha}_i$. From this observation, it is evident that $K = G \oplus G$ inherits the transitivity condition from G.

Next we assume that G is fully transitive. Let a and b be elements of infinite order in G with $\|a\| = \|b\|$. We need to show that $V(a) = V(b)$. Replacing a by an appropriate p-power multiple, we may write $a = x_1 + x_2 + \cdots + x_r$, where $\{x_1, x_2, \ldots, x_r\}$ is an incomparable set of primitive elements. Let $\bar{\alpha}_i = \|x_i\|$ for $i = 1, 2, \ldots, r$. Select endomorphisms ϕ and θ of G such that $\phi(a) = b$ and $\theta(b) = a$. Then, $(\theta\phi)(a) = a$ so that
$$a = x_1 + x_2 + \cdots + x_r = \theta(y_1) + \theta(y_2) + \cdots + \theta(y_r)$$
where $y_i = \phi(x_i)$ for each i. By Lemma 1.2, each $\theta(y_i)$ is primitive and there is a nonnegative integer n such that $\|\theta(p^n y_i)\| = \|p^n x_i\| = p^n \bar{\alpha}_i$. Notice then that $\|p^n y_i\| = \|p^n x_i\|$ because $\|x_i\| \leq \|y_i\| \leq \|\theta(y_i)\|$ due to the fact that endomorphisms cannot decrease heights. We claim that all the $p^n y_i$'s are primitive. If this were not so, there would be a j such that $p^n y_j \in G(p^n \bar{\alpha}_j^*)$. But $G(p^n \bar{\alpha}_j^*)$ is a fully invariant subgroup of G and so $\theta(p^n y_j) \in G(p^n \bar{\alpha}_j^*)$, contrary to the fact that $\theta(p^n y_j)$ is primitive. Since $b = \phi(a)$, $p^n b = p^n y_1 + p^n y_2 + \cdots + p^n y_r$. Moreover, $\{p^n y_1, p^n y_2, \ldots, p^n y_r\}$ is an incomparable set of primitive elements. Therefore, $V(a) = V(b)$ since each of these type vectors is represented by the height-sequence vector $(p^n \bar{\alpha}_1, p^n \bar{\alpha}_2, \ldots, p^n \bar{\alpha}_r)$. □

THEOREM 5.7. *Suppose H is a Σ-isotype subgroup of a local Warfield group G. Then, H is transitive if and only if it is fully transitive.*

PROOF. First assume that H is transitive and that a and b are elements of H with $\|a\| \leq \|b\|$. Because H is a k-group by Theorem 2.6, part (1) of Proposition 5.6 says that $H \oplus H$ satisfies the transitivity condition. Since $H \oplus H$ is obviously Σ-isotype in $G \oplus G$, it follows from Theorem 5.1 that $H \oplus H$ is transitive. Now $\|a\| \leq$

$\|b\|$ implies that $\|(a,0)\| = \|(a,b)\|$ in $H \oplus H$ and so there is an automorphism of $H \oplus H$ that maps $(a,0)$ to (a,b). Composing this automorphism with the appropriate canonical maps yields the desired endomorphism of H mapping a to b. Conversely, suppose that H is fully transitive. Since H is weakly transitive by Theorem 5.1 and satisfies the transitivity condition by part (2) of Proposition 5.6, H is indeed transitive. □

The reader should not be left with the false impression that Theorems 5.1 and 5.7 may be true for k-groups in general. Indeed, they even fail in the torsion case. To see this, observe that any p-group is (trivially) a p-local k-group, and that the notions of transitivity and weak transitivity coincide for such groups. However, it is well known that there exist nontransitive p-groups (for example, see [**M**] and [**H2**]). Moreover, as shown in [**C**], there exist fully transitive p-groups (for any prime p) that are nontransitive, and transitive 2-groups that are not fully transitive. For a more honest example, we now construct a nonsplit mixed k-group that is neither weakly transitive nor fully transitive.

EXAMPLE 5.8. Let C be an unbounded p-group with $p^\omega C = 0$ and select a pure subgroup B of C with $C/B \cong \mathbb{Z}(p^\infty)$. Fix a homomorphism $\psi : C \to \mathbb{Q}/\mathbb{Z}_p$, with $\ker \psi = B$, and let L be the subgroup of $C \oplus \mathbb{Q}$ consisting of all pairs (x,y) such that $\psi(x) = y + \mathbb{Z}_p$. It is straightforward to verify the following: $p^\omega L = \mathbb{Z}_p$, $L/p^\omega L \cong C$, the maximal torsion subgroup of L is B, and B is a p^ω-high subgroup of L. Thus, if we take c to be a generator of $p^\omega L$ (as a $\mathbb{Z}_p$-module), then c is primitive with height sequence $\|c\| = \{\omega, \omega+1, \omega+2, \dots\}$.

Now, as the above shows, there exist mixed p-local groups K and H and primitive elements $a \in K$ and $b \in H$ that satisfy the following conditions: $K/\langle a \rangle$ is an unbounded torsion-complete p-group and $H/\langle b \rangle$ is an unbounded direct sum of cyclic p-groups, $p^\omega K = \langle a \rangle$ and $p^\omega H = \langle b \rangle$, and $\|a\| = \|b\| = \{\omega, \omega+1, \omega+2, \dots\}$. Moreover, K and H can be constructed so that $K \oplus H$ does not split.

Finally, set $G = K \oplus H$. Then, G is certainly a k-group since it has the decomposition basis $\{a, b\}$. Also, $\|a\| = \|b\|$ so that $V(a) = V(b)$ due to the fact that both a and b are primitive. However, there is not even an endomorphism of G that maps a to b. Indeed, suppose to the contrary that G has an endomorphism ϕ with $\phi(a) = b$. Then, a slight modification of the proof of Theorem 2.4 in [**M**] yields elements $w, z \in K$ such that $pw = a$ and $\phi(w) - z \in p^\omega G = \langle a \rangle \oplus \langle b \rangle$. It now follows that $b \in K \oplus \langle pb \rangle$, which is absurd.

6. Special Constructions

By Theorem 2.6, a $*$-isotype subgroup H of a k-group G is Σ-isotype in G if and only if H is a k-group. But, as the following theorem demonstrates, $*$-isotype subgroups of k-groups are not necessarily Σ-isotype (and therefore not necessarily k-groups).

THEOREM 6.1. *There exists a simply presented p-local group G that contains a $*$-isotype subgroup H that is not Σ-isotype. Moreover, H can be chosen so that it contains no primitive elements.*

PROOF. First, partition the nonnegative integers into infinitely many infinite sets P_q, $q < \omega$. For each q, let $\mathcal{F}_q$ be the collection of all finite subsets of P_q and select a sequence $\{F_{q,n}\}_{n<\omega}$ satisfying the following conditions.

(a) Each $F_{q,n}$ belongs to $\mathcal{F}_q$.
(b) For any $F \in \mathcal{F}_q$, $F_{q,n} = F$ for infinitely many $n < \omega$.

In accordance with Proposition 6.1 of [**HM3**], for each q we obtain a sequence $\{E_{q,i}\}_{i<\omega}$ such that the following conditions hold whenever $i, j, k < \omega$.

(1) $E_{q,i} \subseteq P_q$.
(2) $E_{q,i}$ is infinite and $F_{q,i} \subseteq E_{q,i}$.
(3) $E_{q,i} \cap E_{q,j}$ is finite whenever $i \neq j$.
(4) $E_{q,k} \cap E_{q,j}$ properly contains $E_{q,i} \cap E_{q,j}$ whenever $i \neq j$ and $k > \max\{i,j\}$.

For each $n < \omega$, now set $E_n = \bigcup_{q<\omega} E_{q,n}$ and define a height sequence $\bar{\alpha}_n = \{\alpha_{n,m}\}_{m<\omega}$ by decreeing that $\alpha_{n,m} = 2m+1$ if $m \in E_n$ and $\alpha_{n,m} = 2m$ if $m \notin E_n$. Associate with each $\bar{\alpha}_n$ a rank 1 simply presented p-local group G_n that contains an element g_n with $\|g_n\| = \bar{\alpha}_n$, and take G to be the direct sum of the G_n's. Now define a map ϕ from G into the additive group of rationals $\mathbb{Q}$ by mapping each g_n to 1. Set $H = \ker \phi$.

Clearly H is isotype in G by virtue of G/H being torsion free. We intend to show that for every $x \in H$ of infinite order, there exists a nonnegative integer e such that $p^e x \in H(\|p^e x\|^*)$. Note that this condition would certainly preclude H from having any primitive elements or from being a k-group. Moreover, this condition together with the fact that H is isotype in G would then imply that H is actually $*$-isotype in G, but an application of Theorem 2.6 would show that H is not Σ-isotype.

Suppose $x \in H$ has infinite order. Then, for some nonnegative integer e,

$$p^e x = c_1 g_1 + c_2 g_2 + \cdots + c_r g_r$$

where $c_i \in \mathbb{Z}_p$. Since $\phi(p^e x) = 0$ and $\phi(g_i) = 1$, $\sum_{i=1}^{r} c_i = 0$ and at least two c_i's are nonzero. By invoking condition (4) and the definition of the E_n's, there exists $k > r$ such that $E_k \cap E_i$ not only properly contains $E_i \cap E_j$ for all $i, j \leq r$, $i \neq j$, but also, $(E_k \cap E_i) \setminus (E_i \cap E_j)$ is infinite. Therefore,

$$p^e x = c_1(g_1 - g_k) + c_2(g_2 - g_k) + \cdots + c_r(g_r - g_k),$$

where, for each $i \leq r$, $c_i(g_i - g_k) \in H$, $\|c_i(g_i - g_k)\| \geq \|p^e x\|$, and the height sequence $\|c_i(g_i - g_k)\|$ differs from $\|p^e x\|$ in infinitely many places (whether computed in H or in G). In other words, $p^e x \in H(\|p^e x\|^*)$, as desired. □

Observe that the $*$-isotype subgroup constructed in the proof of Theorem 6.1 has infinite rank. Thus it does not resolve the following fundamental question.

QUESTION 6.2. Must a finite rank $*$-isotype subgroup of a local Warfield group be Σ-isotype?

It is of interest to note that the torsion-free analogue of the preceding question has an affirmative answer. Indeed, it is proved in [**DR**] that a finite rank $*$-pure subgroup of a completely decomposable torsion-free group is a direct summand and hence a Σ-isotype subgroup.

In Corollary 2.9 it was shown that a countable Σ-isotype subgroup of a k-group is a Warfield group. In our final theorem, we show that the restriction on cardinality cannot be removed. In fact, we actually do more by showing that there exists a Σ-isotype subgroup of a p-local simply presented group that does not have a decomposition basis. In order to prove this, we need a definition and a preliminary proposition. Our definition strengthens the notion of a separable subgroup introduced in [**H3**] by replacing heights by height sequences.

DEFINITION 6.3. Call a subgroup H of a p-local group G *strongly separable* if for each $g \in G$ there is a corresponding countable subset $\{h_n\}_{n<\omega}$ of H with the following property: if $h \in H$, there exists $n < \omega$ such that $\|g+h\| \leq \|g+h_n\|$.

The following result was proved by M. Lane in [**L**] for separable rather than strongly separable subgroups.

PROPOSITION 6.4. *Suppose H is an isotype subgroup of a p-local group G. If H has a decomposition basis X, then $\langle X \rangle$ is a strongly separable subgroup of G.*

PROOF. Suppose to the contrary that $\langle X \rangle$ is not strongly separable in G. Then, there is a $g \in G$ such that for each countable subset C of $\langle X \rangle$, there exists $x_C \in \langle X \rangle$ such that $\|g + x_C\| \leq \|g + h\|$ fails for each $h \in C$.

Write $\langle X \rangle = \bigoplus_{i \in I} \langle x_i \rangle$, a valued coproduct with each x_i of infinite order, and select an ascending chain
$$J(0) \subseteq J(1) \subseteq \cdots \subseteq J(n) \subseteq \ldots \qquad (n < \omega)$$
of countable subsets of I such that $J(0) = \emptyset$ and the following condition holds for all n:

if $h' \in \bigoplus_{i \in J(n)} \langle x_i \rangle$, and if $|p^k(g+h)| \gneq |p^k(g+h')|$ for some $h \in \langle X \rangle$ and $k < \omega$, then there exists $h'' \in \bigoplus_{i \in J(n+1)} \langle x_i \rangle$ such that $|p^k(g+h'')| \gneq |p^k(g+h')|$.

Now, set $J(\omega) = \bigcup_{n < \omega} J(n)$ and $X_\omega = \bigoplus_{i \in J(\omega)} \langle x_i \rangle$. Because X_ω is a countable subgroup of $\langle X \rangle$, there exists $h \in \langle X \rangle$ such that $\|g+h\| \leq \|g+y\|$ fails for all $y \in X_\omega$. In particular, $\|g+h\| \leq \|g+h'\|$ fails, where h' is the projection of h onto X_ω. Thus,
$$|p^k(g+h)| \gneq |p^k(g+h')|$$
for some $k < \omega$. According to the selection of the $J(n)$'s, there exists $h'' \in X_\omega$ such that
$$|p^k(g+h'')| \gneq |p^k(g+h')|.$$
Therefore, we have
$$|p^k(h'-h'')| = |p^k(g+h')| \lneq |p^k(g+h) - p^k(g+h'')| = |p^k(h-h'')|.$$
However, this is absurd. Indeed, as a consequence of the facts that $p^k(h'-h'')$ is the projection of $p^k(h-h'')$ onto X_ω and that $\langle X \rangle = \bigoplus_{i \in I} \langle x_i \rangle$ is a valued coproduct, we obtain $|p^k(h'-h'')| \geq |p^k(h-h'')|$. □

REMARK 6.5. In the preceding proof, the full force of the assumption that X is a decomposition basis for H was not used. All that was required of X is that $\langle X \rangle$ is a valued coproduct of countable subgroups of H.

THEOREM 6.6. *There exists a Σ-isotype subgroup of a simply presented p-local group that does not have a decomposition basis.*

PROOF. Let $\mathcal{S}$ be the set of all height sequences s such that each entry of s is a nonnegative integer. For each $s \in \mathcal{S}$, select a rank 1 simply presented p-local group G_s which contains an element g_s with $\|g_s\| = s$. Set $G = \bigoplus_{s \in \mathcal{S}} G_s$ and define a map $\phi: G \to \mathbb{Q}$ by decreeing that $\phi(g_s) = 1$ for all $s \in \mathcal{S}$. We claim that $H = \ker \phi$ is Σ-isotype in G, but that H does not have a decomposition basis.

To see that H is Σ-isotype, suppose $h \in H \cap \sum_{i=1}^{r} G(\bar{\alpha}_i)$ for some height sequences $\bar{\alpha}_1, \bar{\alpha}_2, \ldots, \bar{\alpha}_r$. Since G/H is torsion free, H is isotype in G. Therefore, by Corollary 2.2, to show that H is Σ-isotype in G, it is enough to show that $p^k h \in \sum_{i=1}^{r} H(p^k \bar{\alpha}_i)$ for some nonnegative integer k.

Write $h = x_1 + x_2 + \cdots + x_r$, where $x_i \in G(\bar{\alpha}_i)$. Thus, there exist a nonnegative integer k and distinct $s_1, s_2, \ldots, s_t \in \mathcal{S}$ such that, for each $i \leq r$,

$$p^k x_i = c_{i,1} g_{s_1} + c_{i,2} g_{s_2} + \cdots + c_{i,t} g_{s_t}$$

for some $c_{i,j} \in \mathbb{Z}_p$. Observe that $\sum_{i \leq r, j \leq t} c_{i,j} = 0$ and that each $c_{i,j} g_{s_j} \in G(p^k \bar{\alpha}_i)$. Now select a height sequence $s \in \mathcal{S}$ such that $s \geq s_j$ for all j, $1 \leq j \leq t$. Then,

$$y_i = c_{i,1}(g_{s_1} - g_s) + c_{i,2}(g_{s_2} - g_s) + \cdots + c_{i,t}(g_{s_t} - g_s)$$

is in $H(p^k \bar{\alpha}_i)$ and $p^k h = y_1 + y_2 + \cdots + y_r \in \sum_{i=1}^{r} H(p^k \bar{\alpha}_i)$. Therefore, H is Σ-isotype in G.

Finally, suppose to the contrary that H has a decomposition basis X. Select and fix a particular $s_0 \in \mathcal{S}$ and set $g = g_{s_0}$. According to Proposition 6.4, $\langle X \rangle$ is strongly separable in G. Therefore, there is a countable subset $\{h_n\}_{n<\omega} \subseteq \langle X \rangle$ such that if $h \in \langle X \rangle$ and if $k < \omega$, there exists an n (depending on h and k) such that

$$\|p^k g + h\| \leq \|p^k g + h_n\|.$$

Since g is fixed, the collection of all height sequences $\|p^k g + h_n\|$ is a countable subset of $\mathcal{S}$. (They are in $\mathcal{S}$ because $p^k g + h_n \notin H$ and H contains all torsion.) Select a bijection γ from the countable collection of ordered pairs $\{(k, n) : k < \omega, n < \omega\}$ to the nonnegative integers, and construct a height sequence $\bar{\alpha} = \{\alpha_l\}_{l<\omega} \in \mathcal{S}$ that satisfies the following condition: if $\gamma(k, n) = l$, then

$$\alpha_l \geq |p^l(p^k g + h_n)| + 1.$$

Now set $h' = g_{\bar{\alpha}} - g = g_{\bar{\alpha}} - g_{s_0} \in H$ and select a nonnegative integer m so that $p^m h' \in \langle X \rangle$. Then, $\|p^m g + p^m h'\| = \|p^m g_{\bar{\alpha}}\|$ and at least one entry of each $\|p^k g + h_n\|$ is strictly smaller than the corresponding entry of $\|p^m g_{\bar{\alpha}}\| = p^m \bar{\alpha}$. In particular, the inequality

$$\|p^m g + p^m h'\| \leq \|p^m g + h_n\|$$

fails for all n. This contradiction shows that H does not have a decomposition basis and completes the proof of the theorem. $\square$

One interpretation of Theorem 6.6 is that the class of Σ-isotype subgroups of Warfield groups is much larger than the class of Warfield groups.

References

[C]   A.L.S. Corner, *The independence of Kaplansky's notions of transitivity and full transitivity*, Quart. J. Math. Oxford Ser.(2), 27, 1976, 15–20

[DR]   M. Dugas and K. M. Rangaswamy, *On torsion-free abelian k-groups*, Proc. Amer. Math. Soc., 99, 1987, 403–408

[F]   S. Files, *Transitivity and full transitivity for nontorsion modules*, J. Algebra, 197, 1997, 468–478

[H1]   P. Hill, *On the classification of abelian groups*, Photocopied manuscript, 1967

[H2]   P. Hill, *On transitive and fully transitive primary groups*, Proc. Amer. Math. Soc., 22, 1969, 414–417

[H3]   P. Hill, *Isotype subgroups of totally projective groups*, in **Abelian Group Theory**, Lecture Notes in Math., 894, Springer-Verlag, New York, 1981, 305–321

[HM1]   P. Hill and C. Megibben, *On the theory and classification of abelian p-groups*, Math. Z., 190, 1985, 17–38
[HM2]   P. Hill and C. Megibben, *Axiom 3 modules*, Trans. Amer. Math. Soc., 295, 1986, 715–734
[HM3]   P. Hill and C. Megibben, *Pure subgroups of torsion-free groups*, Trans. Amer. Math. Soc., 303, 1987, 765–778
[HM4]   P. Hill and C. Megibben, *The local equivalence theorem*, Contemp. Math., 87, 1989, 201–219
[HM5]   P. Hill and C. Megibben, *Mixed groups*, Trans. Amer. Math. Soc., 334, 1992, 121–142
[HU]   P. Hill and W. Ullery, *The transitivity of local Warfield groups*, J. Algebra, 208, 1998, 643–661
[HRW]   R. Hunter, F. Richman and E. Walker, *Warfield modules*, in **Abelian Group Theory**, Lecture Notes in Math., 616, Springer-Verlag, New York, 1977, 87–123
[L]   M. Lane, *On p-local abelian groups with decomposition bases*, Math. Z., 202, 1989, 129–141
[M]   C. Megibben, *Large subgroups and small homomorphisms*, Mich. Math. J., 13, 1966, 153–160

(Hill) DEPARTMENT OF MATHEMATICS, AUBURN UNIVERSITY, AUBURN, ALABAMA 36849
E-mail address: hillpad@math.auburn.edu

(Megibben) DEPARTMENT OF MATHEMATICS, VANDERBILT UNIVERSITY, NASHVILLE, TENNESSEE 37240
E-mail address: megibben@math.vanderbilt.edu

(Ullery) DEPARTMENT OF MATHEMATICS, AUBURN UNIVERSITY, AUBURN, ALABAMA 36849
E-mail address: ullery@math.auburn.edu

Character Modules and Endomorphism Rings of Modules Over Artinian Serial Rings

George Ivanov

Dedicated to Professor Laszlo Fuchs in honour of his 75th birthday.

ABSTRACT. Some methods for studying modules over Artinian serial rings are developed. These are used to show that it is sometimes possible to define "character modules" for modules over noncommutative rings and that these can have nice properties similar to the commutative case.

Endomorphism rings of abelian groups have been studied extensively for more than half a century. In particular, those of torsion abelian groups are well understood. Their study is the study of p–groups. For modules over noncommutative rings the situation is much more sketchy. Since the indecomposable p–groups are uniserial, it would appear that modules over Artinian serial rings might be an appropriate first (partial) generalisation. The structure of these is particularly simple so their study allows us to concentrate on the problems caused by having nonisomorphic indecomposables of the same length. This is one of the main differences with the torsion abelian group case.

We study two classes of Artinian serial rings: the nonsingular ones and those which are Quasi-Frobenius (QF) rings. In a sense these are at the opposite ends of the structure spectrum of Artinain serial rings. The remaining ones are fairly similar to the QF ones. In fact, since all Artinian serial rings are quotients of serial QF rings, results about them can be deduced from the QF case. To each ring we associate a graph which gives us all the information we want about the endomorphisms of its modules.

For commutative rings there is a natural module structure on the group of homomorphisms between two modules. For noncommutative rings that is not the case and it is one of the main reasons that modules over the latter are so much more difficult to study. However, for the rings that we study it is possible, under certain conditions, to impose a module structure on the group of homomorphisms and even one which behaves nicely. It turns out that there are three obstructions to performing this task. The first is that for rings with non-central idempotents the the

---

1991 *Mathematics Subject Classification.* Primary 16D10, 16D70, 16D50; Secondary 20K30, 20K10.

Key words and phrases. character modules, endomorphism rings, uniserial, Quasi-Frobenius.

right ideals are "perpendicular" to the left ones; the second is the noncommutativity of the endomorphism rings of the indecomposable projectives; and the third is the different multiplicites of the indecomposable projectives in a decomposition of the ring. The first one we can overcome because we know the structure of our rings well. For the strong properties that we want, the second one can only be handled by assuming commutativity. For the third, we reduce the study to that of the basic subring. This is a far cry from the commutative situation but a lot of usefull information can still be obtained. Moreover there is hope that similar constructions can be performed for other rings.

Recently Anh [1] has shown that the well known theorem that a finite abelian group is isomorphic to its character group extends to modules over commutative Artinian serial rings. We extend this result to noncommutrative serial rings in the following way. First we show that Anh's definition of character module extends and we get a module, albeit on the opposite side. Secondly, there is a similarity between the structures of the module and its character module. It is, in fact, possible to formalise a notion of *left-right isomorphism* which the two modules satisfy, if the ring has two additional properties. The first is that the endomorphism ring of an indecomposable projective is commutative. The second is that the ring is basic. This shows that the full force of commutativity of the underlying rings is necessary for both the abelian group and Anh's results.

The requirement that the ring be basic can be dropped, if the multiplicities of the principle indecomposables are all the same. But one needs to slightly modify the definition of character module.

1. Basics

Throughout the paper all modules will be unital left modules. *Ring* will always mean indecomposable basic Artinian serial ring and R will always denote such a ring. The letters e and f will always stand for idempotents. The length of a uniserial module U will be denoted by $\ell(U)$. We will use the algebraic convention for homomorphisms. That is, they will be written on the right of their arguments (without brackets) and composition will be represented by concatenation in the order in which the maps act.

We fix a decomposition $R = Re_0 \oplus Re_1 \oplus ... \oplus Re_n$ of R into indecomposables. We know from [4] that it can be assumed that for each i, the maximal subideal We_i of Re_i is a homomorphic image of Re_{i+1}, where addition of subscripts is taken modulo $n + 1$. We call Re_{i+1} the **successor** of Re_i. Only one We_i can be zero, necessarily We_n, in which case we will abuse termino;ogy and say that Re_0 is the successor of Re_n. Three cases occur. The first is when Re_i has length $n + 1 - i$, for each i. Then for each i, We_i is isomorphic to Re_{i+1}. At the other extreme, all the Re_i have the same length and all the epimorphisms $Re_{i+1} \to We_i$ have simple kernels. In the third case the lengths of Re_i and Re_{i+1} vary by at most n.

The first class is the class of nonsingular rings. They are isomorphic to the rings of triangular matrices over skew-fields ([3]). The second is the class of Quasi-Frobenius rings which are those whose indecomposable projectives have the same lengths. That is well known, but we give a proof for completeness. We take the currently most commonly used definition of Quasi-Frobenius rings: self-injective Artinian rings.

PROPOSITION 1.1. *R* is Quasi-Frobenius if, and only if, the composition lengths of its indecomposable projectives are the same.

PROOF. First assume that all the composition lengths are equal. By the Nakayama–Fuller criterion ([**2**], Theorem 3.1) it is sufficient to show that there is a permutation α of $\{0, 1, \ldots, n\}$ such that the socle of Re_i is an image of $Re_{i\alpha}$ and the socle of $e_{i\alpha}R$ is an image of e_iR. Different Re_i have nonisomorphic socles so the function which, associates $\text{Soc}(Re_i)$ with its projective cover, induces a permutation α of $\{0, 1, \ldots, n\}$. That is, $\text{Soc}(Re_i) = e_{i\alpha}W^t e_i$, for some power t. Since $e_{i\alpha}W^{t+1} = 0$ and $e_{i\alpha}R$ is uniserial, $e_{i\alpha}W^t e_i = \text{Soc}(e_{i\alpha}R)$ is a homomorphic image of e_iR — as required.

Now assume that R is Quasi-Frobenius and let $i \in \{0, 1, \ldots, n\}$. Then every indecomposable injective is isomorphic to an Re_k and so $E(Re_i/\text{Soc}(Re_i)) \cong Re_j$, for some j. From [**4**] we know that We_{i-1} is a homomorphic image of maximal length in R. It must be a proper image since Re_i is injective. Therefore $We_{i-1} \cong Re_i/\text{Soc}(Re_i)$ and so $Re_{i-1} \cong E(Re_i/\text{Soc}(Re_i))$. That is, Re_i and Re_{i-1} have the same length — as required. □

It follows from the above Proposition 1.1 and Theorem 11 of [**4**] that a QF serial ring R is determined by a uniserial ring A, the number $n+1$ of nonisomorphic indecomposable projectives (which we call the **breadth** of R) and the length d of these (which we call the **depth** of R). We will therefore denote such a ring by $\text{QFS}(A, n+1, d)$.

It is often useful to draw a graph representing the subideal structure of the ring being studied. One way is on a vertical cylinder. The vertices are placed on integer lattice points as follows. The top (horizontal) row has $n+1$ vertices labelled $Re_0, \ldots, Re_n$. In the (vertical) column headed by Re_i are placed vertices labelled in order by $We_i, W^2 e_i, \ldots, W^{d_i-1} e_i$. Hence a column represents the lattice of submodules of an indecomposable projective. We place an arrow from Re_{i+1} to We_i to show that the latter is an image of the former and from Re_0 to We_n, if Re_n is not simple. Figure 1 shows such a graph for a nonsingular ring, which we denote by $\Gamma_{\text{id}}(n+1)$; and Figure 2 for a QF ring, which we denote by $\Gamma_{\text{id}}(n+1, d)$. Note that the first graph is actually planar. Since $W^j e_{i-1}$ is an image of $W^{j-1} e_i$ we could add additional arrows in the graphs to represent this fact.

The results of [**4**] show that to each such graph there is a serial ring with that subideal structure. We use this to get the following result, specials case of which were proved by Murase [**6**].

THEOREM 1.2. *Every Artinian serial ring is a quotient ring of a QF Artinian serial ring.*

PROOF. Let R be our canonic indecomposable Artinian serial ring with its canonic decomposition into indecomposable projectives. From the Re_i of shortest length we pick one whose successor has maximal length amongst such successors and call it and its successor Re_k and Re_{k+1}, repectively. If an Re_i is simple it is necessarily Re_n so we rename the Re_i so that Re_n becomes Re_k. We want to show that R is a homomorphic image of a serial ring S, with canonic decomposition $S = Sf_0 \oplus Sf_1 \oplus \ldots \oplus Sf_n$, such that $\ell(Sf_i) = \ell(Re_i)$ for $i \neq k$ and $\ell(Sf_k) = \ell(Rf_k) + 1$. By repeating the argument (a finite number of times) it would follow that R is an image of a serial ring whose principal indecomposables have the same length and this ring would therefore be a QF ring (by Proposition 1.1).

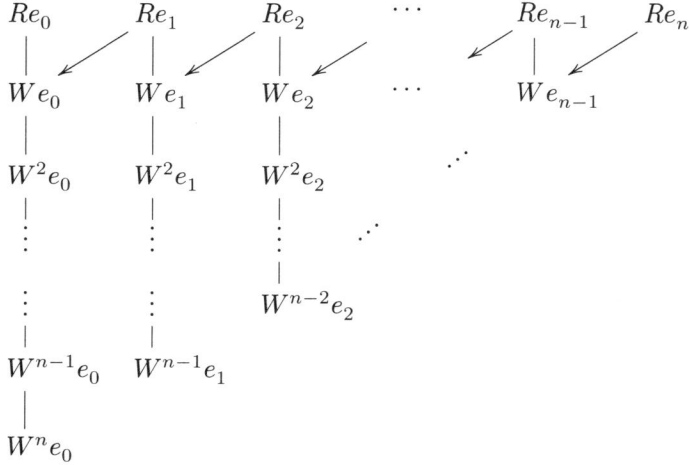

FIGURE 1. The Graph $\Gamma_{\mathrm{id}}(n+1)$

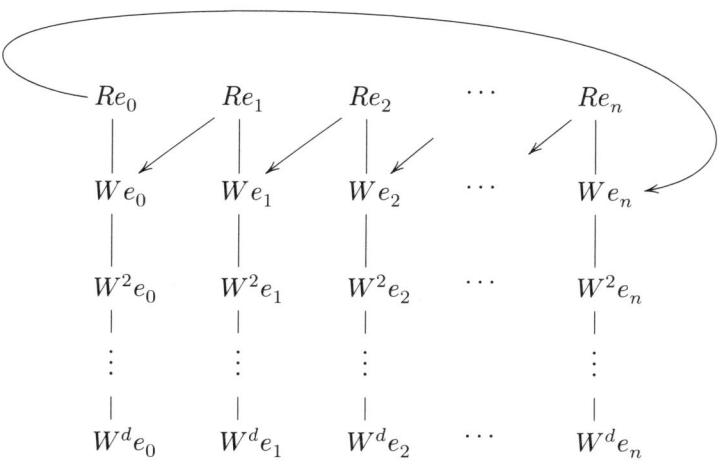

FIGURE 2. The Graph $\Gamma_{\mathrm{id}}(n+1,d)$

If no Re_i is simple then $\mathrm{Soc}(Re_k)$ is an image of We_{k+1}^t (for some t) and We_{k+1}^t is an image of Re_{j-1}, say. Therefore $\mathrm{Soc}(Re_k) \cong Re_{j-1}/We_{j-1}$ and $We_{k+1}^{t+1}/We_{k+1}^{t+2}$ is a non-zero image of Re_j since the length of Re_{k+1} exceeds the length of Re_k. In that case let $V = We_{k+1}^{t+1}/We_{k+1}^{t+2}$ and if Re_k is simple, let $V = Re_{k+1}/We_{k+1}$. Form the the group direct sum $S = R \oplus V$. We want to make S into a ring by constructing a multiplication which extends the ring structure on R. Let $r, r_1, r_2 \in R$ and $v, v_1, v_2 \in V$ be any elements. We define a multiplication on S by $(r_1, 0) * (r_2, v) = (r_1 r_2, r_1 v)$, where the multiplications are those in R or the action of R on its module V. We define $(0, v_1) * (0, v_2) = (0, 0)$ so it only remains to define $(0, v) * (r, 0)$.

We know from Theorem 11 of [4] that all the skew fields $H_{ii} = e_i Re_i / e_i We_i$ are isomorphic, so there is a bilinear action of (H_{jj}, H_{kk}) on V which makes V into an (H_{jj}, H_{kk})-bimodule which is one dimensional on both sides. For $i \neq k$ and any $r \in e_k R e_i$ or $r \in e_i R$ we define $(o,v) * (r, 0) = (0,0)$ for all $v \in V$. If $r \in e_k R e_r$ then we define $(o,v) * (r, 0) = (0, vr)$ where the product vr is that obtained from the (H_{jj}, H_{kk})-bimodule structure on V.

Since $f_i = (e_i, 0)$ we will abuse notation and refer to the f_i's as e_i's and to $0 \oplus V$ as V. It is easy to check that S is a ring under this multiplication, that V is a two sided ideal of S and that $R \approx S/V$. For S to be a serial ring we need V to be contained in every non-zero $J^m e_k$ and $e_j J^m$, where J is the radical of S. That is, we need to ensure that $V = e_j J^u e_q . e_q J^v e_k$ whenever $e_j W^u e_q . e_q W^v e_k$ is non-zero.

When Re_k is simple for there is nothing to prove as then all $e_p W^u e_q . e_q W^v e_k$ are zero. In the other case, $e_j W^u e_q . e_q W^v e_k$ is non-zero if, and only if, $u + v \leq t+1$. Assume that $u+v \leq t+1$. Then $e_j W^u e_q . e_q W^v e_k = e_j W^{u+v} e_k$ and $e_j W^{u+v} e_k / e_j W^{u+v+1} e_k$ is a one dimensional, on both sides, (H_{jj}, H_{kk})-bimodule. Hence there is an (H_{jj}, H_{kk})-bilinear isomorphism $\alpha : e_j W^{u+v} e_k / e_j W^{u+v+1} e_k \to V$ which is induced by an $(e_j R e_j, e_k R e_k)$-bilinear epimorphism $\beta : e_j W^{u+v} e_k \to V$. We extend the multiplication in S as follows. If $a \in e_j W^u e_q$ and $b \in e_q W^v e_k$ then $(a, 0) * (b, 0) = (0, ab\beta)$. Therefore $V = e_j J^{t+2} e_k$ and so $V \subseteq J^s e_k$ and $V \subseteq e_j J^s$ whenever $s \leq t+2$. This shows that S is a serial ring which completes the proof. □

2. Representing endomorphisms as graphs and matrices

Since all R-modules are direct sums of uniserial modules, the $Re_i / W^t e_i$ are the only indecomposables. For simplicity of notation we will denote $Re_i / W^j e_i$ by U_{ij} where, by definition, $W^0 = R$. That is, U_{ij} is the indecomposable of length j which is a quotient of Re_i.

Much of the structure of the endomorphism ring of an R-module can be obtained from that of the **maximal basic** module $B = \oplus_{ij} U_{ij}$. To study the latter we construct a graph which will give us most of the information we seek. We do this for two classes of rings: the nonsingular ones and the Quasi-Frobenius ones. For our purposes these are done slightly differently, so we do them separately.

2.1. Nonsingular rings. We now change the notation to simplify it. $R = T_N(D)$ will now be the ring of upper triangular $N \times N$ matrices over a skew field D, where N is an integer greater than 1. For $d \in D$ and each pair (i, j), $1 \leq i, j \leq N$, $d_{ij} \in R$ will be the matrix whose only nonzero entry is d at the place (i, j). We denote the identity of D by e and for simplicity, and emphasis, we will often write e_i for e_{ii}. Then Re_N is the biggest indecomposable projective and it is injective. Its quotient modules are also injective and they are the only indecomposable injectives.

We study the endomorphisms of an arbitrary module by first studying those of the maximal basic module B. We begin by constructing a graph $\Gamma_{\text{mod}}(N)$ in the real plane whose vertices are at the points (i, j) for $1 \leq i \leq N$ and $1 \leq j \leq N-i+1$. In the ith column we place the vertex $V_{i+j,j}$ at the point (i, j). We will think of, and refer to, these vertices as the modules $U_{i+j,j}$. Therefore the first column will have as "vertices" the projectives Re_i. The edges in $\Gamma_{\text{mod}}(N)$ will represent special homomorphisms between the U_{ij}. At each vertex there will be a loop, representing the identity map on the corresponding module. Along the ith column there will be vertical arrows pointing up $U_{i,j} \to U_{i,j+1}$. These represent the canonic embeddings $e_i + W^j e_i \to e_{i,i+1} + W^{j+1} e_{i+1}$. Along the diagonals will be edges $U_{i,j} \to U_{i,j-1}$ representing the canonic projections $e_i + W^j e_i \to e_i + W^{j-1} e_i$. These maps have simple kernels. The graph is shown in Figure 3, where we have abused notation and put in the modules as vertices. For clarity of the diagram the loops have been omitted.

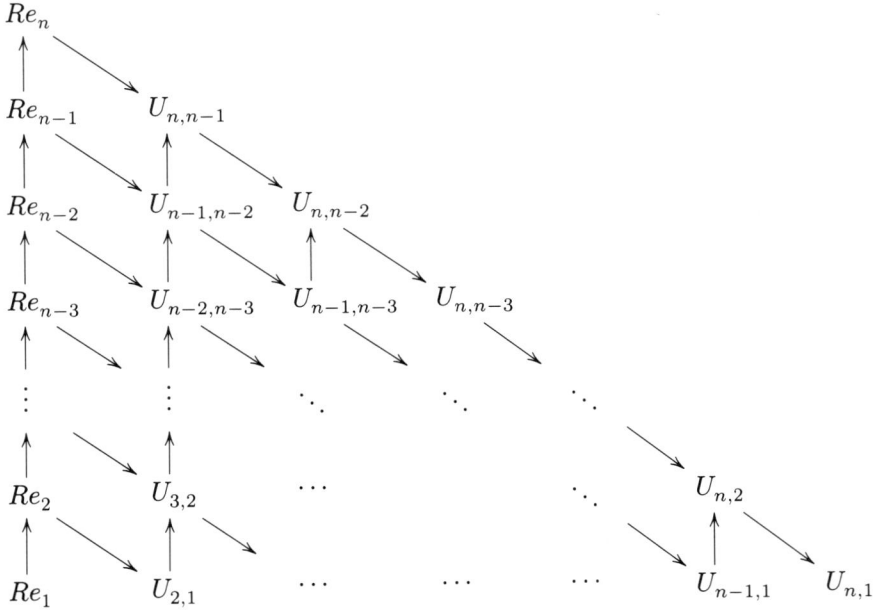

FIGURE 3. The Graph $\Gamma_{\mathrm{mod}}(N)$

The modules on the top (main) diagonal are the indecomposable injectives. Any homomorphism between a uniserial and one immediately vertically above it, or immediately diagonally below it, is simply a d-multiple of one of the above canonic maps, for some $d \in D$. Note also that any simple parrallelogram is commutative in that the maps represented are the same. Therefore all composite parrallelograms are commutative. Hence any homomorphism between two uniserials can be represented by a path in $\Gamma_{\mathrm{mod}}(N)$. A vertical path of m arrows followed by a diagonal path of n arrows is non-zero if, and only if, $n < m$.

The following result is clear but it is useful, for reference purposes, to have it stated.

LEMMA 2.1. If $R = T_N(D)$ then the following describe the maps between the U_{ij}'s.
(a) For each (i,j) the pairs (r,s) such that $\mathrm{Hom}(U_{ij}, U_{rs}) \neq 0$ are those which are indices of vertices in $\Gamma_{\mathrm{mod}}(N)$ which are contained in the parallelogram with corner vertices $U_{i,j}, U_{i,1}, U_{N,N-i+j}$ and $U_{N,N-i+1}$.
(b) For each (r,s) the pairs (i,j) such that $\mathrm{Hom}(U_{ij}, U_{rs}) \neq 0$ are those which are indices of vertices in $\Gamma_{\mathrm{mod}}(N)$ which lie inside the parallelogram whose corners are $U_{r,s}, U_{r-s+1,r-s+1}, U_{r-s+1,1}$ and $U_{r,r}$.

One can regard $\Gamma_{\mathrm{mod}}(N)$ as a category provided the edges represent the full endomorphism groups. Then $\Gamma_{\mathrm{mod}}(N)$ becomes a ring (of maps) and it is clear that

End(B) is isomorphic to this ring. In fact much of the structure of End(B) can be readily read off from $\Gamma_{\text{mod}}(N)$. For some questions it is useful to have a matrix representation of End(B). We obtain it by ordering the U_{ij} as follows. We place U_{11} first, then U_{22} followed by U_{21}, then U_{33} followed by U_{32} and U_{31}, and so on, down each diagonal. The final uniserial is U_{N1}. This leads to the following result.

$$\begin{pmatrix} D_1 & V_{1,2} & & V_{1,4} & & & V_{1,7} & & & & V_{1,11} & & \cdots \\ & D_2 & V_{2,3} & V_{2,4} & V_{2,5} & & V_{2,7} & V_{2,8} & & & V_{2,11} & V_{2,12} & \\ & & D_3 & & V_{3,5} & & & V_{3,8} & & & & V_{3,12} & \\ & & & D_4 & V_{4,5} & V_{4,6} & V_{4,7} & V_{4,8} & V_{4,9} & & V_{4,11} & V_{4,12} & \cdots \\ & & & & D_5 & V_{5,6} & & V_{5,8} & V_{5,9} & & & V_{5,12} & \cdots \\ & & & & & D_6 & & & V_{6,9} & & & & \ddots \\ & & & & & & D_7 & V_{7,8} & V_{7,9} & V_{7,10} & V_{7,11} & V_{7,12} & \cdots \\ & & & & & & & D_8 & V_{8,9} & V_{8,10} & & V_{8,12} & \cdots \\ & & & & & & & & D_9 & V_{9,10} & & & \ddots \\ & & & & & & & & & D_{10} & & & \\ & & & & & & & & & & D_{11} & V_{11,12} & \cdots \\ & & & & & & & & & & & D_{12} & \cdots \\ & & & & & & & & & & & & \ddots \end{pmatrix}$$

FIGURE 4. The ring H

THEOREM 2.2. *Let $R = T_N(D)$ and $m = \frac{1}{2}N(N+1)$. Then End(B) is isomorphic to the ring H of all upper triangular $m \times m$ matrices whose only nonzero entries occur at the places (p, q) which are defined as follows. If i, j are any two integers satisfying $1 \leq i \leq N$ and $1 \leq j \leq i$ then p, q are given by the following equations: $p = \frac{1}{2}i(i-1) + i + j - 1$ and $q = \frac{1}{2}(i+t)(i+t+1) - 1 + l$ where $0 \leq t \leq N - i$ and $1 + t \leq l \leq j + t$. The entries along the diagonals are arbitrary elements of D and at the place (p, q), $p \neq q$, they are arbitrary elements of V_{pq}, a null D-algebra one dimensional on both sides. The V_{ij}'s have distinguished generators δ_{ij}. Multiplication on $\bigcup V_{pq}$ is given by $V_{ij} \otimes V_{jk} \xrightarrow{\times} V_{ik}$ which maps $\delta_{ij} \otimes \delta_{jk}$ onto δ_{ik}.*

PROOF. The ring $\text{End}(R)$ is represented as a ring H of matrices in the usual way. Entries at the place (p,q) correspond to homomorphisms from the pth summand to the qth summand. Remember that we a using a new indexing of the summands and that the endomorphism ring of a U_{ij} is simply (isomorphic to) D. Each $\text{Hom}(U_{ij}, U_{kl})$ is a one dimensional bivector space over $\text{End}(U_{ij}) - \text{End}(U_{kl})$ and we denote it by a V with an appropriate subscript (according to the new ordering of the U's).

There is a nonzero homomorphism from U_{ij} to U_{kl} if, and only if, $U_{k,l}$ is in the parallelogram described in Lemma 2.1 (a). The canonic projection from B onto U_{ij} corresponds to the matrix whose only nonzero entry is at the place (p,p) where $p = \frac{1}{2}i(i-1) + (i-j+1)$. Now $i \leq k \leq N$ and if $k = i+t$, for $t \geq 0$, then $1+t \leq l \leq j+t$. Then $\epsilon_{k,l}$ corresponds to the matrix whose only nonzero entry is at the place (q,q) where

$$q = \tfrac{1}{2}k(k-1) + k - l + 1 = \tfrac{1}{2}(i+t)(i+t-1) + i+t-l+1$$

where $N - i \geq t \geq 0$ and $1+t \leq l \leq j+t$. Therefore the nonzero entries in H occur at the places $\left(\frac{1}{2}i(i-1)+1-j+1, \frac{1}{2}(i+t)(i+t+1)-l+1\right)$ for all $1 \leq i \leq N$, $1 \leq j \leq i$, $0 \leq t \leq N-i$, $1+t \leq l \leq j+t$. Homomorphisms from U_{ij} to U_{kl} are induced by homomorphisms from Re_i to Re_k. The latter can be expressed as products $\alpha\sigma_{ik}$ where α is an endomorphism of Re_i and σ_{ik} maps e_i onto e_{ik}. Note that $\sigma_{ik}\sigma_{kl} = \sigma_{il}$. As each $\text{End}(Re_i) \cong D$ it follows that the composition of homomorphisms can be described as a D-bilinear homomorphism $\text{Hom}(Re_i, Re_k) \otimes \text{Hom}(Re_k, Re_l) \xrightarrow{\times} \text{Hom}(Re_i, Re_l)$ such that $\sigma_{ij} \otimes \sigma_{jk}$ is mapped onto σ_{ik}. The induced homomorphisms also have a D-bilinear homomorphism $\text{Hom}(U_{ij}, U_{kl}) \otimes \text{Hom}(U_{kl}, U_{rs}) \to \text{Hom}(U_{ij}, U_{rs})$. Since the V's are just other names for the $\text{Hom}(U_{ij}, U_{rs})$ and the δ's are new notation for the σ's, this gives a D-bilinear homomorhism $V_{st} \otimes V_{tu} \xrightarrow{\times} V_{su}$ which maps $\delta_{st} \otimes \delta_{tu}$ to δ_{su}, for appropriate s, t and u. □

The structure is best seen by looking at Figure 4. Much about the ideal structure of H can be readily read off the graph $\Gamma_{\text{mod}}(N)$ by applying Lemma 2.1.

PROPOSITION 2.3. Let H be the ring in Theorem 2.2. For any primitive idempotent $\epsilon_{i,j}$ the lattice of primitive cyclic subideals of $\epsilon_{i,j}H$ is isomorphic to the full sublattice of $\Gamma_{\text{mod}}(N)$ consisting of the parrallelogram with corners $U_{i,j}$, $U_{i,1}$, $U_{N,N-i+j}$ and $U_{N,N-i+1}$. The inclusions are obtained by reversing the arrows. Each $U_{k,l}$ in this sublattice corresponds to the unique image of $\epsilon_{k,l}H$ in $\epsilon_{i,j}H$.

For any primitive idempotent $\epsilon_{r,s}$ the lattice of primitive cyclic subideals of $H\epsilon_{r,s}$ is isomorphic to the full sublattice of $\Gamma_{\text{mod}}(N)$ consisting of the parrallelogram with corners $U_{r,s}$, $U_{r-s+1,1}$, $U_{r-s+1,r-s+1}$ and $U_{r,r}$. The inclusions are given by the arrows. Each $U_{k,l}$ in this sublattice corresponds to the unique image of $H\epsilon_{k,l}$ in $H\epsilon_{r,s}$.

COROLLARY 2.4. The $H\epsilon_{Ni}$ are injective left ideals with socles isomorphic to the tops of $H\epsilon_{N-i+1,N-i+1}$. The $\epsilon_{ii}H$ are injective right ideals with socles isomorphic to the tops of $\epsilon_{N,N-i+1}H$.

PROOF. One can see immediately from $\Gamma_{\text{mod}}(N)$ that the statements about the socles and tops are correct. The Nakayama–Fuller criterion ([2], Theorem 3.1) now shows that the ideals are injective. □

We now consider the ring of endomorphisms of B as a right H-module. The next result follows from some more general results about rings of finite representation type, but these are not so readily accessible so we include it because a short simple proof is available.

THEOREM 2.5. $\text{End}_H(B_H) \simeq R$

PROOF. We first show that $B_H = \oplus e_i \epsilon_{ii} H$. Consider the summand $U_{ij} = Re_i/W^j e_i$ of B. Every element of U_{ij} can be expressed as

$$d_i e_i + d_{i-1} e_{i-1,i} + \ldots + d_{i-j+1} e_{i-j+1,i} + W^j e_i = d_{i1} + \ldots d_{i-j+1,i} + W^j e_i.$$

Each $e_{i-k,i} + W^j e_i$ is an image of e_{i-k} under an element of H. Since $e_{i-k} Re_i$ is one dimensional on the right over $e_i Re_i \cong D$, each $d_{i-k} e_{i-k,i}$ is a homomorphic image of e_{i-k}. It follows that every element of B is a sum of homomorphic images of the e_i's. Therefore $B = \Sigma e_i H$ and it is clear that this sum is direct. But $e_i = e_i \epsilon_{ii}$ so $e_i H = e_i \epsilon_{ii} H$ and our assertion follows.

Now $e_i \epsilon_{ii} H \cong \epsilon_{ii} H$ since no element of $\epsilon_{ii} H$ annihilates e_i. From the representation of H given in Theorem 2.2 it can be seen that $\text{Hom}_H(\epsilon_{ii} H, \epsilon_{jj} H) = D$ if $i \leq j$ and it is zero otherwise. It follows that $\text{End}_H(B_H) \simeq T_N(D) = R$. □

The Baer-Kaplansky Theorem (two abelian torsion groups are isomorphic if their endomorphism rings are isomorphic) has generated much interest over the years. It is shown in [5] that it can be generalisaed to modules over arbitrary rings provided the modules decompose nicely into indecomposables. That work leads to some natural questions connecting the ring automorphisms of a ring with its linear automorphisms. The next result extends a theorem in that paper and its proof shows that the graphs we have been studying have many uses.

If M is a module which is a direct sum of indecomposable submodules M_i, then an automorphism ϕ of $\text{End}(M)$ is an **IP-automorphism** if for every projection π from M onto M_i, $M_i \pi \phi$ is isomorphic to M_i.

THEOREM 2.6. *Let H be the ring in Theorem 2.2, J its Jacobson radical and let $I \subseteq J^2$ be a two sided ideal. Every ring automorphism of H/I is an IP-automorpism.*

PROOF. Denote H/I by H^* and $K + I$ by K^*, for any subset K of H. Let $\alpha : H^* \to H^*$ be any ring automorphism. We will constantly use the following facts. Every indecomposable summand of H^* is isomorphic to an $H^* \epsilon_{ij}^*$ and α maps indecomposable summands onto indecomposable summands. Note also that α preserves subideal lattice structure. For simplicity of terminology we call the indecomposable summands of H^* *principle indecomposables*. Our proof is a study of $\Gamma_{\text{mod}}(N)$.

The only $H^* \epsilon_{ij}^*$ which is simple is $H^* \epsilon_{11}^*$ so $H^* \epsilon_{11}^* \alpha \cong_{H^*} H^* \epsilon_{11}^*$. Then $H^* \epsilon_{22}^*$ is the only principle indecomposable which has a copy of $H^* \epsilon_{11}^*$ as its only subideal. Therefore $H^* \epsilon_{22}^* \alpha \cong_{H^*} H^* \epsilon_{22}^*$. Both $H^* \epsilon_{33}^*$ and $H^* \epsilon_{21}^*$ have as their maximal subideals images of $H^* \epsilon_{22}^*$. But $\epsilon_{21}^* H^*$ is uniserial whereas $\epsilon_{33}^* H^*$ is not (Proposition 2.3) therefore $H^* \epsilon_{33} \alpha \cong_{H^*} H^* \epsilon_{33}$. Since $H^* \epsilon_{44}^*$ is the only principle indecomposables whose maximal subideal is an image of $H^* \epsilon_{33}^*$, it follows that it is isomorphic to its image under α. In this way we show that all the principle indecomposables in the first column are isomorphic to their images under α.

Now consider the principle indecomposables in the second column of $\Gamma_{\text{mod}}(N)$. $H^* \epsilon_{32}^*$ is the only principle indecomposable whose maximal subideal is the sum of images of $H^*\epsilon_{33}^*$ and $H^*\epsilon_{21}^*$. Therefore it is isomorphic to its image under α. By repeating this argument we prove the theorem for all the principle indecomposables in the second column. We then repeat the arguments for the third column and so on. $\square$

2.2. Quasi-Frobenius rings. We now consider the case when R is a QF ring. Since every indecomposable module is a submodule of an indecomposable injective and since R is the direct sum of one copy of all the indecomposable injectives, every indecomposable module is isomorphic to one of the $W^j e_i$. Hence our basic generator B is $\oplus_{i,j} W^j e_i$. We can use the graph $\Gamma_{\mathrm{id}}(n+1,d)$ as the graph of homomorphisms of B. We need only put in the arrows between the $W^j e_i$'s and rename it $\Gamma_{\mathrm{mod}}(n+1,d)$. This is shown in Figure 5.

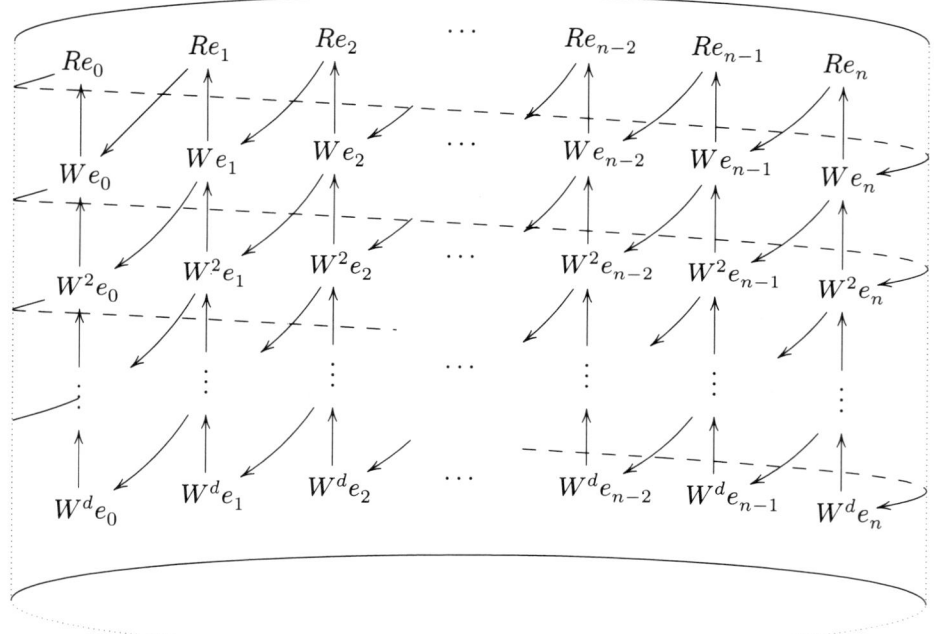

FIGURE 5. $\Gamma_{\mathrm{mod}}(n+1,d)$

The above graph can also be interpreted as the module diagram for the right modules of $\mathrm{QFS}(A, n+1, d)$. The top vertices now represent the ideals $e_i R$ and the other vertices represent the subideals $e_i W^j$ as follows. The vertex that represented Re_i now represents $e_i R$ and the vertex that represents $W^j e_i$ now represents $e_{i+j} W^j$. The canonic maps between the modules are represented by reversing the arrows. So the (reversed) diagonal arrows now represent the inclusions $e_i W^j \to e_i W^{j-1}$ and the (reversed) vertical arrows represent the epimorphisms $e_i W^j \to e_{i+1} W^{j+1}$.

One can, of course, obtain a matrix representation of $G = \mathrm{End}(B)$ using the same approach as before. Unfortunately, there is no simple form into which it can be placed.

3. Character Modules

For commutative rings R, Anh [1] has defined the **character** module M^* of an R-module M to be $\text{Hom}_R(M, E)$, where E is the minimal injective cogenerator. E is the direct sum of the injective hulls of the simple modules (one for each isomorphism class). Since R is a commutative ring M^* is naturally an R-module. He shows that a commutative Noetherian ring has the property that $M^* \cong_R M$, for every module M of finite length, if, and only if the ring is a finite product of Dedekind rings and uniserial Artinian rings.

For noncommutative rings $\text{Hom}_R(M, E)$ has no natural R-module structure. But for the rings we are studying there is a right module structure on $\text{Hom}_R(M, E)$ which is (in a sense) "isomorphic" to the left structure on M.

DEFINITION. Let R and S be rings. An **anti-isomorphism** $\psi : R \to S$ is an additive group isomorphism which satisfies the equation $(r_1 r_2)\psi = (r_2)\psi(r_1)\psi$ for all r_1, r_2 in R. If ψ is such an anti-isomorphism then $\Phi : {}_R M \to N_S$ is an **op-ψ-isomorphism** if it is an isomorphism between the underlying additive groups and if $(rm)\Phi = (m)\Phi(r)\psi$ for all $r \in R, m \in M$. We denote this by $\Phi : {}_R M \cong^{\text{op}}_\psi N_S$. If there is such a map Φ (and ψ) then we say M and N are **op-isomorphic** or **left-right isomorphic**. We denote the existence of such an isomorphism by $M \cong^{\text{op}} N$.

Since the proofs are somewhat different, we will treat the nonsingular and the QF cases separately.

THEOREM 3.1. *Assume D is commutative, let $R = T_N(D)$ and let M be any R-module. Then there is a right R^{op}-module structure on M^* such that M^* is op-isomorphic to M.*

PROOF. Clearly we need only consider the case when M is uniserial. So let $M = U_{pq}$. The minimal injective cogenerator E is $\oplus_k U_{Nk}$. The only maps from U_{pq} to E are the embeddings into its injective hull $U_{N,N-(p-q)}$ and their compositions with the surjections $U_{N,N-(p-q)} \to U_{N,N-(p-q)-i}$, where $0 \leq i \leq p-1$. This can be readily seen from the graph $\Gamma_{\text{mod}}(N)$. It can also be seen that $\text{End}(\oplus_{1 \leq i \leq q} U_{N,N-p+i}) \cong T_q(D)$.

Let R_p denote $\oplus_{1 \leq i \leq p} Re_i$. Note that R_p is a ring with identity. The action of R on U_{pq} is simply the action of R_p. Its kernel is $f_{p-q} R_p$ where $f_{p-q} = e_1 + \ldots + e_{p-q}$ and so U_{pq} is a right module over $T_q(D) \cong R_p / f_{p-q} R_p \cong R_q^*$. There is an anti-isomorphism $\psi : R_q^* \to T_q(D)$ given by $e_{ij} \mapsto e_{ji}$, Note that the later represents an epimorphism $U_{N,N-q+j} \to U_{N,N-q+i}$ whilst the former represents a monomorphism $U_{ii} \to U_{jj}$.

It just remains to construct an op-isomorphism Φ from U_{pq} to $\text{Hom}(U_{pq}, E)$. This can be most easily done by regarding U_{pq} as a set of homomorphisms from $R_p = \oplus_{i \leq p} U_{ii}$ (represented by paths in $\Gamma_{\text{mod}}(N)$). U_{pq} is then just the set of all paths into it modulo the paths which represent the same homomorphism. Φ is the map which takes a path from $U_{p-i,p-i}$ to U_{pq} to the path from U_{pq} to $U_{N,N-p+q-i}$. Algebraically, the former canonic path represents the homomorphism f which takes e_{p-i} to $e_{p-i,p} + W^q e_p$ and the later represents the homomorphism g which takes $e_p + W^q e_p$ to $e_N + W^{N-p+q-i} e_N$. Then Φ is the map which takes df onto $g(d)\psi$, for $d \in D$. $\square$

We now turn to the Quasi-Frobenius case. We will regard the ring as a ring of matrices over the uniserial ring A.

THEOREM 3.2. *Let R be an indecomposable basic QF serial ring with the property that the endomorphim ring of an indecomposable projective is commutative, and let M be an R-module. Then there is a right R^{op}-module structure on M^* such that M^* is op-isomorphic to M.*

PROOF. Assume that R has breadth n and depth d (that is, is one of the rings $\mathrm{QFS}(A, d, n)$). We will regard R as a ring of matrices over the commutative uniserial ring A, along the diagonal, and A or $A/\mathrm{Soc}(A)$ off the diagonal. Multiplication between off diagonal elements of A or $A/\mathrm{Soc}(A)$ is not that in A or $A/\mathrm{Soc}(A)$, but it is given in [4]. The matrix whose only non-zero entry is a at the place (i,j) will be denoted by $a_{i,j}$. Addition of subscripts will always be taken modulo n.

It is sufficient to consider the case when M is uniserial, say $M = U_{pq}$. Now the injective cogenerator is R itself, U_{pq} is one of the ideals $W^t e_i$ and U_{pq}^* is the set of all homomorphisms from that $W^t e_i$ to R. In general, U_{pq} will have nonzero maps into every Re_j and every Re_j will have nonzero maps into U_{pq}.

We first consider the case when $U_{pq} = Re_p$. Then $U_{pq}^* = e_p R$ and the problem becomes to place a right R-module structure on $e_p R$ which supports the required op-isomorphism with Re_p. To do that we define an anti-isomorphism $\psi_p : R \to R$ by the relation: $a_{p+r,p+s} \mapsto a_{p-s,p-r}$, where r, s are positive and addition of subscripts is modulo n. This defines an anti-isomorphism because A is commutative.

It only remains to define a linear op-ψ_p-isomorphism $\Phi_p : Re_p \to e_p R$. For r positive, let $q = p + r$. We define Φ_p to be the additive map which takes $a_{p+r,p}$ to $a_{p,p-r}$. We leave it to the reader to check that Φ_p is bijective. For any $b_{p+s,q}$ and $a_{q,p}$ consider the image of their product under Φ_p. Since $b_{p+s,q} a_{q,p} = (ba)_{p+s,p}$, it follows that $(b_{p+s,q} a_{q,p})\Phi_p = (ba)_{p,p-s}$. But $(b_{p+s,q})\psi_p = b_{p-r,p-s}$ and $a_{p,p-r} b_{p-r,p-s} = (ba)_{p,p-s}$. That is, $((ba)_{q,p})\Phi = (a_{q,p})\Phi_p(b)\psi_p$ which shows that Φ_p is an op-ψ_p-isomorphism, as required.

Now consider the case when U_{pq} is not projective. Then $U_{pq} = W^{d-q} e_{p-q}$ and $U_{pq} \cong Re_p/W^q e_p$. So $\mathrm{Hom}(U_{pq}, R) \cong \{\alpha \in \mathrm{Hom}(Re_p, R) : \ker \alpha \supseteq W^q e_p\}$ and therefore $\mathrm{Hom}(W^{d-q} e_i, R) \subseteq e_p R$, modulo the usual identification. In fact $(W^{d-q} e_{p-q})^* = e_p W^{d-q}$ under this identification. To obtain an op-isomophism between these two modules, we first define an anti-isomorphism ψ_{pq} of R by the formula $a_{p+r,p+s} \mapsto a_{p-q-s,p-q-r}$ where r, s are positive. Then the additive map $\Phi_{pq} : W^{d-q} e_i \to e_p W^{d-q}$ given by $e_{p+r} a e_i \mapsto e_p a e_{p-r}$ defines an op-ψ_{pq}-isomorphism. □

If R is not basic then the above two results will not hold. For example, let R be the matrix ring $\begin{pmatrix} D & D & D \\ D & D & D \\ & & D \end{pmatrix}$ then the simple projective Re_2 has dimension two over D. Its character module is simply $e_2 Re_3$, since Re_3 is the only indecomposable injective into which Re_2 maps, and it has dimension one over D. However, if the multiplicities of the indecomposable projectives in a decomposition of the ring are the same, say m, then this problem can be overcome by replacing, in the definition of character module, the injective cogenerator E by a direct sum of m copies of E.

For arbitrary rings there would appear to be no definition of character module which satisfies the above theorems. One could pass to the category of modules over the basic subring, of the base ring, and use Anh's definition on the corresponding module. Since our modules decompose nicely, determining the corresponding module over the basic ring would not be a problem. We then get a right module structure on the character module.

Acknowledgements

The author began working on modules over nonsingular serial rings while visiting the University of Western Australia in 1994. He would like to thank the university for its hospitality and especially Phillip Schultz for many interesting discussions. He would like to thank Pham Ngoc Anh for engendering the author's interest in some of the questions considered and Ross Moore for drawing the diagrams using XY-pic and for his never-ending help with TeX.

References

[1] Phạm Ngọc Ánh *A note on selfdual modules and Dedekind rings*, JPAA **133** (1998), 23–25.
[2] K. R. Fuller *On indecomposable injectives over Artinian rings*, Pac. J. of Maths **29** (1969), 115–135
[3] A. W. Goldie *Torsion-free modules and rings*, J. of Algebra **1** (1964), 268–287
[4] G. Ivanov *Left generalized uniserial rings*, J. of Algebra **31** (1974), 166–181
[5] G. Ivanov *Generalising the Baer-Kaplansky Theorem*, JPAA **133** (1998), 107–115.
[6] I. Murase *On the structure of generalized uniserial rings, II*, Sci. Papers College of Gen. Ed. Univ. Tokyo., **13** (1963), 131–158.

DEPARTMENT OF MATHEMATICS, MACQUARIE UNIVERSITY, N.S.W. AUSTRALIA 2109.
E-mail address: ivanov@maths.mq.edu.au

Topologically Pure Extensions

Peter Loth

Dedicated to Professor Laszlo Fuchs in honour of his 75th birthday

ABSTRACT. A proper short exact sequence
$$0 \to H \to G \to K \to 0 \quad (*)$$
in the category of locally compact abelian groups is said to be *topologically pure* if the induced sequence
$$0 \to \overline{nH} \to \overline{nG} \to \overline{nK} \to 0$$
is proper short exact for all positive integers n. Some characterizations of topologically pure sequences in terms of direct decompositions, pure extensions and tensor products are established. A simple proof is given for a theorem on pure subgroups by Hartman and Hulanicki. Using topologically pure extensions, we characterize those splitting locally compact abelian groups whose torsion part is a direct sum of a compact group and a discrete group. We determine the compact and discrete groups H with the property that every topologically pure sequence $(*)$ splits. Some structural information on topologically pure injectives and projectives is obtained.

1. Introduction

All groups in this paper are assumed to be Hausdorff abelian topological groups. Let $\mathcal{L}$ denote the category of locally compact abelian (LCA) groups with continuous homomorphisms as morphisms. A morphism is called *proper* if it is open onto its image. An exact sequence
$$G_1 \xrightarrow{\alpha_1} G_2 \xrightarrow{\alpha_2} \ldots \xrightarrow{\alpha_{n-1}} G_n$$
in $\mathcal{L}$ is called *proper exact* if each morphism α_i is proper. A proper short exact sequence $0 \to H \xrightarrow{\alpha} G \xrightarrow{\beta} K \to 0$ in $\mathcal{L}$ is called an extension of H by K (in $\mathcal{L}$), and H may be identified with $\alpha(H)$ and K with $G/\alpha(H)$. The group of extensions of H by K is denoted by $\text{Ext}(K, H)$, and $\text{Pext}(K, H)$ is the subgroup of pure extensions of H by K.

Recall that a proper exact sequence $0 \to H \to G \to K \to 0$ splits exactly if H is a (topological) direct summand of G. If G is a direct sum of closed subgroups A and B, then we write $G = A \oplus B$. We say that G *splits* if the torsion part tG of G

1991 *Mathematics Subject Classification.* Primary 20K35, 22B05; Secondary 20K21, 20K25.

© 2001 American Mathematical Society

is a direct summand of G. For example, G splits if tG is locally compact and G/tG is compact, which can be derived from [1] (6.28). Note that if tG is finite, then G need not split (cf. [8] or Example 2.4).

In [4], Fulp studied pure exact sequences in the category $\mathfrak{L}$. As it was pointed out by Armacost [1], much of the paper is based on [4] Proposition 2 (stating that the dual sequence of a proper pure exact sequence is pure) which is unfortunately not true for all extensions in $\mathfrak{L}$. A closed subgroup H of G is said to be *topologically pure* if

$$\overline{nH} = H \cap \overline{nG}$$

for every positive integer n (cf. [12]). A pure subgroup need not be topologically pure, and a topologically pure subgroup need not be pure. The identity component G_0 and the subgroup bG of all compact elements of an LCA group G are both pure and topologically pure. It turns out that a closed subgroup H of an LCA group G is pure if and only if its annihilator is topologically pure in the dual group and each subgroup $G[n] + H$ is closed in G (cf. Proposition 2.1). Using this result, we give a short proof of the following theorem which is due to Hartman and Hulanicki [6]: If G is an LCA group which is either compactly generated or has no small subgroups, then a closed subgroup of G is pure if and only if its annihilator is pure in the dual group (Theorem 2.3).

Some characterizations of topologically pure sequences in terms of direct decompositions, pure exact sequences and tensor products are contained in Theorem 2.5 and Proposition 2.7. We prove that a topologically pure exact sequence $0 \to B \to G \to K \to 0$ splits if $B = C \oplus D$ is a bounded group where C is compact and D is discrete (Theorem 3.3). Notice that this result generalizes [9, Theorem 1.3] and [10, Theorem 2.4].

In Example 3.6, a non-splitting topologically pure exact sequence $0 \to B \to G \to K \to 0$ with bounded group B is constructed. It is shown that an LCA group G whose torsion part is a direct sum of a compact group and a discrete group, splits if and only if it possesses an ascending sequence $A_1 \subset A_2 \subset \ldots \subset A_n \subset \ldots$ ($n < \omega$) of open subgroups such that (i) $\bigcup_{n<\omega} A_n = G$, (ii) the torsion part of A_n is bounded for each $n < \omega$, (iii) $t(G/A_n) = (tG + A_n)/A_n$ for each $n < \omega$ and (iv) $0 \to tA_n \to A_n \to A_n/tA_n \to 0$ is a topologically pure exact sequence for each $n \leq \omega$ where $A_\omega = G$ (Theorem 3.7). Dually, we obtain a characterization of a certain class of LCA groups G whose subgroup $\mathrm{div} G = \bigcap_{n<\omega} \overline{nG}$ is a direct summand (Theorem 3.8).

If an LCA group has the injective property relative to the class of topologically pure sequences, then its identity component is a direct sum of a vector group and a toral group. It is shown that the groups of the form $R \oplus T \oplus A \oplus B$ (where R is a vector group, T is a toral group, A is a topological direct product of finite cyclic groups and B is a discrete bounded group) are exactly those LCA groups G for which G/G_0 has a pure compact open subgroup and every topologically pure sequence $0 \to G \to X \to Y \to 0$ splits (cf. Theorem 4.3). In particular: A discrete group B has the property that that every topologically pure sequence starting with B splits if and only if B is bounded, and the compact groups C with the property that every topologically pure sequence starting with C splits, are exactly the groups of the form $T \oplus A$ where T is a toral group and A is a topological direct product of finite cyclic groups (see Theorem 4.1). If an LCA group has the projective property relative to the class of topologically pure sequences, then it has the form

$R \oplus F \oplus bG$ where R is a vector group and F is a free group. The projective groups in $\mathfrak{L}$ are precisely those LCA groups G with the property that every topologically pure sequence $0 \to X \to Y \to G \to 0$ splits and bG possesses a topologically pure compact open subgroup (see Theorem 4.4).

The additive topological group of real numbers is denoted by $\mathbf{R}$, $\mathbf{Q}$ is the group of rationals and $\mathbf{Z}$ is the group of integers. By $\mathbf{T}$ we mean the quotient $\mathbf{R}/\mathbf{Z}$, $\mathbf{Z}(n)$ is the cyclic group of order n and $\mathbf{Z}(p^\infty)$ denotes the quasicyclic group. For any groups G and H, $\mathrm{Hom}(G, H)$ is the group of all continuous homomorphisms from G to H, endowed with the compact-open topology. The dual group of G is

$$\hat{G} = \mathrm{Hom}(G, \mathbf{T})$$

and $(\hat{G}, S)$ denotes the annihilator of $S \subset G$ in $\hat{G}$. Throughout this paper, we use the term "isomorphic" for "topologically isomorphic", and "direct summand" for "topological direct summand". For details and fundamental results on locally compact abelian groups and Pontrjagin duality, we may refer to the book [7] by Hewitt and Ross.

2. Pure and topologically pure extensions

The annihilator of a closed pure subgroup of an LCA group G is topologically pure in the dual group if G is a topological p-group (cf. [12] Theorem 22). This restriction of G is redundant:

PROPOSITION 2.1. *Let H be a closed subgroup of $G \in \mathfrak{L}$. Then we have:*
1. *If H is pure in G, then $(\hat{G}, H)$ is topologically pure in $\hat{G}$.*
2. *If H is topologically pure in G such that $\hat{G}[n] + (\hat{G}, H)$ is a closed subgroup of $\hat{G}$ for all positive integers n, then $(\hat{G}, H)$ is pure in $\hat{G}$.*

PROOF. If H is pure in G, then $(G/H)[n] = (G[n] + H)/H$ for all positive integers n by [3] Theorem 28.1. Let $\varphi : G \to G/H$ be the natural map and let $\rho : (G/H)\hat{\;} \to (\hat{G}, H)$ be the induced topological isomorphism. Then ρ maps each group $((G/H)\hat{\;}, (G/H)[n])$ onto $\overline{n(\hat{G}, H)}$. On the other hand, ρ maps $((G/H)\hat{\;}, (G[n] + H)/H)$ onto $(\hat{G}, G[n] + H)$ which is the same as $\overline{n\hat{G}} \cap (\hat{G}, H)$. Therefore $(\hat{G}, H)$ is topologically pure in $\hat{G}$. The proof of the second assertion is similar. $\square$

COROLLARY 2.2. *Suppose H is a compact or an open subgroup of $G \in \mathfrak{L}$. Then H is pure in G if and only if $(\hat{G}, H)$ is topologically pure in $\hat{G}$.*

A topological group is said to have *no small subgroups* if there exists a neighborhood of 0 which does not contain any nontrivial subgroups. Moskowitz [11] proved that LCA groups without small subgroups have the form $\mathbf{R}^n \oplus \mathbf{T}^m \oplus D$ where m and n are nonnegative integers and D is discrete, and that their Pontrjagin duals are exactly the compactly generated LCA groups. Let $\mathfrak{K}$ denote the class consisting of all LCA groups which are either compactly generated or have no small subgroups. Let G be in $\mathfrak{K}$ and suppose H is a closed subgroup of G. Then H is in $\mathfrak{K}$ as well (see [11] Theorem 2.6). But then, H is pure if and only if H is topologically pure. Therefore, the following result of Hartman and Hulanicki [6] is an immediate consequence of Proposition 2.1:

THEOREM 2.3. [6, Hartman and Hulanicki] *Suppose G is an LCA group which is either compactly generated or has no small subgroups. If H is a closed subgroup of G, then H is pure in G if and only if $(\hat{G}, H)$ is pure in $\hat{G}$.*

Recall that a proper exact sequence $E : 0 \to H \to G \to K \to 0$ in $\mathfrak{L}$ is said to be *pure* if the sequence
$$0 \to nH \to nG \to nK \to 0$$
is exact for all positive integers n. We call the sequence E *topologically pure* if
$$0 \to \overline{nH} \to \overline{nG} \to \overline{nK} \to 0$$
is a proper exact sequence for all positive integers n.

EXAMPLE 2.4. Let p be a prime and n a positive integer, and let H be any densely divisible LCA group such that $H/p^n H \neq 0$. [For instance, topologize the group $A = \prod_{\aleph_0} \mathbf{Z}(p^\infty)$ so that it is locally compact and contains the topological direct product $A[p] = \prod_{\aleph_0} \mathbf{Z}(p)$ as an open subgroup, and set $H = A[p] + \bigoplus_{\aleph_0} \mathbf{Z}(p^\infty) \subset A$ (cf. [8]).] Then $\mathrm{Ext}(\mathbf{Z}(p^n), H) \neq 0$, so there is a non-splitting extension
$$E_1 : 0 \to H \to G \to \mathbf{Z}(p^n) \to 0$$
in $\mathfrak{L}$. Notice that E_1 is topologically pure but not pure, and that the dual sequence
$$E_2 : 0 \to \mathbf{Z}(p^n) \to \hat{G} \to \hat{H} \to 0$$
is pure because $\hat{H}$ is torsion-free, but not topologically pure.

Some characterizations of topologically pure sequences are contained in the next theorem.

THEOREM 2.5. *Consider the following conditions for a proper exact sequence $E : 0 \to H \to G \to K \to 0$ in $\mathfrak{L}$:*

1. $0 \to \overline{nH} \to \overline{nG} \to \overline{nK} \to 0$ *is proper exact for all n;*
2. $0 \to H/\overline{nH} \to G/\overline{nG} \to K/\overline{nK} \to 0$ *is proper exact for all n;*
3. $\overline{nG + H}/\overline{nH} = \overline{nG/nH} \oplus H/\overline{nH}$ *for all n;*
4. $0 \to H[n] \to G[n] \to K[n] \to 0$ *is proper exact for all n;*
5. $0 \to H/H[n] \to G/G[n] \to K/K[n] \to 0$ *is proper exact for all n;*
6. $(n^{-1}H)/H[n] = G[n]/H[n] \oplus H/H[n]$ *for all n.*

Then we have: $(1) \Leftrightarrow (2) \Leftrightarrow (3)$ *and* $(4) \Leftrightarrow (5) \Leftrightarrow (6)$. *Further,* $(1) \not\Rightarrow (4) \not\Rightarrow (1)$ *generally.*

PROOF. Suppose the sequence E is topologically pure. Then $\overline{nG + H}$ is a closed subgroup of G, and the topological isomorphism $\overline{(nG + H)}/H \to \overline{nG}/\overline{nH}$ induces a continuous homomorphism
$$\phi : \overline{(nG + H)}/\overline{nH} \to \overline{nG}/\overline{nH}$$
which is the identity on $\overline{nG}/\overline{nH}$ so that the kernel of ϕ is equal to $H/\overline{nH}$. By [7] (6.22), $\overline{(nG + H)}/\overline{nH}$ is a direct sum of $\overline{nG}/\overline{nH}$ and $H/\overline{nH}$, and the equivalence of (1) and (3) follows. Next we assume that $0 \to H[n] \to G[n] \to K[n] \to 0$ is proper exact. Then there is a continuous homomorphism
$$\psi : (n^{-1}H)/H[n] \to G[n]/H[n]$$
which is the identity on $G[n]/H[n]$ so that $\ker \psi = H/H[n]$, hence $(n^{-1}H)/H[n]$ is a direct sum of $G[n]/H[n]$ and $H/H[n]$. Consequently, (4) and (6) are equivalent. To prove $(2) \Leftrightarrow (3)$ and $(5) \Leftrightarrow (6)$, similar arguments can be used. Finally, Example 2.4 shows that $(1) \not\Rightarrow (4) \not\Rightarrow (1)$. □

Let us call a proper exact sequence $0 \to H \to G \to K \to 0$ in $\mathfrak{L}$ *-pure* if it satisfies the (equivalent) conditions (4), (5) and (6) in Theorem 2.5. Then we have:

COROLLARY 2.6. *A proper exact sequence in $\mathfrak{L}$ is $*$-pure if and only if its dual sequence is topologically pure.*

PROOF. The sequence $0 \to H[n] \to G[n] \to K[n] \to 0$ is proper exact if and only if the dual sequence

$$0 \to \hat{K}/\overline{n\hat{K}} \to \hat{G}/\overline{n\hat{G}} \to \hat{H}/\overline{n\hat{H}} \to 0$$

is proper exact. □

Following Fulp [4], we define the tensor product of LCA groups G and H to be the topological group

$$G \otimes H = \text{Hom}(G, \hat{H})\hat{\ }.$$

Note that $G \otimes H$ is locally compact if G is finitely generated, and in this case, our definition coincides with the definition of Moskowitz in [11]. If G and H are discrete, then $G \otimes H$ is the usual tensor product of discrete abelian groups.

PROPOSITION 2.7. *Suppose $E : 0 \to H \to G \to K \to 0$ is a proper exact sequence in $\mathfrak{L}$ such that $\hat{G}$ is σ-compact. Then the following conditions are equivalent:*
  1. *E is a topologically pure sequence;*
  2. *the sequence $0 \to F \otimes H \to F \otimes G$ is proper exact for every finitely generated discrete group F;*
  3. *the sequence $0 \to \mathbf{Z}(n) \otimes H \to \mathbf{Z}(n) \otimes G$ is proper exact for every positive integer n.*

PROOF. (1) implies (2). For any LCA group G, we write G_d for the group G with the discrete topology. Suppose $0 \to H \to G \to K \to 0$ is topologically pure and let F be a finitely generated discrete group. Then the sequence $0 \to \hat{K}_d \to \hat{G}_d \to \hat{H}_d \to 0$ is pure by Corollary 2.6, hence

$$0 \to \text{Hom}(F, \hat{K}_d) \to \text{Hom}(F, \hat{G}_d) \to \text{Hom}(F, \hat{H}_d) \to \text{Pext}(F, \hat{K}_d) = 0$$

is an exact sequence. Notice that $\text{Hom}(F, \hat{G})$ is σ-compact, hence the sequence

$$\text{Hom}(F, \hat{G}) \to \text{Hom}(F, \hat{H}) \to 0$$

is proper exact by the Open Mapping Theorem. Consequently, the sequence in (2) is proper exact.

It is clear that (2) implies (3). Now suppose $0 \to \mathbf{Z}(n) \otimes H \to \mathbf{Z}(n) \otimes G$ is proper exact for all n. Then each induced homomorphism $\text{Hom}(\mathbf{Z}(n), \hat{G}) \to \text{Hom}(\mathbf{Z}(n), \hat{H})$ is surjective. Since $\hat{G}$ is σ-compact, it follows that $0 \to \hat{K} \to \hat{G} \to \hat{H} \to 0$ is $*$-pure. By Corollary 2.6, the sequence E is topologically pure. Therefore, (3) implies (1). □

Notice that Proposition 2.7 fails if "topologically pure" is replaced by "pure" (cf. [4] Proposition 3): For instance, the group G in Example 2.4 possesses a sequence of compact open subgroups H_i whose intersection is trivial, hence

$$\hat{G} = \sum(\hat{G}, H_i)$$

is σ-compact. The tensor map $F \otimes H \to F \otimes G$ is proper and injective for every finitely generated group F, but H is not pure in G.

3. Extensions of torsion groups

A pure subgroup H of a discrete group G is a direct summand if the quotient G/H is a direct sum of cyclic groups (cf. [3] Theorem 28.2). By Corollary 2.2 and duality, we have:

THEOREM 3.1. *Let H be a closed topologically pure subgroup of an LCA group G. Then H is a direct summand if it has the form $\mathbf{T}^\mathfrak{m} \oplus A$, where $\mathfrak{m}$ is a cardinal and A is a topological direct product of finite cyclic groups. In particular, a compact torsion subgroup of an LCA group is a direct summand if it is topologically pure.*

PROPOSITION 3.2. *Let $0 \to H \to G \to K \to 0$ be a proper exact sequence in $\mathfrak{L}$ and suppose H is a discrete group which is a direct sum of cyclic groups of the same order p^k where p is a prime and k is a positive integer. Then the following conditions are equivalent:*

1. $0 \to H \to G \to K \to 0$ *is topologically pure;*
2. $\overline{p^k G + H} = \overline{p^k G} \oplus H$;
3. $0 \to H \to G \to K \to 0$ *splits.*

PROOF. (1) implies (2) because of Theorem 2.4.

(2) implies (3). Suppose $\overline{p^k G + H} = \overline{p^k G} \oplus H$. Since $G/\overline{p^k G}$ is totally disconnected, it contains a compact open subgroup $C/\overline{p^k G}$ such that $C \cap H = 0$. The set of all open subgroups A of G such that $A \cap H = 0$ and $p^k G \subset A$ is partially ordered by inclusion and contains a maximal element K by Zorn's lemma. Now it can be shown that the quotient $G/(H \oplus K)$ is both torsion and torsion-free (see the proof of [3] Proposition 27.1), thus $0 \to H \to G \to K \to 0$ splits.

It is clear that (3) implies (1). □

Recall that a bounded pure subgroup of a discrete group is a direct summand (see [3] Theorem 27.5). This result can be extended as follows:

THEOREM 3.3. *Suppose B is a bounded group which is a direct sum of a compact group and a discrete group. Then:*

1. *A proper exact sequence $0 \to B \to G \to K \to 0$ in $\mathfrak{L}$ splits if and only if it is topologically pure.*
2. *Dually, a proper exact sequence $0 \to H \to G \to B \to 0$ in $\mathfrak{L}$ splits if and only if it is $*$-pure.*

PROOF. By Corollary 2.6 and duality, it suffices to prove the first assertion. Suppose $0 \to B \to G \to K \to 0$ is a topologically pure exact sequence where B is a bounded group.

First, we assume that B is discrete and write $B = H \oplus H'$ where H is a direct sum of cyclic groups of the same prime power p^k and the maximum of orders of the elements of H' is less than the maximum of orders of the elements of B. By Theorem 2.5, we have

$$\overline{p^k G + B}/\overline{p^k B} = \overline{p^k G}/\overline{p^k B} \oplus B/\overline{p^k B},$$

therefore $\overline{p^k G}$ is an open subgroup of $\overline{p^k G + H}$. Now Proposition 3.2 shows that there is a closed subgroup G' of G such that $G = H \oplus G'$. Clearly, we have $B = H \oplus (G' \cap B)$. The sequence

$$0 \to G' \cap B \to G' \to G'/G' \cap B \to 0$$

is topologically pure and splits by induction. Therefore, $0 \to B \to G \to K \to 0$ splits.

Now suppose that B is a direct sum of a compact group C and a discrete group D. By Theorem 3.1, we can write $G = C \oplus X$ and by the first part of this proof, $X \cap B$ is a direct summand of X. Again, the sequence $0 \to B \to G \to K \to 0$ splits. The converse is obvious. □

COROLLARY 3.4. *Let B be a bounded LCA group consisting of elements of square-free order. Then a proper exact sequence $0 \to B \to G \to K \to 0$ in $\mathfrak{L}$ splits if it is topologically pure.*

A discrete bounded topologically pure subgroup of an LCA group need not be a direct summand:

EXAMPLE 3.5. Let p be a prime and let A be any proper dense subgroup (taken discrete) of the topological direct product $H = \prod_{\aleph_0} \mathbf{Z}(p^2)$. Then $B = \{(px, px) : x \in A\}$ is a discrete subgroup of $G = A \times pH$. We have $\overline{pB} = B \cap \overline{pG}$, therefore B is topologically pure in G. Since $\overline{pG} + B$ is not a closed subgroup of G, B is not a direct summand of G.

A topologically pure exact sequence $0 \to B \to G \to K \to 0$ with bounded group B need not split:

EXAMPLE 3.6. Take the locally compact group $A = \prod_{\aleph_0} \mathbf{Z}(p^\infty)$ of Example 2.4, let G be the subgroup $A[p^2]$ and consider the subgroup B consisting of all elements $(x_i) \in G = \prod_{\aleph_0} \mathbf{Z}(p^2)$ such that $px_i = 0$ for almost all i. Then $\overline{p(G/B)} = 0$ and therefore the sequence $0 \to B \to G \to G/B \to 0$ is topologically pure. Since the sequence is not pure, it does not split.

Splitting LCA groups whose torsion part is a direct sum of a compact group and a discrete group, can be characterized using topologically pure extensions:

THEOREM 3.7. *Let G be an LCA group so that tG is a direct sum of a compact group and a discrete group. Then G splits if and only if it possesses an ascending sequence $A_1 \subset A_2 \subset \ldots \subset A_n \subset \ldots$ ($n < \omega$) of open subgroups such that*
1. $\bigcup_{n<\omega} A_n = G$;
2. tA_n *is bounded for all* $n < \omega$;
3. $t(G/A_n) = (tG + A_n)/A_n$ *for all* $n < \omega$;
4. $0 \to tA_n \to A_n \to A_n/tA_n \to 0$ *is a topologically pure sequence for all* $n \leq \omega$ *(we set $A_\omega = G$).*

PROOF. Suppose $tG = C \oplus D$ where C is compact and D is discrete. If $G = tG \oplus F$, then the open subgroups $A_n = C \oplus D[n!] \oplus F$ satisfy conditions (1)–(4).

Conversely, assume that G has open subgroups A_n satisfying (1) - (4). By Theorem 3.1, there is a closed subgroup G' of G such that $G = C \oplus G'$. Since C is compact, we may assume that $C \subset A_n$ for each n. Letting $A'_n = A_n \cap G'$ we have $A_n = C \oplus A'_n$. By Theorem 3.3, we can write $A_n = tA_n \oplus H_n$, hence

$$A'_n = (tA_n \oplus H_n) \cap G' = (tA'_n \oplus C \oplus H_n) \cap G' = tA'_n \oplus [(C \oplus H_n) \cap G'].$$

Since $t(G'/A'_n) = (tG' + A'_n)/A'_n$ for all n, the groups $G_n = (C \oplus H_n) \cap G'$ can be chosen so that $G_1 \subset G_2 \subset \ldots \subset G_n \subset \ldots$ (see the proof of [3] Proposition 100.4). Consequently, we have $G = C \oplus (D \oplus \bigcup_{n<\omega} G_n) = tG \oplus \bigcup_{n<\omega} G_n$, as desired. □

For an LCA group G, a closed subgroup $\operatorname{div}G$ is defined by
$$\operatorname{div}G = \bigcap_{n<\omega} \overline{nG}.$$
Then dualization of Theorem 3.7 yields a characterization of a certain class of LCA groups G whose subgroup $\operatorname{div}G$ is a direct summand:

THEOREM 3.8. *Let G be an LCA group such that $G/\operatorname{div}G$ is a direct sum of a compact totally disconnected group and a discrete bounded group. Then $\operatorname{div}G$ is a direct summand of G if and only if G has a descending sequence $B_1 \supset B_2 \supset \ldots \supset B_n \supset \ldots$ ($n < \omega$) of compact subgroups such that*

1. $\bigcap_{n<\omega} B_n = 0$;
2. $G/(\operatorname{div}G + B_n)$ *is bounded for all* $n < \omega$;
3. $\operatorname{div}B_n = \operatorname{div}G \cap B_n$ *for all* $n < \omega$;
4. $0 \to (\operatorname{div}G + B_n)/B_n \to G/B_n \to G/(\operatorname{div}G + B_n) \to 0$ *is a $*$-pure sequence for all $n \leq \omega$ (we set $B_\omega = 0$).*

PROOF. The subgroup $\operatorname{div}G$ is a direct summand of G exactly if $t\hat{G} = (\hat{G}, \operatorname{div}G)$ is a direct summand of $\hat{G}$. The sequence
$$B_1 \supset B_2 \supset \ldots \supset B_n \supset \ldots$$
satisfies the conditions (1) - (4) in Theorem 3.8 if and only if the sequence
$$(\hat{G}, B_1) \subset (\hat{G}, B_2) \subset \ldots \subset (\hat{G}, B_n) \subset \ldots$$
satisfies the conditions (1) - (4) in Theorem 3.7. □

4. Splitting problems

Let G and X be groups in $\mathfrak{L}$. Recall that $\operatorname{Pext}(X,G) = 0$ if and only if every pure exact sequence $0 \to G \to H \to X \to 0$ splits. We say that
$${}^*\operatorname{Pext}(X,G) = 0$$
if every $*$-pure sequence $0 \to G \to H \to X \to 0$ splits. Similarly, we say that
$${}^t\operatorname{Pext}(X,G) = 0$$
if every topologically pure sequence $0 \to G \to H \to X \to 0$ splits.

THEOREM 4.1. *Let G be an LCA group.*

1. *If G is discrete, then ${}^*\operatorname{Pext}(X,G) = 0$ for each X in $\mathfrak{L}$ if and only if $G = 0$.*
2. *If G is discrete, then ${}^t\operatorname{Pext}(X,G) = 0$ for each X in $\mathfrak{L}$ if and only if G is bounded.*
3. *If G is compact, then ${}^*\operatorname{Pext}(X,G) = 0$ for each X in $\mathfrak{L}$ if and only if $G \cong \mathbf{T}^{\mathfrak{m}}$ where $\mathfrak{m}$ is a cardinal.*
4. *If G is compact, then ${}^t\operatorname{Pext}(X,G) = 0$ for each X in $\mathfrak{L}$ if and only if $G \cong \mathbf{T}^{\mathfrak{m}} \oplus A$ where $\mathfrak{m}$ is a cardinal and A is a topological direct product of finite cyclic groups.*

PROOF. (1) Suppose G is discrete such that ${}^*\operatorname{Pext}(X,G) = 0$ for each X in $\mathfrak{L}$. Since any extension of G by $\hat{\mathbf{Q}}$ is $*$-pure, we have $\operatorname{Ext}(\hat{\mathbf{Q}},G) = 0$. Hence the exactness of
$$\operatorname{Ext}(\hat{\mathbf{Q}},tG) \to \operatorname{Ext}(\hat{\mathbf{Q}},G) \to \operatorname{Ext}(\hat{\mathbf{Q}},G/tG) \to 0$$

implies that $\operatorname{Ext}(\hat{\mathbf{Q}}, G/tG) = 0$. The sequence
$$\operatorname{Hom}((\mathbf{Q}/\mathbf{Z})\hat{}, G/tG) \to \operatorname{Ext}(\hat{\mathbf{Z}}, G/tG) \to \operatorname{Ext}(\hat{\mathbf{Q}}, G/tG)$$
is exact and $\operatorname{Hom}((\mathbf{Q}/\mathbf{Z})\hat{}, G/tG) = 0$. Therefore we have
$$G/tG \cong \operatorname{Ext}(\mathbf{T}, G/tG) = 0,$$
hence G is a torsion group. Since G is also a cotorsion group, it is a direct sum of a bounded group B and a divisible group D (cf. [**3**] Corollary 54.4). Consequently, $^*\operatorname{Pext}(X, B) = 0$ and $^*\operatorname{Pext}(X, D) = 0$ for each X in $\mathfrak{L}$. Since every nontrivial bounded discrete group can be identified with the torsion part of some non-splitting LCA group (see Example 2.4), $B = 0$ follows. Further, $^*\operatorname{Pext}(X, Q) = \operatorname{Ext}(X, Q)$ if Q is quasicyclic, therefore $D = 0$. This proves the first statement.

(2) Now suppose G is discrete such that $^t\operatorname{Pext}(X, G) = 0$ for each X in $\mathfrak{L}$. Then $\operatorname{Ext}(\hat{\mathbf{Q}}, G) = 0$ and again, we conclude that G is a reduced torsion group. Since G is also algebraically compact, G is bounded (cf. [**3**] Corollary 40.3). The converse is true because of Theorem 3.3.

(3) If G is compact where $^*\operatorname{Pext}(X, G) = 0$ for each X in $\mathfrak{L}$, then $\operatorname{Pext}(\hat{G}, C) = 0$ for all discrete groups C, hence $\hat{G}$ is a direct sum of cyclic groups. Consequently, G is of the form $\mathbf{T}^m \oplus A$ where A is a topological direct product of finite cyclic groups. By (1), A is trivial. Since $\operatorname{Ext}(X, \mathbf{T}^m) = 0$ for all X in $\mathfrak{L}$, the third assertion follows.

(4) The last statement follows from the proof of (3) and Theorem 3.1. □

Dually, we have:

THEOREM 4.2. *Let G be an LCA group.*

1. *If G is discrete, then $^*\operatorname{Pext}(G, X) = 0$ for each X in $\mathfrak{L}$ if and only if G is a direct sum of cyclic groups.*
2. *If G is discrete, then $^t\operatorname{Pext}(G, X) = 0$ for each X in $\mathfrak{L}$ if and only if G is free.*
3. *If G is compact, then $^*\operatorname{Pext}(G, X) = 0$ for each X in $\mathfrak{L}$ if and only if G is torsion.*
4. *If G is compact, then $^t\operatorname{Pext}(G, X) = 0$ for each X in $\mathfrak{L}$ if and only if $G = 0$.*

The next theorem contains some structural information on those LCA groups G with the property that every $*$-pure (resp. topologically pure) exact sequence $0 \to G \to H \to X \to 0$ splits.

THEOREM 4.3. *Let G be an LCA group.*

1. *If for each group X in $\mathfrak{L}$, $^*\operatorname{Pext}(X, G) = 0$ or $^t\operatorname{Pext}(X, G) = 0$, then $G \cong \mathbf{R}^n \oplus \mathbf{T}^m \oplus G'$ where G' is totally disconnected.*
2. *$^*\operatorname{Pext}(X, G) = 0$ for each X in $\mathfrak{L}$ and G/G_0 possesses a pure compact open subgroup if and only if $G \cong \mathbf{R}^n \oplus \mathbf{T}^m$.*
3. *$^t\operatorname{Pext}(X, G) = 0$ for each X in $\mathfrak{L}$ and G/G_0 possesses a pure compact open subgroup if and only if $G \cong \mathbf{R}^n \oplus \mathbf{T}^m \oplus A \oplus B$, where A is a topological direct product of finite cyclic groups and B is a discrete bounded group.*

PROOF. (1) Let C be a connected LCA group. Then the proper exact sequence $0 \to G_0 \xrightarrow{\alpha} G \to G/G_0 \to 0$ induces the exact sequence
$$0 = \operatorname{Hom}(C, G/G_0) \to \operatorname{Ext}(C, G_0) \xrightarrow{\alpha_*} \operatorname{Ext}(C, G).$$

If $E : 0 \to G_0 \xrightarrow{\phi_1} X \to C \to 0 \in \text{Ext}(C, G_0)$, then we have

$$\alpha_*(E) = \alpha E : 0 \to G \xrightarrow{\phi_2} X' \to C \to 0$$

where $X' = (G \oplus X)/N$ and $N = \{(-\alpha(g), \phi_1(g)) : g \in G_0\}$. Since G_0 and C are connected, X is connected (cf. [**7**] (7.14)), hence $X[m]$ is a compact group for each positive integer m. But then $X'[m]/\phi_2(G[m])$ is compact as well which implies that the sequence

$$0 \to G[m] \to X'[m] \to C[m] \to 0$$

is proper exact. Therefore, αE is $*$-pure. Since we have a commutative diagram

$$\begin{array}{ccccccccc} 0 & \to & G_0 & \to & X & \to & C & \to & 0 \\ & & \downarrow & & \downarrow & & \| & & \\ 0 & \to & mG & \to & mX' & \to & mC & \to & 0 \end{array}$$

with proper exact bottom row, αE is also topologically pure. It follows that $\text{Ext}(C, G_0) = 0$ if $^*\text{Pext}(C, G) = 0$ or $^t\text{Pext}(C, G) = 0$. By [**5**] Theorem 3.3, the identity component of G is isomorphic to $\mathbf{R}^n \oplus \mathbf{T}^m$ and is therefore a direct summand of G (cf. [**11**] Theorem 3.2).

(2) Suppose $^*\text{Pext}(X, G) = 0$ for each X in $\mathfrak{L}$ such that $D = G/G_0$ has a pure compact open subgroup A. By (1), G_0 is a direct summand of G, therefore A is an algebraically compact open subgroup of D. Then there is a discrete subgroup B of G so that $D = A \oplus B$. It follows that $^*\text{Pext}(X, A) = 0$ and $^*\text{Pext}(X, B) = 0$ for each X in $\mathfrak{L}$ and Theorem 4.1 yields $A = B = 0$.

(3) Finally, assume that $^t\text{Pext}(X, G) = 0$ for all X in $\mathfrak{L}$ such that $D = G/G_0$ has a pure compact open subgroup A. Again, we can write $D = A \oplus B$ and by Theorem 4.1, A is a topological direct product of finite cyclic groups and B is a discrete bounded group. Conversely, every group G of the form $\mathbf{R}^n \oplus \mathbf{T}^m \oplus A \oplus B$ as in (3) satisfies $^t\text{Pext}(X, G) = 0$ for all X in $\mathfrak{L}$. This completes the proof of the theorem. $\square$

Dualization of Theorem 4.3 yields the following result involving the subgroup bG of all compact elements of G:

THEOREM 4.4. *Let G be an LCA group.*

1. *If for each group X in $\mathfrak{L}$, $^*\text{Pext}(G, X) = 0$ or $^t\text{Pext}(G, X) = 0$, then $G \cong \mathbf{R}^n \oplus \bigoplus_m \mathbf{Z} \oplus bG$.*
2. *$^*\text{Pext}(G, X) = 0$ for each X in $\mathfrak{L}$ and bG possesses a topologically pure compact open subgroup if and only if $G \cong \mathbf{R}^n \oplus C \oplus D$ where C is a compact torsion group and D is a discrete group which is a direct sum of cyclic groups.*
3. *$^t\text{Pext}(G, X) = 0$ for each X in $\mathfrak{L}$ and bG possesses a topologically pure compact open subgroup if and only if $G \cong \mathbf{R}^n \oplus \bigoplus_m \mathbf{Z}$.*

References

[1] D. L. Armacost, *The Structure of Locally Compact Abelian Groups*, Marcel Dekker Inc., New York, 1981.

[2] J. Braconnier, *Sur les groupes topologiques localement compacts*, J. Math. Pures Appl., N.S. **27** (1948), 1–85.

[3] L. Fuchs, *Infinite Abelian Groups*, Vols. **I** and **II**, Academic Press, New York, 1970 and 1973.

[4] R. O. Fulp, *Homological study of purity in locally compact groups*, Proc. London Math. Soc. **21** (1970), 501–512.

[5] R. O. Fulp and P. Griffith, *Extensions of locally compact abelian groups II*, Trans. Amer. Math. Soc. **154** (1971), 357–363.

[6] S. Hartman and A. Hulanicki, *Les sous-groupes purs et leurs duals*, Fund. Math. **45** (1957), 71–77.

[7] E. Hewitt and K. Ross, *Abstract Harmonic Analysis*, Vol. **I**, Springer Verlag, Berlin, 1963.

[8] J. A. Khan, *The finite torsion subgroup of an LCA group need not split*, Period. Math. Hungar. **31** (1995), 43–44.

[9] P. Loth, *Direct docompositions of LCA groups*, Abelian Groups and Modules, Trends Math., Birkhäuser, Basel (1999), pp. 301–307.

[10] P. Loth, *When is a discrete bounded subgroup of an LCA group necessarily a topological direct summand?*, Far East J. Math. Sci. (FJMS), to appear.

[11] M. Moskowitz, *Homological algebra in locally compact abelian groups*, Trans. Amer. Math. Soc. **127** (1967), 361–404.

[12] N. Vilenkin, *Direct decompositions of topological groups I*, Mat. Sb., N. S. **19** (61) (1946), 85–154. [English translation from the Russian by E. Hewitt in A.M.S. Translations, Series 1, Volume 23, Providence, Rhode Island (1950).]

DEPARTMENT OF MATHEMATICS, SACRED HEART UNIVERSITY, 5151 PARK AVENUE, FAIRFIELD, CONNECTICUT 06432

E-mail address: `lothp@sacredheart.edu`

Rings having simple adjoint semigroup

N.R. McConnell and T. Stokes

Dedicated to Professor Laszlo Fuchs in honour of his 75th birthday.

ABSTRACT. Jacobson radical rings are precisely those rings which are groups with respect to the adjoint operation a+b+ab. Rings which are simple semigroups under the adjoint operation can also be shown to form a radical class; we consider some of its properties and related matters. We conjecture that this class and the Jacobson radical class are the only two such radical classes defined by an associating expression defined purely in terms of the adjoint operation.

Recall that the adjoint operation on an associative ring is defined by the rule $a \circ b = a + b + ab$; $\circ$ is associative (so that $(R, \circ)$ is a semigroup) with identity 0, and is commutative if and only if the ring is. The following family of identities is easily shown:

$$x_1 \circ x_2 \circ \cdots \circ x_n = (x_1 + 1)(x_2 + 1) \cdots (x_n + 1) - 1,$$

where the identity on the right hand side is formally adjoined if necessary. We use the notation $x^{\circ n}$ to denote $x \circ x \circ \cdots \circ x$ (n times).

For background on radical theory of rings, see [4] or [14]. We will denote the Brown-McCoy radical class (the upper radical of the class of simple rings with identity) by $\mathcal{G}$, and the Behrens radical class (the upper radical of the class of subdirectly irreducible rings with heart containing a non-zero idempotent) by $\mathcal{J}_B$.

We shall be dealing with semigroups throughout; for background, see [6] for instance.

1. Introduction

Let $\mathcal{V}$ be a variety of associative rings. Let Ω_1 be the free ring in $\mathcal{V}$ on the generators $x, y_1, z_1, y_2, z_2, \ldots$ and Ω the free ring on the generators $x, y_1, y_2, \ldots$. We view Ω as a subset of Ω_1.

An element $f(x, y_1, y_2, \ldots, y_n)$ of Ω is said to be *associating* if there exist $g_1, g_2, \ldots, g_n \in \Omega_1$ such that

$$f(f(x, y_1, y_2, \ldots, y_n), z_1, z_2, \ldots, z_n) = f(x, g_1, g_2, \ldots, g_n).$$

This notion is defined in greater generality in [7].

1991 *Mathematics Subject Classification.* Primary 16N80, 16N20; Secondary 16U99.

© 2001 American Mathematical Society

For $f(x,y_1,y_2,\ldots,y_n) \in \Omega$, let $\mathcal{R}_f$ be the class of rings in $\mathcal{V}$ defined as follows: R is in $\mathcal{R}_f$ provided that, for every $r \in R$ there exist $s_1, s_2, \ldots, s_n \in R$ such that $f(r,s_1,s_2,\ldots,s_n) = 0$. By Theorem 4 of [**7**], $\mathcal{R}_f$ is a radical class in $\mathcal{V}$ provided f is associating.

The classes of quasiregular and von Neumann regular rings have the form $\mathcal{R}_f$ for some associating $f \in \Omega$: in the former case, we may let $f = x + y + xy$ and $g = y + z + yz$; in the latter, $f = x - xyx$ and $g = y + z - zxy - yxz + yxzxy$. These examples have a common generalisation, explored in [**11**].

The study of radical classes defined purely additively is essentially an exercise in abelian group theory. Radical classes defined only in terms of the multiplicative semigroup are trivial: any product of two or more distinct elements is zero provided at least one of the elements is zero, so the only interesting cases are those featuring only one variable, of which there is only one example, the class of nil rings. However, it is a non-trivial problem to determine all radical classes of rings of the form $\mathcal{R}_f$, where f is associating and defined only in terms of the adjoint operation $\circ$.

We solved this problem for the variety of commutative rings in [**8**]. The distinct radical classes of commutative rings defined by associating monomials in the adjoint operation are exactly the $\mathcal{J}_m = \mathcal{R}_{f_m}$'s, $f_m = x \circ y^{\circ m}$, where m is a product of distinct primes, and where $\mathcal{J}_m \cap \mathcal{J}_n = \mathcal{J}_{lcm(m,n)}$.

The general associative case is a quite distinct and apparently far harder problem. So far we have discovered only one example other than $\mathcal{J}$ itself, and we conjecture that there are no others, although our attempts to prove this have so far been fruitless. Here we discuss this possibly unique non-Jacobson example.

2. The radical class $\mathcal{K}$

Let $f(x,y,z) = y \circ x \circ z$. Then

$$f(f(x,y,z),u,v) = u \circ y \circ x \circ z \circ v = f(x,g,h),$$

where $g = u \circ y$ and $h = z \circ v$. Hence f is associating and so $\mathcal{R}_f$ is a radical class. Henceforth we refer to this class as $\mathcal{K}$.

Quasiregular rings are exactly those for which the adjoint semigroup is a group. There is a similar characterisation of rings in $\mathcal{K}$. Recall that a semigroup S is *simple* if it has no non-trivial ideals (subsets I for which $i \in I$ implies $is, si \in I$ for all $s \in S$). Hence every group is a simple semigroup.

PROPOSITION 2.1. $\mathcal{K}$ is the class of rings R for which $(R, \circ)$ is a simple semigroup.

PROOF. If S is a monoid, the principal ideal generated by $a \in S$ is the subset $(a) = \{bac | b, c \in S\}$; clearly, (a) is contained in any ideal of S containing a. Thus, for any ring R, $(R, \circ)$ is simple if and only if, for every $a \in R$, there exist $b, c \in R$ for which $b \circ a \circ c = 0$, that is, if and only if $R \in \mathcal{K}$. □

It is shown in [**3**] that if L is the ring of all linear transformations on a vector space of dimension $\aleph_\alpha$, where α is a non-zero ordinal, and S_α is the subring consisting of all transformations $T \in L$, with $\dim((T)) < \aleph_\alpha$, then $S_\alpha \in \mathcal{K}$. In fact, the non-trivial proper ideals of L are precisely the rings

$$S_\mu = \{T : V \to V | \dim((T)) < \aleph_\mu\}$$

(so $R = S_\alpha$). All these rings are primitive, and hence are $\mathcal{J}$-semisimple. For $\mu \neq 0$, the proof in [**3**] can be adapted to show that $S_\mu \in \mathcal{K}$. The case $\mu = 0$ is dealt with by the following result:

LEMMA 2.2. *If L is any dense ring of linear transformations over a vector space V and S_0 is the ideal of L consisting of all finite rank transformations, then $S_0 \notin \mathcal{K}$ or $S_0 = 0$.*

PROOF. If $S_0 = 0$, the result is immediate. Consider $S_0 \neq 0$, and let x be the element of S_0 which maps the basis vector e to $-e$ and maps all other basis vectors to zero. Suppose there exist $y, z \in S_0$ for which $y \circ x \circ z = 0$. Let W be the subspace of V generated by the images of x, y, z; let $m = dim(W)$, and note that each of x, y, z maps W into itself and moreover $y \circ x \circ z = 0$ restricted to W also. Let B be an ordered basis of W containing e, and let X, Y, Z be the representing matrices of x, y, z respectively, relative to B, with e listed first. Because $y \circ x \circ z = 0$, it follows that $Y \circ X \circ Z = 0$ in the algebra of $m \times m$ matrices, so $(I+Y)(I+X)(I+Z) = I$. Now X has -1 in the top left position and zeros elsewhere, so $I + X$ has a column of zeros and hence is singular, a contradiction. Thus no such y, z exist and so $S_0 \notin \mathcal{K}$. □

Thus we have
$$0 \subset S_0 \subset S_1 \subset \cdots \subset S_\alpha \subset L,$$
where each containment is strict, $S_0 \notin \mathcal{K}$, $S_\mu \in \mathcal{K}$ for $\mu \neq 0$, and $\mathcal{G}(L/S_\alpha) = 0$ (as L/S_α is a simple ring with identity). Also, S_0 is simple (as an idempotent minimal ideal of L), and has no identity, so $S_0 \in \mathcal{G}$.

We summarise some properties of the radical class $\mathcal{K}$.

THEOREM 2.3. *$\mathcal{K}$ is non-hereditary, $\mathcal{J} \subset \mathcal{K} \subset \mathcal{G}$ with both inclusions strict, and $\mathcal{J} = \mathcal{K} \cap \mathcal{J}_B$.*

PROOF. Obviously $\mathcal{J} \subset \mathcal{K}$ since every group is a simple semigroup.

If R is a ring with identity, then
$$x \circ (-1) \circ z = (x+1)(-1+1)(z+1) - 1 = -1,$$
so $R \notin \mathcal{K}$ and $\mathcal{K} \subset \mathcal{G}$. Clearly, S_0 and S_α above provide examples to show that the containments are strict, and that $\mathcal{K}$ is not hereditary.

Now suppose $R \in \mathcal{K} \cap \mathcal{J}_B$. Then R has no non-zero idempotents, so if $a \circ a = a$, then $2a + a^2 = a$, and so $(-a)^2 = a^2 = -a$, so $-a$ is an idempotent and therefore is 0. Hence $(R, \circ)$ has no non-zero idempotents. However, $(R, \circ)$ is simple, so for all $a \in R$ there exist $b, c \in R$ for which $b \circ a \circ c = 0$, so
$$c \circ b \circ a = c \circ (b \circ a \circ c) \circ b \circ a = (c \circ b \circ a) \circ (c \circ b \circ a),$$
and so $c \circ b \circ a = 0$; similarly one can show that $a \circ c \circ b = 0$, so $c \circ b$ is the inverse of a in $(R, \circ)$, which is therefore a group. Thus $R \in \mathcal{J}$, and so $\mathcal{J} = \mathcal{K} \cap \mathcal{J}_B$. □

Köthe's problem, the question of whether the sum of two nil left ideals need be nil, is equivalent to whether the polynomial ring in one indeterminate $R[X]$ over a nil ring R is always Jacobson radical. Beidar, Fong and Puczyłowski have recently shown ([**2**], Corollary 3.6) that $R[X]$ is Behrens radical; thus we have the alternative formulation:

COROLLARY 2.4. *Köthe's problem has a positive solution if and only if, for any nil ring R, $R[X] \in \mathcal{K}$*

Note that with S_0 and S_α as above, S_0 is a simple ring with no identity, contains idempotents and is $\mathcal{K}$-semisimple as shown above. Hence $(\mathcal{J}_B \vee \mathcal{K})(S_0) = 0$. But as also shown, $S_0 \in \mathcal{G}$, so $\mathcal{J}_B \vee \mathcal{K}$ is strictly contained in $\mathcal{G}$.

Finite primitive rings are matrix rings over finite fields, so finite $\mathcal{J}$-semisimple rings are subdirect products of these; however all such rings are $\mathcal{K}$-semisimple also. Hence for finite rings, $\mathcal{K} = \mathcal{J}$. Similarly, $\mathcal{K} = \mathcal{J}$ for commutative rings, as is obvious.

We can see that $\mathcal{K}$ is not left or right hereditary, and

$$\mathcal{K} \ni \begin{bmatrix} 0 & 0 \\ \mathbb{Q} & 0 \end{bmatrix} \triangleleft \begin{bmatrix} \mathbb{Q} & 0 \\ \mathbb{Q} & 0 \end{bmatrix} \triangleleft_l \begin{bmatrix} \mathbb{Q} & \mathbb{Q} \\ \mathbb{Q} & \mathbb{Q} \end{bmatrix}$$

(where $\triangleleft_l$ denotes left ideal) shows that $\mathcal{K}$ is not left (and similarly not right) stable. It is not clear whether or not $\mathcal{K}$ is left and right strong; certainly its symmetry indicates it will be both or neither.

3. The lattice of radicals near $\mathcal{K}$

The placement of $\mathcal{K}$ in the lattice of radical classes indicates that the most interesting area of study is how $\mathcal{K}$ divides the Jacobson semisimple but Brown-McCoy radical rings. With this in mind, we will consider some related radical classes.

1. The pseudocomplement of the Jacobson radical, $\mathcal{J}^*$.

This has been characterised as follows (see [**13**], [**9**]). The following are equivalent for a ring A:

- A is in $\mathcal{J}^*$;
- A is in $(0; \mathcal{J})$, that is, is strongly $\mathcal{J}$-semisimple (i.e. every homomorphic image of A is $\mathcal{J}$-semisimple);
- A is in the upper radical $U(\mathcal{J})$ of $\mathcal{J}$.

From our previous discussion, the rings $S_{\mu+1}/S_\mu$ for $\mu \neq 0$ are simple $\mathcal{K}$-rings containing idempotents, so they are all in $(0; \mathcal{J})$. Thus, as the proper homomorphic images of S_μ are S_μ/S_ν for $\nu < \mu$, by closure under extensions of $\mathcal{J}^*$, each S_μ is in $(0; \mathcal{J})$. Thus $\mathcal{K} \cap \mathcal{J}^*$ contains all our examples so far of $\mathcal{J}$-semisimple $\mathcal{K}$-rings.

2. The "special closure" of $\mathcal{K}$, $\mathcal{K}_\phi$.

Andrunakievic introduced this construction in [**1**]: for hereditary supernilpotent radical classes $\mathcal{R}$, $\mathcal{R}_\phi$ is the upper radical of the class of subdirectly irreducible rings with $\mathcal{R}$-semisimple hearts. For these radicals, $\mathcal{R}$ is contained in $\mathcal{R}_\phi$, and $\mathcal{R}_\phi$ and the class of strongly $\mathcal{R}$-semisimple rings $\bar{\mathcal{R}}$ are upper radicals of each other. If we define $\mathcal{K}_\phi$ to be the upper radical of the class of subdirectly irreducible rings with $\mathcal{K}$-semisimple hearts, then by a result in [**1**], $\mathcal{K}_\phi$ is a special radical (and hence hereditary), but $\mathcal{K}$ is not contained in $\mathcal{K}_\phi$. In fact, the ring S_0 is $\mathcal{K}$-semisimple and is the heart of each S_μ above, so $S_\mu \notin \mathcal{K}_\phi$. Clearly, $\mathcal{J}_\phi \subseteq \mathcal{K}_\phi$, however.

3. The largest hereditary subradical of $\mathcal{K}$.

This is the class $T_\mathcal{K}$ of rings for which every accessible subring is in $\mathcal{K}$ (see [**12**]). By Lemma 2.2, any primitive ring containing finite image transformations will have an ideal which is not in $\mathcal{K}$, so none of these can be in $T_\mathcal{K}$.

4. The strongly $\mathcal{K}$-semisimple rings, $(0;\mathcal{K})$.

This may or may not be a radical class; by Theorems 6 and 7 of [**5**], it is a hereditary radical class if and only if $\mathcal{K}$ is not mutagenic. By Theorem 8 of [**5**], $\mathcal{K}$ is mutagenic if and only if it contains a ring which is the union of a chain of ideals which are strongly $\mathcal{K}$-semisimple rings. Example 3 of [**5**] shows that the lower radical class generated by S_{ω_0} plus the class of nilpotent rings is mutagenic; however, this uses the fact that the rings S_μ for finite μ are not in that radical class. Clearly, $(0;\mathcal{K}) \subset (0;\mathcal{J})$ strictly, since the first class considered above shows the existence of strongly $\mathcal{J}$-semisimple $\mathcal{K}$-rings.

5. The base radical class of $\mathcal{J}$-semisimple $\mathcal{K}$-rings, $\mathcal{L}_b(\mathcal{K}\backslash\mathcal{J})$.

Introduced in [**10**], $\mathcal{L}_b(\mathcal{X})$ is the class of all rings for which every non-zero homomorphic image has a non-zero accessible subring in the class $\mathcal{X}$; it is always a radical class, and coincides with the lower radical when $\mathcal{X}$ is homomorphically closed. By Corollary 5 of [**10**], $\mathcal{L}_b(\mathcal{K}\backslash\mathcal{J})$ contains all rings for which every homomorphic image is in $\mathcal{K}\backslash\mathcal{J}$; this includes the rings S_μ, so $\mathcal{L}_b(\mathcal{K}\backslash\mathcal{J}) \neq 0$.

From all this, a number of questions arise:
1. Does $\mathcal{K} \cap \mathcal{J}^*$ contain all $\mathcal{J}$-semisimple $\mathcal{K}$-rings?
2. Does $\mathcal{K} \cap \mathcal{K}_\phi = \mathcal{J}$?
3. Does $T_\mathcal{K} = \mathcal{J}$?
4. Is $\mathcal{K}$ mutagenic, or equivalently, is $(0;\mathcal{K})$ a radical class?
5. How are $\mathcal{L}_b(\mathcal{K}\backslash\mathcal{J})$ and $\mathcal{K} \cap (0;\mathcal{J})$ related?

These questions are closely related; further study of some of the properties of primitive rings may be fruitful in answering them.

Acknowledgement

The authors would like to acknowledge Dr. B.J. Gardner for asking the question that started all this.

References

[1] V.A. Andrunakievic, *Radicals of associative rings* I, Amer. Math. Soc. Transl. **52** (1966), 95–128.

[2] K.I.Beidar, Y.Fong and E.Puczyłowski, *Polynomial rings over nil rings can not be homomorphically mapped onto rings with non-zero idempotents*, submitted to J. Algebra.

[3] W.E. Clark and J. Lewin, *On minimal ideals in the circle composition semigroup of a ring*, Publ. Math. Debrecen **14** (1967), 99–104.

[4] N. Divinsky, *Rings and radicals*, Allen and Unwin, 1965.

[5] N. Divinsky and A. Sulinski, *Radical pairs*, Can. J. Math. **29** no.5 (1977), 1086–1091.

[6] J.M. Howie, *Fundamentals of Semigroup Theory*, London Mathematical Society Monographs, Oxford University Press, 2nd ed., 1995.

[7] N.R. McConnell and T. Stokes, *Equationally defined radical classes*, Bull. Aust. Math. Soc. **47** (1993), 217–220.

[8] N.R. McConnell and T. Stokes, *Radical classes of commutative rings defined in terms of the adjoint operation*, Comm. Alg. **22** (1994), 5533–5548.

[9] R.G. McDougall, *On the lattice of all radical classes, part 1: examples of pseudocomplements*, Comm. Alg. **27** no.5 (1999), 2441–2465.

[10] R.G. McDougall, *A generalisation of the lower radical class*, Bull. Aust. Maths Soc. **59** (1999), 139–146.

[11] G.L. Musser, *Linear semiprime $(p;q)$ radicals*, Pacific Journal of Mathematics **37** (1971), 749–757.

[12] E.R. Puczylowski, *A note on hereditary radicals*, Acta Sci. Math. **44** (1982), 133–135.

[13] R.L. Snider, *Lattices of radicals*, Pacific J. Math. **40** no.1 (1971), 307–320.

[14] R. Wiegandt, *Radical and semisimple classes of rings*, Queen's Papers in Pure and Applied Mathematics vol. 37, Kingston, Ontario, 1974.

DEPARTMENT OF MATHEMATICS AND COMPUTING, CENTRAL QUEENSLAND UNIVERSITY, ROCKHAMPTON, QUEENSLAND, 4702, AUSTRALIA.
Current address: Department of Defence, Locked Bag 5076, Kingston, ACT 2604, Australia
E-mail address: `nickm@defcen.gov.au`

SCHOOL OF MATHEMATICAL AND PHYSICAL SCIENCES, MURDOCH UNIVERSITY, MURDOCH, W.A. 6150, AUSTRALIA.
E-mail address: `stokes@prodigal.murdoch.edu.au`

Invariants of Global crq–Groups

A. Mader, L.G. Nongxa and M.A. Ould–Beddi

This paper is dedicated to Laszlo Fuchs on his 75th birthday.

ABSTRACT. Almost completely decomposable groups with a cyclic regulating quotient, the crq–groups, are a reasonably accessible class of groups for arbitrary critical typesets and have been studied intensively. Previous results were based on special, convenient representations of the groups. In this paper invariants are introduced that are independent of any particular representations and lead to a new classification theorem. Local–global relations are detailed and heavily utilized in order to reduce to the relatively easy local case when the regulating index is a prime power.

1. Introduction

An **almost completely decomposable group** X is a finite essential (abelian) extension of a completely decomposable group A of finite rank. In 1974 E.L. Lady [**Lad74**] initiated a systematic theory of such groups based on the fundamental concept of *regulating subgroup*. The regulating subgroups can be defined as the completely decomposable subgroups of least index in an almost completely decomposable group. Details on the beginnings and subsequent developments of the theory of almost completely decomposable groups can be found in the monograph [**Mad00**]. Rather than citing the original sources we will quote [**Mad00**] which contains an extensive bibliography of journal articles.

Almost completely decomposable groups are easily written down. In fact, one traditionally starts with a completely decomposable group A inside a divisible hull $\mathbb{Q}A$, and adjoins a finite number of elements x_i of $\mathbb{Q}A$ to get the almost completely decomposable group $X = A \dotplus \sum_i \mathbb{Z}x_i$. In [**MV95**], [**BM96**] and [**DMM97**] the simpler but important subclass of almost completely decomposable groups with a cyclic regulating quotient was studied. These crq–*groups* thus are almost completely decomposable groups X containing a regulating subgroup A such that X/A

1991 *Mathematics Subject Classification.* 20K15, 20K35.

Key words and phrases. socle, radical, near–isomorphism, almost completely decomposable group, cyclic regulating quotient.

The third author thanks the University of the Western Cape for her kind hospitality and support during the preparation of this paper.

is cyclic. Later Campagna (([**Cam95**], [**Cam00**])) showed that any cyclic extension of a completely decomposable group is a crq–group.

The extensive results on clipped local crq–groups obtained in [**MV95**] ([**Mad00**, Theorem 6.4.1]) are based on a special representation of such groups detailed in the following Structure Theorem. Recall that a group is **clipped** if it has no completely decomposable direct summands, and **local** if its regulating index is a prime power.

THEOREM 1.1. *The almost completely decomposable group X is a clipped crq–group with $\mathrm{rgi}(X) = p^n > 1$ if and only if X contains a completely decomposable subgroup $A = \bigoplus_{\rho \in \mathrm{T}_{\mathrm{cr}}(X)} A_\rho$ and an element $a \in A$ such that the following hold.*

1. $X = A + \mathbb{Z}p^{-n}a$;
2. $a = \sum_{\rho \in \mathrm{T}_{\mathrm{cr}}(X)} s_\rho a_\rho$ *for certain p–powers s_τ and certain elements $a_\tau \in A_\tau$ with $\gcd^A(p, a_\tau) = 1$;*
3. *for each $\tau \in \mathrm{T}_{\mathrm{cr}}(X)$, $A_\tau = \tau a_\tau$;*
4. *the p–powers s_τ satisfy*
   (StrA) *For each $\tau \in \mathrm{T}_{\mathrm{cr}}(X)$, $s_\tau < p^n$.*
   (StrB) $s_\tau = 1$ *for some $\tau \in \mathrm{T}_{\mathrm{cr}}(X)$.*
   (StrC) *For each $\tau \in \mathrm{T}_{\mathrm{cr}}(X)$, there is $\rho \not\geq \tau$ such that $s_\rho \leq s_\tau$.*
   (StrD) *For $\sigma, \tau \in \mathrm{T}_{\mathrm{cr}}(X)$, $s_\tau < s_\sigma$ whenever $\sigma < \tau$.*

This strategy was continued in [**DMM97**] ([**Mad00**, Section 6.3]) for global crq–groups. In contrast the emphasis in this article is on results for global crq–groups that are independent of any particular representation. Local–global principles ([**MMV99**], [**Mad00**, Section 5]) play a major role as well as radicals that have been conspicuously rare in the theory of almost completely decomposable groups. The key is that in the situation of Theorem 1.1 it is true that $s_\tau = \mathrm{rgi}\, X[\tau]$. Main results are Proposition 3.6, Theorem 3.10 on local–global relations, Theorem 4.1 characterizing crq–groups, Theorem 4.3 stating that the class of crq–groups is closed under direct summands, and finally Theorem 4.5 containing a complete system of near–isomorphism invariants for global crq–groups.

The discussion of direct decompositions of global crq–groups is left to a subsequent paper.

2. General Background

The set of all prime number is denoted by $\mathbb{P}$. The purification of a subgroup H in a torsion–free group G is denoted by H_*^G. We take it for granted that the reader is familiar with the usual type subgroups, namely the **socles** $G(\tau)$, $G^*(\tau)$, $G^\sharp(\tau) = G^*(\tau)_*^G$, and the **radicals** $G[\tau]$, $G^\sharp[\tau] = \bigcap_{\rho < \tau} G[\rho]$. A type τ is **critical** for G if $G(\tau)/G^\sharp(\tau) \neq 0$. The **critical typeset** $\mathrm{T}_{\mathrm{cr}}(G)$ is the set of all critical types of G. If A is a completely decomposable group, then $A = \bigoplus_{\rho \in \mathrm{T}_{\mathrm{cr}}(A)} A_\rho$ is always assumed to be a decomposition of A into ρ–homogeneous components $A_\rho (\neq 0)$. The **typeset** of a group G is denoted by $\mathrm{Tst}(G)$. If G is an almost completely decomposable group, then its typeset is the meet closure of its critical typeset.

Every almost completely decomposable group X can be decomposed as $X = X_{cd} \oplus X_{cl}$ where X_{cd} is completely decomposable and X_{cl} is **clipped**, i.e., X_{cl} has no completely decomposable direct summands. The existence of such a **main decomposition** is immediate since X has finite rank but it is not unique. Lady showed that X_{cd} is unique up to isomorphism and consequently X_{cl} is unique up to near–isomorphism ([**Mad00**, Theorem 9.2.7]).

The Purification Lemma [**Mad00**, Lemma 11.4.1] and [**Mad00**, Corollary 11.2.5] will be an important tool in calculating the invariants of purifications of direct summands of regulating subgroups of crq–groups. For future use we add a variant of part of the Purification Lemma. It involves a generalized greatest common divisor $\gcd^A(N, a^\downarrow)$ of a non–singular integral $k \times k$ matrix N and a column vector $a^\downarrow$ of k elements of A ([**Mad00**, Chapter 11]), but for the purposes of this article $k = 1$ and the reader may think of N as a positive integer and $a^\downarrow$ as a single element of A.

LEMMA 2.1. *Let X be an almost completely decomposable group and A a subgroup of finite index in X. Suppose that $A = B \oplus C$ and $X = A + \vec{\mathbb{Z}} N^{-1} a^\downarrow$ with $a^\downarrow = b^\downarrow + c^\downarrow$ where $a^\downarrow \in A^\downarrow$, $b^\downarrow \in B^\downarrow$ and $c^\downarrow \in C^\downarrow$. Then $B_*^X = B + \vec{\mathbb{Z}} N_B^{-1} b^\downarrow$ where $N_B = \gcd^A(N, c^\downarrow)$. If A is a regulating subgroup of X, then B is regulating in B_*^X.*

PROOF. The only result that is not part of the Purification Lemma is the claim that B is regulating in B_*^X if A is regulating in X. We will verify that B is regulating in B_*^X by showing that it is a completely decomposable subgroup of minimal index. Let B' be a regulating subgroup of B_*^X and let $A' = B' \oplus C$ which is completely decomposable. We have

$$[X : A'] = [X : (B_*^X \oplus C)][(B_*^X \oplus C) : (B' \oplus C)] = [X : (B_*^X \oplus C)][B_*^X : B'],$$

and similarly

$$[X : A] = [X : (B_*^X \oplus C)][B_*^X : B].$$

By a theorem of Lady ([**Mad00**, Theorem 4.2.13]) the regulating index $\mathrm{rgi}(X) = [X : A]$ divides $[X : A']$, hence by the above index identities $[B_*^X : B]$ divides $[B_*^X : B']$, and $\mathrm{rgi}(B_*^X) = [B_*^X : B']$ divides $[B_*^X : B]$ again by Lady. Thus $\mathrm{rgi}(B_*^X) = [B_*^X : B]$ which says that B is regulating in B_*^X. □

Let A be a regulating subgroup of the almost completely decomposable group X. Lemma 2.1 implies the well–known facts that $A(\tau)$ is regulating in $X(\tau)$, $A^\sharp(\tau)$ is regulating in $X^\sharp(\tau)$, $A[\tau]$ is regulating in $X[\tau]$, $A^\sharp[\tau]$ is regulating in $X^\sharp[\tau]$ but also less evidently that $A[\tau](\sigma)$ is regulating in $X[\tau](\sigma)$, $A^\sharp(\tau)[\tau]$ is regulating in $X^\sharp(\tau)[\tau]$ and more. If X is a crq–group, then every one of these canonical subgroups are crq–groups also.

Recall that a subgroup $A = \bigoplus_{\rho \in \mathrm{T}_{\mathrm{cr}}(A)} A_\rho$ of an almost completely decomposable group X is regulating in X if and only if $X(\tau) = A_\tau \oplus X^\sharp(\tau)$ for every $\tau \in \mathrm{T}_{\mathrm{cr}}(X)$. In particular, the homogeneous components A_τ of a regulating subgroup are pure in X and therefore $\mathrm{hgt}_p^{A_\tau}(a) = \mathrm{hgt}_p^A(a) = \mathrm{hgt}_p^X(a)$ for any $a \in A_\tau$. We also mention in this context that for a subset T of $\mathrm{T}_{\mathrm{cr}}(A)$ and $a = \sum_{\rho \in T} a_\rho$, where $a_\tau \in A_\tau$, it is true that $\gcd^A(n, a) = \gcd\{\gcd^A(n, a_\rho) : \rho \in T\}$. This well–known and easily verified fact ([**Mad00**, Lemma 11.2.9]) will be used freely and without reference in the sequel.

For general background on torsion–free abelian groups, and for almost completely decomposable groups in particular, we refer to [**Mad00**]. See also [**Fuc73**] and [**Arn82**].

3. Local–Global Issues

We recall some basic results that can be found in [**Mad00**, Chapter 5]. The **regulator** $\mathrm{R}(X)$ of an almost completely decomposable group is the intersection of all regulating subgroups of X and was shown by Burkhardt to be itself a completely decomposable subgroup of finite index in X. If $A = \bigoplus_{\rho \in \mathrm{T}_{\mathrm{cr}}(A)} A_\rho$ is any regulating subgroup of X, then $\mathrm{R}(X) = \bigoplus_{\rho \in \mathrm{T}_{\mathrm{cr}}(A)} \beta_\rho^X A_\rho$ where the **Burkhardt invariants** are given by $\beta_\tau^X = \exp(X^\sharp(\tau)/\mathrm{R}(X^\sharp(\tau)))$. By definition $X_{lp}/\mathrm{R}(X) = (X/\mathrm{R}(X))_p$, the p–primary component of the finite abelian group $X/\mathrm{R}(X)$, and for an integer n, n_{lp} is the highest p–power dividing n. In the following we consider a fixed almost completely decomposable group X and various of its subgroups that are themselves almost completely decomposable groups and have localizations as such. The reader must therefore be cautioned that we consider exclusively localization with respect to the regulator of the fixed group X. Explicitly, if H is a subgroup of X, its **p–local constituent** H_{lp} is given by

$$H_{lp} = H \cap \mathbb{Z}[p^{-1}]\mathrm{R}(X), \qquad \frac{H_{lp} + \mathrm{R}(X)}{\mathrm{R}(X)} = \left(\frac{H + \mathrm{R}(X)}{\mathrm{R}(X)}\right)_p.$$

We recall some important facts ([**Mad00**, Theorem 5.1.3]).

LEMMA 3.1. *Let X be an almost completely decomposable group.*
1. $\mathrm{R}(X) = \mathrm{R}(X_{lp})$.
2. $\beta_\tau^{X_{lp}} = (\beta_\tau^X)_{lp}$.
3. $\mathrm{rgi}(X_{lp}) = (\mathrm{rgi}(X))_{lp}$.
4. *If* $A = \bigoplus_{\rho \in \mathrm{T}_{\mathrm{cr}}(A)} A_\rho \in \mathrm{Regg}(X)$, *then* $A_{lp} = \bigoplus_{\rho \in \mathrm{T}_{\mathrm{cr}}(A)} (A_\rho)_{lp} \in \mathrm{Regg}(X_{lp})$ *where* $(A_\tau)_{lp} = (\beta_\tau^X/(\beta_\tau^X)_{lp})A_\tau$.
5. $X(\tau)_{lp} = X_{lp}(\tau);\ X^\sharp(\tau)_{lp} = X_{lp}^\sharp(\tau);\ X[\tau]_{lp} = X_{lp}[\tau];\ X^\sharp[\tau]_{lp} = X_{lp}^\sharp[\tau]$.

The next lemma is a general fact in local–global affairs.

LEMMA 3.2. *Let X be any almost completely decomposable group and suppose that $X = Y \oplus Z$. Then $Y_{lp} = Y \cap X_{lp}$, $Z_{lp} = Z \cap X_{lp}$, $X_{lp} = Y_{lp} \oplus Z_{lp}$, $Y = \sum_{p \in \mathbb{P}} Y \cap X_{lp}$ and $Z = \sum_{p \in \mathbb{P}} Z \cap X_{lp}$.*

PROOF. Since $\mathrm{R}(X)$ is a fully invariant subgroup of X, we have that $\mathrm{R}(X) = Y \cap \mathrm{R}(X) \oplus Z \cap \mathrm{R}(X)$. It follows that $Y_{lp} = Y \cap X_{lp}$, $Z_{lp} = Z \cap X_{lp}$, that $X_{lp} = Y_{lp} \oplus Z_{lp}$, and that

$$\frac{X}{\mathrm{R}(X)} \cong \frac{Y}{Y \cap \mathrm{R}(X)} \oplus \frac{Z}{Z \cap \mathrm{R}(X)}.$$

Hence $Y/(Y \cap \mathrm{R}(X))$ is finite. It is easily checked that

$$\left(\frac{Y}{(Y \cap \mathrm{R}(X))}\right)_p = \frac{Y \cap X_{lp}}{Y \cap \mathrm{R}(X)}.$$

We then have

$$\frac{Y}{Y \cap \mathrm{R}(X)} = \bigoplus_{p \in \mathbb{P}} \frac{Y \cap X_{lp}}{Y \cap \mathrm{R}(X)}$$

and this implies that $Y = \sum_{p \in \mathbb{P}} Y \cap X_{lp}$. The proof for Z is the same. □

Given an almost completely decomposable group X, one can pass to the localizations X_{lp} that are considerably easier to deal with than the original global group and then draw conclusions for X. It turns out that it is also easy to start with a family of local groups and construct a global group that has the original groups as localizations. The only restriction is that the given groups have a common regulator, a condition that is necessary by Lemma 3.1.1. Proposition 3.3 can be used to construct global groups from local groups with known properties. For example, if just one of the local groups is indecomposable, then so is the global group.

PROPOSITION 3.3. *Let P be a finite set of primes and let R be a completely decomposable group of finite rank. Suppose that for each $p \in P$, X_{lpl} is a group such that $R \leq X_{lpl} \leq \mathbb{Q}R$, X_{lpl}/R is a finite p-group and $\mathrm{R}(X_{lpl}) = R$. Then $X := \sum_{p \in P} X_{lpl}$ is an almost completely decomposable group such that $\mathrm{R}(X) = R$ and $X_{lp} = X_{lpl}$ for each $p \in P$ where X_{lp} is the p-local constituent of X (with respect to $\mathrm{R}(X)$).*

PROOF. It is clear that X is an almost completely decomposable group and that

$$(3.4) \qquad \frac{X}{R} = \frac{\sum_{p \in P} X_{lpl}}{R} = \bigoplus_{p \in P} \frac{X_{lpl}}{R}.$$

It follows from (3.4) that $X_{lp} = X_{lpl}$. The localization is with respect to R but it remains to show that $R = \mathrm{R}(X)$. We show next the following auxiliary fact.

(3.5)

If $R = U \oplus V$, then $U_*^X = \sum_{p \in P} U_*^{X_{lpl}}$, $\dfrac{U_*^X}{U} = \bigoplus_{p \in P} \dfrac{U_*^{X_{lpl}}}{U}$, and $\left(U_*^X\right)_{lp} = U_*^{X_{lpl}}$.

The localization of U_*^X is with respect to R (or equivalently U). It is clear that $U_*^{X_{lpl}} \subseteq U_*^X$ and hence $\sum_{p \in P} U_*^{X_{lpl}} \subseteq U_*^X$. Conversely, let $x \in U_*^X$. Then $nx \in U$ for some $0 \neq n \in \mathbb{Z}$. Since $\dfrac{U_*^X}{U} \cong \dfrac{U_*^X \oplus V}{U \oplus V} \leq \dfrac{X}{R}$ we assume without loss of generality that n is a divisor of $|X/R| = \prod_{p \in P} |X_{lpl}/R|$. By partial fraction decomposition ([**Mad00**, Lemma 1.1.4]) we can write $1 = \sum_{p \in P} u_p \frac{n}{n_{lp}}$. Hence $x = \sum_{p \in P} u_p \left(\frac{n}{n_{lp}} x\right)$ with $\frac{n}{n_{lp}} x \in U_*^{X_{lpl}}$. So $U_*^X \subseteq \sum_{p \in P} U_*^{X_{lpl}}$ and equality is established. It is easy to see that $\dfrac{U_*^X}{U} = \bigoplus_{p \in P} \dfrac{U_*^{X_{lpl}}}{U}$ and hence $\left(U_*^X\right)_{lp} = \{x \in U_*^X : p^k x \in R \cap U_*^X = U \text{ for some } k\} = U_*^{X_{lpl}}$. This proves (3.5).

It follows from (3.5) that

$$X(\tau) = R(\tau)_*^X = \sum_{p \in P} R(\tau)_*^{X_{lpl}} = \sum_{p \in P} X_{lpl}(\tau)$$

and

$$\left(X^\sharp(\tau)\right)_{lp} = \left(R^\sharp(\tau)_*^X\right)_{lp} = R^\sharp(\tau)_*^{X_{lpl}} = X_{lpl}^\sharp(\tau).$$

We show that $\mathrm{R}(X) = R$ by using Burkhardt's Regulator Criterion ([**Mad00**, Theorem 4.4.6]). Let $\beta_\tau := \prod_{p \in P} \beta_\tau^{X_{lpl}}$ for $\tau \in \mathrm{T}_{\mathrm{cr}}(X) = \mathrm{T}_{\mathrm{cr}}(R)$. Choose a Butler decomposition $X_{lpl}(\tau) = A_{p\tau} \oplus X_{lpl}^\sharp(\tau)$. Then $R(\tau) = \beta_\tau^{X_{lpl}} A_{p\tau} \oplus R^\sharp(\tau)$ ([**Mad00**,

Theorem 4.4.4]). By [**Mad00**, Lemma 5.2.1] there is a decomposition $R(\tau) = R_\tau \oplus R^\sharp(\tau)$ such that $R_\tau \subseteq \beta_\tau X(\tau)$. Also $\beta_\tau X(\tau) = \beta_\tau \sum_{p \in P} X_{lpl}(\tau) \subseteq R(\tau)$ and

$$\exp \frac{X^\sharp(\tau)}{R^\sharp(\tau)} = \prod_{p \in P} \exp \frac{X^\sharp_{lpl}(\tau)}{R^\sharp(\tau)} = \prod_{p \in P} \exp \frac{X^\sharp_{lpl}(\tau)}{R(X_{lpl})^\sharp(\tau)} = \prod_{p \in P} \beta_\tau^{X_{lpl}} = \beta_\tau.$$

By [**Mad00**, Theorem 4.4.6] we have $R(X) = R$. □

We show next that the crq–property behaves as expected in local–global affairs.

PROPOSITION 3.6. *An almost completely decomposable group X is a crq–group if and only if for every prime p the local constituent X_{lp} is a crq–group.*

PROOF. Suppose first that X is a crq–group, and that A is a regulating subgroup such that X/A is cyclic. Since

$$(3.7) \qquad \frac{X}{A} \cong \frac{X/R(X)}{A/R(X)} \cong \bigoplus_{p \mid \mathrm{rgi}(X)} \frac{X_{lp}/R(X)}{A_{lp}/R(X)} \cong \bigoplus_{p \mid \mathrm{rgi}(X)} \frac{X_{lp}}{A_{lp}}$$

every group X_{lp}/A_{lp} is cyclic and $A_{lp} \in \mathrm{Regg}(X_{lp})$ by Lemma 3.1.4.

In order to see the converse, choose, for each prime $p \mid \mathrm{rgi}(X)$, a group $A(p) \in \mathrm{Regg}(X_{lp})$ such that $X_{lp}/A(p)$ is cyclic. Then, by [**Mad00**, Theorem 5.3.2], $A = \sum_{p \mid \mathrm{rgi}(X)} A(p) \in \mathrm{Regg}(X)$ and $A_{lp} = A(p)$. It now follows from (3.7) that X/A is cyclic. □

We will need some information on localization of base group purifications. Special cases of these are again the various type subgroups.

LEMMA 3.8. *Let X be an almost completely decomposable group with regulator $R = \mathrm{R}(X)$. Let $A = B \oplus C$ be a regulating subgroup of X. Then B_{lp} is a regulating subgroup of $(B^X_*)_{lp}$ and $(\mathrm{rgi}\, B^X_*)_{lp} = \mathrm{rgi}\left((B^X_*)_{lp}\right)$.*

PROOF. By Burkhardt's regulator theorem ([**Mad00**, Theorem 4.4.4]) we have that $R = R_B \oplus R_C$ where $R_B = B \cap R$ and $R_C = C \cap R$. For any prime p, it follows that

$$\begin{aligned}
\mathbb{Z}[p^{-1}]R &= \mathbb{Z}[p^{-1}]R_B \oplus \mathbb{Z}[p^{-1}]R_C, \\
B_{lp} &= B \cap \mathbb{Z}[p^{-1}]R = B \cap \mathbb{Z}[p^{-1}]R_B \geq R_B, \\
C_{lp} &= C \cap \mathbb{Z}[p^{-1}]R = C \cap \mathbb{Z}[p^{-1}]R_C \geq R_C, \\
(B^X_*)_{lp} &= B^X_* \cap \mathbb{Z}[p^{-1}]R = B^X_* \cap \mathbb{Z}[p^{-1}]R_B, \\
(C^X_*)_{lp} &= C^X_* \cap \mathbb{Z}[p^{-1}]R = C^X_* \cap \mathbb{Z}[p^{-1}]R_C.
\end{aligned}$$

By Lemma 3.1.4 the group $A_{lp} = B_{lp} \oplus C_{lp}$ (Lemma 3.2) is regulating in X_{lp} and hence in $(B^X_*)_{lp} \oplus (C^X_*)_{lp}$. This shows that B_{lp} is a regulating subgroup of $(B^X_*)_{lp}$. From the above identities we conclude that

$$\frac{B^X_*}{R_B} = \bigoplus_{p \in \mathbb{P}} \frac{(B^X_*)_{lp}}{R_B} \geq \bigoplus_{p \in \mathbb{P}} \frac{B_{lp}}{R_B} = \frac{B}{R_B}.$$

It follows that

$$\frac{B^X_*}{B} \cong \bigoplus_{p \in \mathbb{P}} \frac{(B^X_*)_{lp}}{B_{lp}},$$

and, B_{lp} being regulating in $\left(B_*^X\right)_{lp}$ it follows that $\mathrm{rgi}((B_*^X)_{lp}) = (\mathrm{rgi}(B))_{lp}$. □

We also need to clarify what the localization of a crq–group looks like and relate top decompositions with main decompositions of localizations. We recall the definition of top decomposition for easy reference.

DEFINITION 3.9. *Let $X = A + \mathbb{Z}n^{-1}a$ be an almost completely decomposable group with completely decomposable subgroup $A = \bigoplus_{\rho \in \mathrm{T_{cr}}(A)} A_\rho$ of finite index. Write $a = \sum_{\rho \in \mathrm{T_{cr}}(A)} a_\rho$ with $a_\tau \in A_\tau$. A critical type τ is called **p-movable** if p is a prime divisor of n and either $\mathrm{hgt}_p^A(a_\tau) \geq \mathrm{hgt}_p^{\mathbb{Z}}(n)$ or there exists some critical type $\sigma < \tau$ with $\mathrm{hgt}_p^A(a_\sigma) \leq \mathrm{hgt}_p^A(a_\tau)$. The homogeneous decomposition $D: A = \bigoplus_{\rho \in \mathrm{T_{cr}}(A)} A_\rho$ is called a **top decomposition** of A, if, for all prime divisors p of n, $\mathrm{hgt}_p^A(a_\tau) \geq \mathrm{hgt}_p^{\mathbb{Z}}(n)$ for each p-movable type τ.*

It was shown in [**DMM97**] ([**Mad00**, Section 6.3]) that a top decomposition can always be achieved, and after doing so the completely decomposable direct summands of the crq–group under consideration become visible. The latter is also true when passing to localizations.

THEOREM 3.10. *Let $X = A + \mathbb{Z}n^{-1}a$ be a crq–group where $A \in \mathrm{Regg}(X)$, $a \in A$ and $\gcd^A(n, a) = 1$ (or equivalently, $n = \mathrm{rgi}(X)$). Let $A = \bigoplus_{\rho \in \mathrm{T_{cr}}(X)} A_\rho$ be a homogenous decomposition of A. Then the following hold.*

1. *$X_{lp} = A_{lp} + \mathbb{Z}n_{lp}^{-1}a$ where $A_{lp} \in \mathrm{Regg}(X_{lp})$, $a \in A_{lp}$, and $\gcd^{A_{lp}}(n_{lp}, a) = 1$ (or equivalently, $n_{lp} = \mathrm{rgi}(X_{lp})$).*
2. *$A_{lp} = \bigoplus_{\rho \in \mathrm{T_{cr}}(X)} (A_\rho)_{lp}$ and $(A_\tau)_{lp} = \left(\beta_\tau^X / \left(\beta_\tau^X\right)_{lp}\right) A_\tau$.*
3. *$\gcd^{A_{lp}}(n_{lp}, a_\tau) = \gcd^A(n_{lp}, a_\tau) = \gcd^A(n, a_\tau)_{lp}$.*
4. *$A = \bigoplus_{\rho \in \mathrm{T_{cr}}(A)} A_\rho$ is a top decomposition for X if and only if for all $p \in \mathbb{P}$,*

$$X_{lp} = \left(\bigoplus_\rho \{(A_\rho)_{lp} : n_{lp} = \gcd^A(n, a_\rho)_{lp}\}\right)$$
$$\oplus \left(\bigoplus_\rho \{(A_\rho)_{lp} : n_{lp} > \gcd^A(n, a_\rho)_{lp}\} + \mathbb{Z}n_{lp}^{-1}\sum_\rho \{a_\rho : n_{lp} > \gcd^A(n, a_\rho)_{lp}\}\right)$$

is a main decomposition of X_{lp}.

PROOF. 1. By Lemma 3.1.4 A_{lp} is a regulating subgroup of X_{lp} and $n_{lp} = \mathrm{rgi}(X_{lp})$. Note first that $a = n(n^{-1}a) \in R(X) \subseteq A_{lp}$. It is also clear that $A_{lp} + \mathbb{Z}n_{lp}^{-1}a \subseteq X_{lp}$. To show the reverse inclusion let $x \in X_{lp}$. Write $x = b + mn^{-1}a$ where $b \in A$ and $m \in \mathbb{Z}$. By hypothesis there is a p-power p^s with $p^s \leq n_{lp}$ such that $p^s x \in R(X)$. It follows that $p^s m n^{-1}a = p^s x - p^s b \in A$. But $\mathrm{ord}(n^{-1}a + A) = n$ by assumption, so

(3.11) $\qquad\qquad p^s m = n k_1 \quad$ for some integer $\ k_1$.

We conclude that $p^s b = p^s x - k_1 n(n^{-1}a) \in R(X)$ and hence $b \in A_{lp}$. Factoring n as $n = n_{lp}\check{n}_{lp}$, so that p and $\check{n}_{lp}$ are relatively prime, (3.11) implies that

(3.12) $\qquad\qquad k_2 = m\check{n}_{lp}^{-1} \in \mathbb{Z}$.

Altogether we now have that $x = b + k_2 n_{lp}^{-1} a \in A_{lp} + \mathbb{Z}n_{lp}^{-1}a$ as desired.

2. Recall that $R(X) = \bigoplus_{\rho \in T_{cr}(X)} \beta_\rho^X A_\rho$. It follows easily that $a_\tau \in \beta_\tau^X A_\tau \subseteq (A_\tau)_{\wr p} = \left(\beta_\tau^X / \left(\beta_\tau^X\right)_{\wr p}\right) A_\tau$.

3. The first identity follows since $\gcd(p, \beta_\tau^X / \left(\beta_\tau^X\right)_{\wr p}) = 1$, and the second equality follows by a direct check of the greatest common divisors.

4. **Necessity.** $A = \bigoplus_{\rho \in T_{cr}(A)} A_\rho$ being in top decomposition means, by definition, that for every prime divisor p of n and every two critical types $\sigma < \tau$, we have ([**Mad00**, Definition 6.3.4])

$$\mathrm{hgt}_p^A(a_\sigma) > \mathrm{hgt}_p^A(a_\tau) \quad \text{unless} \quad \mathrm{hgt}_p^A(a_\tau) \geq \mathrm{hgt}_p^{\mathbb{Z}}(n).$$

By the first item of the current theorem,

$$X_{\wr p} = A_{\wr p} + \mathbb{Z} n_{\wr p}^{-1} \sum_{\rho \in T_{cr}(A)} a_\rho$$

where $A_{\wr p} = \bigoplus_{\rho \in T_{cr}(A)} (A_\rho)_{\wr p} \in \mathrm{Regg}(X_{\wr p})$. It is clear that

$$\begin{aligned}
X_{\wr p} &= Y \oplus Z \quad \text{where} \\
Y &= \bigoplus_\rho \{(A_\rho)_{\wr p} : n_{\wr p} = \gcd{}^A(n, a_\rho)_{\wr p}\}, \quad \text{and} \\
Z &= \bigoplus_\rho \{(A_\rho)_{\wr p} : n_{\wr p} > \gcd{}^A(n, a_\rho)_{\wr p}\} + \mathbb{Z} n_{\wr p}^{-1} \sum_\rho \{a_\rho : n_{\wr p} > \gcd{}^A(n, a_\rho)_{\wr p}\}.
\end{aligned}$$

Here Y is completely decomposable. We claim that Z is clipped. For $\tau \in T_{cr}(Z)$, let $s_\tau = \gcd{}^{A_{\wr p}}(n_{\wr p}, a_\tau)$ and write $a_\tau = s_\tau a'_\tau$. Then $\mathrm{hgt}_p^{A_{\wr p}}(a'_\tau) = 0$, and one checks that $Z = \bigoplus_{\rho \in T_{cr}(Z)} (A_\rho)_{\wr p} + \mathbb{Z} n_{\wr p}^{-1} \sum_{\rho \in T_{cr}(Z)} s_\rho a'_\rho$ satisfies all the conditions of Theorem 1.1.

Sufficiency. Use the structure theorem for local clipped crq–groups (Theorem 1.1) and the fact that $\mathrm{hgt}_p(s_\tau a_\tau) = \log_p(s_\tau)$ to see that $A = \bigoplus_{\rho \in T_{cr}(A)} A_\rho$ is a top decomposition of A. □

We illustrate Theorem 3.10 with a simple example.

EXAMPLE 3.13. 1. *Let*

$$\tau_1 = \mathbb{Z}[7^{-1}], \quad \tau_2 = \mathbb{Z}[11^{-1}], \quad \tau_3 = \mathbb{Z}[7^{-1}, 13^{-1}], \quad \tau_4 = \mathbb{Z}[7^{-1}, 17^{-1}].$$

The poset of these types is as shown.

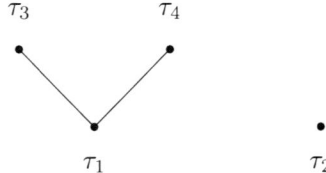

2. $A = \tau_1 v_1 \oplus \tau_2 v_2 \oplus \tau_3 v_3 \oplus \tau_4 v_4$ *is the regulating regulator of* $X = A + \mathbb{Z}\frac{1}{30}(2v_1 + v_2 + 15v_3 + 15v_4)$. *Furthermore* $\mathrm{rgi}(X) = 30$, *and* $A = \tau_1 v_1 \oplus \tau_2 v_2 \oplus \tau_3 v_3 \oplus \tau_4 v_4$ *is in top decomposition.*

3. *The primary constituents of X are*

$$X_{l2} = \sigma_1 v_1 \oplus [(\sigma_2 v_2 \oplus \sigma_3 v_3 \oplus \sigma_4 v_4) + \mathbb{Z}\frac{1}{2}(v_2 + v_3 + v_4)]$$

$$X_{l3} = \sigma_3 v_3 \oplus \sigma_4 v_4 \oplus [(\sigma_1 v_1 \oplus \sigma_2 v_2) + \mathbb{Z}\frac{1}{3}(2v_1 + v_2)]$$

$$X_{l5} = \sigma_3 v_3 \oplus \sigma_4 v_4 \oplus [(\sigma_1 v_1 \oplus \sigma_2 v_2) + \mathbb{Z}\frac{1}{5}(2v_1 + v_2)].$$

In each case, this is a main decomposition with indecomposable clipped part. Each localization has completely decomposable direct summands.

4. *X is directly indecomposable.*

4. Invariants and Classification of Global crq–Groups

Our first result gives criterion for an almost completely decomposable group to be a crq–group.

THEOREM 4.1. *An almost completely decomposable group X is a crq–group if and only if $\exp(X/\operatorname{R}(X)) = \operatorname{rgi}(X)$.*

PROOF. We first observe that, in general,

$$\exp(X/\operatorname{R}(X)) = \operatorname{lcm}\{\exp(X/A) : A \in \operatorname{Regg}(X)\}.$$

It is immediate that $\exp(X/\operatorname{R}(X)) = \operatorname{rgi}(X)$ if X is a crq–group.

Conversely, suppose that $\exp(X/\operatorname{R}(X)) = \operatorname{rgi}(X)$ and we first consider the case where X is p-local for some prime p. We have

$$\operatorname{rgi}(X) = \exp\frac{X}{\operatorname{R}(X)} = \max\left\{\exp\frac{X}{A} : A \in \operatorname{Regg}(X)\right\} = \exp\frac{X}{B}$$

for some $B \in \operatorname{Regg}(X)$. This implies that X/B is a cyclic group. In the general case, by Proposition 3.6, it suffices to show that X_{lp} is a crq–group for every prime p dividing $\operatorname{rgi}(X)$. Now, by Lemma 3.1.3, $\operatorname{rgi}(X) = \prod_{p|\operatorname{rgi}(X)} \operatorname{rgi}(X_{lp})$. Furthermore, since $\operatorname{rgi}(X_{lp})$ is the highest p–power dividing $\operatorname{rgi}(X)$,

$$\operatorname{rgi}(X_{lp}) = \max\left\{\left(\exp\left(\bigoplus_{q|\operatorname{rgi}(X)} \frac{X_{lq}}{A_{lq}}\right)\right)_{lp} : A \in \operatorname{Regg}(X)\right\}$$

$$= \max\left\{\exp\frac{X_{lp}}{A_{lp}} : A \in \operatorname{Regg}(X)\right\}$$

and this will be $\exp(X_{lp}/A_{lp})$ for some $A \in \operatorname{Regg}(X)$. By the first part of the proof, X_{lp} is a crq–group. Thus X must be a crq–group. □

The following result extends Theorem 4.1. The latter is obtained by the choosing τ to be the least type in $\operatorname{Tst}(X)$.

PROPOSITION 4.2. *Let X be an almost completely decomposable group, and assume that for some type $\tau \in \operatorname{Tst}(X)$ it is true that $\beta_\tau^X = \operatorname{rgi}(X)$. Then X is a crq–group. Furthermore, if τ is not the least type of $\operatorname{Tst}(X)$, then X has a rank–one summand of type σ whenever $\sigma \not\geq \tau$. Thus, if X is clipped and σ is a critical type, then β_σ^X is a proper divisor of $rgi(X)$.*

PROOF. Let τ_0 be the least type of $\mathrm{Tst}(X)$, i.e., τ_0 is the so-called inner type of X. Then $\beta_\tau^X \leq \beta_{\tau_0}^X \leq \mathrm{rgi}(X)$, and therefore $\beta_{\tau_0}^X = \mathrm{rgi}(X)$. Hence

$$\mathrm{rgi}(X) = \beta_{\tau_0}^X = \exp\left(\frac{X^\sharp(\tau_0)}{\mathrm{R}(X^\sharp(\tau_0))}\right) = \exp\left(\frac{X}{\mathrm{R}(X)}\right).$$

By Theorem 4.1 X is a crq–group.

Now suppose that $\beta_\tau^X = \mathrm{rgi}(X)$ and $\tau_0 < \tau$. Choose a regulating subgroup $A = \bigoplus_{\rho \in \mathrm{T}_{\mathrm{cr}}(A)} A_\rho$ of X such that X/A is cyclic. Write $X = A + \mathbb{Z}n^{-1}a$ with $a \in A$ and $n = [X : A] = \mathrm{rgi}(X)$. Further write $a = \sum_{\rho \in \mathrm{T}_{\mathrm{cr}}(X)} a_\rho$ with $a_\rho \in A_\rho$. Then $\mathrm{rgi}(X) = \beta_\tau^X = \gcd\{\gcd^A(n, a_\rho) : \rho \in \mathrm{T}_{\mathrm{cr}}(X), \rho \not\geq \tau\}$ ([**DMM97**, Corollary 3.8]). This implies that $\mathrm{rgi}(X)$ divides each a_σ in A provided that $\sigma \not\geq \tau$ and by [**DMM97**, Lemma 4.1] X has a rank–one summand of type σ for every type $\sigma \not\geq \tau$.

Since X is clipped the inner type cannot be critical and no Burkhardt invariant β_τ^X can equal $\mathrm{rgi}(X)$ for a critical type τ. □

It was noted in [**Mad00**, Theorem 6.5.15] that every direct summand of a local clipped crq–group is again a crq–group. We will show next that this is true for arbitrary crq–groups.

THEOREM 4.3. *Let X be a global* crq*–group. Then any direct summand of X is again a* crq*–group.*

PROOF. Assume first that X is a local crq–group and that $X = Y \oplus Z$. Choose main decompositions of these three groups, $X = X_{cd} \oplus X_{cl}$, $Y = Y_{cd} \oplus Y_{cl}$, $Z = Z_{cd} \oplus Z_{cl}$. By [**Mad00**, Corollary 7.3.14] the sum $Y_{cl} \oplus Z_{cl}$ is again clipped, and the equality $X = X_{cd} \oplus X_{cl} = (Y_{cd} \oplus Z_{cd}) \oplus (Z_{cl} \oplus Y_{cl})$ implies by [**Mad00**, Theorem 9.2.7] that $X_{cl} \cong_{\mathrm{nr}} Y_{cl} \oplus Z_{cl}$. Now Y_{cl} is nearly isomorphic to a summand of X_{cl} by [**Mad00**, Theorem 10.2.5] and the latter is a crq–group by [**Mad00**, Theorem 6.5.15]. Finally, Y_{cl} is a crq–group being nearly isomorphic to a crq–group ([**Mad00**, Theorem 9.2.6.5]) and so is $Y = Y_{cd} \oplus Y_{cl}$. By symmetry, Z is also a crq–group.

We now assume that X is an arbitrary global crq–group and $X = Y \oplus Z$. By Lemma 3.2 $Y = \sum_{p \in \mathbb{P}} Y \cap X_{lp}$ and by the first part of the proof $Y \cap X_{lp}$ is a crq–group. Choose $A_Y(p) \in \mathrm{Regg}(Y \cap X_{lp})$ such that $(Y \cap X_{lp})/A_Y(p)$ is cyclic. Then $\bigoplus_{p \in \mathbb{P}} (Y \cap X_{lp})/A_Y(p)$ is cyclic and there exists an epimorphism

$$\Sigma : \bigoplus_{p \in \mathbb{P}} \frac{Y \cap X_{lp}}{A_Y(p)} \twoheadrightarrow \frac{\sum_{p \in \mathbb{P}} Y \cap X_{lp}}{\sum_{p \in \mathbb{P}} A_Y(p)} = \frac{Y}{\sum_{p \in \mathbb{P}} A_Y(p)} :$$

$$[\ldots, x_{lp} + A_Y(p), \ldots]\Sigma = \sum_{p \in \mathbb{P}} x_{lp} + \sum_{p \in \mathbb{P}} A_Y(p).$$

It follows that $Y/\sum_{p \in \mathbb{P}} A_Y(p)$ is cyclic and, since $\sum_{p \in \mathbb{P}} A_Y(p)$ is regulating in Y by [**Mad00**, Theorem 5.3.2], the summand Y is a crq–group. □

The following proposition contains the information that shows that the invariants $\mathrm{rgi}\,X[\tau]$ largely determine the crq–group X.

PROPOSITION 4.4. *Let $X = A + \mathbb{Z}n^{-1}a$ be a* crq*–group with $\gcd^A(n, a) = 1$ and $A = \bigoplus_{\rho \in \mathrm{T}_{\mathrm{cr}}(A)} A_\rho \in \mathrm{Regg}\,X$. Write $a = \sum_{\rho \in \mathrm{T}_{\mathrm{cr}}(A)} a_\rho$ such that $a_\tau \in A_\tau$ and set $s_\tau = \gcd^A(n, a_\tau)$. Then the following hold for any type τ (critical or not).*

1. $\operatorname{rgi} X[\tau] = \gcd\{s_\sigma : \sigma \leq \tau\}$, $\operatorname{rgi} X^\sharp[\tau] = \gcd\{s_\sigma : \sigma < \tau\}$, and $\operatorname{rgi} X[\tau] = \operatorname{rgi} X^\sharp[\tau]$ if and only if A_τ is a summand of X.
2. $\tau \in \operatorname{T_{cr}}(X_{cl})$ if and only if $\operatorname{rgi} X^\sharp[\tau] \neq \operatorname{rgi} X[\tau]$.
3. $\gcd\{\operatorname{rgi} X[\sigma] : \sigma \in \operatorname{T_{cr}}(X), \sigma \neq \tau\} = 1$.
4. For any two types τ, σ, if $\sigma \leq \tau$, then $\operatorname{rgi}(X[\tau])$ divides $\operatorname{rgi}(X[\sigma])$.
5. $\operatorname{rgi} X^\sharp[\tau] = \gcd\{\operatorname{rgi} X[\sigma] : \sigma \in \operatorname{T_{cr}}(X), \sigma < \tau\}$ for every type τ for which $\{\sigma : \sigma \in \operatorname{T_{cr}}(X), \sigma < \tau\}$ is not empty.
6. If X is a clipped local group given in the form of Theorem 1.1, then $s_\tau = \operatorname{rgi} X[\tau]$.
7. $\beta_\tau^X = \gcd\{\operatorname{rgi} X[\sigma] : \sigma \in \operatorname{T_{cr}}(X), \sigma \not\geq \tau\} = \gcd\{\operatorname{rgi} X[\sigma] : \sigma \in \operatorname{T_{cr}}(X), \sigma \not> \tau\}$.

PROOF. 1. $\operatorname{rgi} X[\tau] = \gcd\{s_\sigma : \sigma \leq \tau\}$ and $\operatorname{rgi} X^\sharp[\tau] = \gcd\{s_\sigma : \sigma < \tau\}$ by the Purification Lemma. Both sides of the equivalence are trivially true if τ is not a critical type of X. Hence suppose that $\tau \in \operatorname{T_{cr}}(X)$. Suppose that B_τ is a τ-Butler complement of X and $X = B_\tau \oplus Y$. Then $A_\tau = B_\tau(1 + \phi)$ for some $\phi \in \operatorname{Hom}(B_\tau, X^\sharp(\tau))$ and $X^\sharp(\tau) \subseteq Y$. Hence $X = A_\tau \oplus Y$ also. By [**Mad00**, Corollary 7.3.16]

$$A_\tau \sqsubseteq X \Leftrightarrow \operatorname{rwd}_\tau(X) = \operatorname{rk} A_\tau$$
$$\Leftrightarrow \operatorname{rk} A_\tau - \operatorname{width} \frac{X^\sharp[\tau]}{X(\tau) + X[\tau]} = \operatorname{rk} A_\tau$$
$$\Leftrightarrow X^\sharp[\tau] = X(\tau) + X[\tau] = A_\tau \oplus X[\tau]$$
$$\Leftrightarrow \operatorname{rgi} X^\sharp[\tau] = \operatorname{rgi} X[\tau].$$

For the last equivalence note that $A^\sharp[\tau] = A_\tau \oplus A[\tau]$ is a regulating subgroup of both $X^\sharp(\tau)$ and $X(\tau) + X[\tau]$.

2. The validity of either side of the equivalence implies that τ is a critical type. So let τ be critical. Then 2. is the equivalence obtained by negating both sides of the equivalence in 1.

3. $\gcd\{\operatorname{rgi} X[\sigma] : \tau \neq \sigma \in \operatorname{T_{cr}}(X)\} = \gcd\{\gcd\{s_\rho : \rho \leq \sigma\} : \tau \neq \sigma \in \operatorname{T_{cr}}(X)\} \leq \gcd\{s_\rho : \rho \neq \tau\} = 1$ by the Purification Lemma and the fact that A_τ is pure in X.

4. Let $\sigma \leq \tau$, then $\{s_\rho : \rho \leq \sigma\} \subseteq \{s_\rho : \rho \leq \tau\}$ and this implies that $\operatorname{rgi} X[\tau] = \gcd\{s_\rho : \rho \leq \tau\}$ divides $\gcd\{s_\rho : \rho \leq \sigma\} = \operatorname{rgi} X[\sigma]$.

5. Without loss of generality assume that $A \in \operatorname{Regg}(X)$. Then

$$\gcd\{\operatorname{rgi} X[\sigma] : \sigma < \tau\} = \gcd\{\gcd\{s_\rho : \rho \leq \sigma\} : \sigma \in \operatorname{T_{cr}}(X), \sigma < \tau\}$$
$$= \gcd\{s_\rho : \rho \in \operatorname{T_{cr}}(X), \rho < \tau\}$$
$$= \operatorname{rgi} X^\sharp[\tau].$$

6. In this case the s_τ are all p–powers for some prime p. By Theorem 1.1(StrD) $s_\tau < s_\sigma$ whenever $\sigma < \tau$. Hence $\operatorname{rgi}(X[\tau]) = \min\{s_\sigma : \sigma \leq \tau\} = s_\tau$ for every $\tau \in \operatorname{T_{cr}}(X)$.

7. Firstly, suppose that X is a clipped p–local crq–group and let $X = A + \mathbb{Z}p^{-n} \sum_{\rho \in \operatorname{T_{cr}}(X)} s_\rho a_\rho$ be a cyclic representation of X satisfying Theorem 1.1. Then $s_\tau = \operatorname{rgi} X[\tau]$ by 6., and by [**Mad00**, Theorem 6.2.2]

$$\beta_\tau^X = \gcd\{s_\sigma : \sigma \not\geq \tau\} = \gcd\{s_\sigma : \sigma \not> \tau\}.$$

Therefore the proposition holds for clipped local crq–groups. If $X = X_{cd} \oplus X_{cl}$ is a main decomposition of an arbitrary p-local crq–group X, then $\beta_\tau^X = \beta_\tau^{X_{cl}}$ for all

$\tau \in \mathrm{T}_{\mathrm{cr}}(X)$ ([**Mad00**, Proposition 4.4.9]) and $\mathrm{rgi}(X[\tau]) = \mathrm{rgi}(X_{cl}[\tau])$ for all types τ. By the first part of the proof we can conclude that the proposition holds for all local crq–groups.

If X is a global crq–group, then X_{lp} is a local crq–group for every prime p dividing $\mathrm{rgi}(X)$. Consequently, for every type τ

$$\beta_\tau^{X_{lp}} = \gcd\{\mathrm{rgi}(X_{lp}[\sigma]) : \sigma \not\geq \tau\} = \gcd\{\mathrm{rgi}(X_{lp}[\sigma]) : \sigma \not> \tau\}.$$

We now have (using Lemma 3.8 again)

$$\gcd\{\mathrm{rgi}(X[\sigma]) : \sigma \not> \tau\} = \prod_{p|\mathrm{rgi}(X)} \gcd\{(\mathrm{rgi}(X[\sigma]))_{lp} : \sigma \not> \tau\}$$

$$= \prod_{p|\mathrm{rgi}(X)} \gcd\{(\mathrm{rgi}(X_{lp}[\sigma]) : \sigma \not> \tau\} = \prod_{p|\mathrm{rgi}(X)} \beta_\tau^{X_{lp}} = \beta_\tau^X.$$

Similarly,

$$\gcd\{\mathrm{rgi}(X[\sigma]) : \sigma \not\geq \tau\} = \prod_{p|\mathrm{rgi}(X)} \gcd\{(\mathrm{rgi}(X[\sigma]))_{lp} : \sigma \neq \tau\}$$

$$= \prod_{p|\mathrm{rgi}(X)} \gcd\{(\mathrm{rgi}(X_{lp}[\sigma]) : \sigma \not\geq \tau\} = \prod_{p|\mathrm{rgi}(X)} \beta_\tau^{X_{lp}} = \beta_\tau^X$$

and the proof is complete. $\square$

We are now in a position to prove the near–classification theorem for global crq–groups.

THEOREM 4.5. *Two* crq*–groups* X *and* Y *are nearly isomorphic if and only if*
1. $\mathrm{R}(X) \cong \mathrm{R}(Y)$
2. $\mathrm{rgi}(X) = \mathrm{rgi}(Y)$
3. $\mathrm{rgi}\, X[\tau] = \mathrm{rgi}\, Y[\tau]$ *for every* $\tau \in \mathrm{T}_{\mathrm{cr}}(X) = \mathrm{T}_{\mathrm{cr}}(Y)$.

PROOF. Necessity being clear, suppose that the three conditions are satisfied for the groups X and Y.

We observe first that (by Assumption 1.) $\mathrm{T}_{\mathrm{cr}}(X) = \mathrm{T}_{\mathrm{cr}}(Y)$ and $\mathrm{rk}\, X(\tau)/X^\sharp(\tau) = \mathrm{rk}\, \mathrm{R}(X)(\tau)/\mathrm{R}(X)^\sharp(\tau) = \mathrm{rk}\, \mathrm{R}(Y)(\tau)/\mathrm{R}(Y)^\sharp(\tau) = \mathrm{rk}\, Y(\tau)/Y^\sharp(\tau)$.

Secondly, let μ be minimal in $\mathrm{T}_{\mathrm{cr}}(X) = \mathrm{T}_{\mathrm{cr}}(Y)$. Then $X^\sharp[\mu] = X$ and $Y^\sharp[\mu] = Y$. Hence, by Assumption 2., $\mathrm{rgi}\, X^\sharp[\mu] = \mathrm{rgi}\, Y^\sharp[\mu]$. Together with Assumption 3. and Proposition 4.4.5 we now have that

(4.6) $\quad \mathrm{rgi}\, X^\sharp[\tau] = \mathrm{rgi}\, Y^\sharp[\tau]$ for all critical types $\tau \in \mathrm{T}_{\mathrm{cr}}(X) = \mathrm{T}_{\mathrm{cr}}(Y)$.

For starters, assume that

$$X = A + \mathbb{Z}\frac{1}{p^n} \sum_{\rho \in \mathrm{T}_{\mathrm{cr}}(X)} s_\rho a_\rho \quad \text{and} \quad Y = B + \mathbb{Z}\frac{1}{p^n} \sum_{\rho \in \mathrm{T}_{\mathrm{cr}}(Y)} t_\rho b_\rho$$

are clipped p–local crq–groups given as in the Structure Theorem 1.1. Then $s_\tau = \mathrm{rgi}(X[\tau]) = \mathrm{rgi}\, Y[\tau] = t_\tau$ for every $\tau \in \mathrm{T}_{\mathrm{cr}}(X) = \mathrm{T}_{\mathrm{cr}}(Y)$. Consequently, by [**Mad00**, Theorem 12.1.2], $X \cong_{\mathrm{nr}} Y$.

We now drop the assumption that X and Y are clipped but still assume that X and Y are p–local. Let $X = X_{cd} \oplus X_{cl}$ and $Y = Y_{cd} \oplus Y_{cl}$ be main decompositions of X and Y respectively. By Proposition 4.4.2, Assumption 2. and (4.6) we have that $\mathrm{T}_{\mathrm{cr}}(X_{cl}) = \mathrm{T}_{\mathrm{cr}}(Y_{cl})$. It follows, X_{cl}, Y_{cl} being slim, that $\mathrm{R}(X_{cl}) \cong \mathrm{R}(Y_{cl})$.

This together with 1., by counting ranks of the homogeneous components, implies that $X_{cd} \cong Y_{cd}$. Furthermore

$$\mathrm{rgi}(X_{cl}[\tau]) = \mathrm{rgi}(X[\tau]) = \mathrm{rgi}(Y[\tau]) = \mathrm{rgi}(Y_{cl}[\tau])$$

which implies by the previous case that $X_{cl} \cong_{\mathrm{nr}} Y_{cl}$. Combining this with the fact that the completely decomposable parts of X and Y are isomorphic we obtain that $X \cong_{\mathrm{nr}} Y$.

Finally we consider the general case. For every prime p we have $\mathrm{R}(X_{\wr p}) = \mathrm{R}(X) \cong \mathrm{R}(Y) = \mathrm{R}(Y_{\wr p})$ and, for every $\tau \in \mathrm{T}_{\mathrm{cr}}(X) = \mathrm{T}_{\mathrm{cr}}(X_{\wr p})$, we have

$$\mathrm{rgi}(X_{\wr p}[\tau]) = (\mathrm{rgi}(X[\tau]))_{\wr p} = (\mathrm{rgi}(Y[\tau]))_{\wr p} = \mathrm{rgi}(Y_{\wr p}[\tau]).$$

Similarly,

$$\mathrm{rgi}(X_{\wr p}^{\sharp}[\tau]) = (\mathrm{rgi}(X^{\sharp}[\tau]))_{\wr p} = (\mathrm{rgi}(Y^{\sharp}[\tau]))_{\wr p} = \mathrm{rgi}(Y_{\wr p}^{\sharp}[\tau]).$$

By the local result we conclude that $X_{\wr p}$ and $Y_{\wr p}$ are nearly isomorphic for every prime p and, by the Primary Reduction Theorem ([**Mad00**, Theorem 9.2.8]) we conclude that X and Y are nearly isomorphic. $\square$

There is an easy corollary which simplifies the description of crq–groups by passing to a nearly isomorphic group.

COROLLARY 4.7. *Let* $A = \bigoplus_{\rho \in \mathrm{T}_{\mathrm{cr}}(A)} A_\rho$ *be a homogeneous decomposition of the completely decomposable group* A, *and let* $a_\tau \in A_\tau$. *Consider the two crq–groups* $X = A + \mathbb{Z}\frac{1}{n} \sum_{\rho \in \mathrm{T}_{\mathrm{cr}}(A)} t_\rho a_\rho$ *and* $Y = A + \mathbb{Z}\frac{1}{n} \sum_{\rho \in \mathrm{T}_{\mathrm{cr}}(A)} a_\rho$ *where the* t_τ *are integers. If* $\gcd(n, t_\tau) = 1$ *for every* $\tau \in \mathrm{T}_{\mathrm{cr}}(A)$, *then* $X \cong_{\mathrm{nr}} Y$.

PROOF. It is routine to check that $\gcd^A(n, t_\tau a_\tau) = \gcd^A(n, a_\tau)$. The Purification Lemma tells how various regulating indices are computed from these greatest common divisors and it follows in particular that $\mathrm{rgi}\, X = \mathrm{rgi}\, Y$ and $\mathrm{rgi}\, X[\tau] = \mathrm{rgi}\, Y[\tau]$. By Theorem 4.5 the groups X and Y are nearly isomorphic. $\square$

The following proposition gives an invariant description of the main decomposition of a slim crq–group.

PROPOSITION 4.8. *Let* $X = A + \mathbb{Z}n^{-1}a$ *be a slim crq-group,* $A = \bigoplus_{\rho \in \mathrm{T}_{\mathrm{cr}}(A)} A_\rho$, $a = \sum_{\rho \in \mathrm{T}_{\mathrm{cr}}(A)} a_\rho$, *where* $a_\tau \in A_\tau$. *Then*

$$\begin{aligned}
X_{\wr p} &= \left(\bigoplus_\rho \{(A_\rho)_{\wr p} : (\mathrm{rgi}\, X[\rho])_{\wr p} = (\mathrm{rgi}\, X^{\sharp}[\rho])_{\wr p}\} \right) \\
&\oplus \left(\bigoplus_\rho \{(A_\rho)_{\wr p} : (\mathrm{rgi}\, X[\rho])_{\wr p} \neq (\mathrm{rgi}\, X^{\sharp}[\rho])_{\wr p}\} \right. \\
&\left. + \mathbb{Z}n_{\wr p}^{-1} \sum_\rho \{a_\rho : (\mathrm{rgi}\, X[\rho])_{\wr p} \neq (\mathrm{rgi}\, X^{\sharp}[\rho])_{\wr p}\} \right)
\end{aligned}$$

is a main decomposition of X.

References

[Arn82] D. M. Arnold, *Finite Rank Torsion Free Abelian Groups and Rings, Lecture Notes in Mathematics*, Vol. 931, Springer Verlag, 1982.

[BM96] R. Burkhardt and O. Mutzbauer, Almost completely decomposable groups with primary cyclic regulating quotient, *Rend. Sem. Mat. Univ. Padova*, 95, 81–93, 1996.

[Cam95] M. J. Campagna, *Single-relation almost completely decomposable groups*. PhD thesis, Wesleyan University, 1995.

[Cam00] M. J. Campagna, Single-relation almost completely decomposable groups, *Communications in Algebra*, 28, 83–92, 2000.

[DMM97] U. Dittmann, A. Mader, and O. Mutzbauer, Almost completely decomposable groups with a cyclic regulating quotient. *Communications in Algebra*, 25, 769–784, 1997.

[Fuc73] L. Fuchs, *Infinite Abelian Groups, Vol. I, II*. Academic Press, 1970 and 1973.

[Lad74] E. L. Lady, Almost completely decomposable torsion-free abelian groups. *Proc. Amer. Math. Soc.*, 45, 41–47, 1974.

[Mad00] A. Mader, *Almost Completely Decomposable Groups, Algebra, Logic and Applications*, Vol. 13, Gordon and Breach Science Publishers, 2000.

[MMV99] A. Mader, O. Mutzbauer, and C. Vinsonhaler, Local–global relations for almost completely decomposable groups, *Rocky Mountain J. Math.*, 29, 1429–1453, 1999.

[MV95] A. Mader and C. Vinsonhaler, Almost completely decomposable groups with a cyclic regulating quotient, *J. Algebra*, 177, 463–492, 1995.

(Mader) Department of Mathematics, University of Hawaii, 2565 The Mall, Honolulu, HI 96822, USA
E-mail address: adolf@math.hawaii.edu

(Nongxa) Department of Mathematics, University of the Western Cape, 7535 Bellville, South Africa
E-mail address: gnongxa@uwc.ac.za

(Ould–Beddi) Faculté des Sciences et Techniques, Université de Nouakchott, B. P. 5026, Nouakchott, Mauritania
E-mail address: beddima@univ-nkc.mr

On Varieties of Groups Generated by Wreath Products of Abelian Groups

Vahagn H. Mikaelian

To Marine Mikaelian on her birthday

ABSTRACT. Generalizing results of Higman and Houghton on varieties generated by wreath products of finite cycles, we prove that the (direct or cartesian) wreath product of *arbitrary* abelian groups A and B generates the product variety $\text{var}(A) \cdot \text{var}(B)$ if and only if one of the groups A and B is not of finite exponent, or if A and B are of finite exponents m and n respectively and for all primes p dividing both m and n, the factors $B[p^k]/B[p^{k-1}]$ are infinite, where $B[s] = \langle b \in B \mid b^s = 1 \rangle$ and where p^k is the highest power of p dividing n.

Introduction

The problem, whether the standard wreath product $A \text{ wr } B$ of abelian groups A and B generates the product variety $\text{var}(A) \cdot \text{var}(B)$, is solved by Higman for the case when $A = C_p$ and $B = C_n$ are finite cycles of orders p and n, where p is a prime [H59], and by Houghton for the case of arbitrary finite cycles $A = C_m$ and $B = C_n$. Namely, equality $\text{var}(C_m \text{ wr } C_n) = \text{var}(C_m) \cdot \text{var}(C_n) = \mathfrak{A}_m \cdot \mathfrak{A}_n$ holds if and only if m and n are coprime. As we are informed by Professor C. Houghton, his result never was published. However this theorem is frequently cited in the literature and can be found, say, in [N68]: this is not only a well-known result of independent interest, but also an argument frequently used in other constructions of the theory of varieties of groups: descriptions of lattices of subvarieties of certain product varieties, basis ranks of varieties (see for example [N68]).

The aim of this paper is to generalize Houghton's result for the case of *arbitrary* abelian groups A and B. Namely, *for arbitrary abelian groups A and B the (direct or cartesian) wreath product of groups A and B generates the product variety* $\text{var}(A) \cdot \text{var}(B)$ *if and only if at least one of the groups A and B is not of finite exponent, or if A and B are of finite exponents m and n respectively and for all primes p dividing both m and n, the factors $B[p^k]/B[p^{k-1}]$ are infinite, where $B[s]$ is defined as $B[s] = \langle b \in B \mid b^s = 1 \rangle$ and where p^k is the highest power of p dividing n* (Theorem 6.1). As we will see below these factors $B[p^k]/B[p^{k-1}]$ have very "understandable" structure and our criterion is easily applicable in concrete situations.

2000 Mathematics Subject Classification. 20E22, 20E10, 20K01, 20K25.

© 2001 American Mathematical Society

The structure of infinitely generated abelian groups of non-finite exponent is complicated and at first sight such a generalization may demand techniques very different from those of critical groups, of Cross varieties or of finite nilpotent p-groups. The main idea that enables us to deal with the case of infinitely generated groups is the following main dichotomy: *each abelian group is either of finite exponent and, thus, is a direct sum of (possibly infinitely many) copies of some finitely many cycles of prime power orders, or is a discriminating group* (see definitions and notations below). The point is that if the "active" group B is a discriminating group, then the cartesian or direct wreath product of A and B always generates (and discriminates) the variety var $(A) \cdot$ var (B) and we can restrict ourselves to the first case of the dichotomy. In this case if p is a prime dividing n, then the p-primary component B_p of B is simply a direct sum of some (possibly infinitely many) copies of cycles $C_p, C_{p^2}, \ldots, C_{p^k}$, and our condition $|B[p^k]/B[p^{k-1}]| = \infty$ simply means that the direct decomposition of B_p contains infinitely many summands isoporphic to the cycle C_{p^k}.

Since the proof of Theorem 6.1 consists of consideration of several cases and subcases, we have divided it into parts which occupy Sections 2–5 and each one of them is presented as an independent result (Theorems 2.5, 3.3, 4.5, 5.5 closing corresponding sections).

For arbitrary groups A and B the cartesian wreath product $A \operatorname{Wr} B$ and direct wreath product $A \operatorname{wr} B$ generate the very same variety of groups. We build our construction for the case of *cartesian* wreath products, bearing in mind, that our proofs are also true for *direct* wreath products of groups. Only in a few cases do we consider direct wreath products for some specific details of the proofs.

For general information on the theory of groups we refer to [R96, KM96]. Following [N68, N64] we denote by $A \operatorname{Wr} B$ the *cartesian* wreath product of groups A and B and by $A \operatorname{wr} B$ the *direct* wreath product of these groups. The *base groups* of the cartesian and direct wreath products of A and B will be denoted by A^B and by $A^{(B)}$ respectively. Detail information on wreath products can be found in [N68, M95, N64, KM96]. For general information on varieties of groups we refer to the book of Hanna Neumann [N68]. We reserve notations $\mathfrak{A}, \mathfrak{A}_n, \mathfrak{N}_c$ and $\mathfrak{B}_e$ for varieties of all abelian groups, of all abelian groups of exponent dividing n, of all nilpotent groups of class at most c, and of all groups of exponents dividing e respectively. For a set $\mathfrak{X}$ of groups we denote by var $\mathfrak{X}$, as usual, the variety *generated* by $\mathfrak{X}$. Information on the notion of *discriminating group* can be found in [BNNN64, N68]. See also the articles of Bryce [B70, B76] and of Kovács and Newman [KN94] for results of more general nature related to the material of this paper. We write abelian groups *additively*, all other groups will be written *multiplicatively*. Background information on abelian groups used in this paper can be found in [F70, R96, KM96].

I am extremely grateful to Professor Alexander Yurievich Ol'shanskii, who introduced me to varieties of groups and guided and encouraged me in all parts of my work during my post-graduate study at the Lomonosov Moscow State University, where most of this investigation was done. I am very pleased that this paper appears in the volume dedicated to Professor Laszlo Fuchs.

1. Wreath products and operations Q, S, C

As usual, for a given set $\mathfrak{X}$ of groups we denote by $\mathsf{Q}\mathfrak{X}$, $\mathsf{S}\mathfrak{X}$ and $\mathsf{C}\mathfrak{X}$, the sets of all homomorphic images, subgroups and cartesian products of groups of $\mathfrak{X}$ respectively. According to Birkhoff's Theorem [B35, N68], for the given set $\mathfrak{X}$ of groups the variety var $\mathfrak{X}$ can be realized as: var $\mathfrak{X}$ = QSC $\mathfrak{X}$.

For given $\mathfrak{X}$ and $\mathfrak{Y}$ denote $\mathfrak{X}\operatorname{Wr}\mathfrak{Y} = \{X\operatorname{Wr}Y \mid X \in \mathfrak{X}, Y \in \mathfrak{Y}\}$ and $\mathfrak{X}\operatorname{wr}\mathfrak{Y} = \{X\operatorname{wr}Y \mid X \in \mathfrak{X}, Y \in \mathfrak{Y}\}$. Since the product variety var $\mathfrak{X}$ · var $\mathfrak{Y}$ consists of all extensions of groups $X^* \in \operatorname{var}\mathfrak{X}$ by groups $Y^* \in \operatorname{var}\mathfrak{Y}$ and, since by Kaloujnine and Krasner Theorem [KK51] each extension of such a type can be embedded into the appropriate wreath product $X^*\operatorname{Wr}Y^*$, we get that the set $\mathfrak{X}\operatorname{Wr}\mathfrak{Y}$ generates the variety var $\mathfrak{X}$ · var $\mathfrak{Y}$ if and only if for each pair $X^* \in \mathsf{QSC}\mathfrak{X}$ and $Y^* \in \mathsf{QSC}\mathfrak{Y}$ variety var $(\mathfrak{X}\operatorname{Wr}\mathfrak{Y})$ contains $X^*\operatorname{Wr}Y^*$

The following lemmas, however, show that, to see whether var $(\mathfrak{X}\operatorname{Wr}\mathfrak{Y})$ = var $\mathfrak{X}$ · var $\mathfrak{Y}$, for purposes of the current paper we have to check just *one* of six conditions assumed, namely, whether for abelian sets $\mathfrak{X}$ and $\mathfrak{Y}$ of groups the variety var $(\mathfrak{X}\operatorname{Wr}\mathfrak{Y})$ contains wreath products $X\operatorname{Wr}Y^*$ for every $X \in \mathfrak{X}$ and $Y^* \in \mathsf{C}\mathfrak{Y}$.

LEMMA 1.1. *For arbitrary sets $\mathfrak{X}$ and $\mathfrak{Y}$ of groups and arbitrary groups X^* and Y, where $X^* \in \mathsf{Q}\mathfrak{X}$, $X^* \in \mathsf{S}\mathfrak{X}$ or $X^* \in \mathsf{C}\mathfrak{X}$ and where $Y \in \mathfrak{Y}$, the group $X^*\operatorname{Wr}Y$ belongs to variety* var $(\mathfrak{X}\operatorname{Wr}\mathfrak{Y})$.

PROOF. If X^* is the homomorphic image of some $X \in \mathfrak{X}$ under a homomorphism f, then for arbitrary $Y \in \mathfrak{Y}$ the group $X^*\operatorname{Wr}Y$ is the homomorphic image of the group $X\operatorname{Wr}Y$ under the homomorphism f_W defined as: $f_W\colon y\varphi \mapsto y\varphi_f$, where $y \in Y$, $\varphi \in X^Y$ and where $\varphi_f \in (X^*)^Y$ is set as: $\varphi_f(g) = f(\varphi(g))$ for each $g \in Y$ [N68, 22.11].

If X^* is the subgroup of some $X \in \mathfrak{X}$, then, clearly, $X^*\operatorname{Wr}Y$ is the subgroup of $X\operatorname{Wr}Y$ [N68, 22.12].

If X^* is cartesian product of some groups $X_i \in \mathfrak{X}$, $i \in I$, then we can define an embedding of $X^*\operatorname{Wr}Y$ into the cartesian product $W^* = \prod_{i \in I}(X_i\operatorname{Wr}Y)$ by the following rule: $y\varphi^* \mapsto \theta \in W^*$, where $y \in Y$, $\varphi^* \in (X^*)^Y$ and θ it defined as $\theta(i) = y\varphi_i$ with $\varphi_i \in X_i^Y$, $\varphi_i(g) = [\varphi(g)](i)$, $g \in Y$, $i \in I$. □

LEMMA 1.2. *For arbitrary sets $\mathfrak{X}$ and $\mathfrak{Y}$ of groups and arbitrary groups X and Y^*, where $X \in \mathfrak{X}$ and where $Y^* \in \mathsf{S}\mathfrak{Y}$, the group $X\operatorname{Wr}Y^*$ belongs to variety* var $(\mathfrak{X}\operatorname{Wr}\mathfrak{Y})$. *Moreover, if $\mathfrak{X}$ is a set of abelian groups, then for each $Y^* \in \mathsf{Q}\mathfrak{Y}$ the group $X\operatorname{Wr}Y^*$ also belongs to* var $(\mathfrak{X}\operatorname{Wr}\mathfrak{Y})$.

PROOF. The first statement of the lemma is obvious (see [N68, 22.13]).

Since the cartesian and direct wreath products $H\operatorname{Wr}G$ and $H\operatorname{wr}G$ of arbitrary groups H and G generate the same variety [N68, 22.31, 22.32], it is sufficient to show that, if Y^* is a homomorphic image of some $Y \in \mathfrak{Y}$ under some homomorphism h, then the direct wreath product $X\operatorname{wr}Y^*$ is the homomorphic image of $X\operatorname{wr}Y$ under some homomorphism h_W, provided that, X is abelian. h_W is defined by its values $h_W(y)$ and $h_W(\varphi_{x,y})$ over the following set of elements generating $X\operatorname{wr}Y$:

$$\{y \in Y, \varphi_{x,y} \in X^{(Y)} \mid \varphi_{x,y}(y) = x \text{ and } \varphi_{x,y}(g) = 1 \text{ for } g \in Y\setminus\{y\}, x \in X\}.$$

Namely:

$$h_W\colon y \mapsto h(y) \quad \text{and} \quad h_W\colon \varphi_{x,y} \mapsto \varphi_{x,h(y)} \in X^{(Y^*)}.$$

See also [S65, B63]. □

In particular, if each of $\mathfrak{X}$ and $\mathfrak{Y}$ consist of one group only, it follows from the previous two lemmas that:

LEMMA 1.3. *For arbitrary groups A and B, if $A^* \cong A/N$ (N is any normal subgroup of A), $A^* \leq A$ or $A^* = \prod_{i \in I} A$ (I is any index set), then $A^* \operatorname{Wr} B \in \operatorname{var}(A \operatorname{Wr} B)$.*

On the other hand, if $B^* \cong B/K$ (A is abelian and K is any normal subgroup of B) or if $B^* \leq B$, then $A \operatorname{Wr} B^* \in \operatorname{var}(A \operatorname{Wr} B)$.

COROLLARY 1.4. *If A and B are abelian groups, then $\operatorname{var}(A \operatorname{Wr} B) = \operatorname{var}(A) \cdot \operatorname{var}(B)$ if and only if $\operatorname{var}(A \operatorname{Wr} B)$ contains the wreath product $A \operatorname{Wr} (\prod_{i \in I} B)$ for every index set I.*

REMARK 1.5. As we will see in Section 7, an even stronger result of independent interest can be proved: $\operatorname{var}(A \operatorname{Wr} B) = \operatorname{var}(A) \cdot \operatorname{var}(B)$ holds for abelian groups A and B if and only if $\operatorname{var}(A \operatorname{Wr} B)$ contains the wreath product $A \operatorname{Wr}(B \oplus B)$ of A and of the direct sum of *two copies* of B (see Theorem 7.1).

2. Discriminating sets of groups and the case of abelian groups of non-finite exponents

Let us begin by considering the case of wreath products of abelian groups A and B, where at least one of these groups is *not* of finite exponent. As we will see, this situation can be described via properties of discriminating sets of groups.

DEFINITION 2.1 (see [BNNN64]). The set $\mathfrak{D}$ of groups is said to be *discriminating*, if for arbitrary finite word set V with the property that, for each $w \in V$ there exists a homomorphism δ_w of a free group F_n into some group of $\mathfrak{D}$, such that $\delta_w(w) \neq 1$, there exist a group $D \in \mathfrak{D}$ and a single homomorphism δ of F_n into D, such that $\delta(w) \neq 1$ for all $w \in V$.

A discriminating set of groups $\mathfrak{D}$ can be described by the following property: every finite set of identities $\{w \equiv 1 \,|\, w \in V\}$ that can be separately falsified in some groups $\{D_w \in \mathfrak{D} \,|\, w \in V\}$ can also be *simultaneously* falsified in a group $D = D_V \in \mathfrak{D}$ for certain choice of values $d_1, d_2, \ldots, d_n \in D$. Every discriminating set $\mathfrak{D}$ *discriminates* the variety $\operatorname{var}\mathfrak{D}$ generated by $\mathfrak{D}$, that is, $\mathfrak{D} \subseteq \operatorname{var}\mathfrak{D}$ and for every finite set V of words in, say, n variables, none of which is identically 1 in $\operatorname{var}\mathfrak{D}$ there is a group $D \in \mathfrak{D}$ and elements $d_1, d_2, \ldots, d_n \in D$ such that for all $w \in V$ $w(d_1, d_2, \ldots, d_n) \neq 1$ holds [BNNN64]. Discriminating set $\mathfrak{D}$ always generates $\operatorname{var}\mathfrak{D}$. If a discriminating set consists of one group D we term *discriminating group* D.

LEMMA 2.2 (see [BNNN64]). *If $\mathfrak{D}$ discriminates the variety $\mathfrak{U} = \operatorname{var}\mathfrak{D}$ and for the given set $\mathfrak{D}_1$ the relations $\mathfrak{D} \subseteq \operatorname{QS}\mathfrak{D}_1$ and $\mathfrak{D}_1 \subseteq \mathfrak{U}$ hold, then $\mathfrak{D}_1$ also discriminates $\mathfrak{U}$.*

Now we can prove the following:

LEMMA 2.3. *If the group B is not of finite exponent, then the cartesian wreath product $A \operatorname{Wr} B$ generates the variety $\operatorname{var}(A) \cdot \operatorname{var}(B) = \operatorname{var}(A) \cdot \mathfrak{A}$.*

PROOF. Assume, firstly, that B contains an element c of infinite order. Then B contains an infinite cycle $C = \langle c \rangle$ which is a discriminating group for the variety $\mathfrak{A}$ [BNNN64]. According to Lemma 2.2, B also discriminates $\mathfrak{A}$. Therefore for an *arbitrary* (and not only abelian) group A the wreath product $A \operatorname{Wr} B$ discriminates var $(A) \cdot \mathfrak{A}$ because in this situation the *direct* wreath product $A \operatorname{wr} B$ discriminates var $(A) \cdot \mathfrak{A}$ [BNNN64]: $A \operatorname{wr} B \in \mathsf{S}(A \operatorname{Wr} B)$.

Assume now that B is not of finite exponent but that B contains no element of infinite order. Then there exists a sequence of elements

(2.1) $$c_1, c_2, \ldots \in B$$

such that for arbitrary $l \in \mathbb{N}$ there is such a $c_{i(l)}$ whose $\exp c_{i(l)} \geq l$. Let

(2.2) $$w_1(x_1, \ldots, x_d), \ldots, w_k(x_1, \ldots, x_d)$$

be a finite set of words none of which is an identically 1 for all abelian groups (we can assume all these words to contain the same variables $x_1, \ldots, x_d$ because the number of these words is finite). Since the infinite cycle $C = \langle c \rangle$ discriminates $\mathfrak{A}$, there exist elements $c^{i_1}, \ldots, c^{i_d} \in C$ ($i_1, \ldots, i_d \in \mathbb{Z}$) such that

(2.3) $$w_1(c^{i_1}, \ldots, c^{i_d}) \neq 1, \ldots, w_k(c^{i_1}, \ldots, c^{i_d}) \neq 1.$$

Now let us consider all these values (2.3) together with all values of all *subwords* of words (2.2) over elements $c^{i_1}, \ldots, c^{i_d}$. Since in this way we will get only *finitely many* values, that is, only finitely many elements $c^j \in C$, we can choose a power c^{j_0}, such that the absolute value of j_0 is greater than that of all j's obtained. Now take such a $c_{i(j_0)}$ in (2.1) that $\exp c_{i(j_0)}$ is greater than $|j_0|$. Since, clearly, all values (2.3) remain unchanged if they are calculated not in C but in $\langle c_{i(j_0)} \rangle$ (that is, modulo $\exp c_{i(j_0)}$), we get that cycles

(2.4) $$\langle c_1 \rangle, \langle c_2 \rangle, \ldots \leq B$$

form a discriminating set for the variety $\mathfrak{A}$. And since $\{\langle c_1 \rangle, \langle c_1 \rangle, \ldots\} \subseteq \mathsf{S}B$, the group B discriminates $\mathfrak{A}$ according to Lemma 2.2. □

An analog of this lemma for the "passive" group A is also true:

LEMMA 2.4. *If the group A is not of finite exponent, then cartesian wreath product $A \operatorname{Wr} B$ generates the variety* var $(A) \cdot$ var $(B) = \mathfrak{A} \cdot$ var (B).

PROOF. Taking into account Lemma 2.3 we assume, without loss of generality, that B is a group of finite exponent n. Since the infinite cycle C belongs to var $(A) = \mathfrak{A}$, it sufficient, according to Lemma 1.3, to prove that

$$\operatorname{var}(C \operatorname{Wr} C_n) = \operatorname{var}(A) \cdot \operatorname{var}(B) = \mathfrak{A} \cdot \mathfrak{A}_n,$$

where C_n is a finite cycle of order n. We can choose infinitely many cycles

$$C_{p_1}, C_{p_2}, \ldots \in \mathsf{Q}(C)$$

of orders $p_1, p_2, \ldots$ all coprime to n and to each other. Thus var $(C \operatorname{Wr} C_n)$ contains each one of the wreath products $C_{p_1} \operatorname{Wr} C_n, C_{p_2} \operatorname{Wr} C_n, \ldots$ and, thus, varieties $\mathfrak{A}_{p_1} \cdot \mathfrak{A}_n, \mathfrak{A}_{p_2} \cdot \mathfrak{A}_n$ generated by these wreath products according to the result of Houghton [N68]. It remains to use the fact that:

$$(\mathfrak{A}_{p_1} \cdot \mathfrak{A}_n) \cup (\mathfrak{A}_{p_2} \cdot \mathfrak{A}_n) \cup \cdots = (\mathfrak{A}_{p_1} \cup \mathfrak{A}_{p_2} \cup \cdots) \cdot \mathfrak{A}_n = \mathfrak{A} \cdot \mathfrak{A}_n.$$

□

We collect the information of Lemmas 2.3 and 2.4 below:

THEOREM 2.5. *If at least one of the groups A and B is not of finite exponent, then cartesian or direct wreath product of groups A and B generates the variety* $\text{var}(A) \cdot \text{var}(B)$.

3. The case of finitely generated abelian groups

Assume A and B to be arbitrary *finitely generated* abelian groups. If in a direct decomposition of A or B an infinite cycle is present, then we apply the construction of Section 2. So what we have to deal with are merely *finite* abelian groups A and B.

LEMMA 3.1. *For finite abelian groups A and B of exponents m and n respectively the wreath product of A and B generates the variety* $\text{var}(A) \cdot \text{var}(B) = \mathfrak{A}_m \cdot \mathfrak{A}_n$ *if and only if the exponents m and n are coprime.*

REMARK 3.2. So, as we see, *the results of Higman* [H59] *and Houghton* [N68] *remain true not only for finite cycles, but also for arbitrary finite groups.* On the other hand, as the familiar example $\text{var}(C_p \text{Wr} \prod_{i=1}^{\infty} C_p) = \mathfrak{A}_p \cdot \mathfrak{A}_p$ shows (p is a prime), the criterion of Higman and Houghton has no direct analog for the case of infinite groups, even for the case of infinite groups of finite exponent.

PROOF OF LEMMA 3.1. If $\text{var}(A) \cdot \text{var}(B) = \mathfrak{A}_m \cdot \mathfrak{A}_n = \text{var}(A \text{Wr} B)$, then $\mathfrak{A}_m \cdot \mathfrak{A}_n$ is a Cross variety [N68] and m and n are coprime according to result of Šmelkin on product varieties of group generated by a finite group [S65].

On the other hand, if the condition of the lemma is satisfied, we can choose a cyclic subgroup $\langle a_m \rangle$ of order m in A and a cyclic subgroup $\langle b_n \rangle$ of order n in B. Now according to Lemma 1.3 $\text{var}(A \text{Wr} B) = \text{var}(\langle a_m \rangle \text{Wr} \langle b_n \rangle)$ and the latter is equal $\mathfrak{A}_m \cdot \mathfrak{A}_n$ according to the result of Houghton. □

Let us present the information of Sections 2 and 3 in an "easy-to-use" form:

THEOREM 3.3. *Let A and B be arbitrary finitely generated abelian groups with direct decompositions respectively*

$$A = \underbrace{C \oplus \cdots \oplus C}_{r_A} \oplus C_{p_1^{u_1}} \oplus \cdots \oplus C_{p_s^{u_s}} \text{ and } B = \underbrace{C \oplus \cdots \oplus C}_{r_B} \oplus C_{q_1^{k_1}} \oplus \cdots \oplus C_{q_d^{k_d}}.$$

Then the cartesian or direct wreath product of groups A and B generates the product variety $\text{var}(A) \cdot \text{var}(B)$ *if and only if*

1. *at least one of the decompositions of A and B contains a non trivial infinite cycle, that is, $r_A \neq 0$ or $r_B \neq 0$,*
2. *or all cycles in the decompositions of A and B are finite, that is, $r_A = 0$, $r_B = 0$, and the sets of primes $\{p_1, \ldots, p_s\}$ and $\{q_1, \ldots, q_d\}$ have empty intersection.*

4. The case of arbitrary abelian p-groups

4.1. Some notations. As we mentioned in Remark 3.2, in the case of infinitely generated abelian groups A and B of finite exponent no analogs of Lemma 3.1 and of Theorem 3.3 do exist. What a variety can be generated by wreath product, say, $C_p \text{Wr} (C_{p^2} \oplus \sum_{i=1}^{\infty} C_p)$ of groups of exponent p and p^2? We clear this situation by means of a specially defined function $\lambda(A, B, t)$:

DEFINITION 4.1. For given abelian p-groups A and B of finite exponents and for given $t \in \mathbb{N}$ the value of the function
$$\lambda = \lambda(A, B, t),$$
is defined to be the *maximum of the nilpotency classes of the t-generated groups* of variety $\operatorname{var}(A \operatorname{Wr} B)$.

Firstly we have to observe that this definition is *correct*: t-generated groups of $\operatorname{var}(A \operatorname{Wr} B)$ belong to the variety generated by all t-generated subgroups of the group $A \operatorname{Wr} B$ [N68, 16.31]. The number of non-isomorphic copies of mentioned t-generated subgroups is finite and every one of them is *a finite p-group* and, thus, nilpotent (in the next subsection we will find concrete upper bounds for nilpotency classes of t-generated groups of $\operatorname{var}(A \operatorname{Wr} B)$).

For the purposes of the rest of this paper for the given p-group B of finite exponent p^k let us denote: $k(B, p) = k$. If B is not a p-group but still has finite exponent, then let us denote by $k(B, p)$ the largest k for which p^k divides $\exp B$. Further, for the given B and positive integer s let us denote $B[s] = \langle b \in B \mid b^s = 1 \rangle$. The factor groups
$$B[p^k]/B[p^{k-1}] = B[p^{k(B,p)}]/B[p^{k(B,p)-1}]$$
will play a key role in our construction. So let us clear what do they mean in our situation of p-groups. According to Prüfer's Theorem the group B is a direct sum of finite cycles which can be arranged as:
$$(4.1) \qquad B = C_{p^{k_1}} \oplus C_{p^{k_2}} \oplus \cdots,$$
where $k_1 = k = k(B, p) \geq k_2 \geq \cdots$. From this decomposition it is clear that
$$(4.2) \qquad B[p^k]/B[p^{k-1}] = \underbrace{C_p \oplus \cdots \oplus C_p}_{\mu \text{ times}},$$
where μ is such an ordinal that $k_1 = k, \ldots, k_\mu = k$ and $k_{\mu+1} < k$ (notation is correct for the set of all ordinals greater than μ can be well-ordered).

4.2. An upper bound for the function $\lambda(A, B, t)$. According to the remark following Definition 4.1, $\lambda(A, B, t)$ is bounded by the maximum of the nilpotency classes of t-generated subgroups of $A \operatorname{Wr} B$. Let H be such a subgroup and A_H be its intersection with the base group A^B of $A \operatorname{Wr} B$. So A_H is a group of the variety $\mathfrak{A}_{p^u}$, where $p^u = \exp A$. A_H is normal in H and H/A_H is isomorphic to an, at most, t-generated subgroup A_B of
$$(A \operatorname{Wr} B)/A^B \cong A$$
and, therefore, according to Kaloujnine and Krasner Theorem [KK51], we get, that H is embeddable into the cartesian wreath product $A_H \operatorname{Wr} B_H$, where A_H is of exponent $p^{u'}$ dividing p^u and where
$$B_H = C_{k'_1} \oplus C_{k'_2} \oplus \cdots \oplus C_{k'_t} \qquad (k'_1 \geq k'_2 \geq \cdots \geq k'_t)$$
is certain subgroup of B. So B_H is finite and $A_H \operatorname{Wr} B_H$ is a *direct* wreath product. Thus, applying the result of Liebeck [L62] (on the nilpotency class of the direct wreath product of finite abelian p-groups), we calculate the class c of $A_H \operatorname{Wr} B_H$:
$$c = \sum_{i=1}^{t}(p^{k'_i} - 1) + (u' - 1)(p - 1)p^{k'_1 - 1} + 1.$$

And since $u' \leq u$, $k_1' \leq k_1 = k$, $k_2' \leq k_2, \ldots$, we have:

LEMMA 4.2. *For given abelian p-group A of finite exponent p^u and for the abelian p-group B of form (4.1):*

$$(4.3) \qquad \lambda(A, B, t) \leq \sum_{i=1}^{t}(p^{k_i} - 1) + (u-1)(p-1)p^{k(B,p)-1} + 1.$$

4.3. An example of a t-generated group in $\mathfrak{A}_{p^u} \cdot \mathfrak{A}_{p^k}$, the general criterion for wreath products of abelian p-groups.

EXAMPLE 4.3. The product variety $\mathfrak{A}_{p^u} \cdot \mathfrak{A}_{p^k}$ contains the t-generated group $T(p,t) = C_{p^u} \text{ wr } \sum_{i=1}^{t-1} C_{p^k}$. According to [L62] the nilpotency class of $T(p,t)$ is equal to:

$$\nu(p,t) = \sum_{i=1}^{t-1}(p^k - 1) + (u-1)(p-1)p^{k-1} + 1.$$

LEMMA 4.4. *If $|B[p^k]/B[p^{k-1}]| < \infty$, then for sufficiently large values of t:*

$$\nu(p,t) > \lambda(A, B, t)$$

and, thus, the group $T(p,t)$ does not belong to the variety $\text{var}(A \text{ Wr } B)$.

On the other hand, if $|B[p^k]/B[p^{k-1}]| = \infty$, then $\text{var}(A \text{ Wr } B) = \mathfrak{A}_{p^u} \cdot \mathfrak{A}_{p^k}$.

PROOF. Assume $B[p^k]/B[p^{k-1}]$ is finite and is of order p^μ, according to (4.2). If $t \leq \mu$, then all powers of p on the right side of (4.3) are equal to p^k. But when t becomes greater than μ, then the later summands in sum $\sum_{i=1}^{t}(p^{k_i} - 1)$ become less or equal to $p^{k-1} - 1$. Thus for sufficiently large t:

$$\nu(p,t) - \lambda(A,B,t) > \sum_{i=1}^{t-1}(p^k - 1) - \left(\sum_{i=1}^{\mu}(p^k - 1) + \sum_{i=\mu+1}^{t}(p^{k-1} - 1)\right)$$

$$= \sum_{i=\mu+1}^{t-1}\left[(p^k - 1) - (p^{k-1} - 1)\right] - (p^{k-1} - 1)$$

$$= (t - 1 - \mu)(p^k - p^{k-1}) + 1 - p^{k-1}.$$

So taking an arbitrary positive integer $t_0 > (p^{k-1} - 1)/(p^k - p^{k-1}) + \mu + 1$ we get that $\nu(p, t_0) - \lambda(A, B, t_0) > 0$.

Assume now $B[p^k]/B[p^{k-1}]$ is infinite, that is, μ is an infinite ordinal in (4.2). Thus B contains a subgroup D isomorphic with an infinite direct power of cycle C_{p^k}. D is a discriminating group in the variety $\mathfrak{A}_{p^k}$ [BNNN64]. Therefore by Lemma 2.2 the group B itself is a discriminating group for the variety $\mathfrak{A}_{p^k}$ and so $\text{var}(A \text{ Wr } B) = \mathfrak{A}_{p^u} \cdot \mathfrak{A}_{p^k}$. □

Now let us summarise:

THEOREM 4.5. *Let A and B be arbitrary abelian p-groups. Then:*
1. *if at least one of the groups A and B is not of finite exponent, then*
   $\text{var}(A \text{ Wr } B) = \text{var}(A \text{ wr } B) = \text{var}(A) \cdot \text{var}(B)$,
2. *if A and B are groups of finite exponents p^u and p^k respectively then*
   $\text{var}(A \text{ Wr } B) = \text{var}(A \text{ wr } B) = \text{var}(A) \cdot \text{var}(B) = \mathfrak{A}_{p^u} \cdot \mathfrak{A}_{p^k}$ *if and only if in direct decomposition (4.1) of B infinitely many cycles of order p^k are present, that is, if the factor group $B[p^k]/B[p^{k-1}]$ is infinite.*

EXAMPLE 4.6. Applying this result, we easily find an answer to the question asked at the beginning of this section: the group $C_p \operatorname{Wr}(C_{p^2} \oplus \sum_{i=1}^{\infty} C_p)$ does *not* generate the variety $\mathfrak{A}_p \cdot \mathfrak{A}_{p^2}$ because decomposition of the "active" group of this wreath product contains only one direct summand of order $p^2 = p^k$.

5. The case of abelian groups of finite composite exponents

5.1. p-groups in variety $\operatorname{var}(A \operatorname{Wr} B)$. Assume $\exp A = m$, $\exp B = n$ and denote for a given prime p (not necessarily dividing both m and n) by A_p and B_p the p-primary components of A and B respectively.

LEMMA 5.1. *Using this notation:*

(5.1) $$\operatorname{var}(A \operatorname{Wr} B) \cap \mathfrak{A}_{p^u} \cdot \mathfrak{A}_{p^k} = \operatorname{var}(A_p \operatorname{Wr} B_p),$$

where $u = k(A, p)$, $k = k(B, p)$.

PROOF. The right side of (5.1) lies in the left side. So it is sufficient to prove that every p-group P of $\operatorname{var}(A \operatorname{Wr} B)$ belongs to $\operatorname{var}(A_p \operatorname{Wr} B_p)$. Moreover, since $\operatorname{var}(A \operatorname{Wr} B)$ is a locally finite variety, we can assume P to be a *finite* group [N68]. We omit the trivial case, when p is coprime with m or with n, that is, when $A_p = \{1\}$ or $B_p = \{1\}$. The variety $\operatorname{var}(A \operatorname{Wr} B)$ is generated by the set $\{R_i \mid i \in I\}$ of all finite subgroups of $A \operatorname{Wr} B$. So there is a finite subset $\{R_1, \ldots, R_l\}$ of this set, such that $P \in \operatorname{var}(R_1, \ldots, R_l)$ [N68] and:

$$P \in \operatorname{QSC}(R_1, \ldots, R_l).$$

That is, P is a surjective image of a subgroup R of a direct product $R_1 \times \ldots \times R_l$ under some homomorphism $\varphi : R \to P$:

$$P = \varphi(R), \quad R \leq R_1 \times \ldots \times R_l.$$

P is a p-group and, thus, is an image of *some* Sylow p-subgroup P^* of R. In turn P^* is a sub-direct product of its projections P_i^* on R_i, $i = 1, \ldots, l$. Denote by $P_{i,B}^*$ the intersection of P_i^* with the base group A^B of $A \operatorname{Wr} B$. Clearly, $P_{i,B}^*$ is normal in P_i^* and the factor group $P_i^*/P_{i,B}^*$ is isomorphic to some subgroup $P_{i,A}^*$ of the factor group $(A \operatorname{Wr} B)/A^B \cong A$. Thus P_i^* is isomorphically embeddable into the wreath product $P_{i,A}^* \operatorname{Wr} P_{i,B}^*$. The group $P_{i,B}^*$ is a p-subgroup of B and, thus, lies in the B_p. $P_{i,A}^*$ is a p-subgroup of the base group A^B and, thus, lies in cartesian power $(A_p)^B$ of A_p. So $P_{i,A}^* \in \operatorname{var}(A_p)$ and, according to Lemma 1.3,

$$P_i^* \leq P_{i,A}^* \operatorname{Wr} P_{i,B}^* \in \operatorname{var}(A_p \operatorname{Wr} B_p).$$

□

5.2. Wreath products of abelian groups of finite composite exponents. Assume $\exp A = m$, $\exp B = n$ as above and

$$A = A_{p_1} \oplus \cdots \oplus A_{p_s}, \quad B = B_{q_1} \oplus \cdots \oplus B_{q_d}$$

are direct decompositions of A and B as direct sums of finitely many primary components A_{p_i}, $i = 1, \ldots, s$ and B_{p_j}, $j = 1, \ldots, d$.

Let us begin with the case when the set $\{q_1, \ldots, q_d\}$ is a subset of $\{p_1, \ldots, p_s\}$. Assume $p_1 = q_1$ and denote for brevity $p = p_1 = q_1$. Assume further that $\exp A_p = p^u$ and $\exp B_p = p^k$.

LEMMA 5.2. *If A, B, p are as above, then*
$$\mathrm{var}\,(A\,\mathrm{Wr}\,B) = \mathrm{var}\,(A) \cdot \mathrm{var}\,(B) = \mathfrak{A}_m \cdot \mathfrak{A}_n$$
if and only if for each p dividing n the factor $B[p^k]/B[p^{k-1}]$, where $k = k(B,p)$, is infinite.

PROOF. If $A\,\mathrm{Wr}\,B$ generates $\mathfrak{A}_m \cdot \mathfrak{A}_n$, then according to Lemma 5.1:
$$\mathrm{var}\,(A_p\,\mathrm{Wr}\,B_p) = \mathrm{var}\,(A\,\mathrm{Wr}\,B) \cap \mathfrak{A}_{p^u} \cdot \mathfrak{A}_{p^k} = \mathfrak{A}_{p^u} \cdot \mathfrak{A}_{p^k}.$$
Therefore the group B_p must satisfy the condition of Lemma 4.4:
$$|B_p[p^{k(B_p,p)}]/B_p[p^{k(B_p,p)-1}]| = \infty.$$
But, since B_p is the p-primary component of B, we have:
$$k(B_p, p) = k(B, p) = k \quad \text{and} \quad B_p[p^{k(B_p,p)}] = B_p[p^k] = B[p^k].$$

Now assume, on the other hand, that for all $q_1, \ldots, q_d$ all factors
$$B[q_1^{k_1}]/B[q_1^{k_1-1}], \ldots, B[q_d^{k_d}]/B[q_d^{k_d-1}],$$
where $k_i = k(B_{q_i}, q_i)$ ($i = 1, \ldots, d$) are infinite. Since the cycle C_n is the direct sum of cycles $C_{q^{k_1}}, \ldots, C_{q^{k_d}}$, we get that B contains the infinite direct power D of cycle C_n. And since D discriminates $\mathfrak{A}_n$, the group B also discriminates $\mathfrak{A}_n$ and $A\,\mathrm{Wr}\,B$ discriminates $\mathfrak{A}_m \cdot \mathfrak{A}_n$. □

It remains to consider the case when the exponent n of B has a prime divisor p which does *not* divide the exponent m of A. Let
$$m = p_1^{u_1} \cdots p_s^{u_s} \quad \text{and} \quad n = q_1^{k_1} \cdots q_{s'}^{k_{s'}} q_{s'+1}^{k_{s'+1}} \cdots q_d^{k_d},$$
where primes p_i and q_j are arranged such that $p_1 = q_1, \ldots, p_{s'} = q_{s'}$ and $p_i \neq q_j$ for all $i = 1, \ldots, s$; $j = s'+1, \ldots, d$. Then B has a decomposition $B_1 \oplus B_2$, where $n_1 = \exp B_1 = q_1^{k_1} \cdots q_{s'}^{k_{s'}}$ and $n_2 = \exp B_2 = q_{s'+1}^{k_{s'+1}} \cdots q_d^{k_d}$.

LEMMA 5.3. *In the above notation the wreath product $A\,\mathrm{Wr}\,B$ generates the variety $\mathfrak{A}_m \cdot \mathfrak{A}_n$ if and only if the wreath product $A\,\mathrm{Wr}\,B_1$ generates the variety $\mathfrak{A}_m \cdot \mathfrak{A}_{n_1} = \mathfrak{A}_m \cdot \mathrm{var}\,(B_1)$.*

PROOF. If $\mathrm{var}\,(A\,\mathrm{Wr}\,B) = \mathfrak{A}_m \cdot \mathfrak{A}_n$, then $\mathrm{var}\,(A\,\mathrm{Wr}\,B_1) = \mathfrak{A}_m \cdot \mathfrak{A}_{n_1}$ according to Lemma 5.1 and to the first part of the proof of Lemma 5.2.

Assume, on the other hand, $\mathfrak{A}_m \cdot \mathfrak{A}_{n_1} = \mathrm{var}\,(A\,\mathrm{Wr}\,B_1)$. Then B_1 contains the direct sum B_1^* of infinitely many copies of $C_{q_i^{k_i}}$ for each $i = 1, \ldots, s'$:
$$B_1^* = \sum_{j=1}^{\infty} C_{q_1^{k_1}} \oplus \cdots \oplus \sum_{j=1}^{\infty} C_{q_{s'}^{k_{s'}}} \leq B_1.$$
The groups B_2 contains the cycle
$$B_2^* = C_{n_2} \cong C_{q_{s'+1}^{k_{s'+1}}} \oplus \cdots \oplus C_{q_d^{k_d}}.$$

Thus, according to Lemma 1.3, it is sufficient to prove that $\mathfrak{A}_m \cdot \mathfrak{A}_n = \mathrm{var}\,(A\,\mathrm{Wr}\,B^*)$, where $B^* = B_1^* \oplus B_2^*$.

$\mathfrak{A}_m \cdot \mathfrak{A}_n$ is a locally finite variety generated by its critical groups [N68]. Let Q be such a critical group. Q is an extension of a normal subgroup $H \in \mathfrak{A}_m$ by means of group $G = Q/H \in \mathfrak{A}_n$. So $G = G_1 \oplus G_2$, where $G_1 \in \mathfrak{A}_{n_1}$ and $G_2 \in \mathfrak{A}_{n_2}$. Let M

be the monolith of Q [N68]. M is a finite direct product of, say, r copies of some cycle C_p, $p \in \{p_1, \ldots, p_s\}$. So M is an r-dimensional space over the finite field $\mathbb{F}_p$ and the operation of conjugation of elements of M by elements of Q defines a linear representation of the group G degree r over the field $\mathbb{F}_p$. Since $\exp G_2$ is coprime with p we think of this groups to be isomorphically embedded in Q. Let us use the same notaion G_2 for that isomorphic copy in Q. Since M is a *minimal* normal subgroup of Q, this representation of G is irreducible. Let us apply the theorem of Clifford [CR62] to the representation of *normal subgroup* G_2 of G, that is, to $\mathbb{F}_p G_2$-module M_{G_2}. Our $\mathbb{F}_p G$-module M is, thus, a direct sum of its submodules, each of which is homogenous regarding G_2: $M = M_1 \oplus \cdots \oplus M_l$. Since H and Q/H are both abelian, $M_1, \ldots, M_l$ are normal in Q. Thus $l = 1$ and $M = M_1$.

The representation of G_2 is *faithful*. For, if not, the direct sum $H \oplus K$ of H and of non-trivial kernel K would be normal in Q and, so, D would contain a "second monolith" of Q. Thus, as an abelian group with irreducible faithful representation, G_2 has to be a *cycle*. Since this cycle belongs to $\mathfrak{A}_{n_2}$, we get that G_2 is a subgroup of B_2^*. Further, since G_1 is a finite group in $\mathfrak{A}_{n_1}$, then G_1 is a subgroup of B_2^*. Therefore $G = G_1 \oplus G_2$ is a subgroup of $B_1^* \oplus B_2^*$ and, thus, the extension Q of H by G lies in $H \operatorname{Wr} G$ and

$$H \operatorname{Wr} G \leq H \operatorname{Wr} B^* \in \operatorname{var}(A \operatorname{Wr} B^*).$$

□

EXAMPLE 5.4. Replacing the wreath product of Example 4.6 by the following one: $W = C_p \operatorname{Wr} (C_{p^2} \oplus \sum_{i=1}^{\infty} C_p \oplus \sum_{i \in I} C_q)$, where q is a prime different from p, we get that W does *not* generate the variety $\mathfrak{A}_p \cdot \mathfrak{A}_{p^2 \cdot q}$ for any index set I, in spite of the fact that $C_p \operatorname{Wr} \sum_{i \in I} C_q$ does generate the variety $\mathfrak{A}_p \cdot \mathfrak{A}_q$ for arbitrary non-empty index set I. On the other hand, replacing in W the summand C_{p^2} by $\sum_{i=1}^{\infty} C_{p^2}$, we will obtain a wreath product $C_p \operatorname{Wr} \left[\sum_{i=1}^{\infty} C_{p^2} \oplus \sum_{i=1}^{\infty} C_p \oplus \sum_{i \in I} C_q\right]$ *generating* the variety $\mathfrak{A}_p \cdot \mathfrak{A}_{p^2 \cdot q}$ for any non-empty I. Clearly this wreath product will generate variety $\mathfrak{A}_p \cdot \mathfrak{A}_{p^2 \cdot q}$ even if we remove the summand $\sum_{i=1}^{\infty} C_p$ in the active group.

The information of this section can be collected as:

THEOREM 5.5. *Let A and B be arbitrary abelian groups of finite exponents m and n respectively and let $m = p_1^{u_1} \cdots p_s^{u_s}$ and $n = q_1^{k_1} \cdots q_{s'}^{k_{s'}} q_{s'+1}^{k_{s'+1}} \cdots q_d^{k_d}$, where $s' \leq s$ and where the prime divisors p_i and q_j are grouped such that $p_1 = q_1, \ldots, p_{s'} = q_{s'}$ and $p_i \neq q_j$ for all $i = 1, \ldots, s$; $j = s' + 1, \ldots, d$. Then $\operatorname{var}(A \operatorname{Wr} B) = \operatorname{var}(A \operatorname{wr} B) = \operatorname{var}(A) \cdot \operatorname{var}(B) = \mathfrak{A}_m \cdot \mathfrak{A}_n$ if and only if the factors $B[q^{k(B,q)}]/B[q^{k(B,q)-1}]$ are infinite for all $q = q_1, \ldots, q_{s'}$.*

6. The general case of arbitrary abelian groups and wreath products of groups "near" to abelian ones

6.1. The general criterion for arbitrary abelian groups. Statements of Theorems 2.5, 3.3, 4.5 and 5.5 are constituent parts of the following main criterion for cartesian and direct wreath products of arbitrary abelian groups:

THEOREM 6.1 (MAIN CRITERION). *For arbitrary abelian groups A and B their cartesian wreath product $A \operatorname{Wr} B$ (or direct wreath product $A \operatorname{wr} B$) generates the product variety $\operatorname{var}(A) \cdot \operatorname{var}(B)$ if and only if*

1. *at least one of the groups A and B is not of finite exponent,*

2. or if A and B are of finite exponents m and n respectively and for each prime p dividing both m and n the factors $B[p^k]/B[p^{k-1}]$ are infinite, where $B[s] = \langle b \in B \mid b^s = 1 \rangle$ and p^k is the highest power of p dividing n.

As the mentioned Theorems 2.5, 3.3, 4.5 and 5.5 show, this criterion is *effective* in the sense that in concrete cases the factors $B[p^k]/B[p^{k-1}] \cong \sum_{i \in I} C_p$ (if need of their consideration arises) have simple and "understandable" meaning: we have to consider them only in the case when A and B are of finite exponents and there is a prime p dividing those exponents; in these circumstances the condition $|B[p^k]/B[p^{k-1}]| = \infty$ simply means that in the direct decomposition $G_p = C_{p^{k_1}} \oplus C_{p^{k_2}} \oplus \cdots$ (where $k_1 \geq k_2 \geq \cdots$) of the p-primary component B_p of the group B infinitely many cycles (direct summands) $C_{p^{k_1}} = C_{p^{k(B,p)}}$ are present.

An immediate consequence of Theorem 6.1 is the following:

COROLLARY 6.2. *If for abelian groups A and B the wreath product $A \operatorname{Wr} B$ does not generate the product variety $\operatorname{var}(A) \cdot \operatorname{var}(B)$, then the wreath product $A \operatorname{Wr} \sum_{i=1}^{l} B$ also does not generate $\operatorname{var}(A) \cdot \operatorname{var}(B)$ for any positive integer $l \in \mathbb{N}$.*

6.2. Parallels for nilpotent groups of class 2 and metabelian groups. Problems. The following two examples show, that Theorem 6.1 does *not* have obvious generalizations even for cases of "small" groups of classes of groups "near" to abelian groups.

EXAMPLE 6.3. Let $\mathfrak{V} = \mathfrak{N}_2 \cap \mathfrak{B}_3$ be the variety of all nilpotent groups of class at most 2 and exponent dividing 3. Then for no one group A generating $\mathfrak{V}$ and cycle $B = C_2 \operatorname{var}(A \operatorname{Wr} B) = \mathfrak{V} \cdot \mathfrak{A}_2$. On the other hand $\exp A = 3$ is coprime with $\exp B = 2$. According to Lemma 1.3, it is sufficient to prove this for one group A generating $\mathfrak{V}$. Let $A = F_2(\mathfrak{V}) = \langle x_1, x_2 \mid [x_1, x_2, x_1] = [x_1, x_2, x_2] = x_1^3 = x_2^3 = 1\rangle$ be the $\mathfrak{V}$-free group of rank 2 and let R be the extension of A by means of the group of operators generated by automorphisms $\nu_1, \nu_2 \in \operatorname{Aut}(A)$ defined as: $\nu_1 : x_1 \mapsto x_1^{-1}$, $\nu_1 : x_2 \mapsto x_2$; $\nu_2 : x_1 \mapsto x_1$, $\nu_2 : x_2 \mapsto x_2^{-1}$. Clearly $\langle \nu_1, \nu_2 \rangle \cong C_2 \oplus C_2 \in \mathfrak{A}_2$. As it is shown in [B], R is a critical group. Every one of its proper factors, but not R itself, satisfies $[[x_1, x_2], [x_3, x_4], x_5] \equiv 1$. On the other hand the wreath product $A \operatorname{Wr} B$ satisfies this identity because its second commutator subgroup lies in the center.

EXAMPLE 6.4. Let A and B be arbitrary finite groups generating varieties $\mathfrak{A}_p \cdot \mathfrak{A}_q$ and $\mathfrak{A}_r$ respectively, where p, q, r are arbitrary different primes. Then $\operatorname{var}(A \operatorname{Wr} B) \neq (\mathfrak{A}_p \cdot \mathfrak{A}_q) \cdot \mathfrak{A}_r$, in spite of the fact that $\exp A = pq$ is coprime with $\exp B = r$. For, the product of three non-trivial varieties $(\mathfrak{A}_p \cdot \mathfrak{A}_q) \cdot \mathfrak{A}_r = \mathfrak{A}_p \cdot \mathfrak{A}_q \cdot \mathfrak{A}_r$ cannot, by theorem of Šmelkin [S65], be generated by a single finite group $A \operatorname{Wr} B$.

This examples, and the results of Section 1 proved not only for the wreath products of single groups but also for sets $\mathfrak{X} \operatorname{Wr} \mathfrak{Y}$ set the following problems of generalization of our main criterion (Theorem 6.1) in two possible directions:

PROBLEM 6.5. *Let A and B be arbitrary (nilpotent, metabelian, soluble) groups. Find a criterion under which $\operatorname{var}(A \operatorname{Wr} B) = \operatorname{var}(A) \cdot \operatorname{var}(B)$.*

PROBLEM 6.6. *Let $\mathfrak{X}$ and $\mathfrak{Y}$ be arbitrary sets of (abelian) groups. Find a criterion under which $\operatorname{var}(\mathfrak{X} \operatorname{Wr} \mathfrak{Y}) = \operatorname{var}(\mathfrak{X}) \cdot \operatorname{var}(\mathfrak{Y})$.*

7. Wreath products and finite direct sums of abelian groups

As we saw in Section 1, $\mathrm{var}\,(A\,\mathrm{Wr}\,B) = \mathrm{var}\,(A) \cdot \mathrm{var}\,(B)$ if and only if $\mathrm{var}\,(A\,\mathrm{Wr}\,B)$ contains wreath products $A\,\mathrm{Wr}\,(\prod_{i \in I} B)$ for every index set I. It is a result of independent interest, that here instead of arbitrary set I we can take a two-element set, that is:

THEOREM 7.1. *For arbitrary abelian groups A and B equality $\mathrm{var}\,(A\,\mathrm{Wr}\,B) = \mathrm{var}\,(A) \cdot \mathrm{var}\,(B)$ takes place if and only the variety $\mathrm{var}\,(A\,\mathrm{Wr}\,B)$ contains the group $A\,\mathrm{Wr}\,(B \times B)$. The analogous statement is also true for direct wreath products of groups.*

This theorem follows from a more general result:

THEOREM 7.2. *For arbitrary abelian groups A and B one and only one of the following two alternatives holds:*

1. *the variety $\mathrm{var}\,(A\,\mathrm{Wr}\,B)$ is equal to the variety $\mathrm{var}\,(A) \cdot \mathrm{var}\,(B)$ and, thus, to every subvariety $\mathrm{var}\,\bigl(A\,\mathrm{Wr}\,(\prod_{i \in I} B)\bigr)$ of the latter for any index set I;*
2. *the variety $\mathrm{var}\,(A\,\mathrm{Wr}\,B)$ is a proper subvariety of variety $\mathrm{var}\,(A) \cdot \mathrm{var}\,(B)$. Then for any positive integer $s \in \mathbb{N}$ the variety $\mathrm{var}\,(A\,\mathrm{Wr}\,(\prod_{i=1}^{s} B))$ is a proper subvariety of variety $\mathrm{var}\,(A) \cdot \mathrm{var}\,(B)$ and, moreover, of the variety $\mathrm{var}\,(A\,\mathrm{Wr}\,(\prod_{i=1}^{u} B))$ for any integer $u > s$. On the other hand for any infinite index set I: $\mathrm{var}\,\bigl(A\,\mathrm{Wr}\,(\prod_{i \in I} B)\bigr) = \mathrm{var}\,(A) \cdot \mathrm{var}\,(B)$.*

The first alternative holds for any abelian A and B apart from the following case: A and B are of finite exponents m and n respectively and for some prime p dividing both m and n the corresponding factor $B[p^k]/B[p^{k-1}]$ is finite.

The analogous statement is also true for direct wreath products of groups.

REMARK 7.3. Applying this theorem, having any wreath product $A\,\mathrm{Wr}\,B$ which does not generate the variety $\mathrm{var}\,(A) \cdot \mathrm{var}\,(B)$, we get a countably infinite set of linearly ordered proper subvarieties of $\mathrm{var}\,(A) \cdot \mathrm{var}\,(B)$, namely, varieties generated by wreath products: $\{A\,\mathrm{Wr}\,(\underbrace{B \oplus \cdots \oplus B}_{s\ \text{times}}) \mid s \in \mathbb{N}\}$ (see examples below).

PROOF OF THEOREM 7.2. First let us consider the cases, when the statement of this theorem easily follows from one of the results already established. If one of the groups A and B is not of finite exponent, then the first alternative holds. Thus assume $\exp A = m$, $\exp B = n$. If the index set I is infinite, then for any B the direct sum $\sum_{i \in I} B$ always discriminates $\mathrm{var}\,(B)$ and, thus, $A\,\mathrm{Wr}\,(\sum_{i \in I} B)$ discriminates $\mathrm{var}\,(A) \cdot \mathrm{var}\,(B)$. So assume I to be finite. If $\mathrm{var}(A\,\mathrm{Wr}\,\sum_{i \in I} B) \neq \mathrm{var}\,(A) \cdot \mathrm{var}\,(B)$, then by Theorem 6.1 there is a prime p such that the corresponding factor group $B[p^k]/B[p^{k-1}]$ is finite. Denote for brevity by $B_{s,p}$ the p-primary component of direct sum $\sum_{i=1}^{s} B$. If for given $r > s$ $\mathrm{var}\,(A\,\mathrm{Wr}\,\sum_{i=1}^{s} B) = \mathrm{var}\,(A\,\mathrm{Wr}\,\sum_{i=1}^{r} B)$ holds, then following the proof of Lemma 5.1:

(7.1) $$\mathrm{var}\,(A_p\,\mathrm{Wr}\,B_{s,p}) = \mathrm{var}\,(A_p\,\mathrm{Wr}\,B_{r,p}).$$

So to complete the proof it is sufficient to show that (7.1) leads to a contradiction to the fact that $B[p^k]/B[p^{k-1}]$ is finite. We can also omit the case when the group $B_{s,p}$ (and, therefore, the group $B_{r,p}$) is finite, for in such a case by Liebeck's Theorem [L62] the groups $A_p\,\mathrm{Wr}\,B_{s,p}$ and $A_p\,\mathrm{Wr}\,B_{r,p}$ have different nilpotency classes.

For the rest of proof the concrete form of B_p is essential. Since $B[p^k]/B[p^{k-1}]$ is finite, B_p contains in its direct decomposition only *finitely many*, say l_0, summands C_{p^k}. It may turn out that $B[p^{k-1}]/B[p^{k-2}]$ is also finite and the number of summands $C_{p^{k-1}}$ is finite, say l_1. But since the group B_p is infinite, there exists the *first* number d such that the corresponding factor $B[p^{k-d+1}]/B[p^{k-d}]$ is finite, it consists of, say, l_{d-1} summands $C_{p^{k-d+1}}$, and the factor $B[p^{k-d}]/B[p^{k-d-1}]$ is *infinite*. Thus B_p can be presented as:

$$(7.2) \quad B_p = \underbrace{C_{p^k} \oplus \cdots \oplus C_{p^k}}_{l_0} \oplus \cdots \oplus \underbrace{C_{p^{k-d+1}} \oplus \cdots \oplus C_{p^{k-d+1}}}_{l_{d-1}}$$
$$\oplus \underbrace{C_{p^{k-d}} \oplus \cdots \oplus C_{p^{k-d}} \oplus \cdots}_{\infty} \oplus \hat{B},$$

where $\exp \hat{B} \leq p^{k-d-1}$ and where some of $l_1, \ldots, l_{d-1}$ may be equal to 0. Let $\lambda = \lambda(A_p, B_{s,p}, t)$ be the function defined in Section 4 and let $\exp A_p = p^u$. It follows from the proof in Subsection 4.2 and from decomposition (7.2) that, for any $t > s \cdot \sum_{i=0}^{d-1} l_i$ the function $\lambda(A_p, B_{s,p}, t)$ is bounded by

$$(7.3) \quad s \cdot \sum_{i=0}^{d-1} l_i(p^{k-i} - 1) + (t - s \cdot \sum_{i=0}^{d-1} l_i) \cdot (p^{k-d} - 1) + (u-1)(p-1)p^{k-1} + 1.$$

On the other hand for each $t > r \cdot \sum_{i=0}^{d-1} l_i + 1$ the variety $\operatorname{var}(A_p \operatorname{Wr} B_{r,p})$ contains the following t-generated group:

$$T(r,t) = C_{p^u} \operatorname{Wr} \left[\sum_{i=1}^{rl_0} C_{p^k} \oplus \cdots \oplus \sum_{i=1}^{rl_{d-1}} C_{p^{k-d+1}} \oplus \sum_{i=1}^{\omega(t)} C_{p^{k-d}}, \right]$$

where $\omega(t) = t - r \cdot \sum_{i=0}^{d-1} l_i - 1$. The group $T(r,t)$ is nilpotent of class:

$$\nu(p,r,t) = r \cdot \sum_{i=0}^{d-1} l_i(p^{k-i} - 1) + (t - r \cdot \sum_{i=0}^{d-1} l_i - 1) \cdot (p^{k-d} - 1)$$
$$+ (u-1)(p-1)p^{k-1} + 1.$$

It remains to verify that for sufficiently large integers t the value of $\nu(p,r,t)$ is greater than that of $\lambda(A_p, B_{s,p}, t)$. It is sufficient to make calculations for the value $r = s + 1$: the generality of the statement of Theorem 7.2 remains unaffected but our calculations become much shorter.

$$\nu(p, s+1, t) - \lambda(A_p, B_{s,p}, t) = \sum_{i=0}^{d-1} l_i(p^{k-i} - 1)$$
$$+ \left(t - (s+1) \cdot \sum_{i=0}^{d-1} l_i - 1 - t + s \cdot \sum_{i=0}^{d-1} l_i \right) \cdot (p^{k-d} - 1)$$
$$= \sum_{i=0}^{d-1} l_i p^{k-i} - \sum_{i=0}^{d-1} l_i - \left(\sum_{i=0}^{d-1} l_i + 1 \right) \cdot (p^{k-d} - 1)$$
$$= \sum_{i=0}^{d-1} l_i(p^{k-i} - p^{k-d}) - p^{k-d} + 1 > 0.$$

□

Here are two examples concerning subvarieties generated by wreath products in the lattice of all subvarieties of product varieties of abelian groups.

EXAMPLE 7.4. Since $C_p \operatorname{Wr} C_p$ does not generate variety $\mathfrak{A}_p^2$, wreath products $C_p \operatorname{Wr} \sum_{i=1}^{s} C_p$ $(s = 1, 2, \ldots)$ generate infinitely many subvarieties of $\mathfrak{A}_p^2$. We are able to locate them in the lattice of subvarieties of $\mathfrak{A}_p^2$ using its description due to Kovács and Newman [KN71]. For any proper subvariety $\mathfrak{V}$ of $\mathfrak{A}_p^2$ there is a number $s \geq 1$ such that $\operatorname{var}(C_p \operatorname{Wr} \sum_{i=1}^{s} C_p) \subseteq \mathfrak{V} \subseteq \operatorname{var}(C_p \operatorname{Wr} \sum_{i=1}^{s+1} C_p)$ or $\mathfrak{V}$ lies in $\operatorname{var}(C_p \operatorname{Wr} C_p)$. Moreover, for any $s \geq 1$ there are exactly the following $p-2$ subvarieties of $\mathfrak{A}_p^2$ "between" $\operatorname{var}(C_p \operatorname{Wr} \sum_{i=1}^{s} C_p)$ and $\operatorname{var}\left(C_p \operatorname{Wr} \sum_{i=1}^{s+1} C_p\right)$: $\mathfrak{A}_p^2 \cap \mathfrak{B}_{p^2} \cap \mathfrak{N}_j$, $j = s(p-1)+2, \ldots, (s+1)(p-1)$. And there are $2p-3$ subvarieties of $\mathfrak{A}_p^2$ "between" $\operatorname{var}(C_p \operatorname{Wr}\{1\}) = \mathfrak{A}_p$ and $\operatorname{var}(C_p \operatorname{Wr} C_p)$ (see [KN71]).

EXAMPLE 7.5. On the other hand for arbitrary coprime numbers m and n the variety $\mathfrak{A}_m \cdot \mathfrak{A}_n$ contains only finitely many subvarieties [N68]. According to Theorem 7.2 this fact already guarantees, that for an arbitrary pair $A \in \mathfrak{A}_m$, $B \in \mathfrak{A}_n$: $\operatorname{var}(A \operatorname{Wr} B) = \operatorname{var}(A) \cdot \operatorname{var}(B)$.

Closing the current work we would like to announce our recent papers [M00, Msub1, Msub2] as well as our common paper with Professor H. Heineken [HM00], where some related properties of wreath products and their verbal subgroups are considered.

References

[BNNN64] G. Baumslag, B. H. Neumann, Hanna Neumann, P. M. Neumann *On varieties generated by finitely generated group*, Math. Z., 86 (1964), 93–122.

[B35] G. Birkhoff, *On the structure of abstract algebras*, Proc. Cambridge Phil. Soc., 31 (1935), 433–454.

[B70] R. A. Bryce, *Metabelian groups and varieties*, Phil. Trans. Roy. Soc. 266 (1970), 281–355.

[B76] R. A. Bryce, *Varieties of metabelian p-groups*, J. London Math. Soc., 13 (1976), 363–380.

[B63] N. R. Brumberg, *Connection of wreath product with other operations on groups*, Sib. Mat. Zh., 4 (1963), 6, 1221–1234 (Russian).

[B] R. G. Burns, Ph.D. Thesis.

[CR62] C. W. Curtis, I. R. Reiner, *Representation Theory of finite Groups and associative Algebras*, Wiley, New York, Lodon, 1962.

[F70] L. Fuchs, *Infinite Abelian Groups*, 2 vols. Academic Press, New York 1970–73.

[HM00] H. Heineken, V. H. Mikaelian, *On normal verbal embeddings of groups*, J. Math. Sci., New York, 100 (2000), 1, 1915–1924.

[H59] G. Higman, *Some remarks on varieties of groups*, Quart. J. Math. Oxford, (2) 10 (1959), 165–178.

[HNN47] G. Higman, B. Neumann, Hanna Neumann, *Embedding theorems for groups*, J. London Math. Soc. (3), 24 (1947), 247–254.

[KK51] L. Kaloujnine, M. Krasner, *Produit complete des groupes de permutations et le problème d'extension des groupes, III*, Acta Sci. Math. Szeged, 14 (1951), 69–82.

[KM96] M. I. Kargapolov, Ju. I. Merzlyakov, *Fundamentals of the Theory of Groups*, fourth edition, Nauka, Moscow 1996 (Russian). English translation of the second edition by R. G. Burns, Springer Verlag, New York 1979.

[KN71] L. G. Kovács, B. H. Neumann, *On non-Cross varieties of groups*, J. Austral. Math. Soc., 12 (1971), 2, 129–144.

[KN94] L. G. Kovács, B. H. Neumann, *Infinite groups* 1994 (Ravello), 125–128, de Gruyter, Berlin, 1996.

[L62]   H. Liebeck, *Concerning nilpotent wreath products*, Proc. Cambridge Phil. Soc., 58 (1962), 443–451.

[M95]   J. D. R. Meldrum, *Wreath products of Groups and Semigroups*, Longman, Harlow 1995.

[M00]   V. H. Mikaelian, *Subnormal embedding theorems for groups*, J. London Math. Soc., 62 (2000), 398–406.

[Msub1] V. H. Mikaelian, *Über die normalen verbalen Einbettungen einiger Klassen der Gruppen*, submitted (German).

[Msub2] V. H. Mikaelian, *An embedding construction for ordered groups*, submitted.

[NN59] B. H. Neumann, Hanna Neumann, *Embedding theorems for groups*, J. London Math. Soc., 34 (1959), 465–479.

[N68]   Hanna Neumann, *Varieties of Groups*, Springer Verlag, Berlin 1967. Russian translation of A. L. Šmelkin, Mir, Moscow 1969.

[N64]   P. M. Neumann, *On the structure of standard wreath products of groups*, Math. Zeitschr. 84 (1964), 343–373.

[R96]   D. J. S. Robinson, *A Course in the Theory of Groups*, second edition, Springer-Verlag, New York, Berlin, Heidelberg 1996.

[S65]   A. L. Šmelkin, *Wreath products and varieties of groups*, Izv. AN SSSR, ser. matem., 29 (1965), 149–170 (Russian).

DEPARTMENT OF COMPUTER SCIENCE AND APPLIED MATHEMATICS, YEREVAN STATE UNIVERSITY, 375025 YEREVAN, ARMENIA

E-mail address: mikaelian@e-math.ams.org, v.mikaelian@usa.net

Internet: http://www.crosswinds.net/~mikaelian/home.html

Existence of Rigid Indecomposable Almost Completely Decomposable Groups

Otto Mutzbauer

Dedicated to Professor Laszlo Fuchs in honour of his 75th birthday.

ABSTRACT. The existence of almost completely decomposable groups with given isomorphism types of the regulator and the regulator quotient is an open problem in this generality. Here we consider indecomposable rigid groups with a regulator quotient of fixed exponent.

Introduction

We construct indecomposable rigid almost completely decomposable groups (of finite rank) with a regulator quotient of fixed (finite) exponent. Since the regulator quotient is not necessarily a p-group, this is the global case. So we may apply the local-global principle, established in [4] for almost completely decomposable groups to obtain the regulator quotient and a regulator criterion. There is no help concerning indecomposability, cf. [4, 6.2], since there are indecomposable groups with all local constituents decomposable.

1. Preliminaries

Let R be a completely decomposable group of finite rank and G a finite group of exponent m. Then

$$\mathfrak{H}(R,G) = \{X \subset \mathbb{Q}R \mid R \subset X \subset m^{-1}R,\ \mathrm{R}(X) = R,\ X/R \cong G\}$$

denotes the set of almost completely decomposable groups contained in the divisible hull $\mathbb{Q}R$ with regulator R and regulator quotient isomorphic to G. All almost completely decomposable groups with regulator isomorphic to R and regulator quotient isomorphic to G are in $\mathfrak{H}(R,G)$ up to isomorphism.

For later use we collect here some lemmata.

LEMMA 1.1. *If n is a natural number, $X = H \oplus L$ a torsion-free abelian group and U a fully invariant subgroup of X, then*

$$n^{-1}U \cap X = (n^{-1}U \cap H) \oplus (n^{-1}U \cap L).$$

1991 *Mathematics Subject Classification.* Primary 20K15; Secondary 20K20.

PROOF. It is enough to consider $n^{-1}u = h + l \in n^{-1}U \cap X$. Since U is fully invariant, $u = nh + nl \in U = (U \cap H) \oplus (U \cap L)$ by [1, 9.3]. Consequently $h \in n^{-1}U \cap H$ and $l \in n^{-1}U \cap L$. □

LEMMA 1.2. *Let U, V be subgroups of the torsion-free abelian group X, not necessarily of finite rank. Let Y be pure in U and in V. If there is a set P of primes such that $U/(U \cap V)$ is a P-group and $V/(U \cap V)$ is a P'-group, where P' is the complement of P in the set of primes, then Y is pure in $U + V$.*

PROOF. Let $u + v \in U + V$ such that $p(u+v) \in Y \subset U \cap V$ for some prime p. Then $pu, pv \in U \cap V$. Now, if $p \in P$, then $v \in U \cap V$ because $p(V/(U \cap V)) = V/(U \cap V)$. Hence $u + v \in U$, and even $u + v \in Y$ since Y is pure in U, i. e., Y is p-pure in $U + V$. For $p' \notin P$ we obtain by the same arguments that Y is p'-pure in $U + V$. So Y is pure in $U + V$. □

2. Indecomposability

For $m \in \mathbb{N}$ a torsion-free group X is called *properly m-decomposable* if there are nonzero subgroups $U, V \subset X$ such that $X = hU \oplus kV$, for $m = hk$. This implies $hU = U$ and $kV = V$ because $hU \subseteq U \subseteq \langle hU \rangle_*^X = hU$, hence $X = \gcd(h,k)X$. If $m = p^s$ is a prime power, $s > 0$, then either U or V is p-divisible. We give some preliminary consequences implied by m-decomposability.

LEMMA 2.1. *Let A be a subgroup of the torsion-free abelian group X, not necessarily of finite rank. Let m be a natural number such that $\exp(X/A)$ divides m. Then A is properly m-decomposable if and only if X is properly m-decomposable.*

More precisely, if $m = hk$ and $\exp(X/A)$ divides m, then $A = hU \oplus kV$ implies $X = hU_*^X \oplus kV_*^X$ and $X = hU \oplus kV$ implies $A = h(U \cap A) \oplus k(V \cap A)$.

PROOF. If $A = hU \oplus kV$, then $A \subseteq U_*^X \oplus V_*^X \subseteq X$ where U_*^X is h-divisible and V_*^X is k-divisible. Moreover, X/U_*^X is k-divisible, since V is isomorphic to a full subgroup. By the same argument X/V_*^X is h-divisible. Thus $X/(U_*^X \oplus V_*^X)$ is m-divisible. Finally, $X = U_*^X \oplus V_*^X$ since $\exp(X/A)|m$.

Conversely, if $X = hU \oplus kV$, then $mX = h(kU) \oplus k(hV) \subset A$, and the pair (A, mX) satisfies exactly the same hypothesis as the pair (X, A). Thus by the above argument $A = h\langle kU \rangle_*^A \oplus k\langle hV \rangle_*^A = h(U \cap A) \oplus k(V \cap A)$. □

REMARK 2.2. Let m be a natural number. A completely decomposable group A with critical typeset T, i.e., $A = \sum_{\tau \in T} A_\tau$ is a homogeneous decomposition, is properly m-decomposable if and only if there is a partition $P \cup Q$ of the prime divisors of m such that $\tau \in T$ is either divisible by all $p \in P$ or by all $q \in Q$. This induces a partition of the critical typeset $T = T_P \cup T_Q$, where all types in T_P have infinity for all $p \in P$, and all types in T_Q have infinity for all $q \in Q$. Hence $A = \sum_{\tau \in T_P} A_\tau \oplus \sum_{\tau \in T_Q} A_\tau$ is the direct sum of a P-divisible group and a Q-divisible group. In particular, the partition of $T = P \cup Q$ can be trivial, i. e., the sets P or Q can be empty. In this case there is an m-divisible direct summand and A is properly m-decomposable if not of rank 1.

LEMMA 2.3. *A completely decomposable subgroup A of a torsion-free abelian group X, not necessarily of finite rank, is not a maximal completely decomposable subgroup of X if $\dim(X/A)[p] \neq 0$ is finite and equal to $\dim A/pA$ for some prime p.*

PROOF. The finiteness of $\dim(X/A)[p] = \dim A/pA \neq 0$ and $(X/A)[p] \subset p^{-1}A/A$ imply equality. Hence $A \neq p^{-1}A \subset X$, and $p^{-1}A$ is completely decomposable, too. □

A. Mader made me aware of the following general fact. If $A = B \oplus C \subset X$, where X/A is torsion and $B \neq 0$ is p-pure in X, then

$$\dim(X/A)[p] \leq \operatorname{rk}_p C = \dim C/pC \leq \operatorname{rk} C < \operatorname{rk} A.$$

If $B = 0$, then $\dim(X/A)[p] \leq \operatorname{rk} A$. Thus, only the equality $\dim(X/A)[p] = \dim A/pA$ is the real condition in Lemma 2.3. Moreover, the p-rank of the quotient X/A of a torsion-free abelian group X is bounded by the difference $\operatorname{rk}_p X - \operatorname{rk} S$ of the p-rank of X and the rank of the maximal p-divisible subgroup S of X.

To illustrate both Lemmata above, consider

$$A = \mathbb{Z}[2^{-1}]a \oplus \mathbb{Z}[3^{-1}]b \subset X = A + \mathbb{Z}\frac{a+b}{6} = \mathbb{Z}[2^{-1}]\frac{a}{3} \oplus \mathbb{Z}[3^{-1}]\frac{b}{2}.$$

Since A is 6-decomposable there is no indecomposable extension of A whose exponent is 6.

3. Existence

In this section we collect necessary and sufficient conditions for the existence of indecomposable groups in $\mathfrak{H}(R, G)$, where R is rigid and G is of exponent m.

The following Lemmata are of interest in themselves, since they deal with almost completely decomposable groups without a priori restrictions on the typeset.

LEMMA 3.1. *Let X be an almost completely decomposable group with regulator R and X/R a p-group. Then R is the regulator of all subgroups $p^{-i}R \cap X$ for $i \geq 0$.*

PROOF. Let $B_i = p^{-i}R \cap X$ for $i \geq 0$. Then for any type τ, $B_i(\tau) = p^{-i}R \cap X(\tau)$ and $B_i^\sharp(\tau) = p^{-i}R \cap X^\sharp(\tau)$, cf. [**2**, 2.3.10].

Let i and some critical type τ of X be fixed. Let $R = \bigoplus_\rho R_\rho$ be a homogeneous decomposition of the regulator R.

Recall that $\beta_\tau^X = \exp(X^\sharp(\tau)/R^\sharp(\tau)) = \exp(X(\tau)/R(\tau))$ denotes the Burkhardt invariant of X at the type τ, cf. [**2**, 4.4.5]. Here $\beta_\tau^X = p^{h(\tau)}$ is a p power, since $p^e X^\sharp(\tau) \subseteq R \cap X^\sharp(\tau) = R^\sharp(\tau)$ and $\beta_\tau^X | p^e$.

Assume first that $p^i \leq \beta_\tau^X$. Let $b \in X^\sharp(\tau)$ with order $\operatorname{ord}(b + R^\sharp(\tau)) = \beta_\tau^X$. Then $p^{-i}\beta_\tau^X b \in p^{-i}R \cap X^\sharp(\tau)$ has the order p^i relative to $R^\sharp(\tau)$. Thus $\exp(B_i^\sharp(\tau)/R^\sharp(\tau)) = \exp[(p^{-i}R \cap X^\sharp(\tau))/R^\sharp(\tau)] = p^i$ and

$$R_\tau \subseteq \beta_\tau^X X(\tau) = p^i(\beta_\tau^X p^{-i}X(\tau)) \subseteq p^i(p^{-i}R \cap X(\tau)) = p^i B_i(\tau) \subseteq R(\tau),$$

where $p^i = \exp(B_i^\sharp(\tau)/R^\sharp(\tau))$ so that Burkhardt's Regulator Criterion, cf. [**2**, 4.4.6], is satisfied for B_i and R at the type τ.

Secondly let $p^i > \beta_\tau^X$. Then $X^\sharp(\tau) \subseteq (\beta_\tau^X)^{-1}R^\sharp(\tau) \subseteq (\beta_\tau^X)^{-1}R \subseteq p^{-i}R$ and

$$\frac{X^\sharp(\tau)}{R^\sharp(\tau)} = \frac{p^{-i}R \cap X^\sharp(\tau)}{R^\sharp(\tau)} = \frac{B_i^\sharp(\tau)}{R^\sharp(\tau)}$$

yield $\exp(B_i^\sharp(\tau)/R^\sharp(\tau)) = \exp(X^\sharp(\tau)/R^\sharp(\tau)) = \beta_\tau^X$. We get

$$R_\tau \subseteq \beta_\tau^X X(\tau) = \beta_\tau^X (p^{-i}R \cap X(\tau)) = \beta_\tau^X B_i(\tau) \subseteq R(\tau).$$

This is Burkhardt's Regulator Criterion for B_i and R in the second case. So R is the regulator of all B_i. □

A necessary and sufficient condition is already known ([**4**, Lemma 3.7] or [**2**, 11.8.2]) for the existence of (possibly decomposable) groups in $\mathfrak{H}(R,G)$ if R is rigid and G is a finite p-group.

LEMMA 3.2. *An almost completely decomposable group X with regulator quotient of exponent m, which is not properly m-decomposable, is indecomposable if the following two conditions hold:*
  (1) *The p-reduced part of the p-local constituent of X is indecomposable for all $p|m$.*
  (2) *The p-reduced part of the regulator is fully invariant in the regulator for all $p|m$.*

PROOF. Let $X \in \mathfrak{H}(R,G)$ with $\exp G = m$ and $R = \bigoplus_{i=1}^{n} R_i$. For a prime $p|m$ let p^l be the maximal power of p dividing m. Then $R_p^* = \bigoplus \{R_i \mid pR_i \neq R_i, 1 \leq i \leq n\}$ is a maximal p-reduced direct summand of R, hence fully invariant in R by (2). Moreover, $W_p = p^{-l}R_p^* \cap X$ is a maximal p-reduced part of the p-local constituent of X. Since X is not properly m-decomposable it has no properly m-divisible subgroup. Thus $X = \sum_{p|m} X_{lp} = \sum_{p|m} W_p$, cf. [**4**, 2.2], is the sum of all of the maximal p-reduced parts of its p-constituents for all p dividing m.

Now assume that $X = U \oplus V$ is decomposable. The R_p^* are fully invariant in R, and since the regulator is fully invariant in X, the R_p^* are fully invariant in X, too. Then by Lemma 1.1 we obtain $W_p = p^{-l}R_p^* \cap X = (W_p \cap U) \oplus (W_p \cap V)$ and since W_p is indecomposable by (1), $W_p \subset U$ or $W_p \subset V$. Thus there is a partition $P_U \cup P_V = \{p \text{ prime} \mid p|m\}$ of the set of prime divisors of m such that $P_U = \{p \mid W_p \subset U\}$ and $P_V = \{p \mid W_p \subset V\}$. Let $W_U = \sum_{p \in P_U} W_p \subset U$ and $W_V = \sum_{p \in P_V} W_p \subset V$. Then $X = W_U \oplus W_V = U \oplus V$, hence $W_U = U$ and $W_V = V$. This implies $pU = U$ for all $p \in P_V$ and $pV = V$ for all $p \in P_U$. Since $P_U \cup P_V$ is a partition of the set of primes dividing m, there is a decomposition $m = h \cdot k$ in some P_U-number k and some P_V-number h, so that $X = hU \oplus kV$ is a proper m-decomposition, contradicting that X is not properly m-decomposable. Consequently X is indecomposable. □

Comparing the following theorem with constructions used in [**4**] shows that there is no overlap since here we have the additional property indecomposability. Moreover, since we have rigid groups the parts of the proof which could be done by citation of the local-global principle are so easy that we prefer to show things directly.

It is clear in view of Lemma 2.1 that indecomposable almost completely decomposable groups with regulator quotient of exponent m cannot have a proper m-decomposition, cf. the example after Lemma 2.3.

THEOREM 3.3. *Let R be a rigid completely decomposable group of finite rank and G a finite abelian group of exponent $m \neq 1$. There is an indecomposable group in $\mathfrak{H}(R,G)$, if and only if $\dim(G[p]) < \dim(R/pR)$ for all p dividing m and R is not properly m-decomposable.*

PROOF. By the Lemmata 2.1 and 2.3 the conditions are necessary, since in the rigid case the regulating regulator is a maximal completely decomposable subgroup of groups in $\mathfrak{H}(R,G)$.

For the converse we have to construct some indecomposable group in $\mathfrak{H}(R,G)$. First let us consider the special case $G \cong \mathbb{Z}_p^r$ for some prime p, i. e., $\exp(G) = p$ and R is p-reduced, since R is not properly p-decomposable. Hence $R/pR \cong \mathbb{Z}_p^n$ and $r < n = \operatorname{rk} R$. Let $\{a_1,\ldots,a_n\}$ be a p-basis of $R = \bigoplus_{i=1}^n R_i = \bigoplus_{i=1}^n \langle a_i \rangle_*^R$, i. e., $\chi_p^R(a_i) = 0$ for all i, and let

$$(1) \qquad W = R + \sum_{j=1}^r \mathbb{Z} g_j, \quad g_j = p^{-1}\left(a_j + \sum_{k=r+1}^n a_k\right), \; j = 1,\ldots,r.$$

Certainly $W/R = \langle g_j + R \mid j \rangle$ is a subspace of $V = p^{-1}R/R = \langle p^{-1}a_j + R \mid j \rangle$ with $\dim W/R = r$. Moreover, $\langle p^{-1}a_i + R \rangle \cap W/R = ((p^{-1}R_i + R)/R) \cap (W/R) = 0$ for all i. Hence all summands R_i of R are pure in W and R is the regulator of W. Thus $W \in \mathfrak{H}(R,\mathbb{Z}_p^r)$. By a simple linear algebraic argument there is no non–empty subset $I \subsetneq \{1,\ldots,n\}$, such that

$$W/R = (\langle p^{-1}a_i + R \mid i \in I \rangle \cap W/R) \oplus (\langle p^{-1}a_i + R \mid i \in \complement I \rangle \cap W/R).$$

By [**3**, Theorem 3.1] and using rigidity, W is indecomposable.

Let us consider the more general case that $G \cong \bigoplus_{j=1}^r \mathbb{Z}_{p^{t_j}}$ is a finite p-group. Again R is p-reduced, since it is not properly p-decomposable. Thus $r = \dim G[p] < \operatorname{rk} R$. Let

$$(2) \qquad Y = R + \sum_{j=1}^r \mathbb{Z} f_j, \quad f_j = p^{-t_j}\left(a_j + \sum_{k=r+1}^n a_k\right), \; j = 1,\ldots,r.$$

We have to show that $Y \in \mathfrak{H}(R, \bigoplus_{j=1}^r \mathbb{Z}_{p^{t_j}})$ is indecomposable.

Since

$$(Y/R)[p] = \langle p^{t_j-1} f_j + R \mid j = 1,\ldots,r \rangle = \langle g_j + R \mid j = 1,\ldots,r \rangle = W/R \cong \mathbb{Z}_p^r,$$

the group $Y \cap p^{-1}R$ is indecomposable as in the elementary case above with the same elements g_j. Thus Y is indecomposable. Moreover, R is the regulator of Y by Lemma 3.1. Finally the definition of the generators f_j guarantees that $Y/R \cong \bigoplus_{j=1}^r \mathbb{Z}_{p^{t_j}}$.

Now we deal with the general case of a finite abelian group G with $\exp(G) = m$. Let $G = \bigoplus_{p \mid m} G_p$ with primary components G_p. Let $R_p^* = \bigoplus \{R_i \mid pR_i \neq R_i, 1 \leq i \leq n\}$ for $p \mid m$. This is the uniquely determined maximal p-reduced subgroup of R. We have $\sum_{p\mid m} R_p^* = R$, since $mR_i = R_i$ for some i would lead to a proper m-decomposition $R = mR_i \oplus (\bigoplus_{k \neq i} R_k)$ of R and of X, cf. Lemma 2.1. Moreover, $\dim G[p] < \operatorname{rk}(R_p^*)$ by hypothesis.

As in the first part of the proof there is an indecomposable group Y_p in $\mathfrak{H}(R_p^*, G_p)$ for all $p \mid m$ satisfying (2). We will show that $X = \sum_{p \mid m} Y_p$ is in $\mathfrak{H}(R,G)$. For the quotient we get

$$\frac{X}{R} = \frac{\sum_{p\mid m} Y_p}{R} = \bigoplus_{p\mid m} \frac{Y_p + R}{R} \cong \bigoplus_{p\mid m} \frac{Y_p}{Y_p \cap R} = \bigoplus_{p\mid m} \frac{Y_p}{R_p^*} \cong \bigoplus_{p\mid m} G_p = G.$$

Since all R_i are pure in $Y_p + R \subseteq X$ and $\dfrac{Y_p + R}{(Y_p + R) \cap (Y_q + R)} = \dfrac{Y_p + R}{R} \cong G_p$, we have by Lemma 1.2 that all R_i are pure in $X = \sum_{p\mid m}(Y_p + R)$. Thus $X \in \mathfrak{H}(R,G)$ and it remains to show that X is indecomposable.

The Y_p are the maximal p-reduced parts of the local constituents of X. They are fully invariant since X is rigid, and they are indecomposable by construction. Hence X is indecomposable by Lemma 3.2. □

References

[1] L. Fuchs, *Infinite Abelian Groups* Vols. I and II, Academic Press (1970, 1973).

[2] A. Mader, *Almost Completely Decomposable Groups*, Gordon Breach (2000).

[3] K.-J. Krapf und O. Mutzbauer, *Classification of almost completely decomposable groups*, Abelian Groups and Modules, CISM Courses and Lecture Notes, vol. 287, Springer Verlag, 1984, S. 151–161.

[4] A. Mader, O. Mutzbauer und C. Vinsonhaler, *Local-global relations for almost completely decomposable groups*, Rocky Mountain J. Math. **29** (1999), 1429 – 1453.

MATHEMATISCHES INSTITUT, UNIVERSITÄT WÜRZBURG, AM HUBLAND, 97074 WÜRZBURG, GERMANY

E-mail address: `mutzbauer@mathematik.uni-wuerzburg.de`

C2–rings and the FGF–conjecture

W. K. Nicholson and M. F. Yousif

This paper is dedicated to Laszlo Fuchs on his 75th birthday.

ABSTRACT. In 1965 Utumi isolated two module-theoretic properties, C1 and C2, that hold in any injective module. He called a ring continuous if it satisfies both of these conditions as a module over itself, and proved several important results about these rings. While the C1-property has been extensively studied, the C2-property has been neglected. In this paper we survey some recent work on the C2-condition, and show how it can be used to obtain new information about the FGF-conjecture (that a ring for which every finitely generated module embeds in a free module is necessarily quasi-Frobenius), and to simplify what is required to settle the conjecture.

Suppose a ring R has the property that every right module can be embedded in a free right module. It follows immediately that every injective right module is a direct summand of a free right module, and so is projective. But then a theorem of Faith and Walker [5] asserts that R is necessarily quasi-Frobenius. A ring R is called a right *FGF-ring* if every finitely generated right R-module can be embedded in a free module, and the *FGF-conjecture* asserts that every right FGF-ring is quasi-Frobenius. The truth of this conjecture has been an open question since the 1960's.

This paper is a self-contained survey of C2-rings and how they relate to the FGF-conjecture. Here a ring R is called a *right C2-ring* if it satisfies Utumi's C2-condition as a right module over itself, that is if every right ideal isomorphic to a direct summand of R is itself a direct summand. After some preliminaries, the main theorem characterizes when a matrix ring over a right C2-ring is again a C2-ring. This is then used to prove that a ring R is quasi-Frobenius if the ring of $n \times n$ matrices is a right C2-ring for each $n \geq 1$ and every 2-generated right module embeds in a free right module. This result extends two classical results about the FGF-conjecture, and yields several applications, among them that every right FP-injective, right FGF-ring is quasi-Frobenius. We give an example showing that it is not sufficient to be able to embed 1-generated modules in a free module.

1991 *Mathematics Subject Classification.* Primary: 16D50, 16L60, 16E50.
Key words and phrases. C2-condition, quasi-Frobenius rings, FGF-rings.
This research was supported by NSERC Grant A8075, by the University of Calgary and by the Ohio State University.

© 2001 American Mathematical Society

The paper concludes by showing that to prove the FGF-conjecture it is sufficient to show that every right FGF-ring R is a right C2-ring.

Throughout this paper R always denotes an associative ring with unity, and all R-modules are unital. The notations $N \subseteq^{ess} M$ and $N \subseteq^{max} M$ mean that N is an essential (respectively, maximal) submodule of a module M. We write $J = J(R)$ for the Jacobson radical of R, and we write $M_n(R)$ for the ring of $n \times n$ matrices over R. Right annihilators will be denoted as

$$\mathbf{r}_X(Y) = \{x \in X \mid yx = 0 \text{ for all } y \in Y\},$$

with a similar definition of left annihilators $\mathbf{l}_X(Y)$.

1. C2-Rings

A module M_R over a ring R is said to satisfy the *C2-condition* (and is called a *C2-module*) if every submodule that is isomorphic to a direct summand of M is itself a direct summand. This condition was introduced in 1965 by Utumi [14] in his pioneering study of continuous modules and rings. We call a ring R a right **C2-ring** if the right module R_R satisfies the C2-condition, equivalently if T is a right ideal of R such that $T \cong eR$ where $e^2 = e \in R$, then $T = fR$ for some $f^2 = f \in R$.

Clearly, every right selfinjective ring is a right C2-ring, and every (von Neumann) regular ring is both a right and a left C2-ring. The only domains that are right C2-rings are the division rings. More generally, if 0 and 1 are the only idempotents in a ring R, then R is a right C2-ring if and only if all monomorphisms $R_R \to R_R$ are epic (that is, $\mathbf{r}(a) = 0$ implies that a is a unit). In fact, if R is a C2-ring which contains no infinite orthogonal family of idempotents, then monomorphisms from $R_R \to R_R$ are epic (if $\mathbf{r}(a) = 0$, $a \in R$, then $a^n R$ is a summand of R_R for each $n \geq 1$ (being isomorphic to R_R) so the series $aR \supseteq a^2R \supseteq \cdots$ terminates by hypothesis and it follows that a is a unit).

If F is a field and $R = \begin{bmatrix} F & F \\ 0 & F \end{bmatrix}$, write $a = \begin{bmatrix} 0 & 1 \\ 0 & 0 \end{bmatrix}$ and $e = \begin{bmatrix} 1 & 0 \\ 0 & 0 \end{bmatrix}$. Then R is right and left artinian, but R is neither right nor left C2 because $J = Ra \cong Re$ and $J = aR \cong (1-e)R$, and J is not a summand of R_R or $_RR$. Incidentally, this shows that a direct sum of C2-modules need not be C2 because both eR and $(1-e)R$ are C2-modules (eR has exactly one proper submodule and $(1-e)R$ is simple).

Before providing more examples of C2-rings, we give a characterization of these rings that will be used frequently in what follows.

LEMMA 1. *A ring R is a right C2-ring if and only if whenever $\mathbf{r}(a) = \mathbf{r}(e)$, $a \in R$, $e^2 = e \in R$, then necessarily $e \in Ra$.*

PROOF. Suppose that R is a right C2-ring. If $\mathbf{r}(a) = \mathbf{r}(e)$ where a and $e = e^2$ are in R, then $\gamma(ar) = er$ defines an isomorphism $\gamma : aR \to eR$ so the C2-condition forces $aR = fR$ for some $f^2 = f \in R$. Hence $a = fa$, and so $e = \gamma(a) = \gamma(f)a \in Ra$, as required.

Conversely, let $\gamma : aR \to fR$ be an isomorphism where $a \in R$ and $f^2 = f \in R$. We must show that aR is a direct summand of R_R. If we write $c = \gamma(a)$ then $\mathbf{r}(a) = \mathbf{r}(c)$ because γ is one-to-one, and $cR = fR$ because γ is onto. Hence the left multiplication map $c \cdot : R_R \to fR$ is onto, so the kernel $\mathbf{r}(c) = \mathbf{r}(a)$ is a direct summand of R_R, say $\mathbf{r}(a) = (1-e)R$ where $e^2 = e \in R$. Thus $\mathbf{r}(a) = \mathbf{r}(e)$ so $e \in Ra$

by hypothesis. But also $a \in Re$ because $1 - e \in \mathbf{r}(a)$, so $Ra = Re$ is a summand of $_RR$, whence aR is a summand of R_R. □

EXAMPLE 1. Lemma 1 shows that a direct product $R = \Pi_i R_i$ of rings R_i is a right C2-ring if and only if each factor R_i is a right C2-ring.

EXAMPLE 2. The following conditions are equivalent for a local ring R:
1. R is a right C2-ring.
2. Every monomorphism $R_R \to R_R$ is epic.
3. $J = \{a \in R \mid \mathbf{r}(a) \neq 0\}$.

In particular, any local ring with nil radical is a right and left C2-ring.

PROOF. We have already observed that 1. ⇒ 2., and 2. ⇒ 3. is clear in a local ring. Finally, if 3. holds, suppose $\mathbf{r}(a) = \mathbf{r}(e)$, $a \in R$, $e^2 = e \in R$; by Lemma 1 we must show that $e \in Ra$. This is clear if $e = 0$. If $e = 1$ then $\mathbf{r}(a) = 0$ so $a \notin J$ by 3. Hence $Ra = R$ because R is local, and so $e \in Ra$, proving 3. ⇒ 1. The last statement follows from 3. because R is local. □

A ring R is called right *Kasch* if every simple right R-module embeds in R, equivalently if $\mathbf{l}(T) \neq 0$ for every maximal right ideal T of R.

EXAMPLE 3. Every left Kasch ring is a right C2-ring, but the converse is false.

PROOF. The first assertion appears in [**15**, Lemma 1.15], but we include a proof for completeness. Let $\mathbf{r}(a) = \mathbf{r}(f)$, $a \in R$, $f^2 = f \in R$; we must show that $f \in Ra$ by Lemma 1. We have $a = af \in Rf$ because $1 - f \in \mathbf{r}(a)$. If $f \notin Ra$ choose $Ra \subseteq M \subseteq^{max} Rf$, and by the Kasch hypothesis let $\sigma : Rf/M \to {_RR}$ be monic. Write $c = \sigma(f + M)$. Since $af = a$, we have $ac = \sigma(af + M) = \sigma(a + M) = 0$ because $a \in M$. Hence $c \in \mathbf{r}(a) = \mathbf{r}(f)$, so $0 = fc = \sigma(f^2 + M) = \sigma(f + M)$. But then $f \in M$, a contradiction. So R is a right C2-ring. The converse fails because any regular, right selfinjective ring is a right C2-ring, but it is not left Kasch if it is not artinian. A specific example is an infinite product of division rings. □

A ring is called right *P-injective* (for right *principally* injective) if each R-linear map $\gamma : aR \to R_R$, $a \in R$, extends to $R_R \to R_R$, equivalently if $\mathbf{lr}(a) = Ra$ for each a in R. Hence Lemma 1 shows that every right P-injective ring is a right C2-ring (this is [**10**, Theorem 1.2]).

EXAMPLE 4. Every right P-injective ring is a right C2-ring, but not conversely.

PROOF. To disprove the converse, let $S = \left\{ \begin{bmatrix} a & v \\ 0 & a \end{bmatrix} \mid a \in F, v \in V \right\}$ be the trivial extension of a field F by the two-dimensional vector space V. Then it is routine to verify that S is a commutative, local, artinian ring, and S is a C2-ring by Example 2 because $J(S)^2 = 0$. To see that S is not P-injective, let $V = uF \oplus wF$, and write $\bar{u} = \begin{bmatrix} 0 & u \\ 0 & 0 \end{bmatrix}$. Then $\bar{u}S = \begin{bmatrix} 0 & uF \\ 0 & 0 \end{bmatrix}$ and the map $\begin{bmatrix} 0 & ua \\ 0 & 0 \end{bmatrix} \mapsto \begin{bmatrix} 0 & wa \\ 0 & 0 \end{bmatrix}$ is an S-linear map from $\bar{u}S \to S$ which does not extend to $S \to S$ because $w \notin Fu$. Hence S is not P-injective. □

EXAMPLE 5. There exists a left C2-ring that is not right C2.

PROOF. In [4] Faith and Menal construct a right noetherian ring R in which every right ideal is an annihilator and J is nilpotent, but which is not right artinian. In particular, R is left P-injective (every principal right ideal is an annihilator) and so is a left C2-ring by Example 4. Suppose that R is a right C2-ring. Since R has no infinite sets of orthogonal idempotents every monomorphism $R_R \to R_R$ is epic (see the discussion preceding Lemma 1). Hence R is semilocal by the Camps-Dicks theorem [2] (it is right finite-dimensional), and so R is right artinian by Hopkin's theorem, a contradiction. Hence R is not a right C2-ring. □

LEMMA 2. *If R is a right C2-ring, so is fRf for any $f^2 = f \in R$ such that $RfR = R$.*

PROOF. Write $S = fRf$ and let $\mathbf{r}_S(a) = \mathbf{r}_S(e)$, $a \in S$, $e^2 = e \in S$; by Lemma 1 we must show that $e \in Sa$. Since it suffices to show that $e \in Ra$, we show that $\mathbf{r}_R(a) = \mathbf{r}_R(e)$. If $r \in \mathbf{r}_R(a)$ then, for all $x \in R$, $a(frxf) = arxf = 0$ so $frxf \in \mathbf{r}_S(a) = \mathbf{r}_S(e)$. Thus $erxf = 0$ for all $x \in R$ so $er = 0$ because $RfR = R$. Thus $\mathbf{r}_R(a) \subseteq \mathbf{r}_R(e)$; the other inclusion is proved in the same way. □

We do not know if the C2-rings form a Morita invariant class. Lemma 2 is "half" of the proof, and the question comes down to whether the C2-property passes to matrix rings. Our main result about C2-rings is a useful characterization of when this happens.

THEOREM 1. *If R is a ring, $M_n(R)$ is a right C2-ring if and only if R^n satisfies the C2-condition as a right R-module.*

PROOF. It is convenient to view $M_n(R)$ as isomorphic to the ring $E = \text{End}(M)$, where we write $M = R^n$.

Assume first that E is a right C2-ring. If $\alpha : P \to M$ is R-monic where P is a direct summand of M, we must show that $\alpha(P)$ is also a summand of M. Let $\pi^2 = \pi \in E$ satisfy $\pi(M) = P$, and write $ker(\pi) = Q$. Since $M = P \oplus Q$, extend α to $\bar{\alpha} \in E$ by defining $\bar{\alpha}(p+q) = \alpha(p)$. We have $ker(\bar{\alpha}) = Q = ker(\pi)$ because α is monic, and it follows easily that $\mathbf{r}_E(\bar{\alpha}) = \{\lambda \in E \mid \lambda(M) \subseteq Q\} = \mathbf{r}_E(\pi)$. Hence $\pi \in E\bar{\alpha}$ by hypothesis (and Lemma 1), say $\pi = \beta \circ \bar{\alpha}$ with $\beta \in E$. Then $\pi \circ \beta \circ \alpha = \iota_P$ where ι_P is the inclusion $P \hookrightarrow M$, and so $\theta = \alpha \circ \pi \circ \beta$ is an idempotent in E with $\theta(M) \subseteq \alpha(P)$. But $\alpha(P) \subseteq \theta(M)$ because $\theta \circ \alpha = \alpha$ so $\alpha(P) = \theta(M)$ is a summand of M, as required.

Conversely, assume that $M = R^n$ satisfies the right C2-condition, and let $\mathbf{r}_E(\alpha) = \mathbf{r}_E(\pi)$ where α and $\pi = \pi^2$ are in E. By Lemma 1 we must show that $\pi \in E\alpha$.

CLAIM. $ker(\alpha) = ker(\pi)$.

PROOF. $1 - \pi \in \mathbf{r}_E(\pi) = \mathbf{r}_E(\alpha)$, so $\alpha = \alpha \circ \pi$, whence $ker(\pi) \subseteq ker(\alpha)$. On the other hand, if $m \in ker(\alpha)$ then mR is an image of R, so there exists $\theta \in E$ such that $\theta(M) = mR$. It follows that $ker(\alpha) = \Sigma\{\theta(M) \mid \theta \in E, \theta(M) \subseteq ker(\alpha)\}$. But if $\theta(M) \subseteq ker(\alpha)$ then $\theta \in \mathbf{r}_E(\alpha) = \mathbf{r}_E(\pi)$, and it follows that $\theta(M) \subseteq ker(\pi)$. Hence $ker(\alpha) \subseteq ker(\pi)$, proving the Claim.

Now write $\pi(M) = P$ and $ker(\pi) = Q$. Then $P \cap ker(\alpha) = 0$ by the Claim so $\alpha_{|P} : P \to M$ is monic. Since M is a C2-module, let $\alpha(P) \oplus K = M$ and define $\beta \in E$ by $\beta[\alpha(p) + k] = p$. We have $(\beta \circ \alpha)(p) = p = \pi(p)$ for all $p \in P$, and if $q \in Q$ then $(\beta \circ \alpha)(q) = 0 = \pi(q)$ by the Claim. Since $M = P \oplus Q$ this shows that $\pi = \beta \circ \alpha \in E\alpha$, as required. □

It follows that the right C2-rings form a Morita invariant class if and only if $(R \oplus R)_R$ has the C2-condition whenever R is a right C2-ring. However we do not know if these statements are true.

2. Applications

If a ring R has the property that every right module can be embedded in a free right module, then R is quasi-Frobenius by a theorem of Faith and Walker [5]. A ring R is called a right *FGF-ring* if every finitely generated right R-module can be embedded in a free module, and it is an open question whether every right FGF-ring is quasi-Frobenius (the *FGF-Conjecture*). The conjecture grew out of a question of Levy [8] in 1963, and was formulated in its present form by Faith [3] in 1982. The FGF-conjecture is known to be true for the following classes of rings:

(1) Left Kasch rings.   Kato [7], 1968.
(2) Right perfect rings.   Rutter [12], 1969.
(3) Right selfinjective rings. Björk [1], 1972.
(4) Right CS-rings.   Gómez Pardo and Guil Asensio [6], 1997.

(Recall that a ring R is called a right *CS-ring* if each right ideal is essential in a direct summand of R.) Our next theorem unifies and extends (1) and (3).

Call a ring R a **strongly right C2-ring** if $M_n(R)$ is a right C2-ring for every $n \geq 1$, equivalently by Theorem 1 if R^n is a right C2-module for every $n \geq 1$. It is not difficult to check (using Lemma 2) that the strongly right C2-rings form a Morita invariant class. For convenience, call a ring R a right n**GF-ring** if every n-generated right ideal embeds in a free right module.

THEOREM 2. *Every right 2GF, strongly right C2-ring is quasi-Frobenius.*

PROOF. Let $a \in E(R_R)$, where $E(M)$ denotes the injective hull of a module M. If R is right 2GF, let $\sigma : R + aR \to R^n$ be monic. Then $\sigma(R)$ is a summand of R^n by hypothesis because $\sigma(R) \cong R$, and so $\sigma(R)$ is a summand of $\sigma(R + aR)$. But $\sigma(R) \subseteq^{ess} \sigma(R + aR)$ because $R \subseteq^{ess} R + aR$. Since σ is monic, this implies that $a \in R$, and hence that $R = E(R_R)$. Thus R is right selfinjective so, since every principal right module embeds in a free module, R is quasi-Frobenius by [1, Theorem 2.5], using results of Osofsky [13]. □

COROLLARY 1. *Let $\mathcal{C}$ denote a Morita invariant class consisting of right C2-rings. Then the only right 2GF-rings in $\mathcal{C}$ are the quasi-Frobenius rings.*

PROOF. This is immediate from Theorem 2 because every ring in $\mathcal{C}$ is a strongly right C2-ring. □

Corollary 1 implies that the FGF-conjecture is true for right selfinjective rings [7] and left Kasch rings [1], as mentioned in (1) and (3) above. In fact we get:

COROLLARY 2. *If R is a right selfinjective, right 2GF-ring then R is quasi-Frobenius.*

COROLLARY 3. *If R is a left Kasch, right 2GF-ring then R is quasi-Frobenius.*

PROOF. Since "left Kasch" is a Morita invariant, Corollary 1 applies. □

We remark in passing that the right Kasch rings are exactly those rings for which every simple right module embeds in a free module.

As another application of Corollary 1, recall that a ring R is called right *FP-injective* if, whenever K is a finitely generated submodule of a free right R-module M, every R-linear mapping $K \to R_R$ extends to M. Examples include regular rings and right selfinjective rings. FP-injectivity is a Morita invariant, and every right FP-injective ring is right C2 by Example 4 (being right P-injective), so the next result follows from Corollary 1.

COROLLARY 4. *Every right FP-injective, right 2GF-ring is quasi-Frobenius.*

It is known (see [**11**, Theorem 1]) that a ring R is right FP-injective if and only if $M_n(R)$ is right P-injective for all $n \geq 1$. Hence the right FP-injective rings are the largest Morita invariant class contained in the class of right P-injective rings. We do not know whether every right P-injective, right FGF-ring is quasi-Frobenius.

A ring R is called *semiregular* if R/J is regular and idempotents can be lifted modulo J, equivalently [**9**, Theorem 2.9] if every finitely generated right ideal T has the form $T = eR \oplus S$ where $e^2 = e$ and $S \subseteq J$ is a right ideal. It follows that $Z(R_R) \subseteq J$ in every semiregular ring where $Z(R_R)$ denotes the right singular ideal of R, and Utumi showed that every right continuous ring R is semiregular with $J = Z(R_R)$.

COROLLARY 5. *Every semiregular ring R with $J = Z(R_R)$ that is a right 2GF ring is quasi-Frobenius.*

PROOF. Semiregularity and the condition that $J = Z(R_R)$ are Morita invariants (by [**9**, Corollary 2.8] and [**16**, Lemma 1] respectively). Hence by Corollary 1 it remains to show that if R is semiregular with $J = Z(R_R)$ then R is a right C2-ring. So let T be a right ideal of R which is isomorphic to a summand. Then T is principal, so $T = eR \oplus S$ where $e^2 = e$ and $S \subseteq J = Z(R_R)$ is a right ideal. Thus S is both singular and projective (since T is projective), so $S = 0$ and $T = eR$ is a summand of R. □

The following result is known, but the simple proof is worth recording.

COROLLARY 6. *Every right and left FGF-ring is quasi-Frobenius.*

PROOF. If R is a left FGF-ring so is $M_n(R)$ because "left FGF-ring" is a Morita invariant. Hence $M_n(R)$ is left Kasch (as every left ideal is an annihilator) and so is a right C2-ring by Example 3. If R is also a right FGF-ring, it is quasi-Frobenius by Theorem 2. □

The right 1GF-rings are commonly referred to as right CF-rings (cyclic modules embed in free modules), and they are characterized as the rings R in which every right ideal T has the form $T = \mathbf{r}\{a_1, a_2, \cdots, a_k\}$ for finitely many $a_i \in R$. We note in passing that Björk [**1**] actually proves that every right selfinjective, right CF-ring is quasi-Frobenius, and Gómez Pardo and Guil Asensio [**6**] show that every right CS, right CF-ring is quasi-Frobenius. However, Theorem 2 is false if the 2GF condition is replaced by the CF condition.

EXAMPLE 6. There exists a local, left and right artinian, left and right strongly C2-ring in which every right ideal is an annihilator (and so is a right CF-ring), but which is not quasi-Frobenius.

PROOF. Let F be a field and let $a \mapsto \bar{a}$ be an isomorphism $F \to \bar{F} \subseteq F$ where the subfield $\bar{F} \neq F$. The right vector space R on basis $\{1, t\}$ is an F-algebra if we set $t^2 = 0$ and $at = t\bar{a}$ for all $a \in F$. Then R is local, $R/J \cong F$, $J^2 = 0$ and $J = tR = tF$ is the only proper right ideal of R. Hence R is right artinian and right CF (since $J = \mathbf{r}(t)$). Moreover, R is left artinian if and only if $_{\bar{F}}F$ is finite dimensional (in fact $X \leftrightarrow tX$ is a lattice isomorphism between the $\bar{F}$-subspaces X of F and the left ideals of R contained in J). Furthermore, if we take $F = \mathbb{Z}_p(x)$ to be the field of rational forms over the field $\mathbb{Z}_p$ of p elements where p is a prime, then $w \longmapsto \bar{w} = w^p$ is an isomorphism $F \to \bar{F} = \{w^p \mid w \in F\}$ and $dim(_{\bar{F}}F) = p$. Finally, R is left and right Kasch (being local with both socles nonzero), and hence is a left and right strongly C2-ring (again by Example 3 and the fact that "right Kasch" is a Morita invariant). However, R is not quasi-Frobenius because it is not left selfinjective (if x and y in F are $\bar{F}$-independent, the projection $Rtx \oplus Rty \to Rtx$ does not extend to $R \to R$). □

Note that, while Example 6 shows that the "CF-conjecture" fails—a right CF–ring need not be quasi–Frobenius—it remains an open question whether every right CF-ring is right artinian.

Since being a right FGF-ring is a Morita invariant, Theorem 2 immediately gives the following simplification of what is required to prove the FGF-conjecture.

THEOREM 3. *The FGF-conjecture is true if and only if every right FGF-ring is a right C2-ring.*

References

[1] J.-E. Björk, *Radical properties of perfect modules*, J. Reine Angew. Math. **253** (1972), 78-86.
[2] R. Camps and W. Dicks, *On semi-local rings*, Israel J. Math. **81** (1993), 203-211.
[3] C. Faith, *Embedding modules in projectives. A report on a problem*, Lect. Notes in Mathematics, **951**, Springer, New York, 1982.
[4] C. Faith and P. Menal, *A counter example to a conjecture of Johns*, Proc. Amer. Math. Soc. **116** (1992), 21-26.
[5] C. Faith and E.A. Walker, *Direct-sum representations of injective modules*, J. Algebra **5** (1967), 203-221.
[6] J.L. Gómez Pardo and P.A. Guil Asensio, *Rings with finite essential socle*, Proc. Amer. Math. Soc. **125** (1997), 971-977.
[7] T. Kato, *Torsionless modules*, Tôhoku Math. J. **20** (1968), 234-243.
[8] L.S. Levy, *Torsionfree and divisible modules over non-integral domains*, Canad. J. Math. **15** (1963), 132-151.
[9] W.K. Nicholson, *Semiregular modules and rings*, Canad. J. Math. **28** (1976), 1105-1120.
[10] W.K. Nicholson and M.F. Yousif, *Principally injective rings*, J. Algebra **174** (1995), 77-93.
[11] W.K. Nicholson and M.F. Yousif, *FP-rings*, to appear in Comm. in Algebra.
[12] E.A. Rutter, *Two characterizations of quasi-Frobenius rings*, Pacific J. Math. **30** (1969), 777-784.
[13] B. Osofsky, *A generalization of quasi-Frobenius rings*, J. Algebra **4** (1966), 373-387.
[14] Y. Utumi, *On continuous and self-injective rings*, Trans. Amer. Math. Soc. **118** (1965), 158-173.
[15] M.F. Yousif, *On continuous rings*, J. Algebra **191** (1997) 495-509.
[16] M.F. Yousif, *On large FPF-rings*, Comm. in Algebra **26** (1998), 221-224.

DEPARTMENT OF MATHEMATICS, UNIVERSITY OF CALGARY, CALGARY, CANADA T2N 1N4
E-mail address: wknichol@ucalgary.ca

DEPARTMENT OF MATHEMATICS, OHIO STATE UNIVERSITY, LIMA, OHIO, USA 45804
E-mail address: yousif.1@osu.edu

Lifting Direct Sum Decompositions of Bounded Abelian p-Groups

Barbara L. Osofsky

Dedicated to Professor Laszlo Fuchs in honor of his 75th birthday

ABSTRACT. We show the theorem: Let G be a free module over a pid and H a submodule of G such that the annihilators of elements in G/H are powers of a given prime p, where the powers are bounded. Then for any direct sum decomposition $G/H = \bigoplus_{\iota \in \mathcal{I}} K_\iota$, there is a direct sum decomposition $G = \bigoplus_{\iota \in \mathcal{I}} L_\iota$ such that the quotient map takes L_ι to K_ι.

The object of this paper is to prove the theorem:

MAIN THEOREM. *Let G be a free abelian group and H a subgroup of G such that G/H is a bounded p-group where p is a prime. Then for any direct sum decomposition $G/H = \bigoplus_{\iota \in \mathcal{I}} K_\iota$, there is a direct sum decomposition $G = \bigoplus_{\iota \in \mathcal{I}} L_\iota$ such that the quotient map takes L_ι to K_ι.*

The proof works over any principal ideal domain as indicated in the abstract, but for convenience it is done in $\mathbb{Z}$.

The proof consists of a reduction to the countable case, and then an algorithm which will actually produce the desired lifting.

1. Introductory remarks

The Hill-Megibben generalization of Cohen and Gluck's stacked basis theorem (see [1] and [2]) answers the following question:

> Let $0 \to H \to G \to A \to 0$ and $0 \to H' \to G' \to A \longrightarrow 0$ be exact sequences of abelian groups, with G and G' free. When are these resolutions equivalent in the sense that there is an isomorphism Φ from G to G' taking H onto H'?

Their definitive result gives a family of numerical invariants which determine when this occurs, but says nothing about the automorphism α of A induced by an

---

1991 *Mathematics Subject Classification.* Primary 20C38, 15A15; Secondary 05A15, 15A18.
Key words and phrases. p-groups, direct sum decompositions.

© 2001 American Mathematical Society

equivalence Φ of the two presentations of A in the commutative diagram

$$\begin{array}{ccccccccc} 0 & \longrightarrow & H & \longrightarrow & G & \stackrel{\nu}{\longrightarrow} & A & \longrightarrow & 0 \\ & & \downarrow \Phi|_H & & \downarrow \Phi & & \downarrow \alpha & & \\ 0 & \longrightarrow & H' & \longrightarrow & G' & \stackrel{\nu'}{\longrightarrow} & A & \longrightarrow & 0. \end{array}$$

That is not at all surprising. There is almost nothing that can be said about α for a very good reason. Multiplication by an integer > 1 may well induce an automorphism of A, but that automorphism usually is not induced by an isomorphism from G to G'. There is no hope in general of replacing Φ by a different isomorphism so that the resulting α becomes the identity, even in the simplest possible case where G and G' are both isomorphic to $\mathbb{Z}$.

In the special case that A is a bounded abelian p-group for some prime p, our result says that this is essentially the only obstacle to modifying the isomorphism Φ so that α becomes the identity.

NOTATION. Let G be an infinitely generated free abelian group, and H a subgroup of G such that G/H is annihilated by p^ε. We use an overline to denote the natural map from G to $A = G/H$.

Note that every module over $\mathbb{Z}/p^\varepsilon\mathbb{Z}$ is a direct sum of cyclic modules.

DEFINITION. We will use the word basis of a nonzero group to denote an independent generating set consisting of nonzero elements, as in 'the *basis* theorem for finitely generated modules over a pid'. There is no implication that the group involved is free over any ring, just that it is a direct sum of (nonzero) cyclics, and we have taken a a generator of each of the summands.

Let $\overline{\mathfrak{B}}$ be any basis of the quotient group $A = G/H$ as a module over $\mathbb{Z}/p^\varepsilon\mathbb{Z}$. Let $\tilde{G}$ denote the free subgroup of G generated by any lifting $\mathfrak{B}$ of $\overline{\mathfrak{B}}$ to G. A subgroup K of G maximal in the set of those subgroups which have zero intersection with $\tilde{G}$ will be a direct summand of G contained in H. Without loss of generality we can replace G by a direct sum complement of K containing $\tilde{G}$. That is, we may (and will) assume that $H \subseteq pG$.

Now assume we have another presentation $0 \longrightarrow H' \longrightarrow G' \longrightarrow A \longrightarrow 0$ equivalent to $0 \longrightarrow H \longrightarrow G \longrightarrow A \longrightarrow 0$. A basis $\overline{\mathfrak{B}}$ of A induces a direct sum decomposition $A = \bigoplus_{\overline{b} \in \overline{\mathfrak{B}}} \overline{b}\mathbb{Z}$. By our theorem, this decomposition lifts to direct sum decompositions $G = \bigoplus_{\overline{b} \in \overline{\mathfrak{B}}} b\mathbb{Z}$ and $G' = \bigoplus_{\overline{b} \in \overline{\mathfrak{B}}} b'\mathbb{Z}$.

We then have an equivalence

$$\begin{array}{ccccccccc} 0 & \longrightarrow & H & \longrightarrow & G & \longrightarrow & A & \longrightarrow & 0 \\ & & \downarrow \Phi|_H & & \downarrow \Phi & & \downarrow \alpha & & \\ 0 & \longrightarrow & H' & \longrightarrow & G' & \longrightarrow & A & \longrightarrow & 0 \end{array}$$

where Φ is induced by mapping a b in the basis of G to the b' in the basis of G' used in lifting the same $\overline{b}\mathbb{Z} \in \overline{B}$. The corresponding map α takes the image of each $\overline{b}$ in the basis for $\overline{G}$ to a nonzero multiple of itself where the multiplier is a unit modulo the appropriate power of p.

Admittedly this is a long way from the torsion–free case which occupies much of the Hill-Megibben paper, but it is of interest that one can get any result whatsoever. Moreover, the theorem has as a consequence the solution of a thirty year old question on projective dimension of ideals in commutative von Neumann regular rings (see [**4**]). The algorithm which is the major part of the proof we give here is somewhat different than the proof given in [**4**].

Understanding the notation is so important for reading the proofs of the lemmas in Section 2 that we repeat a portion of it.

NOTATION. We let G denote a nonzero free abelian group, and let H and p be as in the statement of the Main Theorem with $H \subseteq pG$. We use the notation $\overline{G}$ to denote the quotient group G/H. We will notationally go back and forth between G and $\overline{G}$ by using an overline for the natural map and removing it to denote a preimage of an element of $\overline{G}$.

We note that any subgroup of $\overline{G}$ has a basis (not necessarily free) over $\mathbb{Z}/p^\varepsilon\mathbb{Z}$, so if we can lift the cyclic subgroups corresponding to a basis, we can lift any direct sum decomposition.

2. Reduction to the countable case

The reduction to the countable case is a standard argument based on Kaplansky [**3**]. This version of it is taken directly from [**4**], where the proofs can be found. The lemmas and corollary are stated so that a reader familiar with the Kaplansky argument might be able to supply his or her own proof.

LEMMA 2.1. *Let G be a nonzero free abelian group with free basis $\mathfrak{X}$, and let $\mathfrak{B} = \{b_\alpha : \alpha \in \mathfrak{J}\}$ be a basis of $\overline{G}$. Let $\mathfrak{c}$ be any countable subset of $\mathfrak{B}$. Then there exists a nonzero countably generated direct summand H of G such that*

$$\overline{H} = \sum_{i=0}^{\infty} \overline{b_{\alpha_i}} \mathbb{Z}$$

for $\{\overline{b_{\alpha_i}} : i \in \omega\}$ some countable subset of $\mathfrak{B}$ containing $\mathfrak{c}$. Moreover, H itself is generated by a countable subset of $\mathfrak{X}$. □

LEMMA 2.2. *Let G be a nonzero free abelian group, and let $\mathfrak{B}$ be a basis of $\overline{G}$. Then G is the union of a well-ordered (by inclusion) family $\{H_\mu : \mu < \Omega\}$ of subgroups such that: H_μ and $\bigcup_{\kappa < \mu} H_\kappa$ are direct summands of G for every μ in the ordinal Ω; for each μ, $H_\mu / \bigcup_{\kappa < \mu} H_\kappa$ is countable; and each $\overline{H_\mu}$ is generated by some subset of the $\{\overline{b_\alpha} : \alpha \in \mathfrak{J}\}$.* □

COROLLARY 2.3. *Assume that, for any countably generated free abelian group G with $\mathfrak{B}$ a basis for $\overline{G}$, there is a direct decomposition lifting of*

$$\overline{G} = \bigoplus_{\alpha \in \mathfrak{J}} \overline{b_\alpha} \mathbb{Z}$$

to the direct decomposition

$$G = \bigoplus_{\alpha \in \mathfrak{J}} y_\alpha \mathbb{Z}.$$

Then the Main Theorem is true for any free abelian group G. □

3. The proof of the Main Theorem

In the previous section we reduced the problem of proving the Main Theorem to the problem of showing it in the countable case. The algorithm used is relatively straightforward in the finite case, so we will write the proof assuming the basis is infinite. So now G will have a countable basis $\{x_i : i \in \omega\}$, and we have a lifting $\mathfrak{B}$ of the countable basis of $\overline{G}$. Each $b_i \in \mathfrak{B}$ can be represented as a row vector indexed by ω with nonzero entries corresponding to the coefficient of x_i in the representation of b_i as a sum of the x_j. We put all of these row vectors together in an $\omega \times \omega$ row finite matrix $\mathbf{A}$ with entries in $\mathbb{Z}$, and identify the row space of this matrix with G. For each basis element x_j, there is a power of p, say p^{ε_j}, which annihilates it in $\overline{G}$. A lifting of the decomposition induced by the basis $\overline{\mathfrak{B}}$ over $\sum_{\iota \in \omega} p^{\varepsilon_j} x_j \mathbb{Z}$ will also lift it over H, so we will assume that $H = \sum_{\iota \in \omega} p^{\varepsilon_j} x_j \mathbb{Z}$. We do elementary row operations on this matrix to try and reduce $\mathbf{A}$ modulo H to a row permutation of the identity matrix. Simultaneously we produce a square matrix $\mathbf{C}$ over $\mathbb{Z}$ of determinant 1 whose size increases with every new pivot row. $\mathbf{C}$ agrees with an appropriate submatrix of $\mathbf{A}$ modulo H, and will eventually hold our lifting.

The reader is assumed thoroughly familiar with the Gaussian and Gauss-Jordan elimination algorithms and elementary row operations as found in a linear algebra course. Since $\overline{\mathfrak{B}}$ is a basis modulo H, the matrix $\overline{\mathbf{A}}$ can be row reduced to a matrix which is a permutation matrix over $\mathbb{Z}/p^\varepsilon \mathbb{Z}$, that is, a matrix representing a re-ordering of the basis $\overline{\mathfrak{B}}$. For technical reasons, we look for a pivot in each successive row rather than in each successive column as is usually done in elementary or numerical linear algebra. There are three obstacles to the same Gaussian elimination that works modulo H working in G itself.

- The only integers which are units are ± 1. In $\overline{G}$ any integer relatively prime to p is a unit. We get around this problem by lifting subgroups, not elements. That is, when we find a pivot in row i which is relatively prime to p, say in column $j(i)$, we multiply row i by the inverse of the pivot modulo $p^{\varepsilon_{J(i)}}$ and then treat the pivot in the Gaussian elimination as though it is 1 in $\mathbb{Z}$ so that the candidate matrix $\mathbf{C}$ is what would be obtained if it really were 1.
- The row reduction that makes $\mathbf{A}$ congruent to a permutation of the identity modulo H may produce entries off the diagonal which are nonzero integers divisible by the appropriate power of p. To overcome this problem, we treat such entries as zero in $\mathbb{Z}$ when they occur in a pivot row or pivot column of the reduced matrix so that the candidate matrix $\mathbf{C}$ is what would be obtained if they really were 0.
- The multiples of the p^{ε_i} which are ignored in the row reduction may cause entries in $\mathbf{C}$ which are zero modulo p^{ε_i} not to be zero in $\mathbb{Z}$. This can lead to a row of $\mathbf{C}$ not being row finite when the algorithm has run to infinity, and hence not corresponding to an element of G. To get around this problem, we first observe that if a row of $\mathbf{C}$ has an infinite number of nonzero entries, the powers of p which occur in these nonzero entries congruent to zero modulo p must be unbounded. This is so because the quotient of G modulo the row space of $\mathbf{A}$ is p-divisible and an infinite number of entries in a row of $\mathbf{C}$ when the algorithm has run its course means that row is not zero modulo the row space of $\mathbf{A}$.

After Gaussian elimination has been used to clear the current pivot column $j(i)$ above the pivot row, beginning at the top row, look down column i of $\mathbf{C}$ to see if there is a first row, say row k, that satisfies the following: it has a nonzero entry $c_{k,i}$ congruent to 0 mod $p^{\varepsilon_{j(i)}}$ in column $j(i)$; $S = \{c_{k,h} : c_{h',h} = 0 \text{ if } 1 \leq h' < h\} \neq \emptyset$; and the power of $p^{\varepsilon_{j(i)}}$ dividing $c_{k,i}$ is larger than the maximum power of $p^{\varepsilon_{j(i)}}$ dividing all of the entries in S. If so, proceed as follows. Let $c_{k,j} = p^{\varepsilon_{j(i)}} n$ for some $n \neq 0$, and let d denote the greatest common divisor of the elements of S. We have $d \mid n$ and $p^{\varepsilon_{j(i)}} \mid \frac{c_{k,h}}{d}$. There exist integers $\{m_h\}$ with $\sum_S c_{k,h} m_h = d$. For each h with $c_{k,h} \in S$, subtract $\frac{c_{k,j} m_h}{d}$ times column h of $\mathbf{C}$ from column j of $\mathbf{C}$. The nonzero entry congruent to 0 mod $p^{\varepsilon_{j(i)}}$ in the k,j position of $\mathbf{C}$ becomes 0 and all entries above it stay 0. The top row of $\mathbf{C}$ will have nonzero multiples of the appropriate p^{ε_j} made 0 and so be 0 from some point on, and once that happens, from the point of view of this correction, the next row behaves like the top row, so it too will cease to have these 'false zeros' from some point on.

Here is a step by step description of the algorithm which will lift a countable direct sum of cyclic p-groups to a basis of G. It can be adapted to also compute the inverse of the matrix $\mathbf{C}$ being computed by appending the identity to $\mathbf{R}$.

ALGORITHM (Infinite Gaussian Elimination modulo p). The input to this algorithm is a $\omega \times \omega$ (or $n \times m$) matrix $\overline{\mathbf{A}}$ with integer entries considered as elements of $\overline{G}$, a prime p, and an $1 \times \omega$ (or $1 \times m$) vector of powers of p. The output of a successful completion is a triple of matrices $\mathbf{R}$, $\mathbf{C}$, and $\mathbf{U}$ with integer entries. $\mathbf{R}$ has the same size as $\overline{\mathbf{A}}$, $\mathbf{C}$ and $\mathbf{U}$ are square with the same number of rows as $\overline{\mathbf{A}}$, and $\mathbf{U}$ is diagonal with integers relatively prime to p on the diagonal. Each row of $\mathbf{R}$ has a pivot column with 1, and that is the only entry in the row not congruent to 0 mod the appropriate power of p. Each row of $\mathbf{C}$, when the permutation caused by pivoting off the diagonal is undone, spans exactly the same subgroup modulo H as the corresponding portion of the same row of $\overline{\mathbf{A}}$; indeed each row of $\mathbf{C}$ is congruent to the corresponding part of the same row of $\mathbf{U}\overline{\mathbf{A}}$ modulo p. We also use some temporary locations for various operations.

STEP 1: **Initialize.** Set up the matrices $\mathbf{C} = \mathbf{I}$ initially, $\mathbf{R}$, $\mathbf{U}$. Set up a row vector J to hold pivot columns. Set your row index I to 1. Read in the first row of $\overline{\mathbf{A}}$ to the first row of $\mathbf{R}$. Set up a temporary vector to check for the determinant, and read the first row of $\overline{\mathbf{A}}$ into it.

STEP 2: **Set up to search for a pivot along row I.** For K going from 1 to I − 1, if $\mathbf{R}_{I,J(K)}$ is not zero modulo p, then subtract $\mathbf{R}_{I,J(K)}$ times row K of $\mathbf{R}$ from the temporary vector. (This step is not done when I = 1.)

STEP 3: **Search for the pivot.** Search the temporary vector for the first entry which is a unit modulo p. If no unit is found then STOP. The rows of $\overline{\mathbf{A}}$ are not linearly independent modulo p. Otherwise, let the first entry which is a unit modulo p be in column J(I), and call column J(I) the Ith pivot column.

STEP 4: **Make the pivot 1 modulo $p^{\varepsilon_{J(I)}}$.** Set $\mathbf{U}_{I,I}$ equal to an integer u_i such that $u_i \mathbf{R}_{I,J(I)} \equiv 1 \mod p^{\varepsilon_{J(I)}}$. Multiply row I of $\overline{\mathbf{A}}$ by u_i. Insert the result as row I in $\mathbf{R}$.

STEP 5: **Clear below the diagonal (forward pass).** For K going from 0 to I − 1, if $\mathbf{R}_{I,J(K)}$ is not zero in $\mathbb{Z}/p^{\varepsilon_{J(K)}}\mathbb{Z}$ then subtract $\mathbf{R}_{I,J(K)}$ times row K of

R from row I of **R** and add $\mathbf{R}_{I,J(K)}$ times column I of **C** to column K of **C**. (This forward pass uses our new row I which has been adjusted to make the pivot equal 1 in $\mathbb{Z}/p^{\varepsilon J(I)}\mathbb{Z}$. Every elementary row operation on **R** is accompanied by the inverse column operation on **C**. The pivot is treated as though it is 1 in $\mathbb{Z}$, and an entry in row I, column J of **R** which is 0 modulo $p^{\varepsilon J}$ is treated as though it is 0 in $\mathbb{Z}$.)

STEP 6: **Clear above the diagonal (back substitution).** For K going from 0 to $I-1$, if $\mathbf{R}_{K,J(I)}$ is not zero in $\mathbb{Z}/p^{\varepsilon J(I)}\mathbb{Z}$ then subtract $\mathbf{R}_{K,J(I)}$ times row I of **R** from row K of **R** to clear every entry in column J(I) above the Ith row, and add $\mathbf{R}_{K,J(I)}$ times column K to column I of **C**. This back substitution treats entries of **R** in column J(I) which are 0 modulo $p^{\varepsilon J(I)}$ as though they are 0 in $\mathbb{Z}$.

STEP 7: **Starting from the top and working down, prevent unwanted multiples of powers of p from propagating an infinite number of times in a row.** For K starting at 0 and going to (at most) $I-1$, test if the first nonzero entry, if it exists, is $\mathbf{C}_{K,I}$ where $\mathbf{C}_{K,I} \equiv 0$ modulo $p^{\varepsilon J(I)}$. If so, test if the highest power of $p^{\varepsilon J(I)}$ dividing $\mathbf{C}_{K,I}$ is greater than the highest power of $p^{\varepsilon J(I)}$ dividing the greatest common divisor

$$d = \gcd\{\mathbf{C}_{K,L} : \mathbf{C}_{H,L} = 0 \text{ for } 1 \leq H < K\}.$$

If this test also has a positive answer, find elements $\{m_L\}$ such that $\sum \mathbf{C}_{K,L} m_L = d$. Note that $\mathbf{C}_{K,I}/d$ is necessarily divisible by $p^{\varepsilon J(I)}$ by our second test. For each L in our set whose gcd is d, subtract $m_L \cdot \mathbf{C}_{K,I}/d$ times column L of **C** from column I of **C**. Add the corresponding rows of **R** to row I of **R**.

STEP 8: If you have not looked at all the rows of $\overline{\mathbf{A}}$, increment I by 1, read row I of $\overline{\mathbf{A}}$ into the temporary vector, and GOTO STEP 2.

If you have looked at all of the rows, it is time to unpermute the columns of **C** if necessary. Pad **C** with extra symbolic columns corresponding to pivot columns with index not a row index, and extra rows to make these columns look like pivot columns for those rows. Then undo the permutation $I \longmapsto J(I)$ of the columns of **C**. Then RETURN **C**, **R**, and **U**.

END

It is clear from the algorithm and elementary linear algebra that **C** is a matrix of determinant 1 and so invertible over $\mathbb{Z}$, and its inverse (even if not computed) is the inverse of an appropriate submatrix of **A** modulo H and so **C** is congruent to the inverse of that inverse, namely the submatrix of **A**, modulo H. The algorithm itself does not use the bounded hypothesis. However, if we wish a lifting of the cyclic summands of $\overline{G}$, every row of the computed **C** must eventually contain zeros from some point on. In row i, once every nonzero column is a pivot column, the only way the row can change is by adding nonzero multiples of p^{ε_j} in column j. Assume that the entries in all rows above row i are zero beyond some column. By p-divisibility of $\overline{G}$, somewhere beyond that column there will be an entry in row i divisible by an arbitrarily high power of p and hence by an arbitrarily high power of the exponent of $\overline{G}$. The greatest common divisor of the earlier zeros mod H in row i will not change, so STEP 7 will make any entry in row i divisible by that $\gcd \times p^{\varepsilon}$ equal to 0, and row i will become 0 from some point on. This is where the bounded hypothesis is used. It may not be necessary to show that, if we retain the property that $\overline{G}$ is a direct sum of cyclics, decompositions lift, but I cannot yet show that. I would conjecture that the theorem is true even without the bounded hypothesis.

To observe the output of this algorithm, we will look at two examples of its application.

EXAMPLE. Here we have $\mathbf{A} = \begin{bmatrix} 5 & 2 & -1 \\ 3 & -1 & 2 \\ 2 & 3 & -1 \end{bmatrix}$, $p = 5$, and all ε_i are 1.

A Maple V implementation of the algorithm in this paper, which may be found at URL

http://www.math.rutgers.edu/pub/~osofsky/getbasisfin.html

computes:

$$\mathbf{C} = \begin{bmatrix} 0 & 1 & 2 \\ 1 & -2 & -6 \\ 1 & -1 & -3 \end{bmatrix}$$

$$\mathbf{U} = \begin{bmatrix} -2 & 0 & 0 \\ 0 & 2 & 0 \\ 0 & 0 & -2 \end{bmatrix}$$

$$\mathbf{UA} - \mathbf{C} = \begin{bmatrix} -10 & -5 & 0 \\ 5 & 0 & 10 \\ -5 & -5 & 5 \end{bmatrix}$$

$$\mathbf{C}^{-1} = \begin{bmatrix} 0 & -1 & 2 \\ 3 & 2 & -2 \\ -1 & -1 & 1 \end{bmatrix}$$

Since the implementation used in reducing $\mathbf{C}$ to the identity keeps track of its inverse, we get this computation of $\mathbf{C}^{-1}$ with no additional effort.

EXAMPLE. Here $\mathbf{A} = \begin{bmatrix} 1 & 1 & 1 & 1 & 0 & 0 & 0 & 0 & 0 & 0 \\ 0 & 1 & 1 & 1 & 0 & 0 & 0 & 0 & 0 & 0 \\ 0 & 0 & 1 & 0 & 1 & 1 & 1 & 0 & 0 & 0 \\ 0 & 0 & 0 & 1 & 1 & 1 & 1 & 0 & 0 & 0 \\ 0 & 0 & 0 & 0 & 1 & 1 & 1 & 0 & 0 & 0 \\ 0 & 0 & 0 & 0 & 0 & 1 & 0 & 1 & 1 & 1 \\ 0 & 0 & 0 & 0 & 0 & 0 & 1 & 1 & 1 & 1 \end{bmatrix}$,

$p = 2$, and all ε_i are 1. From the form of $\mathbf{A}$, we see that $\mathbf{U}$ will be the identity

and $\mathbf{A}$ as it stands has pivots on the diagonal. If the pattern continues to an $\omega \times \omega$ version of $\mathbf{A}$, that infinite version cannot be row reduced to the identity over $\mathbb{Z}$. The quotient group of the free group modulo the group generated by the rows of $\mathbf{A}$ is $\mathbb{Z}_{2^\infty}$. Now apply the algorithm.

Up to row 4, **C** looks like $\begin{bmatrix} 1 & 1 & 1 & 1 \\ 0 & 1 & 1 & 1 \\ 0 & 0 & 1 & 0 \\ 0 & 0 & 0 & 1 \end{bmatrix}$ but then at row 5 **C** temporarily looks like

$$\begin{bmatrix} 1 & 1 & 1 & 1 & 2 \\ 0 & 1 & 1 & 1 & 2 \\ 0 & 0 & 1 & 0 & 1 \\ 0 & 0 & 0 & 1 & 1 \\ 0 & 0 & 0 & 0 & 1 \end{bmatrix}$$

and STEP 7 modifies this by subtracting 2 times column 4 from the 5th column of **C** to get the corrected matrix

$$\mathbf{C} = \begin{bmatrix} 1 & 1 & 1 & 1 & 0 \\ 0 & 1 & 1 & 1 & 0 \\ 0 & 0 & 1 & 0 & 1 \\ 0 & 0 & 0 & 1 & -1 \\ 0 & 0 & 0 & 0 & 1 \end{bmatrix}.$$

Our final

$$\mathbf{C} = \begin{bmatrix} 1 & 1 & 1 & 1 & 0 & 0 & 0 \\ 0 & 1 & 1 & 1 & 0 & 0 & 0 \\ 0 & 0 & 1 & 0 & 1 & 1 & 1 \\ 0 & 0 & 0 & 1 & -1 & -1 & -1 \\ 0 & 0 & 0 & 0 & 1 & 1 & 1 \\ 0 & 0 & 0 & 0 & 0 & 1 & 0 \\ 0 & 0 & 0 & 0 & 0 & 0 & 1 \end{bmatrix},$$

and **C** will remain row finite if the pattern of **A** is continued to infinity because in this case we have eliminated all nonzero even numbers from **C**. Without the correction of STEP 7, **C** would have a first row of $\begin{bmatrix} 1 & 1 & 1 & 1 & 2 & 2 & 2 \end{bmatrix}$ and the next entry would be a 4. The powers of 2 increase every 3 columns.

There is an example which shows that rows may become zero before rows lying above them, and that the power of p dividing a nonzero entry of a row may increase without STEP 7 zeroing out the entry. It may be found in the Maple V worksheet implementing the algorithm mentioned above. There are links to this algorithm and the related algorithm used in [4] at URL

http://www.math.rutgers.edu/pub/~osofsky/index.html

References

[1] Joel M. Cohen and Herman Gluck, Stacked bases for modules over principal ideal domains, *J. Algebra* **14**(1970), pp 493–505.
[2] Paul Hill and Charles Megibben, Generalizations of the stacked bases theorem, *Trans. Amer. Math. Soc.* **312**(1989), pp 377–402.
[3] I. Kaplansky, *Projective modules*, Ann. of Math (2) **68** (1958), 372–377.
[4] Barbara L. Osofsky, Projective dimension is a lattice invariant, *J. Pure and Applied Alg.*, to appear.

DEPARTMENT OF MATHEMATICS, RUTGERS, THE STATE UNIVERSITY OF NEW JERSEY, 110 FRELINGHUYSEN ROAD, PISCATAWAY, NJ 08854-8019
E-mail address: osofsky@math.rutgers.edu

On Modules and Submodules with Finite Projective Dimension

K.M. Rangaswamy

Dedicated to Professor Laszlo Fuchs in honor of his 75th birthday

ABSTRACT. Using the idea of a relative projective resolution (which is an adaptation of a corresponding concept in the theory of infinite rank Butler groups), it is shown that a module M with finite projective dimension n over a ring R with some coherency condition has a $G(\aleph_n)$-family of tight submodules. This generalizes a theorem of Kaplansky on projective modules. Conditions are given under which a submodule has the same projective dimension as the original module M.

1. Introduction

This note indicates how some of the recent ideas and techniques from the theory of infinite rank Butler groups can be used to consider questions about modules with finite projective dimension over general rings. A well-known theorem of Kaplansky states that every projective module over a ring R is the direct sum of countably generated (projective) modules. Since projective modules have projective dimension 0, it is natural to inquire whether Kaplansky's theorem can be generalized to modules M with arbitrary projective dimension n (in symbols, $pd(M) = n$), where n is an integer ≥ 0. We begin with a theorem that gives an interesting condition for a submodule S of a projective module P over an integral domain to be projective: There must be a smooth chain of *projective* submodules $S = S_0 < S_1 < \ldots < S_\alpha < S_{\alpha+1} < \ldots < S_\tau = P$ where, for each $\alpha < \tau$, $S_{\alpha+1}/S_\alpha$ is countably generated. To obtain a generalization of Kaplansky's theorem, we require some coherency condition on the ring R. Specifically, let R be an $\aleph_k$-noetherian ring where, to avoid trivial cases, we may assume that $\aleph_k \leq |R| \cdot \aleph_0$. Then it is shown that an R-module M with projective dimension n admits a $G(\aleph_k)$-family $\Im$ of submodules which are *tight* in the sense for each $A \in \Im$, both A and M/A have

---

1991 *Mathematics Subject Classification.* Primary 16D40, 16E10, 13C05; Secondary 13D05, 18G05, 18G20.

Key words and phrases. Projective dimension, relative projective resolution, $G(\aleph_k)$-family of submodules.

projective dimension $\leq n$. Observe that, in this case, M is the union of a smooth ascending chain of submodules $0 = M_0 < \ldots < M_\alpha < M_{\alpha+1} < \ldots < M_\gamma = M$ where γ is some ordinal and, for each $\alpha < \gamma$, $M_\alpha \in \Im$ (with $pd(M_\alpha) \leq n$) and $M_{\alpha+1}/M_\alpha$ is $\aleph_k$-generated with projective dimension $\leq n$. This provides a stronger converse of the classical theorem of Auslander on projective dimension of modules. Moreover, for $n = 0$, the above result becomes Kaplansky's theorem on projective modules. As an application, we provide a criterion for a submodule of a module to have the same projective dimension as the module. Our approach also yields similar results on pure-projective modules. The key tools are the concept of a $G(\aleph_n)$-family introduced by P.Hill [1] and the idea of a relative projective resolution which is an adaptation of a corresponding concept used by Bican and Fuchs ([2], [3]) in the study of infinite rank Butller groups.

This work initially started as a joint research project with Professor L. Fuchs. I wish to thank him for many useful and helpful conversations and for suggesting to submit the paper in my name to this proceedings commemorating his 75th birthday.

2. Preliminaries

All the modules that we consider are assumed to be left modules over an associative ring R with identity. For the general notation and terminology, we refer the reader to [3], [4] and [5]. Let n be an integer ≥ 0 and let A be a submodule of an R-module B. Following Hill [1], we define a $G(\aleph_n)$-*family of submodules of B over A* to be a family $\mathcal{C}$ of submodules of B containing A and satisfying the following conditions: (a) $A, B \in \mathcal{C}$, (b) $\mathcal{C}$ is closed with respect to taking the unions of ascending chains, and (c) For any $C \in \mathcal{C}$ and any subset X of B having cardinality not exceeding $\aleph_n$, there exists a $D \in \mathcal{C}$ such that $(C \cup X) \subset D$ and D/C is $\aleph_n$-generated, that is D/C has a generating set of cardinality not exceeding $\aleph_n$. In the above definition, if $A = 0$, then the corresponding family $\mathcal{C}$ is called a $G(\aleph_n)$-*family of B*. Note that a $G(\aleph_n)$-family is also a $G(\aleph_m)$-family for any $m \geq n$. A ring R is said to be $\aleph_n$-*noetherian* if every ideal of R is $\aleph_n$-generated. Given a submodule A of an R-module B, we define *a relative projective resolution of B with respect to A* to be an exact sequence of the form

$$0 \to K \to A \oplus P \xrightarrow{\phi} B \to 0$$

where P is a projective R-module and ϕ, when restricted to A, acts as the inclusion map. Such relative projective resolutions always exist. For example, if P is a projective R-module for which there is an epimorphism $\gamma : P \to B$, then $\phi = 1_A \oplus \gamma$ is the needed map from $A \oplus P$ to B. If

$$0 \to K' \to A \oplus P' \xrightarrow{\phi'} B \to 0$$

is another relative projective resolution of B with respect to A with P' projective and ϕ' acting as identity on A, then the classical Schanuel's Lemma argument, together with the fact that both ϕ, ϕ' act as identity on A, yields that $K \oplus P' \cong K' \oplus P$. For the general definition and results on projective dimension, we refer to [4] and [5]. Let S be a submodule of an R-module M. S is said to be a *tight submodule* if both $pd(S) \leq pd(M)$ and $pd(M/S) \leq pd(M)$. S is said to be an *RD-submodule* of M if, for all $r \in R$, $S \cap rM = rS$. An ascending chain of submodules

$S_0 < S_1 < ... < S_\alpha < ...$ indexed by ordinals α less than some fixed infinite ordinal τ is said to be *smooth* if for every limit ordinal $\gamma \leq \tau$, $S_\gamma = \cup_{\alpha < \gamma} S_\alpha$.

3. The Results

Suppose $0 \to K \to A \oplus P \to B \to 0$ is a relative projective resolution of an R-module B with respect to a submodule A. Often module-theoretic properties of K seem to influence the way A embeds in B (see [2], [3], [6]). With this in mind, we wish to investigate whether the existence of a $G(\aleph_n)$-family of special submodules in K implies any embedding relations between A and B. Our first theorem examines this.

Theorem 1. *Let $0 \to K \to A \oplus P \xrightarrow{\phi} B \to 0$ be a relative projective resolution of an R-module B with respect to a submodule A where P is projective and ϕ acts as the identity map on A. Let n be an arbitrary integer ≥ 0.*

(i) Let R be an integral domain. If $\mathcal{K}$ is a $G(\aleph_n)$-family of RD-submodules of K and $\mathcal{P}$ is a $G(\aleph_n)$-family of submodules of P, then the family

$$\mathcal{B} = \{((A \oplus Y) + K)/K \mid Y \in \mathcal{P} \text{ and } (A \oplus Y) \cap K \in \mathcal{K}\}$$

is a $G(\aleph_n)$-family of submodules of B over A.

(ii) If R is $\aleph_n$-noetherian, then the statement (i) holds with $\mathcal{K}$ a $G(\aleph_n)$-family of arbitary submodules of K.

Proof. (i) For convenience of notation, consider K as a submodule of $A \oplus P$ with $K \cap A = 0$ and ϕ the natural quotient map with the property that $\phi \mid A = 1_A$. Let $\pi: A \oplus P \to P$ be the cordinate projection with $\ker \pi = A$. We first show that the family $\mathcal{C} = \{Y \mid Y \in \mathcal{P}, (A \oplus Y) \cap K \in \mathcal{K}\}$ is a $G(\aleph_n)$-family of P. We shall only verify the condition (c) of the definition. Let $Y \in \mathcal{C}$ so that $(A \oplus Y) \cap K = K^{(0)} \in \mathcal{K}$ and let X be a subset of P of cardinality at most $\aleph_n$. Then there exists $Y^{(1)} \in \mathcal{P}$ containing $Y \cup X$ such that $Y^{(1)}/Y$ is $\aleph_n$-generated. Now $(A \oplus Y^{(1)}) \cap K = K_1$ has rank $\leq \aleph_n$ over $K^{(0)}$. Select a $K^{(1)} \in \mathcal{K}$ containing $K^{(0)}$ and a maximal independent subset of K_1 modulo $K^{(0)}$ such that $K^{(1)}/K^{(0)}$ is $\aleph_n$-generated. Note that $K^{(1)} \supset K_1$ since K is torsion-free and $K^{(1)}$ is an RD-submodule. Now there exists $Y^{(2)} \in \mathcal{P}$ such that $Y^{(2)} \supset Y^{(1)} + \pi(K^{(1)})$ and $Y^{(2)}/Y^{(1)}$ is $\aleph_n$-generated. Proceeding like this, we obtain ascending chains $Y^{(1)} \subset ... \subset Y^{(n)} \subset ....$, and $K^{(1)} \subset ... \subset K^{(n)} \subset ....$ such that, for each $i < \omega$, $(A \oplus Y^{(i)}) \cap K \subset K^{(i)} \subset (A \oplus Y^{(i+1)}) \cap K$. Then $K^* = \cup_{i < \omega} K^{(i)} \in \mathcal{K}$, $Y^* = \cup_{i < \omega} Y^{(i)} \in \mathcal{P}$ and $(A \oplus Y^*) \cap K = K^*$. Hence $Y^* \in \mathcal{C}$. Moreover, Y^* contains $Y \cup X$ and Y^*/Y is $\aleph_n$-generated. Thus $\mathcal{C}$ is a $G(\aleph_n)$-family. Then it is readily seen that $\mathcal{B} = \{((A \oplus Y) + K)/K \mid Y \in \mathcal{C}\}$ is a $G(\aleph_n)$-family of submodules of B over A. Also $\mathcal{K}' = \{V \in \mathcal{K} \mid V = (A \oplus Y) \cap K \text{ for some } Y \in \mathcal{C}\}$ is a $G(\aleph_n)$-subfamily of $\mathcal{K}$. We may then view $\mathcal{B}$ as being "induced" by $\mathcal{P}$ and $\mathcal{K}'$ in the sense that each $L \in \mathcal{B}$ fits into an exact sequence $0 \to X \to A \oplus Y \to L \to 0$ where $Y \in \mathcal{P}$ and $X \in \mathcal{K}'$.

(ii) Let R be $\aleph_n$-noetherian. In the proof of (i), we now have $\mathcal{K}$ as a $G(\aleph_n)$-family of (not necessarily RD) submodules of K and repeat the same proof. Continuing the notation of (i), observe that $K_1/K^{(0)} = ((A \oplus Y^{(1)}) \cap K)/((A \oplus Y) \cap K)$ is now $\aleph_n$-generated, since $K_1/K^{(0)} \cong \pi(K_1)/\pi(K^{(0)}) = \pi(K_1)/(\pi(K_1) \cap Y) \cong (\pi(K_1) + Y)/Y \subset Y^{(1)}/Y$, an $\aleph_n$-generated R-module, noting that π is monic on K and R is $\aleph_n$-noetherian. Then select $K^{(1)} \in \mathcal{K}$ containing K_1 so that $K^{(1)}/K^{(0)}$

is $\aleph_n$-generated and proceed as in (i) to reach the desired $G(\aleph_n)$-family $\mathcal{B}$ of B over A.

Theorem 1 has a number of applications. The first one considers when a submodule of a projective module will again be projective.

Theorem 2. *Let R be an integral domain or an $\aleph_0$-noetherian ring and let B be a projective R-module. For a submodule A of B the following are equivalent:*

(i) A is projective.

(ii) There is a $G(\aleph_0)$-family of projective submodules of B over A.

(iii) There is a smooth well-ordered ascending chain of projective submodules

$$A = A_0 < \ldots\ldots < A_\alpha < A_{\alpha+1} < \ldots\ldots < A_\tau = \cup_{\alpha<\tau} A_\alpha = B$$

where τ is some ordinal and, for each $\alpha < \tau$, $A_{\alpha+1}/A_\alpha$ is countably generated.

Proof. (i) $\Longrightarrow$ (ii). Consider a relative projective resolution of B over A, say, $0 \to K \to A \oplus P \xrightarrow{\phi} B \to 0$ where P is projective and ϕ acts as the identity map on A. Since B and A are projective, the exact sequence splits and we conclude that K is projective. Let $\mathcal{K}$ and $\mathcal{P}$ be $G(\aleph_0)$-families of projective direct summands of K and P respectively. By Theorem 1, $\mathcal{K}$ induces a $G(\aleph_0)$-family $\mathcal{B}$ of submodules of B over A, namely, $\mathcal{B} = \{((A \oplus Y) + K)/K \mid Y \in \mathcal{P} \text{ and } (A \oplus Y) \cap K \in \mathcal{K}\}$. Observe that $((A \oplus Y) + K)/K$ is projective, since $((A \oplus Y) + K)/K \cong (A \oplus Y)/((A \oplus Y) \cap K)$ and $(A \oplus Y) \cap K$, being a direct summand of K, is also a direct summand of the projective module $A \oplus Y$.

(ii) $\Longrightarrow$ (iii) Let $\mathcal{B}$ be a $G(\aleph_0)$-family of projective submodules of B over A. Define $A_0 = A$. Suppose $A_\alpha \in \mathcal{B}$ has been defined for some $\alpha \geq 0$. If $X \not\subseteq A_\alpha$ is a countable suset, then select $A_{\alpha+1} \in \mathcal{B}$ with $A_\alpha \cup X \subset A_{\alpha+1}$ such that $A_{\alpha+1}/A_\alpha$ is countably generated. Proceeding like this and taking unions at the limit ordinals, we could then construct the desired smooth chain of projective modules connecting A to B.

(iii) $\Longrightarrow$ (i) Obvious.

Corollary 3. *Every R-module M with projective dimension ≤ 1 has a $G(\aleph_0)$-family of tight submodules if*

(i) (L. Fuchs; see [4], Ch. IV, 4.1) R is an integral domain, or

(ii) R is an $\aleph_0$-noetherian ring.

Proof. Let $M = B/A$ where both B and A are projective. By Theorem 2, there is a $G(\aleph_0)$-family $\mathcal{B}$ of projective submodules of B over A. Then $\mathcal{M} = \{L/A \mid L \in \mathcal{B}\}$ is a $G(\aleph_0)$-family of submodules of M which are readily seen to be tight.

As another consequence of Theorem 1, we obtain a generalization of Kaplansky's theorem when the ring R is $\aleph_k$-noetherian. This result can also be viewed as providing a stronger converse of the classical theorem of Auslander on modules with projective dimension n.

Theorem 4. *Suppose n is an integer ≥ 0 and R is an $\aleph_k$-noetherian ring where k is an integer such that $\aleph_k \leq |R| \cdot \aleph_0$. Then every R-module B with projective dimension n admits a $G(\aleph_k)$-family of tight submodules. In particular, B is the union of a smooth well-ordered ascending chain of submodules $0 = B_0 < \ldots.. < B_\alpha < B_{\alpha+1} < \ldots. (\alpha < \tau)$, where, for each $\alpha < \tau$, $B_{\alpha+1}/B_\alpha$ is $\aleph_k$-generated and both B_α and B/B_α have projective dimension $\leq n$.*

Proof. Apply induction on n. Since the result is known (Kaplansky's theorem) for $n = 0$, assume that $n > 1$ and that the result holds for R-modules with projective dimension $n-1$. Taking $A = 0$ in Theorem 1(ii), we now have a projective resolution $0 \to K \to P \to B \to 0$ of B so that $pd(K) = n-1$. Let $\mathcal{K}$ be a $G(\aleph_{k-1})$-family (and hence a $G(\aleph_k)$-family) of tight submodules of K and let $\mathcal{C} = \{Y \in \mathcal{P} \mid Y \cap K \in \mathcal{K}\}$ where $\mathcal{P}$ is a $G(\aleph_0)$-family of (projective) direct summands of P. Then the induced family $\mathcal{B} = \{(Y+K)/K \mid Y \in \mathcal{C}\}$ is the needed $G(\aleph_k)$-family. We only need to show that each $L = (Y+K)/K \in \mathcal{B}$ is tight in B. Now $pd((Y+K)/K) \leq n$, since Y is projective and $Y \cap K$, being tight in K, has projective dimension $\leq n - 1$. Also $pd(Y+K) \leq n-1$, since $(Y+K)/Y \cong K/(Y \cap K)$ has projective dimension $\leq n-1$ and Y is projective. This implies that $B/L \cong P/(Y+K)$ has projective dimension $\leq n$.

We next give conditions under which a submodule of a module will have the same finite projective dimension, thus generalizing Theorem 2.

Theorem 5. *Let $n \geq 0$. Suppose R is $\aleph_n$-noetherian and B is an R-module with $pd(B) = n$. Then the following are equivalent for a submodule A of B:*
(i) $pd(A) \leq n$.
(ii) There is a $G(\aleph_n)$-family $\mathcal{B}^*$ of submodules of B over A, where each $L \in \mathcal{B}^*$ has projective dimension $\leq n$.
(iii) There is a smooth well-ordered increasing chain of submodules

$$A = A_0 < \ldots < A_\alpha < A_{\alpha+1} < \ldots < A_\tau = \cup_{\alpha < \tau} A_\alpha = B$$

where τ is some ordinal number and, for each $\alpha < \tau$, $pd(A_\alpha) \leq n$ and $A_{\alpha+1}/A_\alpha$ is $\aleph_n$-generated.

Proof. (i) $\Longrightarrow$ (ii). Let $0 \to K \to A \oplus P \to B \to 0$ be a relative projective resolution of B with respect to A. Clearly $pd(K) \leq n$. By Theorem 4, K, A and B will have $G(\aleph_n)$-families $\mathcal{K}$, $\mathcal{A}$ and $\mathcal{B}$ respectively consisting of tight submodules. From the proof of Theorem 1, $\mathcal{C} = \{Y \in \mathcal{P} \mid (A \oplus Y) \cap K \in \mathcal{K}\}$ is a $G(\aleph_n)$-family, where $\mathcal{P}$ is a $G(\aleph_0)$-family of (projective) direct summands of P. It is then readily seen, using the usual back-and-forth argument, that $\mathcal{C}' = \{Y \in \mathcal{C} \mid (Y+K)/K \in \mathcal{B}$ and $(Y+K) \cap A \in \mathcal{A}\}$ is also a $G(\aleph_n)$-family. Then $\mathcal{B}^* = \{(A+Y+K)/K \mid Y \in \mathcal{C}'\}$ is the needed $G(\aleph_n)$-family of B over A. Also from the exact sequence

$$0 \to (Y+K)/K \to (A+Y+K)/K \to (A+Y+K)/(Y+K)$$
$$\cong A/((Y+K) \cap A) \to 0$$

we conclude that each member of $\mathcal{B}^*$ has projective dimension $\leq n$.

To prove (ii) $\Longrightarrow$ (iii), proceed as the proof of Theorem 2 and (iii) $\Longrightarrow$ (i) is obvious.

The next theorem gives conditions for a submodule to be a tight submodule.

Theorem 6. *Let R be $\aleph_k$-noetherian for some integer $k \geq 0$ and let B be an R-module with $pd(B) = n \geq 1$. Then the following are equivalent for any submodule A of B:*
(i) A is a tight submodule of B.
(ii) In any relative projective resolution $0 \to K \to A \oplus P \to B \to 0$, $pd(K) \leq n-1$.
(iii) There is a $G(\aleph_k)$-family of tight submodules of B over A.

(iv) There is a continuous well-ordered increasing chain of tight submodules

$$A = A_0 < \ldots < A_\alpha < A_{\alpha+1} < \ldots < A_\tau = B$$

where $A_{\alpha+1}/A_\alpha$ is $\aleph_k$-generated for all $\alpha < \tau$.

Proof. To prove (i) $\Rightarrow$ (ii), just note that $K \cong \pi(K)$, where $\pi : A \oplus P \to P$ is the coordinate projection and that $P/\pi(K) \cong B/A$. To show that (ii) $\Rightarrow$ (iii), use Theorem 4 and the proof of Theorem 1.

REMARK: Note that conditions (i) and (ii) of Theorem 6 are equivalent over any ring R, since $pd(A) = pd(A \oplus P) \leq \max\{pd(K), pd(B)\}$ and $B/A \cong P/\pi(K)$.

Kaplansky's theorem can be extended by considering a different kind of restriction on the ring R. We begin with the following definition:

DEFINITION: Let n be an integer ≥ 0. An integral domain R is said to have *Property P_n* if, in a torsion-free R-module of projective dimension n, any RD-submodule of rank $\leq \aleph_n$ is $\aleph_n$-generated.

EXAMPLES: Any countable domain and also any valuation domain ([**4**], Ch. IV, Th. 5.1) has the Property P_n for every $n \geq 0$.

Theorem 7. *Suppose $n \geq 0$ and R is an integral domain with the Property P_n. Then any R-module B with projective dimension n has a $G(\aleph_n)$-family of tight submodules.*

Proof. We apply induction on n, the result being true when $n = 0$ or 1 (by Corollary 3). Consider a projective resolution of B, say $0 \to K \to P \to B \to 0$, where P is projective and $pd(K) = n - 1$. By induction hypothesis, K has a $G(\aleph_{n-1})$-family $\mathcal{K}'$ of tight submodules. Observe that $\mathcal{K}'$ is also a $G(\aleph_n)$-family. Let $\mathcal{K}''$ be the subfamily of $\mathcal{K}'$ consisting of those $S \in \mathcal{K}'$ which are RD-submodules of K. Then using the hypothesis that R has the Property P_n and applying the usual back-and-forth argument, one can show that $\mathcal{K}''$ is a $G(\aleph_n)$-family (of tight RD-submodules) for K. Then, in Theorem 1 (i), taking $A = 0$ and using $\mathcal{K}''$ in place of the family $\mathcal{K}$, we obtain a $G(\aleph_n)$-family $\mathcal{B}$ of submodules of B each of which, as was shown in the proof of Theorem 4, is tight in B.

The Pure-projective Case.

Now a pure-projective (RD-projective) module over a domain R is a direct summand of a direct sum of finitely (cyclicly) presented modules ([**4**], page 41). By Kaplansky's theorem, it is a direct sum of countably generated modules and hence has a $G(\aleph_0)$-family of direct summands. Using the fact that there are enough pure-projectives (RD-projectives), one could define a *relative pure(RD)-projective resolution* of an R-module B relative to a submodule A to be a pure(RD)-exact sequence of the form $0 \to K \to A \oplus P \xrightarrow{\phi} B \to 0$ where P is a pure(RD)-projective R-module and ϕ acts as the identity map on A. Then the direct analogs of Theorems 2 through 6 hold for pure(RD)-projective modules and modules with finite pure(RD)-projective dimension.

References

[1] U. Albrecht and P. Hill, *Butler groups of infinite rank and axiom-3*, Czech. Math. J. **37** (1987), 293 - 309.

[2] L. Bican and L. Fuchs, *Subgroups of Butler groups*, Comm. Algebra, **22** (1994), 1037 - 1047.

[3] L. Fuchs, *Infinite rank Butler groups*, J. Pure and Applied Algebra, **98** (1995), 25 - 44.

[4] L. Fuchs and L. Salce, Modules over Valuation domains, Lecture Notes in Pure and Applied Math., vol. 97, Marcel-Dekker, New York, 1985.

[5] I. Kaplansky, Fields and Rings, University of Chicago Press, Chicago, 1972.

[6] L. Nongxa, K. M. Rangaswamy and C. Vinsonhaler, *Torsion-free modules of finite balanced-projective dimension over valuation domains*, J. Pure and Applied Algebra, (2000)(to appear).

DEPARTMENT OF MATHEMATICS, UNIVERSITY OF COLORADO, COLORADO SPRINGS, CO. 80933-7150, USA

E-mail address: `ranga@math.uccs.edu`

On the torsion groups in cotorsion classes

Lutz Strüngmann and Simone L. Wallutis

Dedicated to Professor Laszlo Fuchs in honour of his 75th birthday

ABSTRACT. For torsion–free abelian groups G we discuss the class $\mathcal{TC}(G)$ of all torsion groups T satisfying $\operatorname{Ext}(G,T) = 0$, that is the subclass of all torsion elements of the cotorsion class cogenerated by G. For rank–1 groups G, and so also for completely decomposable groups G, we characterize $\mathcal{TC}(G)$. Moreover, for a class $\mathfrak{C}$ of torsion groups a criterion is given to decide whether or not $\mathfrak{C}$ equals $\mathcal{TC}(C)$ for some completely decomposable group C. Using this criterion we can show that, for any countable torsion–free group G, there is a completely decomposable group C such that $\mathcal{TC}(G) = \mathcal{TC}(C)$. For uncountable torsion–free groups, however, this statement is undecidable in ZFC. Finally, we seek torsion–free groups G with either $\mathcal{TC}(G) = \mathcal{TC}(R)$ for some rank–1 group R or $\mathcal{TC}(G) = \mathcal{TC}(\bigoplus_{\tau \in Tst(G)} R_\tau)$ where R_τ is a rank–1 group of type τ and $Tst(G)$ denotes the typeset of G; the latter is proven to be true for B_1–groups.

Introduction

Throughout this paper we work in the category Mod–$\mathbb{Z}$ of abelian groups. All terminology used here can be found in [**F1**], [**F2**] and [**EM**].

Cotorsion theories for abelian groups have been introduced by Salce in 1979 [**S**]. Following his notation we call a pair $(\mathcal{F}, \mathcal{C})$ a cotorsion theory if $\mathcal{F}$ and $\mathcal{C}$ are classes of abelian groups which are maximal with respect to the property that $\operatorname{Ext}(F, C) = 0$ for all $F \in \mathcal{F}$, $C \in \mathcal{C}$.

Salce [**S**] has shown that every cotorsion theory is cogenerated by a class of torsion and torsion–free groups where $(\mathcal{F}, \mathcal{C})$ is said to be cogenerated by the class $\mathcal{A}$ if $\mathcal{C} = \mathcal{A}^\perp = \{X \in \text{Mod–}\mathbb{Z} \mid \operatorname{Ext}(A, X) = 0 \text{ for all } A \in \mathcal{A}\}$ and $\mathcal{F} =^\perp (\mathcal{A}^\perp) = \{Y \in \text{Mod–}\mathbb{Z} \mid \operatorname{Ext}(Y, X) = 0 \text{ for all } X \in \mathcal{A}^\perp\}$. Examples for cotorsion theories are: (Mod–$\mathbb{Z}$, $\mathcal{D}$) = $(^\perp(G^\perp), G^\perp)$ with $G = \bigoplus_{p \in \Pi} \mathbb{Z}(p)$ where $\mathcal{D}$ is the class of all divisible groups and Π is the set of all prime numbers, $(\mathcal{L}, \text{Mod–}\mathbb{Z}) = (^\perp(\mathbb{Z}^\perp), \mathbb{Z}^\perp)$ where

---

1991 *Mathematics Subject Classification.* Primary 20K15, 20K20, 20K35, 20K40; Secondary 13C99, 18E99, 20J05.

The first author was supported by the Graduiertenkolleg *Theoretische und Experimentelle Methoden der Reinen Mathematik* of Essen University.

The second author was supported by the Deutsche Forschungsgemeinschaft.

$\mathcal{L}$ is the class of all free groups, and the classical one $(\mathcal{TF}, \mathcal{CO}) = (^\perp(\mathbb{Q}^\perp), \mathbb{Q}^\perp)$ where $\mathcal{TF}$ is the class of all torsion–free groups and $\mathcal{CO}$ is the class of all (classical) cotorsion groups. In view of the last example the classes $\mathcal{F}$ and $\mathcal{C}$ of a cotorsion theory $(\mathcal{F}, \mathcal{C})$ are said to be the torsion–free class and the cotorsion class of this cotorsion theory.

In this paper we shall restrict our attention to cotorsion classes cogenerated by a single group G. When ordering these classes by inclusion we obviously have that $\mathbb{Z}^\perp$ is maximal and $(\bigoplus_{p\in\Pi} \mathbb{Z}(p))^\perp$ is minimal among these classes.

In fact, all cotorsion classes between $(\bigoplus_{p\in\Pi} \mathbb{Z}(p))^\perp$ and $\mathbb{Q}^\perp$ have been characterized (see [**S**, Proposition 2.8.]) and thus it makes sense to consider the cotorsion classes containing the classical one $\mathbb{Q}^\perp$, i.e. we only consider cotorsion classes cogenerated by a torsion–free group. Well, this has actually been done in [**GSW**] where Göbel, Shelah and the second author have shown that any partially ordered set can be embedded into the lattice of all cotorsion classes. Hence there is no hope at all to characterize these classes. When looking at the groups constructed in [**GSW**] to obtain this result we realize, however, that all these groups are in fact torsion–free. Therefore we turn our attention to the subclass of all torsion groups of the cotorsion class cogenerated by a torsion–free group G; we shall denote this class by $\mathcal{TC}(G)$.

A characterization of these classes is obviously closely related to the solution of the Baer problem (e.g. see [**F2**]); put into our context it says that $\mathcal{TC}(G)$ is maximal, i.e. $\mathcal{TC}(G) = \mathfrak{T}$, where $\mathfrak{T}$ is the class of all torsion groups, if and only if G is free. Although we will not obtain a full characterization of the torsion subclasses of all cotorsion classes cogenerated by a torsion–free group we do achieve some interesting results.

After having established some basic facts in Section 1 we shall first consider easy cases of torsion–free groups, namely the rank–1 groups. As rank–1 groups can be identified with subgroups of the rationals $\mathbb{Q}$ they are also called rational groups. Correspondingly, we call a cotorsion class (theory) cogenerated by a rank–1 group a rational cotorsion class (theory). Using a characterization of Salce [**S**] we obtain a very nice criterion for a torsion group T to belong to $R^\perp$ for some given rank–1 group R. Introducing the quasi–reduced type $t^{qr}(R)$ of an rank–1 group R we will show that $\mathcal{TC}(R) = \mathcal{TC}(R')$ if and only if $t^{qr}(R) = t^{qr}(R')$ for any rank–1 groups R, R'; when representing the type of R by $t(R) = (r_p)_{p\in\Pi}$ the quasi–reduced type of R can be represented by $t^{qr}(R) = (s_p)_{p\in\Pi}$ with $s_p = r_p$ whenever $r_p = 0$ or $r_p = \infty$ and $s_p = 1$ otherwise. This characterization of the classes $\mathcal{TC}(R)$ for rank–1 groups has immediate consequences for completely decomposable groups and others.

In Section 3 we shall use this result to find necessary and sufficient conditions on a class of torsion groups to equal $\mathcal{TC}(C)$ for some completely decomposable group C. Moreover, we show that for any countable group G, $\mathcal{TC}(G)$ satisfies these conditions and so $\mathcal{TC}(G) = \mathcal{TC}(C)$ for some completely decomposable group C. This, however, does not remain true if we omit the countability condition. In fact, we will see that, already in the local case (i.e. for modules over $\mathbb{Z}_{(p)}$), it is undecidable in ZFC whether or not for all groups (modules) G, $\mathcal{TC}(G) = \mathcal{TC}(C)$ for some completely decomposable group C.

Finally, in the last Section we shall alter our original question by asking if there are groups G such that either (a): $\mathcal{TC}(G) = \mathcal{TC}(R)$ for some rational group R or (b):

$\mathcal{TC}(G) = \mathcal{TC}(\bigoplus_{\tau \in Tst(G)} R_\tau)$ where $Tst(G)$ denotes the typeset of G and R_τ is a rank-1 group of type τ. We shall show that completely decomposable groups of finite rank satisfy both (a) and (b), while there are many non–isomorphic completely decomposable groups of infinite rank which do not satisfy (a).

However, in both cases we obtain large classes satisfying the required condition: all TEP–groups satisfy (a) (for definition see 4.3), and all B_1–groups satisfy (b). Note that finite rank Butler groups are both TEP and B_1, and hence satisfy (a) and (b). Let us begin with some known facts.

1. Preliminaries

In this section we introduce the class $\mathcal{TC}(G)$ for a given group G and recall known results adjusted to this notation.

For an arbitrary group G the classes $G^\perp$ and $^\perp(G^\perp)$ are defined by $G^\perp = \{X \in \text{Mod-}\mathbb{Z} \mid \text{Ext}(G, X) = 0\}$ and $^\perp(G^\perp) = \{Y \in \text{Mod-}\mathbb{Z} \mid \text{Ext}(Y, X) = 0 \,\forall X \in G^\perp\}$, respectively. Following Salce [**S**], the pair $(^\perp(G^\perp), G^\perp)$ is called the *cotorsion theory cogenerated by* G. In view of the classical cotorsion theory $(\mathcal{TF}, \mathcal{CO})$ where $\mathcal{CO} = \mathbb{Q}^\perp$ is the class of all cotorsion groups and $\mathcal{TF} = {}^\perp(\mathbb{Q}^\perp)$ is the class of all torsion–free groups, we also call $G^\perp$ the *cotorsion class cogenerated by* G. As we have seen in the introduction it makes sense to restrict our attention to the cotorsion classes cogenerated by torsion–free groups and furthermore to its subclasses of all torsion elements.

Let $\mathfrak{T}$ be the class of all torsion groups. For any group G let $\mathcal{TC}(G) = G^\perp \cap \mathfrak{T}$ be the class of all torsion groups belonging to the cotorsion class $G^\perp$ cogenerated by G. We clearly have that $\mathcal{TC}(\mathbb{Z}) = \mathfrak{T}$ is maximal among these classes and that the class $\mathcal{TC}(\mathbb{Q}) = \mathcal{CO} \cap \mathfrak{T}$ consisting of all torsion cotorsion groups, i.e. of all direct sums of a bounded and a divisible torsion group, is minimal when restricting to torsion–free groups G.

Moreover, it is well known that $G^\perp$ as well as $\mathfrak{T}$ are closed under epimorphic images and extensions (see [**F1**] and [**S**]). Hence we immediately have the following:

LEMMA 1.1. *For any group G the following are true:*
  (i) $\mathcal{TC}(G)$ *is closed under epimorphic images;*
  (ii) $\mathcal{TC}(G)$ *is closed under extensions, especially under finite direct sums;*
  (iii) *if G is torsion–free, then $\mathcal{TC}(G)$ contains all torsion cotorsion groups, i.e.* $\mathcal{TC}(\mathbb{Q}) = \mathcal{CO} \cap \mathfrak{T} \subseteq \mathcal{TC}(G)$. □

For our investigations we shall need the following lemma which is well known. However, we include the proof for the convenience of the reader.

First recall that the basic subgroup B of a torsion group T is the direct sum $B = \bigoplus_{p \in \Pi} B_p$ of the basic subgroups B_p of the p–components T_p; for each prime p, B_p is a direct sum of cyclic p–groups, B_p is a pure subgroup of T_p and the quotient T_p/B_p is divisible (see [**F1**]).

LEMMA 1.2. *Let T be a torsion group and $B \subseteq T$ a basic subgroup of T. Then, for any group G, T is an element of $\mathcal{TC}(G)$ if and only if B is.*

PROOF. The short exact sequence $0 \to B \to T \to T/B \to 0$ induces the exact sequence $\text{Ext}(G, B) \to \text{Ext}(G, T) \to \text{Ext}(G, T/B) = 0$ where the last term is zero since T/B is divisible. Thus $\text{Ext}(G, B) = 0$ implies $\text{Ext}(G, T) = 0$, i.e. if $B \in \mathcal{TC}(G)$ then $T \in \mathcal{TC}(G)$.

Conversely, assume $T \in \mathcal{TC}(G)$. By [**F1**, Theorem 36.1] B is an epimorphic image of T and thus, by Lemma 1.1(ii), B also belongs to $\mathcal{TC}(G)$. □

Let us finish this section by restricting to torsion–free groups G. If $\mathcal{TC}(G)$ is maximal, i.e. $\mathcal{TC}(G) = \mathfrak{T} = \mathcal{TC}(\mathbb{Z})$, then G has to be free by Griffith's solution of the Baer problem (see [**G**]). On the other hand, however, if $\mathcal{TC}(G) = \mathcal{TC}(\mathbb{Q})$ is minimal then G need not to be divisible as the following example shows:

EXAMPLE 1.3. Let $G = P = \prod_{n\in\omega} \mathbb{Z}$ be the Baer–Specker group. Obviously, P is not divisible; in fact, P is homogeneous of type $\mathbb{Z}$.
However, by [**GT**, Lemma 1.3], P contains a subgroup $\mathbb{D}_\omega$ such that $\mathcal{TC}(\mathbb{D}_\omega) = \mathcal{TC}(\mathbb{Q})$ and thus $\mathcal{TC}(P) = \mathcal{TC}(\mathbb{Q})$ since $\mathcal{CO} = \mathbb{Q}^\perp \subseteq P^\perp \subseteq \mathbb{D}_\omega^\perp$. □

We are now ready for the next Section in which we shall consider easy cases of torsion–free groups, namely the rank–1 groups.

2. The torsion elements of rational cotorsion classes

In this Section we characterize $\mathcal{TC}(R)$ for torsion–free groups R of rank 1. Since rank–1 groups can be identified with the subgroups of the rationals $\mathbb{Q}$ they are also called rational groups. Due to this terminology a cotorsion theory $(^\perp(R^\perp), R^\perp)$ cogenerated by a rank–1 group R is called a *rational cotorsion theory* (see [**S**]). Hence $R^\perp$ is said to be a rational cotorsion class. To determine the torsion elements of the rational cotorsion classes we employ a criterion due to Salce (see [**S**, Theorem 3.5]). Before we can state this criterion we need some notation.

Let $R \subseteq \mathbb{Q}$ be any rank–1 group containing $\mathbb{Z}$ and let $\chi(R) = \chi_R(1) = (r_p)_{p\in\Pi}$ be the characteristic of R where $r_p \in \mathbb{N} \cup \{\infty\}$ and Π is the set of all prime numbers. Recall that rank–1 groups are isomorphic if and only if their characteristics only differ in finitely many finite entries, i.e. if they belong to the same type; see [**F2**] for details. For an arbitrary group G we put $G_R = \bigcap_{p\in\Pi} p^{r_p} G$. Then the criterion reads as follows:

LEMMA 2.1 (Salce). *For any group G and any rational group R the following are equivalent:*
  (i) $G \in R^\perp$;
  (ii) G/G_R is cotorsion. □

Note that condition (ii) in the above lemma only depends on the type of R not on the chosen representative although the groups G_R and hence G/G_R might be non–isomorphic for different characteristics.

Using the criterion of Lemma 2.1 it has been shown in [**GSW**] that rational cotorsion classes coincide if and only if the corresponding cogenerating rank–1 groups are of the same type. This will not remain true when we restrict to the torsion subclasses as follows from the next proposition. Note that the result is a direct consequence of [**EH**, Theorem 1.2] but we include the straightforward proof for the convenience of the reader.

PROPOSITION 2.2. *Let R be a rational group with $\chi(R) = (r_p)_{p\in\Pi}$ and let $T = \bigoplus_{p\in\Pi} T_p$ be a reduced torsion group with p-components T_p.*
Then $\operatorname{Ext}(R, T) = 0$ if and only if the following conditions are satisfied:
  (i) T_p *is bounded for all p such that $r_p = \infty$;*

(ii) $T_p = 0$ for almost all p such that $r_p \neq 0$.

PROOF. By Lemma 1.2 we may assume, without loss of generality, that for each prime p, T_p is a direct sum of cyclic p–groups.
We decompose T into $T = T_0 \oplus T_\infty \oplus T_f$ where $T_0 = \bigoplus\limits_{r_p=0} T_p$, $T_\infty = \bigoplus\limits_{r_p=\infty} T_p$ and $T_f = \bigoplus\limits_{0<r_p<\infty} T_p$. Obviously, $\text{Ext}(R,T) = 0$ if and only if $\text{Ext}(R,T_0) = \text{Ext}(R,T_\infty) = \text{Ext}(R,T_f) = 0$. But $\text{Ext}(R,T_0) = 0$ anyway by Lemma 2.1 since $(T_0)_R = \bigcap\limits_{r_q \neq 0} q^{r_q} T_0 = T_0$ as T_p is q–divisible for $p \neq q$ and so $T_0/(T_0)_R = 0$ is cotorsion. Therefore we have $\text{Ext}(R,T) = 0$ if and only if $\text{Ext}(R,T_\infty) = \text{Ext}(R,T_f) = 0$. First assume that the conditions (i) and (ii) are satisfied. Then T_∞ is a finite direct sum of bounded groups and so it is itself bounded. Hence $\text{Ext}(R,T_\infty) = 0$ since R is torsion–free and T_∞ is cotorsion. Moreover, it follows from the assumption that $T_p = 0$ for all but finitely many components of T_f, i.e. $T_f = \bigoplus\limits_{i=1}^{n} T_{p_i}$ where $0 < r_{p_i} < \infty$. For each $i \leq n$ we thus have $(T_{p_i})_R = \bigcap\limits_{r_q \neq 0} q^{r_q} T_{p_i} = p_i^{r_{p_i}} T_{p_i}$ and so $T_{p_i}/(T_{p_i})_R$ is bounded and hence cotorsion. Therefore, by Lemma 2.1, $\text{Ext}(R,T_{p_i}) = 0$ for all $i \leq n$ and so $\text{Ext}(R,T_f) = 0$. This now implies $\text{Ext}(R,T) = 0$.
Conversely, assume that $\text{Ext}(R,T_\infty) = \text{Ext}(R,T_f) = 0$. Then $T_\infty/(T_\infty)_R$ and $T_f/(T_f)_R$ are cotorsion (and torsion) and hence a direct sum of a bounded and a divisible group. Now, $(T_\infty)_R = \bigcap\limits_{0 \neq r_q} q^{r_q} T_\infty = \bigcap\limits_{r_q=\infty} q^\infty T_\infty = 0$ since T_∞ is a direct sum of cyclic groups and so $p^\omega T_\infty$ is either zero (if $(T_\infty)_p \neq 0$) or T_∞ (if $(T_\infty)_p = 0$). Therfore T_∞ is itself cotorsion and hence bounded, i.e. T_p is bounded whenever $r_p = \infty$ and $T_p = 0$ for almost all p with $r_p = \infty$.
Moreover, $(T_f)_R = \bigcap\limits_{0<r_q<\infty} q^{r_q} T_f = \bigcap\limits_{0<r_q<\infty} q^{r_q} \left(\bigoplus\limits_{0<r_p<\infty} T_p \right) = \bigcap\limits_{0<r_q<\infty} (q^{r_q} T_q \oplus \bigoplus\limits_{0<r_p<\infty; p \neq q} T_p) = \bigoplus\limits_{0<r_p<\infty} p^{r_p} T_p$. Thus $T_f/(T_f)_R = \bigoplus\limits_{0<r_p<\infty} T_p/p^{r_p} T_p$ is bounded and so $T_p = 0$ for almost all p such that $0 < r_p < \infty$ since $p^{r_p} T_p \neq T_p$ unless $T_p = 0$. Finally, the conditions (i) and (ii) follow from the above. $\square$

Obviously, in the above proposition it makes no difference if the finite non–zero entries are 1 or any other number. Thus we introduce the following terminology.

DEFINITION 2.3. *Let R be a rational group and $\chi_R(1) = (r_p)_{p \in \Pi}$ the characteristic of R. For each prime p we define s_p by*

$$s_p = \begin{cases} r_p & \text{if } r_p = 0 \text{ or } r_p = \infty, \\ 1 & \text{otherwise.} \end{cases}$$

Then $(s_p)_{p \in \Pi}$ is called the quasi–reduced characteristic *of R and the corresponding type is said to be the* quasi–reduced type; *notation: $\chi^{qr}(R) = (s_p)_{p \in \Pi}$, $t^{qr}(R) = t(\chi^{qr}(R))$. By R_{qr} we denote a rational group of type $t^{qr}(R)$.*

Note that the quasi–reduced characteristic and the quasi–reduced type are well defined and that the quasi–reduced types could be the same even for rational groups R and R' of incomparable types $t(R)$, $t(R')$. In fact, the quasi–reduced type is exactly what we need to characterize the torsion subclasses of the rational cotorsion classes.

THEOREM 2.4. *Let R be a rational group and $\chi^{qr}(R) = (r_p)_{p \in \Pi}$. Then $\mathcal{TC}(R) = \mathcal{TC}(R_{qr})$ is the class of all torsion groups $T' = T \oplus D$ (D divisible, T reduced) such that the reduced part $T = \bigoplus_{p \in \Pi} T_p$ satisfies the following two conditions:*

(i) *T_p is bounded for all p such that $r_p = \infty$;*
(ii) *$T_p = 0$ for almost all p such that $r_p \neq 0$.*

PROOF. The proof is straightforward using Proposition 2.2 and the fact that divisible (torsion) groups T always satisfy $\text{Ext}(R,T) = 0$. □

As an immediate consequence of the above theorem we have that $\mathcal{TC}(R) = \mathcal{TC}(R')$ for rational groups R, R' if and only if $t^{qr}(R) = t^{qr}(R')$. So, by what we have said before, the torsion subclasses of rational cotorsion classes can be the same even for incomparable rational groups. This contrasts with the result from [**GSW**] mentioned earlier, where it was shown that two rational groups R and R' are isomorphic if and only if their cotorsion classes $R^\perp$ and $(R')^\perp$ coincide; in fact it was proven that they are isomorphic if and only if $R^\perp \cap \mathcal{TF} = (R')^\perp \cap \mathcal{TF}$ where $\mathcal{TF}$ denotes the class of all torsion–free groups.

Finally note that the characterization of $\mathcal{TC}(R)$ for rank–1 groups R immediately induces a characterization of $\mathcal{TC}(C)$ for completely decomposable groups $C = \bigoplus_{i \in I} R_i$ with $R_i \subseteq \mathbb{Q}$ since $\mathcal{TC}(C) = \bigcap_{i \in I} \mathcal{TC}(R_i)$ although it is not very explicit. However, this criterion will be useful for tackling the problems in the next Section.

3. A characterization of $\mathcal{TC}(C)$ for completely decomposable groups C

In this section we will derive necessary and sufficient conditions for a class $\mathfrak{C}$ of torsion groups to coincide with $\mathcal{TC}(C)$ for some completely decomposable group C. First recall from the last section that $\mathcal{TC}(C) = \bigcap_{i \in I} \mathcal{TC}(R_i)$ for any completely decomposable group $C = \bigoplus_{i \in I} R_i$ where $\mathcal{TC}(R_i)$ is uniquely determined by the quasi–reduced type of R_i. Moreover, $\mathcal{TC}(C)$ contains all bounded (and all divisible torsion) groups as, in general, $\mathcal{TC}(G)$ does for any torsion–free group G. So, we consider unbounded reduced torsion groups and ask if the fact that a certain unbounded group belongs to $\mathcal{TC}(C)$ ($\mathcal{TC}(G)$) has any consequences for other groups. We begin with considering unbounded p–groups.

PROPOSITION 3.1. *Let p be a prime. The following conditions are equivalent for any group C which is either completely decomposable or countable or p-divisible:*

(i) *$\mathcal{TC}(C)$ contains an unbounded reduced p–group T;*
(ii) *$\mathcal{TC}(C)$ contains all p–groups;*
(iii) *$\mathcal{TC}(C)$ contains $B = \bigoplus_{n \in \omega} \mathbb{Z}(p^n)$, where $\mathbb{Z}(p^n)$ denotes the cyclic group of order p^n.*

PROOF. Note that obviously (ii) implies both (i) and (iii) and also (iii) clearly implies (i). Moreover, it is easily seen that (i) implies (iii): T is an element of $\mathcal{TC}(C)$ if and only if its (also unbounded) basic subgroup B' belongs to $\mathcal{TC}(C)$ by Lemma 1.2. Now, B is obviously an epimorphic image of B' and so, by Lemma 1.1(i), B is an element of $\mathcal{TC}(C)$ if an unbounded reduced p–group T is. Therefore it remains to show that (iii) implies (ii).

First we consider a completely decomposable group. Let $C = \bigoplus_{i \in I} R_i$ where each R_i is a rational group. Assume that B is an element of $\mathcal{TC}(C)$. Then $B \in \mathcal{TC}(R_i)$ for all $i \in I$ and thus $\chi(R_i)_p < \infty$ for all $i \in I$ by Proposition 2.2. Applying the same proposition again we deduce that any p–group is an element of $\mathcal{TC}(R_i)$ for each $i \in I$ and hence the same is true for $\mathcal{TC}(C)$, i.e. (ii) is satisfied in this case. Next, let C be a countable group and assume $B \in \mathcal{TC}(C)$. Since $B^{(\omega)}$ is an epimorphic image of B we also have that $B^{(\omega)} \in \mathcal{TC}(C)$, again by Lemma 1.1(i). We show that the same holds true for any cardinal κ, i.e. we prove $\mathrm{Ext}(C, B^{(\kappa)}) = 0$ for any κ. Therefore let

$$0 \to K \to F \to C \to 0$$

be a free resolution of C with F and K countable. This induces the exact sequence

$$0 \to \mathrm{Hom}(C, B^{(\kappa)}) \to \mathrm{Hom}(F, B^{(\kappa)}) \to \mathrm{Hom}(K, B^{(\kappa)}) \to \mathrm{Ext}(C, B^{(\kappa)}) \to 0.$$

Now any homomorphism from K to $B^{(\kappa)}$ is actually a homomorphism f from K to $B^{(\omega)}$ since K is countable and thus f lifts to a homomorphism from F to $B^{(\omega)} \subseteq B^{(\kappa)}$ as $\mathrm{Ext}(C, B^{(\omega)}) = 0$ by the above. Therefore, the map $\mathrm{Hom}(F, B^{(\kappa)}) \to \mathrm{Hom}(K, B^{(\kappa)})$ is surjective and so $\mathrm{Ext}(C, B^{(\kappa)}) = 0$, i.e. $B^{(\kappa)} \in \mathcal{TC}(C)$. Hence (ii) holds since, for any p–group, its basic subgroup is an epimorphic image of $B^{(\kappa)}$ for some cardinal κ.

Finally, let C be p–divisible. We proceed similarly to the countable case. Of course, K and F in the free resolution cannot be chosen to be countable. However, the induced sequence reduces to

$$0 \to \mathrm{Hom}(F, B^{(\kappa)}) \to \mathrm{Hom}(K, B^{(\kappa)}) \to \mathrm{Ext}(G, B^{(\kappa)}) \to 0$$

since $\mathrm{Hom}(C, B^{(\kappa)}) = 0$ as C is divisible and B is reduced. We prove (ii) by induction on the cardinality λ of C. Obviously, the result is true for $\lambda = \omega$ by the above. So let $C = \bigcup_{\alpha < \lambda} C_\alpha$ be the union of a smooth ascending chain of pure subgroups C_α of smaller cardinality. By the induction hypothesis we have that $\mathrm{Ext}(C_0, B^{(\kappa)}) = 0$ and also $\mathrm{Ext}(C_{\alpha+1}/C_\alpha, B^{(\kappa)}) = 0$ for any $\alpha < \lambda$ and for any cardinal κ. Applying [**ET**][Lemma 1] we conclude $\mathrm{Ext}(C, B^{(\kappa)}) = 0$ for any cardinal κ.

Using exactly the same arguments as in the countable case we achieve that (ii) also holds for any p–divisible group and this finishes the proof. $\square$

As an immediate consequence from the above proposition we have:

COROLLARY 3.2. *Let G be a countable torsion–free $\mathbb{Z}_{(p)}$–module. Then G is free if and only if $\mathrm{Ext}(G, \bigoplus_{n \in \omega} \mathbb{Z}(p^n)) = 0$.*

PROOF. If $\mathrm{Ext}(G, \bigoplus_{n \in \omega} \mathbb{Z}(p^n)) = 0$, then $\mathrm{Ext}(G, T) = 0$ for all p–groups T by Proposition 3.1. Hence G is free by Griffith's result [**G**]. The converse implication is trivial. $\square$

Note that the proof of Proposition 3.1 shows that, for any group G, if $\mathcal{TC}(G)$ contains some unbounded reduced p–group then it contains all countable p–groups. In general one can show, by having a closer look at Griffith's solution of the Baer problem (see [**G**]), that $\mathrm{Ext}(G, B^{(|G|)}) = 0$ if and only if $\mathrm{Ext}(G, T) = 0$ for all p–groups T, where $B = \bigoplus_{n \in \omega} \mathbb{Z}(p^n)$.

In order to prove our next result we need a result of Baer which characterizes the torsion elements of a cotorsion class cogenerated by a countable group. The proof can be found in [**B**].

LEMMA 3.3 (Baer). *Let T be a torsion and G a torsion-free group such that $\operatorname{Ext}(G, T) = 0$. Then the following hold:*

(i) *if $\{p_1, \cdots, p_i, \cdots\}$ is an infinite set of different primes for which each $p_i T \neq T$, then G contains no pure subgroup S of finite rank such that G/S has non-zero elements divisible by all p_i;*

(ii) *if for some prime p, the reduced part of the p-component of T is unbounded, then G contains no pure subgroup S of finite rank such that G/S has non-zero elements divisible by all powers of p.*

Moreover, if G is countable, then (i) and (ii) suffice for $\operatorname{Ext}(G, T)$ to be zero. □

We can now prove the following:

PROPOSITION 3.4. *Let C be completely decomposable or countable. Then the following are equivalent for an infinite set P of primes:*

(i) $\mathcal{TC}(C)$ *contains* $\bigoplus_{p \in P} \mathbb{Z}(p)$;

(ii) $\mathcal{TC}(C)$ *contains* $\bigoplus_{p \in P} T_p$ *for all p-groups $T_p \in \mathcal{TC}(C)$.*

PROOF. Trivially, condition (ii) implies (i) since $\mathcal{TC}(C)$ contains all bounded groups. Hence we assume (i).

If $C = \bigoplus_{i \in I} R_i$ is completely decomposable with $R_i \subseteq \mathbb{Q}$ for all $i \in I$, then (i) implies that $\chi(R_i)_p = 0$ for almost all $p \in P$ and all $i \in I$. By Proposition 2.2 it therefore follows that $\bigoplus_{p \in P} T_p \in \mathcal{TC}(C)$ for all $T_p \in \mathcal{TC}(C)$.

If C is countable, then we use the Baer Criterion 3.3. Let $T_p \in \mathcal{TC}(C)$ be a p-group for any $p \in P$ and assume that $\bigoplus_{p \in P} T_p \notin \mathcal{TC}(C)$. Then the Baer criterion implies that there exist a countable set $P' \subseteq P$ and a pure finite rank subgroup $S \subseteq C$ such that G/S contains non-zero elements which are divisible by all $p \in P'$. Hence $\bigoplus_{p \in P'} \mathbb{Z}(p) \notin \mathcal{TC}(C)$ by Lemma 3.3, contradicting the assumption since $\bigoplus_{p \in P'} \mathbb{Z}(p)$ is an epimorphic image of $\bigoplus_{p \in P} \mathbb{Z}(p)$ (see Lemma 1.1). □

Before we can prove the main result of this section we need:

PROPOSITION 3.5. *Let C be a completely decomposable or a countable group and let P be an infinite set of primes such that $\bigoplus_{p \in P} \mathbb{Z}(p) \notin \mathcal{TC}(C)$. Then there exists an infinite subset P' of P such that $\bigoplus_{p \in X} \mathbb{Z}(p) \notin \mathcal{TC}(C)$ for all infinite $X \subseteq P'$.*

PROOF. First let us assume that $C = \bigoplus_{i \in I} R_i$ is completely decomposable with $R_i \subseteq \mathbb{Q}$ for $i \in I$. Then $\bigoplus_{p \in P} \mathbb{Z}(p) \notin \mathcal{TC}(C)$ implies that $\bigoplus_{p \in P} \mathbb{Z}(p) \notin \mathcal{TC}(R_i)$ for some $i \in I$. By Proposition 2.2 there exists an infinite subset P' of P such that $\chi(R_i)_p \neq 0$ for all $p \in P'$. Using Proposition 2.2 again we conclude that, for all infinite $X \subseteq P'$, we have $\bigoplus_{p \in X} \mathbb{Z}(p) \notin \mathcal{TC}(R_i)$ and hence $\bigoplus_{p \in X} \mathbb{Z}(p) \notin \mathcal{TC}(C)$.

If C is countable then $\bigoplus_{p \in P} \mathbb{Z}(p) \notin \mathcal{TC}(C)$ implies the existence of an infinite subset P' of P and of a finite rank pure subgroup S of C such that C/S contains non–zero elements divisible by all $p \in P'$ (using Lemma 3.3). Hence, for all infinite subsets $X \subseteq P'$, the quotient C/S contains non–zero elements divisible by all $p \in X$. Thus Lemma 3.3 implies that $\bigoplus_{p \in X} \mathbb{Z}(p) \notin \mathcal{TC}(C)$. □

We are now able to characterize classes of torsion groups which can be realized as the torsion subclass of the cotorsion class cogenerated by some completely decomposable group.

THEOREM 3.6. *Let $\mathfrak{C}$ be a class of torsion groups. Then $\mathfrak{C} = \mathcal{TC}(C)$ for some completely decomposable group C if and only if the following conditions are satisfied:*

  (i) *$\mathfrak{C}$ contains all torsion cotorsion groups;*
  (ii) *$\mathfrak{C}$ is closed under epimorphic images;*
  (iii) *$\bigoplus_{n \in \omega} \mathbb{Z}(p^n) \in \mathfrak{C}$ if and only if $\mathfrak{C}$ contains all p–groups for all primes p;*
  (iv) *If P is an infinite set of primes then, $\bigoplus_{p \in P} \mathbb{Z}(p) \in \mathfrak{C}$ if and only if $\bigoplus_{p \in P} T_p \in \mathfrak{C}$ for all p–groups $T_p \in \mathfrak{C}$;*
  (v) *If P is an infinite set of primes such that $\bigoplus_{p \in P} \mathbb{Z}(p) \notin \mathfrak{C}$ then there exists an infinite subset P' of P such that $\bigoplus_{p \in X} \mathbb{Z}(p) \notin \mathfrak{C}$ for all infinite $X \subseteq P'$.*

PROOF. Obviously, for any completely decomposable group C, $\mathcal{TC}(C)$ satisfies conditions (i) to (v) by the Propositions 3.1, 3.4 and 3.5.

Conversely, let $\mathfrak{C}$ be a class of torsion groups with the properties (i) to (v). In order to obtain a suitable completely decomposable group we define the following sets:

$$M = \{p \in \Pi \mid \bigoplus_{n \in \omega} \mathbb{Z}(p^n) \notin \mathfrak{C}\}$$

and

$$\mathcal{P} = \{P \subseteq \Pi \mid \bigoplus_{p \in X} \mathbb{Z}(p) \notin \mathfrak{C} \text{ for all infinite } X \subseteq P\}.$$

For each $P \in \mathcal{P}$ let $R_P \subseteq \mathbb{Q}$ be the rank–1 group generated by all fractions $\frac{1}{p}$ with $p \in P$, i.e. $R_P = \left\langle \frac{1}{p} \mid p \in P \right\rangle$. Note that R_P has only finite entries in its characteristic. Also, recall that $\mathbb{Q}^{(p)} \subseteq \mathbb{Q}$ is defined by $\mathbb{Q}^{(p)} = \{\frac{m}{p^n} \mid m, n \in \mathbb{Z}\}$. We put $C = \bigoplus_{p \in M} \mathbb{Q}^{(p)} \oplus \bigoplus_{P \in \mathcal{P}} R_P$; this completely decomposable group will, in fact, turn out to be the required one.

First we show that $\mathfrak{C} \subseteq \mathcal{TC}(C)$. Suppose that $B = \bigoplus_{n \in \omega} \mathbb{Z}(p^n) \in \mathfrak{C}$. Then $p \notin M$ and hence $\text{Ext}(\mathbb{Q}^{(q)}, B) = 0$ for all $q \in M$ by Proposition 2.2. Moreover, $\text{Ext}(R_P, B) = 0$ for all $P \in \mathcal{P}$, again by Proposition 2.2. Therefore $B \in \mathcal{TC}(C)$ follows in this case. Now consider $B = \bigoplus_{p \in P'} \mathbb{Z}(p)$ for some infinite set P' of primes and assume $B \in \mathfrak{C}$. Clearly, $\text{Ext}(\mathbb{Q}^{(q)}, B) = 0$ for all $q \in M$. We also have, for each $P \in \mathcal{P}$, that $\bigoplus_{p \in P \cap P'} \mathbb{Z}(p) \in \mathfrak{C}$ and thus $P' \cap P$ is finite which implies $\text{Ext}(R_P, B) = 0$. Henceforem

$B \in \mathcal{TC}(C)$, also in this case. Now, the above implies $\mathfrak{C} \subseteq \mathcal{TC}(C)$ by the properties (iii) and (iv).

Finally we show that also $\mathcal{TC}(C) \subseteq \mathfrak{C}$ again by using (iii) and (iv). Suppose first that $B = \bigoplus_{n \in \omega} \mathbb{Z}(p^n) \in \mathcal{TC}(C)$. Then $\text{Ext}(\mathbb{Q}^{(q)}, B) = 0$ for all $q \in M$ and so $p \notin M$, i.e. $B \in \mathfrak{C}$ as required. Now consider $B = \bigoplus_{p \in P} \mathbb{Z}(p)$ for some infinite set P of primes. Suppose, for contradiction, that B is an element of $\mathcal{TC}(C)$ but not of $\mathfrak{C}$. Then, by condition (v), there exists an infinite subset P' of P which is an element of $\mathcal{P}$. Hence, $R_{P'}$ is a direct summand of C and so, by assumption, $\text{Ext}(R_{P'}, B) = 0$. But $\mathbb{Z}(p) \subseteq B$ for all $p \in P'$ contradicting Proposition 2.2. □

Applying some of the former results and the above theorem we deduce:

COROLLARY 3.7. *Let G be a countable torsion-free group. Then $\mathcal{TC}(G) = \mathcal{TC}(C)$ for some completely decomposable group C.*

PROOF. The result follows immediately from the Propositions 3.1, 3.4, 3.5 and Theorem 3.6. □

Note that the above corollary holds for all groups satisfying the Baer criterion (Lemma 3.3). A necessary but not very convenient condition for a group to satisfy 3.3 can be found in [**B**].

We finish this section by discussing the uncountable case. In fact, we show that in the local case it is undecidable in ZFC whether or not any torsion subclass $\mathcal{TC}(G)$ of a cotorsion class cogenerated by a torsion-free (uncountable) module G equals $\mathcal{TC}(C)$ for some completely decomposable module C.
For the notation and unexplained terminology we refer to [**EM**].

PROPOSITION 3.8 ($V = L$). *Let p be any prime number and let G be a torsion-free $\mathbb{Z}_{(p)}$-module where $\mathbb{Z}_{(p)}$ denotes the localization of the integers $\mathbb{Z}$ at the prime p. Then G is free if and only if $\text{Ext}(G, \bigoplus_{n \in \omega} \mathbb{Z}(p^n)) = 0$.*

PROOF. Trivially, if G is free then $\text{Ext}(G, B) = 0$ for any module B. Conversely, if $\text{Ext}(G, \bigoplus_{n \in \omega} \mathbb{Z}(p^n)) = 0$ then G is a Baer-module by [**EFS**, Theorem C]. This implies that G is a free $\mathbb{Z}_{(p)}$-module (e.g. see [**G**]). □

Note that the proposition above implies that, for any $\mathbb{Z}_{(p)}$-module G, there are only two possibilities for $\mathcal{TC}(G)$: either $\mathcal{TC}(G)$ contains all torsion $\mathbb{Z}_{(p)}$-modules, i.e. all p-groups (and then G is free) or $\mathcal{TC}(G)$ contains no unbounded reduced p-group (and then $\mathcal{TC}(G) = \mathcal{TC}(\mathbb{Q})$), assuming $V = L$.

As an immediate consequence of the above we have:

COROLLARY 3.9 ($V = L$). *Let G be a torsion-free group and let p be any prime number. Then the following are equivalent:*
  (i) $\mathcal{TC}(G)$ *contains all p-groups;*
  (ii) $\mathcal{TC}(G)$ *contains* $\bigoplus_{n \in \omega} \mathbb{Z}(p^n)$;
  (iii) $\mathbb{Z}_{(p)} \otimes G$ *is a free $\mathbb{Z}_{(p)}$-module.*

PROOF. The claim follows immediately by Proposition 3.8 since $\text{Ext}_{\mathbb{Z}}(G, \bigoplus_{n \in \omega} \mathbb{Z}(p^n)) = 0$ if and only if $\text{Ext}_{\mathbb{Z}_{(p)}}(\mathbb{Z}_{(p)} \otimes G, \bigoplus_{n \in \omega} \mathbb{Z}(p^n)) = 0$. □

A result contrary to Proposition 3.8, assuming $MA + \neg CH$, follows immediately from some known facts which we state here. A more general form of the same result can be found in [**E**]. However, here we want to include an easier version for the convenience of the reader. First we need:

LEMMA 3.10.

(a) *There exists a strongly $\aleph_1$-free $\mathbb{Z}_{(p)}$-module of cardinality $\aleph_1$ which is not free.*
(b) *A strongly $\aleph_1$-free $\mathbb{Z}_{(p)}$-module is a Shelah module.*
(c) *($MA+\neg CH$) Let M be a countable module (over $\mathbb{Z}_{(p)}$) and let A be a Shelah module of cardinality less than $2^{\aleph_0}$. Then $\text{Ext}(A, M) = 0$.*

PROOF. All claims of the lemma can be found in [**EM**]; for (a) see VII, Theorem 1.3, for (b) see XII, Proposition 1.10, and for (c) see XII, Theorem 1.11. □

As a consequence of the above we now obtain the following:

PROPOSITION 3.11. *($MA+\neg CH$)) For any prime number p there exists a non-free $\mathbb{Z}_{(p)}$-module G of cardinality $\aleph_1$ such that $\text{Ext}(G, \bigoplus_{n\in\omega} \mathbb{Z}(p^n)) = 0$.*

PROOF. By Lemma 3.10(a),(b) there exists a Shelah module G over $\mathbb{Z}_{(p)}$ which is not free. Moreover, 3.10(c) implies $\text{Ext}(G, \bigoplus_{n\in\omega} \mathbb{Z}(p^n)) = 0$ since $\bigoplus_{n\in\omega} \mathbb{Z}(p^n))$ is countable. □

Note that, assuming $MA+\neg CH$, we hence have an $\mathbb{Z}_{(p)}$-module G such that $\mathcal{TC}(G)$ contains an unbounded reduced p-group, namely $\bigoplus_{n\in\omega} \mathbb{Z}(p^n)) \in \mathcal{TC}(G)$ but not all p-groups as it is non-free, i.e. $\mathfrak{C} = \mathcal{TC}(G)$ with G as in Propostion 3.11 does not satisfy condition (iii) in Theorem 3.6 and so $\mathcal{TC}(G) \neq \mathcal{TC}(C)$ for all completely decomposable modules C.

Finally it seems worth mentioning that, if we view a torsion-free $\mathbb{Z}_{(p)}$-module G as an abelian group then $\mathcal{TC}(G)$ consists of all torsion groups such that the p-primary component belongs to $\mathcal{TC}(G)$ when G is viewed as a module and the complementary component is cotorsion (see Theorem 2.4). Therefore, the above results suggest that, also in the abelian group case, it is undecidable in ZFC whether or not all classes $\mathcal{TC}(G)$ for G torsion-free coincide with $\mathcal{TC}(C)$ for some completely decomposable group C. Note that, while Proposition 3.11 also gives a negative answer in Mod-$\mathbb{Z}$, it is open whether there is an affirmative answer in $V = L$ (or in some other model).

4. $\mathcal{TC}(G)$ for some groups G with certain nice properties

In the previous section we have seen that the torsion elements of the cotorsion class cogenerated by certain torsion-free groups G coincide with $\mathcal{TC}(C)$ for a completely decomposable group C. We shall now consider some slightly altered questions. We ask if there are any groups G (besides the obvious ones) such that $\mathcal{TC}(G) = \mathcal{TC}(R)$ for some rank-1 group R. Obviously, G needs to satisfy the conditions of Theorem 3.6 in this case but we shall see that these conditions are not sufficient. Moreover, we discuss the problem whether, for certain torsion-free groups G with $\mathcal{TC}(G)$ satisfying the conditions of Theorem 3.6, the completely decomposable group can be chosen as the one determined by the typeset $Tst(G)$ of G, i.e. if $\mathcal{TC}(G) = \mathcal{TC}\left(\bigoplus_{\tau \in Tst(G)} R_\tau\right)$ where R_τ is a rank–1 group of type τ (unique up to isomorphism). We begin with a straightforward result.

LEMMA 4.1. *Let C be a completely decomposable group of finite rank. Then there is a rank–1 group $R \subseteq \mathbb{Q}$ such that $\mathcal{TC}(C) = \mathcal{TC}\left(\bigoplus_{\tau \in Tst(C)} R_\tau\right) = \mathcal{TC}(R)$.*

PROOF. Let $C = \bigoplus_{i=1}^{n} R_i$ with $R_i \subseteq \mathbb{Q}$ rank–1 groups. Then we clearly have
$$\mathcal{TC}(C) = \bigcap_{i=1}^{n} \mathcal{TC}(R_i) = \bigcap_{\tau \in Tst(G)} \mathcal{TC}(R_\tau) = \mathcal{TC}\left(\bigoplus_{\tau \in Tst(C)} R_\tau\right).$$
Now we define R as the rank–1 group determined by $type(R) = sup\{type(R_i) \mid 1 \leq i \leq n\}$. Then $\mathcal{TC}(C) = \mathcal{TC}(R)$ follows easily by Theorem 2.4. □

If we omit the finite rank condition then, surprisingly, the above result does not remain true. In fact, in the infinite rank case we can prove something quite different.

LEMMA 4.2. *There exist $2^{\aleph_0}$ countable non–isomorphic completely decomposable groups G such that $\mathcal{TC}(G)$ does not coincide with $\mathcal{TC}(R)$ for any rank-1 group R.*

PROOF. It is well known that there exist $2^{\aleph_0}$ pairwise almost disjoint subsets Π_κ of the set of primes Π ($\kappa < 2^{\aleph_0}$). We divide the set $K := \{\Pi_\kappa \mid \kappa < 2^{\aleph_0}\}$ into $2^{\aleph_0}$ countable disjoint sets, say $K = \bigcup_{\kappa < 2^{\aleph_0}} K_\kappa$. For a fixed κ let $K_\kappa = \{\Pi_i^\kappa \mid i \in \omega\}$. Take $R_i^\kappa \subseteq \mathbb{Q}$ of characteristic $(r_p)_{p \in \Pi}$ where $r_p = 1$ if $p \in \Pi_i^\kappa$ and $r_p = 0$ otherwise. Then $C_\kappa = \bigoplus_{i \in \omega} R_i^\kappa$ is a completely decomposable group of infinite rank. Moreover, by the choice of the sets K_κ, all C_κ are pairwise non–isomorphic. So it remains to show that $\mathcal{TC}(C_\kappa) \neq \mathcal{TC}(R)$ for all rank–1 groups $R \subseteq \mathbb{Q}$ and for any $\kappa < 2^{\aleph_0}$. We fix κ and put $C = C_\kappa$, $C_i = C_i^\kappa$ and $\Pi_i = \Pi_i^\kappa$ ($i \in \omega$).
Suppose that $\mathcal{TC}(C) = \mathcal{TC}(R)$ for some rank–1 group R and let $(r_p)_{p \in \Pi}$ represent the quasi–reduced type of R. We apply Proposition 2.2 to deduce a contradiction. Now all r_p are clearly finite since $\mathcal{TC}(C)$ contains all unbounded p–groups for any $p \in \Pi$. Moreover, we have that $r_p = 1$ for almost all $p \in \Pi_i$ ($i \in \omega$) since otherwise the group $\bigoplus_{p \in \Pi_i, r_p = 0} \mathbb{Z}(p)$ would be in $\mathcal{TC}(R)$ but not in $\mathcal{TC}(C)$.
It is easily seen that we can now choose $p_i \in \Pi_i$ such that $r_p = 1$ ($i \in \omega$) and the set $\{p_i \mid i \in \omega\}$ is almost disjoint from each Π_i ($i \in \omega$). But then the group $\bigoplus_{i \in \omega} \mathbb{Z}(p_i)$ is in $\mathcal{TC}(C)$ but not in $\mathcal{TC}(R)$ contradicting the assumption. Thus the claim is true. □

Although the above lemma kills all hope of finding a non–obvious group G of infinite rank with $\mathcal{TC}(G) = \mathcal{TC}(R)$ for some $R \subseteq \mathbb{Q}$, there does exist a large class of finite rank groups with the required property, namely the so–called TEP–groups which are defined as follows:

DEFINITION 4.3. *Let G be a torsion–free group and H a subgroup of G. Then G is said to have the Torsion Extension Property (TEP) with respect to H if, for any torsion group T and any homomorphism $f : H \to T$, there exists a homomorphism*

$g : G \to T$ which extends f; in this case we also say that H is a TEP–subgroup of G. Moreover, G is said to be a TEP–group if G has TEP with respect to all its pure subgroups H.

It is well known (see [**F3**]) that, for instance, any finite rank Butler group satisfies TEP but the converse is not true. This will be used later to show that not all groups satisfy $\mathcal{TC}(G) = \mathcal{TC}\left(\bigoplus_{\tau \in Tst(G)} R_\tau\right)$, even if $\mathcal{TC}(G) = \mathcal{TC}(R)$ for some rank–1 group R. However, next we show that the TEP–groups have indeed the required property.

THEOREM 4.4. *Let G be a torsion–free group of finite rank satisfying TEP. Then $\mathcal{TC}(G) = \mathcal{TC}(R)$ for some rank-1 group $R \subseteq \mathbb{Q}$.*

PROOF. The proof is by induction on the rank n of G. If $n = 1$ then the claim is obviously true. Hence let G be of rank $n > 1$. Choose any pure subgroup H of G of rank $n - 1$. Then $G/H \cong R$ is of rank 1. Now the short exact sequence

$$0 \to H \to G \to R \to 0$$

induces the exact sequence

$$\operatorname{Hom}(G,T) \to \operatorname{Hom}(H,T) \to \operatorname{Ext}(R,T) \to \operatorname{Ext}(G,T) \to \operatorname{Ext}(H,T) \to 0.$$

Since G satisfies TEP the first map is surjective and thus we obtain the short exact sequence

$$0 \to \operatorname{Ext}(R,T) \to \operatorname{Ext}(G,T) \to \operatorname{Ext}(H,T) \to 0.$$

It follows immediately that $\mathcal{TC}(G) = \mathcal{TC}(H) \cap \mathcal{TC}(R)$. By the induction hypothesis $\mathcal{TC}(H) = \mathcal{TC}(S)$ for some rational group S. Therefore, by Lemma 4.1, $\mathcal{TC}(G)$ also coincides with the torsion subclass of the cotorsion class cogenerated by some rank–1 group. □

Next we give an example of a group G which satisfies TEP but is not a Butler group and, moreover, although $\mathcal{TC}(G) = \mathcal{TC}(R)$ for some $R \subseteq \mathbb{Q}$ we do not have $\mathcal{TC}(G) = \mathcal{TC}\left(\bigoplus_{\tau \in Tst(G)} R_\tau\right)$. Note that the same example has been considered in [**F2**, Example 5 on p. 125] in a different context.

EXAMPLE 4.5. Let p be a prime and π a p–adic integer, not a rational; say $\pi = s_0 + s_1 p + \ldots + s_n p^n + \ldots$ with $0 < s_i < p$. Moreover, let x_1, x_2 be linearly independent elements and, for each $n \in \mathbb{N}$, let $y_n = p^{-n}(x_1 + (s_0 + s_1 p + \ldots + s_{n-1}p^{n-1})x_2)$. We define G by

$$G = \langle x_1, x_2, y_1, y_2, \ldots, y_n, \ldots \rangle \subseteq \mathbb{Q}^{(p)} x_1 \oplus \mathbb{Q}^{(p)} x_2.$$

It can be easily verified that G is indecomposable and homogeneous of type $\mathbb{Z}$ (see [**F2**]). Also, G is not a Butler group since homogenous Butler groups are completely decomposable.

Moreover, we have that $G/\mathbb{Z}x_2 \cong \mathbb{Q}^{(p)}$ and that G satisfies TEP since G is of rank 2, i.e. any proper pure subgroup is of rank at most 1. Thus the sequence

$$0 \to \mathbb{Z} \to G \to \mathbb{Q}^{(p)} \to 0$$

induces the exact sequence

$$0 \to \operatorname{Ext}(\mathbb{Q}^{(p)}, T) \to \operatorname{Ext}(G, T) \to 0$$

for any torsion group T. Therefore we conclude $\mathcal{TC}(G) = \mathcal{TC}\left(\mathbb{Q}^{(p)}\right)$ and the latter is obviously different from $\mathcal{TC}(\mathbb{Z}) = \mathcal{TC}\left(\bigoplus_{\tau \in Tst(G)} R_\tau\right)$. □

It is well known that finite rank Butler groups satisfy TEP. Hence we obtain that for any finite rank Butler group G, $\mathcal{TC}(G) = \mathcal{TC}(R)$ for some $R \subseteq \mathbb{Q}$. Actually, for these groups we also get an affirmative answer to the other question raised in the beginning of this section. More generally, we will show that
$$\mathcal{TC}(G) = \mathcal{TC}\left(\bigoplus_{\tau \in Tst(G)} R_\tau\right)$$
for any B_1-group G. Recall that a torsion-free group G is called a B_1-group if $\mathrm{Bext}(G, T) = 0$ for all torsion groups T where Bext consists of all balanced exact sequences; a short exact sequence $0 \to A \to B \to G \to 0$ (with G torsion-free) is called *balanced exact* if, for all rank-1 groups R, the induced homomorphism $\mathrm{Hom}(R, B) \to \mathrm{Hom}(R, G)$ is surjective.

THEOREM 4.6. *Let G be a B_1-group of arbitrary rank. Then $\mathcal{TC}(G) = \mathcal{TC}\left(\bigoplus_{\tau \in Tst(G)} R_\tau\right)$.*

PROOF. Let G be a B_1-group and let $C = \bigoplus_{g \in G} \langle g \rangle_*$. We obtain the following exact sequence
$$0 \to K \to C \to G \to 0$$
where the mapping $C \to G$ is the obvious one and K is the corresponding kernel. In fact, this sequence is easily checked to be balanced exact and hence we obtain the induced sequence
$$\mathrm{Hom}(C, T) \to \mathrm{Hom}(K, T) \to \mathrm{Bext}(G, T) \to \mathrm{Bext}(C, T) = 0$$
for any torsion group T. Also, since G is a B_1-group, we have that $\mathrm{Bext}(G, T) = 0$ and thus the mapping $\mathrm{Hom}(C, T) \to \mathrm{Hom}(K, T)$ is surjective, i.e. K is a TEP-subgroup of C. Therefore we obtain
$$0 \to \mathrm{Ext}(G, T) \to \mathrm{Ext}(C, T) \to \mathrm{Ext}(K, T) \to 0.$$
Thus $\mathcal{TC}(C) \subseteq \mathcal{TC}(G)$.

The converse inclusion is trivial and hencefore $\mathcal{TC}\left(\bigoplus_{\tau \in Tst(G)} R_\tau\right) = \mathcal{TC}(C) = \mathcal{TC}(G)$ as required. □

We finish the paper with an interesting question, the converse of Theorem 4.6.

QUESTION 4.7. *If G is a torsion-free group such that $\mathcal{TC}(G) = \mathcal{TC}\left(\bigoplus_{\tau \in Tst(G)} R_\tau\right)$ is G then necessarily a B_1-group?*

In the finite rank case we actually have some kind of an answer but it is not yet satisfying.

PROPOSITION 4.8. *Let G be a torsion-free group of finite rank. Then G is a Butler group if and only if, for all primes p, we have*

$$(*) \qquad \mathcal{TC}(G/p^\omega G) = \mathcal{TC}\left(\bigoplus_{\tau \in Tst(G/p^\omega G)} R_\tau\right).$$

PROOF. If G is a Butler group then all quotients $G/p^\omega G$ are Butler groups as they are epimorphic images of G. Thus (∗) holds for all primes p by Theorem 4.6. Conversely, if (∗) is satisfied for all primes p then $\text{Ext}(G/p^\omega G, T) = 0$ for all p–groups T by Proposition 2.2 since all types in $Tst(G/p^\omega G)$ are p–reduced. This is equivalent to saying that $G/p^\omega G \otimes \mathbb{Z}_{(p)}$ is a free $\mathbb{Z}_{(p)}$–module. By [**BS**, Theorem 2.1] it follows that G is p–Butler for all primes p and hence it is a Butler group since G is of finite rank (see [**BS**, p. 180]). □

Note that the above gives an affirmative answer to 4.7 for all finite rank groups G such that, for all primes p, G is either p–divisible ($p^\omega G = G$) or p–reduced ($p^\omega G = 0$).

References

[B]   R. Baer, *The subgroup of the elements of finite order of an abelian group*, Ann. of Math. **37** (1936), 766–781.

[BS]   L. Bican and L. Salce, *Butler groups of infinte rank*, Abelian Group Theory, Lecture Notes in Math. **1006** (Springer, 1983), 171–189.

[E]   P.C. Eklof, *Homological algebra and set theory*, Trans. AMS **227** (1977), 207–225.

[EFS] P.C. Eklof, L. Fuchs and S. Shelah, *Baer modules over domains*, Trans. AMS **322** (1990), 547–560.

[EH]   P.C. Eklof and M. Huber, *Abelian group extensions and the axiom of constructibility*, Comment. Math. Helvetici **54** (1979), 440–457.

[EM]   P.C. Eklof and A. Mekler, *Almost Free Modules; Set–Theoretic Methods*, North–Holland (1990).

[ET]   P.C. Eklof and J. Trlifaj, *How to make Ext vanish*, Bull. London Math. Soc. **23** (2000), to appear.

[F1]   L. Fuchs, *Infinite Abelian Groups* **I**, Academic Press (1970).

[F2]   L. Fuchs, *Infinite Abelian Groups* **II**, Academic Press (1973).

[F3]   L. Fuchs, *A survey of Butler Groups of Infinite Rank*, Contemp. Math. **171** (1994), 121–139.

[GSW] R. Göbel, S. Shelah and S.L. Wallutis, *On the lattice of cotorsion theories*, to appear in J. Algebra (2000).

[GT]   R. Göbel and J. Trlifaj, *Cotilting and a hierarchy of almost cotorsion groups*, J. Algebra **224** (2000), 110–122.

[G]   P. Griffith, *A solution to the splitting mixed group problem of Baer*, Trans. Amer. Math. Soc. **139**(1969), 261–269.

[S]   L. Salce, *Cotorsion theories for abelian groups*, Symposia Mathematica **23**(1979), 11–32.

FACHBEREICH 6 – MATHEMATIK, UNIVERSITY OF ESSEN, 45117 ESSEN, GERMANY
Current address: The Hebrew University, Givat Ram, Jerusalem 91904, Israel
E-mail address: lutz@math.huji.ac.il

FACHBEREICH 6 – MATHEMATIK, UNIVERSITY OF ESSEN, 45117 ESSEN, GERMANY
E-mail address: simone.wallutis@uni-essen.de

Cotorsion theories induced by tilting and cotilting modules

Jan Trlifaj

DEDICATED TO PROFESSORS LÁSZLÓ FUCHS AND LADISLAV PROCHÁZKA
IN HONOUR OF THEIR 75TH AND 70TH BIRTHDAYS

ABSTRACT. Let R be a ring and T be an (infinitely generated) tilting module. Then T is known to cogenerate a cotorsion theory, $(\mathcal{X}_T, \text{Gen}(T))$, such that each module has a special $\text{Gen}(T)$-preenvelope and a special $\mathcal{X}_T$-precover.

In §1, we investigate cotorsion theories generated by the classes of all cotorsion modules in the sense of Enochs, Warfield, and Matlis, respectively. Over domains, we prove that each of these cotorsion theories provides for envelopes. In §2, we characterize the class $\mathcal{X}_T$. We show that $\text{Gen}(T)$-envelopes, and hence divisible envelopes and FP-injective envelopes, may not exist in general: this happens when T is the Fuchs' divisible module over a Prüfer domain with proj.dim$(Q) \geq 2$. In §3, we deal with cotilting modules and characterize the cotorsion theories generated by them. For commutative rings, we employ the notion of a dual module to relate the structure of cotilting modules and cotilting torsion-free classes to their tilting counterparts.

Tilting theory was introduced in the finite dimensional case in the works of Brenner-Butler [**BB**], Happel-Ringel [**HR**] and Bongartz [**B**] as a far reaching generalization of the Morita theory of equivalences of module categories. The results were extended to the infinite dimensional case by Miyashita [**Mi**], Colby-Fuller [**CF**], Colpi [**C**] et al. While the key point of the theory, the Tilting Theorem, requires the tilting modules to be finitely generated [**T**], other aspects can be extended to infinitely generated modules [**CT**], [**AC**]. As proved in [**ATT**], the tilting torsion classes always provide for left approximations in the sense of Auslander and Smalo [**AS**], [**AR**], or preenvelopes in the sense of Enochs and Xu [**E**], [**X**].

Despite the fact that there exist no categorial dualities between categories of all modules, the dual notions of a cotilting module and a cotilting torsion-free class can be defined, and shown to provide for right approximations, or precovers, for arbitrary rings [**CDT**], [**ATT**].

1991 *Mathematics Subject Classification.* Primary 16E30, 13C13; Secondary 13G05, 13F05, 16D90, 16G99.

Key words and phrases. cotorsion theory, tilting module, envelope, cotilting module, cover, cotorsion module, divisible module, integral domain.

Research supported by grants GAČR 201/00/0766 and MSM 113200007.

Besides the obvious duality between envelopes and covers, there is a homological relation coming from Salce's notion of a cotorsion theory [S]: If $(\mathcal{A}, \mathcal{B})$ is a cotorsion theory, then every module has a special $\mathcal{A}$-precover iff every module has a special $\mathcal{B}$-preenvelope.

In §1, we investigate the cotorsion theories generated by the classes of all cotorsion modules in the sense of Enochs, Warfield, and Matlis, respectively. We prove that the first two cotorsion theories provide for envelopes and covers over any ring (Theorems 2.3 and 2.6). Moreover, for domains, the third cotorsion theory provides for envelopes (Theorem 2.10).

Each tilting (cotilting) module naturally cogenerates (generates) a cotorsion theory. In this paper, we describe these cotorsion theories in detail - see Theorems 3.3 and 4.1 below. We also show that despite the fact that each tilting torsion class provides for special preenvelopes, there are no envelopes (= minimal versions of preenvelopes) for certain modules over domains with $\text{proj.dim}(Q) \geq 2$ (Theorem 3.5). In particular, though special divisible preenvelopes, and special FP-injective preenvelopes, always exist over any ring, the corresponding envelopes may not exist over Prüfer domains with $\text{proj.dim}(Q) \geq 2$ (Corollary 3.6). In the commutative case, we employ the notion of a dual module to relate the structure of cotilting modules and cotilting torsion-free classes to the tilting ones (Theorem 4.4).

1. Preliminaries

For a ring R, we denote by Mod-R the category of all (right R-) modules. Let S be a commutative ring such that R is an S-algebra. Let E be an injective cogenerator of Mod-S and N be a left R-module. Then N is an R, S-bimodule. Put $N^* = \text{Hom}_S(N, E)$. If $M \cong N^*$ as S, R-bimodules, then M is called a *dual module* (of N). Similarly, we define the R, S-bimodule N^{**} etc.

In the particular case when $S = \mathbb{Z}$, $E = \mathbb{Q}/\mathbb{Z}$, the dual module N^* is called the *character module* of N, and denoted by N^c. If R is a k-algebra over a field k, then any finite k-dimensional module M is dual (since $M \cong M^{**}$ where $S = E = k$). If R is commutative, then the choice of $S = R$ and E an injective cogenerator of Mod-R provides for another instance of a dual module.

For a module M denote by Gen(M) the class of all modules generated by M, i.e., of all homomorphic images of arbitrary direct sums of copies of M. Denote by Add(M) the class of all direct summands of arbitrary direct sums of copies of M. Dually, the class of all modules cogenerated by M, Cog(M), and the class, Prod(M), of all direct summands of arbitrary direct products of copies of M, are defined.

Denote by Q the maximal quotient ring of R. Of course, if R is a (commutative integral) domain, then Q is just the quotient field of R. A ring R is *right coherent* if all finitely generated right ideals of R are finitely presented (equivalently, finitely generated submodules of free modules are finitely presented). If R is a domain, then R is coherent iff the intersections of any pair of finitely generated ideals is finitely generated. For example, any Prüfer domain is coherent, and so is a polynomial ring in any number of commuting variables over a commutative noetherian ring, [W, §26].

Let R be a ring, I be a right ideal of R, and M be a module. Then M is *I-divisible* provided that $\text{Ext}_R(R/I, M) = 0$. If $I = rR$ for some $r \in R$, then the term *r-divisible* will also be used to denote I-divisibility. Following Lam [L, 3.16], we call M *divisible* if M is r-divisible for all $r \in R$. Denote by $\mathcal{DI}$ the class of all divisible modules.

Let I be a left ideal of R. A module M is *I-torsion-free* provided $\text{Tor}_R(M, R/I) = 0$. If $I = Rr$ for some $r \in R$, then the term *r-torsion-free* will also be used to denote I-torsion-freeness. M is *torsion-free* if M is r-torsion-free for all $r \in R$. The class of all torsion-free modules is denoted by $\mathcal{TF}$.

If $r \in R$ with $\text{Ann}_r(r) = 0$, then M is r-divisible iff $Mr = M$. Similarly, if $\text{Ann}_l(r) = 0$, then M is r-torsion-free iff $mr = 0$ implies $m = 0$ for all $m \in M$. In particular, these characterizations hold true for any non-zero $r \in R$ in the case when R is a domain.

Denote by $\mathcal{P}_n$ ($\mathcal{I}_n$) the class of all modules of projective (injective) dimension $\leq n$, and by $\mathcal{FL}$ the class of all flat modules. Note that $\mathcal{P}_0 \subseteq \mathcal{FL} \subseteq \mathcal{TF}$ and $\mathcal{I}_0 \subseteq \mathcal{DI}$ for any ring R.

Let $\mathcal{A}$ be a class of modules. Then ${}^\perp\mathcal{A} = \{B \in \text{Mod-}R \mid \text{Ext}_R(B, A) = 0 \,\forall A \in \mathcal{A}\}$. Similarly, $\mathcal{A}^\perp = \{B \in \text{Mod-}R \mid \text{Ext}_R(A, B) = 0 \,\forall A \in \mathcal{A}\}$. If the class $\mathcal{A}$ consists of a single module A, we simply write ${}^\perp A$ and $A^\perp$. For example, $\mathcal{DI} = A^\perp$ where $A = \oplus_{r \in R} R/rR$.

Following Salce [**S**], a pair of classes of modules $(\mathcal{A}, \mathcal{B})$ is a *cotorsion theory* provided that $\mathcal{A} = {}^\perp\mathcal{B}$ and $\mathcal{B} = \mathcal{A}^\perp$. For a module M denote by $\mathcal{G}_M$ the cotorsion theory *generated* by M (i.e., $\mathcal{G}_M = ({}^\perp M, ({}^\perp M)^\perp)$). Dually, $\mathcal{C}_M$ denotes the cotorsion theory *cogenerated* by M (i.e., $\mathcal{C}_M = ({}^\perp(M^\perp), M^\perp)$). If $\mathcal{C} = (\mathcal{A}, \mathcal{B})$ is a cotorsion theory, then $\text{Ker}_\mathcal{C} = \mathcal{A} \cap \mathcal{B}$ is the *kernel* of $\mathcal{C}$. Note that each element K of the kernel is a *splitter* in the sense of [**GS**] and [**Sc**], i.e., $\text{Ext}_R(K, K) = 0$.

Let $\mathcal{C}$ be a class of modules and M be a module. A homomorphism $f \in \text{Hom}_R(M, C)$ with $C \in \mathcal{C}$ is a *$\mathcal{C}$-preenvelope* of M provided that for each $C' \in \mathcal{C}$ the induced map $\text{Hom}_R(f, C')$ is surjective. The preenvelope f is *special* if it is monic and $C/\text{Im}(f) \in {}^\perp\mathcal{C}$. The preenvelope f is a *$\mathcal{C}$-envelope* of M if $f = gf$ implies g is an automorphism of C, for each $g \in \text{End}_R(C)$. By [**X**, §2.1], if $\mathcal{I}_0 \subseteq \mathcal{C}$ and $\mathcal{C}$ is closed under extensions then each $\mathcal{C}$-envelope is special, and unique up to isomorphism. Dually, a *$\mathcal{C}$-precover*, *special $\mathcal{C}$-precover* and a *$\mathcal{C}$-cover* of M are defined, [**X**, §1.2]. Cotorsion theories naturally tie up the two dual notions:

LEMMA 1.1. [**S**, Lemma 2.2] *Let R be a ring and $(\mathcal{A}, \mathcal{B})$ be a cotorsion theory. The following conditions are equivalent:*

1. *Every module has a special $\mathcal{A}$-precover.*
2. *Every module has a special $\mathcal{B}$-preenvelope.*

A cotorsion theory $(\mathcal{A}, \mathcal{B})$ satisfying the equivalent conditions of Lemma 1.1 is called *complete*. Complete cotorsion theories are abundant:

THEOREM 1.2. *Let R be a ring.*

1. *Let $\mathcal{S}$ be a set of modules and let N be a module. Then there is a module $P \in \mathcal{S}^\perp$ of the form $P = P_\lambda$ where $(P_\alpha \mid \alpha \leq \lambda)$ is a continuous increasing chain of modules satisfying $P_0 = N$ and $P_{\alpha+1}/P_\alpha \cong \oplus_{M \in \mathcal{S}} M$ for all $\alpha < \lambda$. In particular, the cotorsion theory $({}^\perp(\mathcal{S}^\perp), \mathcal{S}^\perp)$ is complete.*
2. *Let $\mathcal{C}$ be any class of pure-injective modules. Then the cotorsion theory $({}^\perp\mathcal{C}, ({}^\perp\mathcal{C})^\perp)$ is complete. Moreover, each module has a ${}^\perp\mathcal{C}$-cover and a $({}^\perp\mathcal{C})^\perp$-envelope.*

PROOF. 1. By [**ET1**, Theorem 2] and [**ET1**, Corollary 10].
2. By [**ET2**, Corollary 10] and [**X**, 2.2.6]. □

2. Cotorsion modules

We start with several consequences of Theorem 1.2.

Let R be a ring. A module M is called *FP-injective* if $\text{Ext}_R(F, M) = 0$ for all finitely presented modules F. Denote by $\mathcal{FI}$ the class of all FP-injective modules. Clearly, $\mathcal{I}_0 \subseteq \mathcal{FI} \subseteq \mathcal{DI}$.

PROPOSITION 2.1. *Let R be a ring.*
1. *Each module has a special $\mathcal{DI}$-preenvelope, and a special $\mathcal{FI}$-preenvelope.*
2. *Let $n < \omega$. Then each module has a special $\mathcal{I}_n$-preenvelope.*

PROOF. 1. By Theorem 1.2.1, since $\mathcal{DI} = A^\perp$ where $A = \oplus_{r \in R} R/rR$ and $\mathcal{FI} = B^\perp$, where B is the direct sum of a representative set of all finitely presented modules.

2. Let N be a module. Then $N \in \mathcal{I}_n$ iff S_n is injective where S_n is the n-th syzygy module of the injective resolution of N [**CE**, Proposition VI.2.1a.]. By Baer's Criterion, this is further equivalent to $\text{Ext}_R(R/I, S_n) = 0$, and hence to $\text{Ext}_R^{n+1}(R/I, N) = 0$, for all right ideals I of R. Denote by C_I the n-th syzygy module of the projective resolution of R/I. Then $\text{Ext}_R^{n+1}(R/I, N) = 0$ iff $\text{Ext}_R(C_I, N) = 0$. It follows that $\mathcal{I}_n = (\oplus_{I \subseteq R} C_I)^\perp$, and Theorem 1.2.1 applies. □

Proposition 2.1.2 extends [**AEJO**, Proposition 3.1] where R was supposed to be noetherian. In Theorem 3.5 and Corollary 3.6, we will show that the special preenvelopes from part 1. may not have minimal versions (that is, there may be no $\mathcal{DI}$-envelopes and $\mathcal{FI}$-envelopes in general).

THEOREM 2.2. *Let R be a ring and M be a module. Denote by $\mathcal{Z}_M$ the class of all direct summands of the modules Z of the form $Z = Z_\lambda$ where $(Z_\alpha \mid \alpha \leq \lambda)$ is a continuous increasing chain of modules, Z_0 is a free module, and $Z_{\alpha+1}/Z_\alpha \cong M$ for all $\alpha < \lambda$. Then $\mathcal{Z}_M = {}^\perp(M^\perp)$.*

PROOF. Since $M \in {}^\perp(M^\perp)$ and $F \in {}^\perp(M^\perp)$ for any free module F, we have $\mathcal{Z}_M \subseteq {}^\perp(M^\perp)$ by [**ET1**, Lemma 1].

Conversely, take $X \in {}^\perp(M^\perp)$ and let $0 \to N \to F \to X \to 0$ be a short exact sequence with F free. Consider the module P corresponding to $\mathcal{S} = \{M\}$ and N by Theorem 1.2.1. Then $P/N = \cup_{\alpha < \lambda}(P_\alpha/N)$. We can form the push-out of the monomorphisms $N \to F$ and $N \to P$:

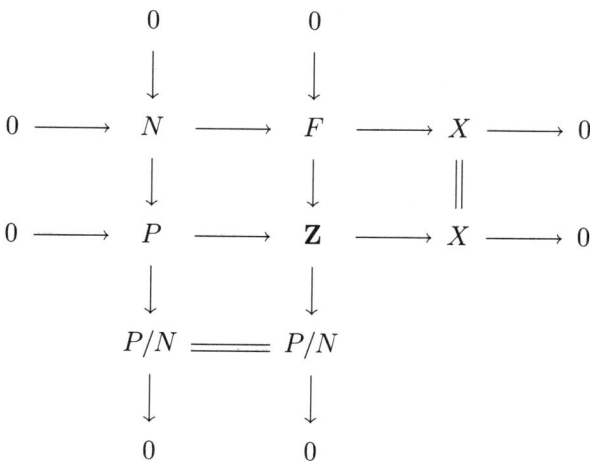

Then $Z = \cup_{\alpha<\lambda} Z_\alpha$, where Z_α are the pre-images of P_α/N in the map $Z \to P/N$. It follows that $Z \in \mathcal{Z}_M$. The second row splits since $P \in M^\perp$ and $X \in {}^\perp(M^\perp)$. So $X \in \mathcal{Z}_M$. □

Note that in the proof of Theorem 2.2, we have $Z \cong X \oplus P$, so that the complement P satisfies $P \in \text{Ker}_{\mathcal{C}_M} = {}^\perp(M^\perp) \cap M^\perp$.

Let R be a ring. Following Enochs, we call a module M *cotorsion* provided that $\text{Ext}_R(F, M) = 0$ for any flat module F, [**X**, §3.3]. For example, any pure-injective module is cotorsion, [**X**, 3.1.1]. Denote by $\mathcal{CO}$ the class of all cotorsion modules. Then $(\mathcal{FL}, \mathcal{CO})$ is a cotorsion theory, [**X**, 3.4.1]. The key facts about this cotorsion theory follow from the recent proof of the Flat Cover Conjecture [**BBE**]:

THEOREM 2.3. *Let R be a ring. Then $\mathcal{C} = (\mathcal{FL}, \mathcal{CO})$ is a complete cotorsion theory. Moreover, every module has a flat cover and a cotorsion envelope. If R is left coherent then $\text{Ker}_\mathcal{C}$ coincides with the class of all pure-injective flat modules.*

PROOF. By [**BBE**] (or by Theorem 1.2.2, since $\mathcal{FL} = {}^\perp\mathcal{PI}$ where $\mathcal{PI}$ denotes the class of all pure-injective modules), and by [**X**, 3.2.3]. □

For left coherent rings, each cotorsion module is a homomorphic image of a pure-injective one:

COROLLARY 2.4. *Let R be a left coherent ring and M be a module.*
1. *M is cotorsion iff F is pure-injective where $F \to M$ is the flat cover of M.*
2. *If M is cotorsion then the minimal flat resolution of M has the form*

$$\cdots \to F_{n+1} \to F_n \to \cdots \to F_0 \to M \to 0$$

where all F_n ($n < \omega$) are (flat and) pure-injective. Conversely, if $\text{fdim} M < \omega$ and all terms of the minimal flat resolution of M are pure-injective, then M is cotorsion.
3. *Assume R is left coherent and right hereditary. Then cotorsion modules coincide with homomorphic images of (flat and) pure-injective modules.*

PROOF. 1. Let $f : F_0 \to M$ be the flat cover of M. Then $C_0 = \text{Ker}(f)$, and hence F_0, is cotorsion. By Theorem 2.3, F_0 is pure-injective. The rest is clear (cf. [**X**, 3.1.2]).

2. We can iterate the construction from part 1., since all the syzygy modules, C_n ($n < \omega$), of the minimal flat resolution of M are cotorsion. Conversely, if $\text{fdim} M = m < \omega$, then $\text{inj.dim}(M^c) = m$, so the character module of the m-th syzygy module, C_m^c, is injective by [**CE**, Proposition VI.2.1a (d)]. Then C_m is flat, so $F_n = 0$ for all $n > m$. By reverse induction, all the syzygy modules C_n ($n < \omega$) of M are cotorsion, hence M is cotorsion.

3. Since Ext_R^2 vanishes on Mod-R, cotorsion modules are closed under homomorphic images, and the assertion follows by part 1. □

Corollary 2.4.3 generalizes the well-known result for $R = \mathbb{Z}$ [**F**, IX.54.1].

EXAMPLE 2.5. 1. Consider a ring R which is right perfect and left coherent (and hence semiprimary, [**W**]). Then flat = projective, so all modules are cotorsion. Moreover, [**X**, 3.2.3] implies a stronger version of Corollary 2.4.1 and 2.4.2: each projective module is pure-injective. Also the claim of 2.4.3 is true (though R need not be right hereditary).

2. Assume R is von Neumann regular (hence left coherent). Then all modules are flat, so cotorsion = pure-injective = injective. A stronger version of Corollary 2.4.3 easily follows: homomorphic images of pure-injectives are cotorsion iff R is right hereditary.

A module M is *strongly cotorsion* (or *Warfield cotorsion*) provided $\operatorname{Ext}_R(T, M) = 0$ for any torsion-free module T. Denote by $\mathcal{SC}$ the class of all strongly cotorsion modules. Clearly, $\mathcal{I}_0 \subseteq \mathcal{SC} \subseteq \mathcal{CO}$.

For example, if R is a domain, then $\mathcal{SC} = \mathcal{CO}$ iff $\mathcal{FL} = \mathcal{TF}$ iff w.gl.dim $R \leq 1$ iff R is Prüfer, cf. [**FS**, IV.1.4 and Ex. 4, p.72]. In this case, all pure-injective modules have injective dimension ≤ 1, [**FS**, XII.3.1].

THEOREM 2.6. .
1. Let R be a ring. Then $\mathcal{D} = (\mathcal{TF}, \mathcal{SC})$ is a complete cotorsion theory. Moreover, every module has a torsion-free cover and a strongly cotorsion envelope.
2. Assume R is a domain. Then $\operatorname{Ker}_{\mathcal{D}}$ coincides with the class of all pure-injective torsion-free modules of injective dimension ≤ 1.

PROOF. 1. By [**CE**, VI.5.1], $\mathcal{TF} = {}^{\perp}(\prod_{r \in R}(R/Rr)^c)$. The module $\prod_{r \in R}(R/Rr)^c$ is pure-injective, so the result follows from Theorem 1.2.2.
2. By [**FS**, XI.3.2 and XII.3.6]. □

Note that Theorem 2.6.1 was proved by Enochs in the particular case when R is a domain (cf. [**X**, 1.3.2]). In fact, the strongly cotorsion envelopes then coincide with the "cotorsion hulls" in the sense of Warfield (cf. [**FS**, XII, §4] and [**X**, 2.2.5]).

Next, we get a result similar to Corollary 2.4.1 for strongly cotorsion modules over domains:

COROLLARY 2.7. *Let R be a domain and M be a module. The following conditions are equivalent:*
1. M is strongly cotorsion.
2. T is pure-injective and $\operatorname{inj.dim}(T) \leq 1$ where $T \to M$ is the torsion-free cover of M.
3. $M \cong T/S$ where $S \subseteq T$, and both S and T are pure-injective torsion-free of injective dimension ≤ 1.

PROOF. 1. implies 3. Let $f : T \to M$ be the torsion-free cover of M. Then $S = \operatorname{Ker}(f)$, and hence T, is strongly cotorsion. Since T and S are torsion-free, they are pure-injective of injective dimension ≤ 1 by Theorem 2.6.2.
3. implies 2. Clear.
2. implies 1. Let S be the kernel of the torsion-free cover. Since $\operatorname{inj.dim}(S) \leq 1$, we have $\operatorname{Ext}_R^2(-, S) = 0$. By Theorem 2.6.2, $\operatorname{Ext}_R(F, T) = 0$ for any torsion-free module F. It follows that $M \in \mathcal{SC}$. □

In contrast with Corollary 2.4.2, the minimal torsion-free resolution of any strongly cotorsion module over a domain has length ≤ 2.

A module M is *weakly cotorsion* (or *Matlis cotorsion* [1]) provided $\operatorname{Ext}_R(Q, M) = 0$. Denote by $\mathcal{MC}$ the class of all weakly cotorsion modules, so $\mathcal{MC} = Q^{\perp}$. Put $\mathcal{SF} = {}^{\perp}(Q^{\perp})$; the elements of $\mathcal{SF}$ will be called *strongly flat*.

---
[1] In [**Ma**], "cotorsion" modules were defined over domains as the h-reduced modules M satisfying $\operatorname{Ext}_R(Q, M) = 0$.

Assume that R is a domain. Since Q is the localization of R at 0, Q is flat. So $\mathcal{CO} \subseteq \mathcal{MC}$. For example, if M is reduced and torsion-free then $M \in \mathcal{MC}$ iff M is R-complete, [**FS**, Proposition V.1.2]. Similarly, if M is *bounded* (i.e., there exists $0 \neq r \in R$ with $Mr = 0$), then $M \in \mathcal{MC}$, [**FS**, XII.3.3].

Warfield proved that $M \in \mathcal{SC}$ iff $M \in \mathcal{MC}$ and $\mathrm{Ext}_R(I, M) = 0$ for any ideal I of R iff $M \in \mathcal{MC}$ and $\mathrm{inj.dim}(M) \leq 1$, cf. [**FS**, XII.3.1-2]. If M is torsion-free, then this is further equivalent to M being pure-injective of $\mathrm{inj.dim} \leq 1$ by Theorem 2.6.2. Matlis proved that all bounded modules have injective dimension ≤ 1 iff R is Dedekind. So $\mathcal{SC} = \mathcal{MC}$ iff R is Dedekind, [**Ma**, Theorem 4.1]. In fact, if M is weakly cotorsion of injective dimension > 1, then there is an ideal I such that $\mathrm{Ext}_R(I, M) \neq 0$, so $I \in \mathcal{TF} \setminus \mathcal{SF}$.

All three notions of cotorsion modules coincide for Dedekind domains. In this case, every module has a $^\perp\mathcal{C}$-cover for any class of cotorsion modules $\mathcal{C}$, cf. [**ET2**, Theorem 16(iii)]. If R is a Prüfer domain which is not Dedekind, then $\mathcal{I}_0 \subsetneq \mathcal{SC} = \mathcal{CO} \subsetneq \mathcal{MC}$, and hence $\mathcal{P}_0 \subsetneq \mathcal{SF} \subsetneq \mathcal{FL} = \mathcal{TF}$.

Theorem 2.2 can be applied to obtain a description of strongly flat modules over domains:

PROPOSITION 2.8. *Let R be a domain and M be a module. The following conditions are equivalent:*

1. *M is strongly flat.*
2. *M is a direct summand of a module N such that there exists an exact sequence $0 \to R^{(\kappa)} \to N \to Q^{(\lambda)} \to 0$ for some cardinals κ and λ.*

In particular, $\mathcal{SF} \subseteq \mathcal{P}_n \cap \mathcal{FL}$, where $n = \mathrm{proj.dim}(Q)$.

PROOF. 1. implies 2. By Theorem 2.2, M is a direct summand in a module N such that N contains a free submodule F with $N/F = Y_\lambda$ where $(Y_\alpha \mid \alpha \leq \lambda)$ is increasing and continuous, $Y_0 = 0$ and $Y_{\alpha+1}/Y_\alpha \cong Q$ for all $\alpha < \lambda$. Since Q is $\sum$-injective [**FS**, Theorem VI.4.1], we infer that $N/F \cong Q^{(\lambda)}$.

2. implies 1. Obvious. $\square$

In general, $\mathcal{SF} \subsetneq \mathcal{P}_n \cap \mathcal{FL}$, where $n = \mathrm{proj.dim}(Q)$:

EXAMPLE 2.9. (Bazzoni) Let R be a Prüfer domain such that $\mathrm{gl.dim}(R) = 2$ and $\mathrm{proj.dim}(Q) = 1$. As mentioned above, there is an ideal I such that $I \in \mathcal{TF} \setminus \mathcal{SF}$. Since R is Prüfer, $\mathcal{TF} = \mathcal{FL}$. Moreover $I \in \mathcal{P}_1$, as $\mathrm{gl.dim}(R) = 2$. So $I \in (\mathcal{P}_1 \cap \mathcal{FL}) \setminus \mathcal{SF}$.

As in Theorem 2.2, in the proof of Proposition 2.8 we get the additional fact that the complement, C, of M in N satisfies $C \in \mathcal{MC} \cap \mathcal{SF}$. The latter observation is essential in proving the second part of our next result.

Let R be a domain and M be a reduced tosion-free module. Denote by $\widehat{M}$ the R-completion of M, so $\widehat{M} \cong \mathrm{Ext}_R(Q/R, M)$ by [**FS**, Theorem V.1.6].

THEOREM 2.10. .

1. *Let R be a ring. Then $\mathcal{E} = (\mathcal{SF}, \mathcal{MC})$ is a complete cotorsion theory.*
2. *Assume R is a domain. Then $\mathrm{Ker}\mathcal{E}$ coincides with the class of all direct summands of the modules of the form $Q^{(\kappa)} \oplus \widehat{R^{(\lambda)}}$ for some cardinals κ and λ.*
3. *Assume R is a domain. Then each module has a weakly cotorsion envelope. Moreover, if M is torsion-free and reduced then the inclusion $M \hookrightarrow \widehat{M}$ is the weakly cotorsion envelope of M.*

PROOF. 1. By Theorem 1.2.1, since $\mathcal{MC} = Q^\perp$.

2. Let $P \in \mathrm{Ker}\,\varepsilon$. By the observation above and by Proposition 2.8, P is a direct summand in a module N such that N is weakly cotorsion and there is an exact sequence

(2.1) $$0 \to R^{(\lambda)} \to N \to Q^{(\lambda')} \to 0$$

for some cardinals λ and λ'. Since N is torsion-free, we have $N \cong Q^{(\kappa)} \oplus N'$ for a cardinal κ and a reduced torsion free module N'. Since $\mathrm{Ext}_R(Q, N') = 0$, we have $N' \cong \mathrm{Hom}_R(R, N') \cong \mathrm{Ext}_R(Q/R, N')$. From (2.1), we get $\mathrm{Ext}_R(Q/R, R^{(\lambda)}) \cong \mathrm{Ext}_R(Q/R, N')$. Since $\mathrm{Ext}_R(Q/R, R^{(\lambda)}) \cong \widehat{R^{(\lambda)}} (\cong \mathrm{Hom}_R(Q/R, (Q/R)^{(\lambda)}))$, we conclude that $N \cong Q^{(\kappa)} \oplus \widehat{R^{(\lambda)}}$.

Conversely, let $N = Q^{(\kappa)} \oplus \widehat{R^{(\lambda)}}$ for some κ and λ. Applying $\mathrm{Hom}_R(-, R^{(\lambda)})$ to the exact sequence $0 \to R \to Q \to Q/R \to 0$, we get

(2.2) $$0 = \mathrm{Hom}_R(Q, R^{(\lambda)}) \to R^{(\lambda)} \to \mathrm{Ext}_R(Q/R, R^{(\lambda)}) \to \mathrm{Ext}_R(Q, R^{(\lambda)}) \to 0.$$

Since $\mathrm{Ext}_R(Q, R^{(\lambda)})$ is a Q-module, we have $\mathrm{Ext}_R(Q, R^{(\lambda)}) \cong Q^{(\kappa')}$ for a cardinal κ'. Since $\mathrm{Ext}_R(Q/R, R^{(\lambda)}) \cong \widehat{R^{(\lambda)}}$, (2.2) induces a presentation of N of the form (2.1). By Proposition 2.8, $N \in \mathcal{SF}$.

Since $(Q/R)^{(\lambda)}$ is h-divisible and torsion, the Matlis duality [**FS**, Theorem V.1.7] gives that $\mathrm{Hom}_R(Q/R, (Q/R)^{(\lambda)})$ is reduced, torsion-free and R-complete, hence $\widehat{R^{(\lambda)}} \cong \mathrm{Hom}_R(Q/R, (Q/R)^{(\lambda)}) \in \mathcal{MC}$ by [**FS**, Proposition V.1.2]. So $N \in \mathcal{MC}$.

It follows that any direct summand of N is in $\mathrm{Ker}\,\varepsilon = \mathcal{MC} \cap \mathcal{SF}$.

3. Let N be a module. Since Q is $\sum$-injective, Theorem 1.2.1 implies the existence of the exact sequence

$$0 \to N \to P \to Q^{(\kappa)} \to 0,$$

where $P \in \mathcal{MC}$ and κ is a cardinal. Note that the map $N \to P$ is a special $\mathcal{MC}$-preenvelope of N. Put $\mathcal{C} = \mathrm{Mod}\text{-}Q$. Then $\mathcal{C}$ is a subclass of $\mathrm{Mod}\text{-}R$ closed under extensions and direct limits, and $\mathcal{C}^\perp = \mathcal{MC}$. By [**X**, 2.2.6], N has an $\mathcal{MC}$-envelope.

Assume M is torsion-free and reduced. There is an exact sequence $0 \to M \hookrightarrow \widehat{M} \to Q^{(\lambda)} \to 0$ for a cardinal λ [**FS**, Lemma V.1.1]. By [**FS**, Proposition V.1.2], the map $i : M \hookrightarrow \widehat{M}$ is a special $\mathcal{MC}$-preenvelope of M. Let f be an endomorphism of $\widehat{M}$ such that $fi = i$. Put $g = \mathrm{id}_{\widehat{M}} - f$. Then $\mathrm{Ker}(g) \supseteq M$, so g induces a morphism $g' : Q^{(\lambda)} \to \widehat{M}$. Since $\widehat{M}$ is reduced, $g' = 0$, hence $g = 0$, and $f = \mathrm{id}_{\widehat{M}}$. It follows that i is the $\mathcal{MC}$-envelope of M. □

COROLLARY 2.11. *Let R be a domain and M be a module. Consider the following two properties:*

1. *M is weakly cotorsion.*
2. *There exists a special strongly flat precover of M, $f : P \to M$, such that $P \cong Q^{(\kappa)} \oplus \widehat{R^{(\lambda)}}$ for some cardinals κ and λ.*

Then 1. implies 2. If $\mathrm{proj.dim}(Q) \leq 1$, then 2. implies 1.

PROOF. 1. implies 2. Let $f : P' \to M$ be any special strongly flat precover of M. Since $\mathrm{Ker}(f) \in \mathcal{MC}$, we have $P' \in \mathcal{MC} \cap \mathcal{SF}$. By Theorem 2.10.2, there exist cardinals κ and λ and a module $C \in \mathcal{MC} \cap \mathcal{SF}$ such that $P = P' \oplus C \cong Q^{(\kappa)} \oplus \widehat{R^{(\lambda)}}$. Then $f \oplus \mathrm{id}_C : P \to M$ is a special strongly flat precover of M.

2. implies 1. f induces an exact sequence $0 \to K \to P \xrightarrow{f} M \to 0$ with P and K weakly cotorsion. If $\mathrm{proj.dim}(Q) \leq 1$, then $\mathrm{Ext}^2_R(Q, K) = 0$, so M is also weakly cotorsion. □

3. Tilting modules and tilting torsion classes

Let R be a ring. A module T is *tilting* provided that $\mathrm{Gen}(T) = T^\perp$ [**CT**]. The (torsion) class $\mathrm{Gen}(T)$ is called the *tilting torsion class*.

Equivalently, T is tilting iff $\mathrm{proj.dim}(T) \leq 1$, $\mathrm{Ext}_R(T, T^{(\kappa)}) = 0$ for all cardinals κ, and there exist $T_1, T_2 \in \mathrm{Add}(T)$ and an exact sequence $0 \to R \to T_1 \to T_2 \to 0$, [**CT**].

We will need the following characterization of tilting torsion classes:

THEOREM 3.1. [**ATT**, Theorem 2.1] *Let R be a ring and $\mathcal{T}$ be a torsion class of modules. Then the following are equivalent:*

1. *$\mathcal{T}$ is a tilting torsion class.*
2. *Every module has a special $\mathcal{T}$-preenvelope.*

For example, any progenerator is a finitely generated tilting module. Many examples of finitely generated tilting modules come from the theory of finite dimensional algebras, cf. [**HR**]. Infinitely generated tilting modules are quite common as well:

EXAMPLE 3.2. Let $R = \mathbb{Z}$. Denote by $\mathbb{P}$ the set of all prime numbers. Take $P \subseteq \mathbb{P}$ and define $\mathcal{T}_P$ as the class of all abelian groups that are p-divisible by all $p \in P$. By [**GT**, Theorem 2.3], $\mathcal{T}_P$ is a tilting torsion class in Mod-R. In fact, $\mathcal{T}_P = \mathrm{Gen}(T_P)$ where $T_P = \oplus_{p \in P} \mathbb{Z}_{p^\infty} \oplus R_P$ is a tilting module. Here, R_P is the subring of $\mathbb{Q}$ generated by R and by $\{p^{-1} \mid p \in P\}$. Assuming Gödel's Axiom of Constructibility, any tilting torsion class in Mod-R is of the form $\mathcal{T}_P$ for a set of primes P, by [**GT**, Theorem 2.3].

In the tilting case, we can refine the description of cotorsion theories cogenerated by modules given in Theorem 2.2 as follows:

THEOREM 3.3. *Let R be a ring and T be a tilting module. Let $0 \to R \to T_1 \to T_2 \to 0$ be a short exact sequence with $T_1, T_2 \in \mathrm{Add}(T)$. Denote by $\mathcal{X}_T$ the class of all direct summands of the modules X such that there exist cardinals κ, λ, and an exact sequence $0 \to R^{(\lambda)} \to X \to T_2^{(\kappa)} \to 0$.*

Then $\mathcal{X}_T = {}^\perp(T^\perp)$, so the cotorsion theory cogenerated by T is $\mathcal{C}_T = (\mathcal{X}_T, \mathrm{Gen}(T))$. Moreover, $\mathrm{Ker}_{\mathcal{C}_T} = \mathrm{Add}(T) \subseteq \mathcal{X}_T \subseteq \mathcal{P}_1 \cap {}^\perp\mathrm{Add}(T)$.

PROOF. Let $M \in {}^\perp\mathrm{Gen}(T)$. Let $0 \to N \to F \to M \to 0$ be a short exact sequence with F free. Put $\kappa = \mathrm{gen}(N)$, and let f be an epimorphism of $R^{(\kappa)}$ on to N. Let ν be the embedding of $R^{(\kappa)}$ into $T_1^{(\kappa)}$. Consider the push-out of ν and f:

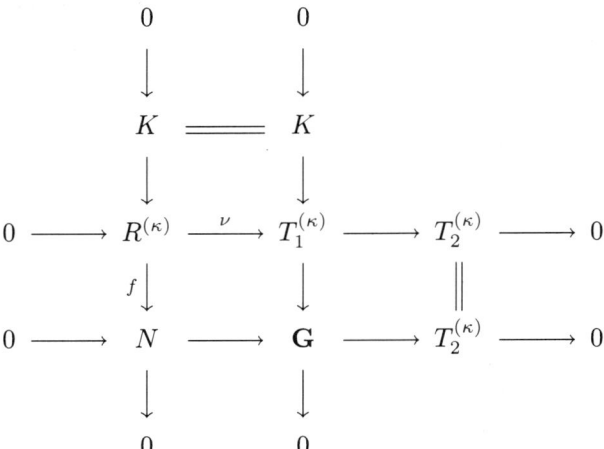

The second column gives $G \in \operatorname{Gen}(T_1) \subseteq \operatorname{Gen}(T)$. Next, consider the push-out of the monomorphisms $N \to F$ and $N \to G$:

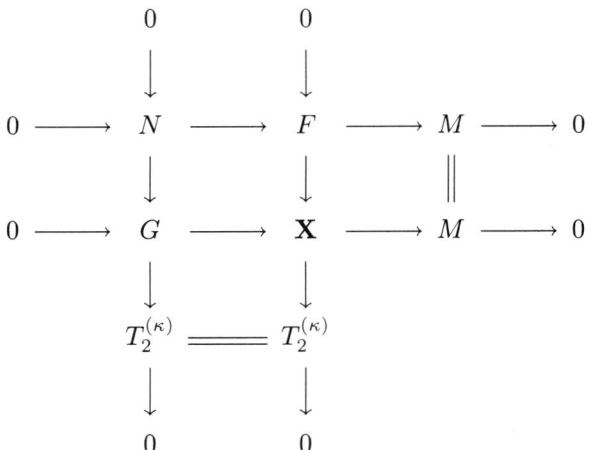

The second column gives $X \in \mathcal{X}_T$. The second row splits since $M \in {}^\perp\operatorname{Gen}(T)$. So $M \in \mathcal{X}_T$.

Conversely, assume that $M \in \mathcal{X}_T$, so M is a summand of some $X \in \mathcal{X}_T$ of the form $0 \to F \to X \to T_2^{(\kappa)} \to 0$ where F is free. Clearly, $F \in {}^\perp\operatorname{Gen}(T)$. Moreover, $T_2^{(\kappa)} \in {}^\perp\operatorname{Gen}(T)$, since $T \in {}^\perp\operatorname{Gen}(T)$. It follows that X, and hence M, belong to ${}^\perp\operatorname{Gen}(T)$.

Since proj.dim$(T) \leq 1$, all elements of $\mathcal{X}_T$ have projective dimension ≤ 1. Since $\operatorname{Add}(T) \subseteq \operatorname{Gen}(T)$, we have also $\mathcal{X}_T \subseteq {}^\perp\operatorname{Add}(T)$. On the other hand, $\operatorname{Add}(T) \subseteq \mathcal{X}_T$ as $T \in {}^\perp\operatorname{Gen}(T)$. It follows that $\operatorname{Add}(T) \subseteq \operatorname{Ker}_{\mathcal{C}_T}$.

Finally, take $M \in \operatorname{Ker}_{\mathcal{C}_T}$. By [**CT**, Lemma 1.2], there is an exact sequence $0 \to K \to T^{(\sigma)} \to M \to 0$, where σ is a cardinal and $K \in \operatorname{Gen}(T)$. Since $M \in {}^\perp\operatorname{Gen}(T)$, the sequence splits, so $M \in \operatorname{Add}(T)$. □

As in Theorem 2.2, we see that the complement G in the proof of Theorem 3.3 satisfies $G \in \operatorname{Add}(T)$.

Assume that R is a domain. Denote by δ the *Fuchs' divisible module*, [**FS**, VI.3], [**Fa**]. So $\delta = F/G$, where F is the free module with the free basis consisting of all k-tuples $(r_1,...,r_k)$ where $k < \omega$ and $0 \neq r_i \in R$ ($i \leq k$), and G is the submodule of F generated by all elements of the form $(r_1,...,r_k)r_k - (r_1,...,r_{k-1})$ ($k \geq 1$). For $k = 0$, we have $w = (\emptyset) + G \in \delta$, $wR \cong R$, and δ/wR is torsion. Clearly, δ is a divisible module.

In fact, δ is tilting, so Theorem 3.3 applies:

THEOREM 3.4. *Let R be a domain and δ be the Fuchs' divisible module.*

1. *δ is a tilting module, the corresponding tilting torsion class being $\mathrm{Gen}(\delta) = \mathcal{DI}$. Every module has a special $\mathrm{Gen}(\delta)$-preenvelope. The cotorsion theory cogenerated by δ has the kernel $\mathrm{Ker}_{\mathcal{C}_\delta} = \mathrm{Add}(\delta)$.*
2. *Assume R is a Prüfer domain. Then $\mathcal{C}_\delta = (\mathcal{P}_1, \mathcal{DI})$. The class $\mathcal{P}_1$ is closed under direct limits iff R is a Dedekind domain. In that case, every module has a $\mathcal{DI}$-envelope.*

PROOF. 1. We have $\mathrm{proj.dim}(\delta) \leq 1$ by [**FS**, Lemma VI.3.1]. Also, for all κ, $\mathrm{Ext}_R(\delta, \delta^{(\kappa)}) = 0$ by [**FS**, Proposition VI.3.4]. Consider the exact sequence $0 \to wR \to \delta \to \delta/wR \to 0$. Since $wR \cong R$ and δ/wR is isomorphic to a summand of δ by [**FS**, p. 124], we infer from [**CT**, Proposition 1.3(i)] that δ is a tilting module.

By [**FS**, Proposition VI.3.4], $\delta^\perp = \mathrm{Gen}(\delta)$ contains all divisible modules. On the other hand, δ, and hence every module in $\mathrm{Gen}(\delta)$, is divisible. So Theorem 1.2.1 (or Theorem 3.1) applies. Finally, $\mathrm{Ker}_{\mathcal{C}_\delta} = \mathrm{Add}(\delta)$ by Theorem 3.3.

2. By part 1, $\mathrm{Gen}(\delta) = \mathcal{DI}$. By Theorem 3.3, we are left to prove that $\mathcal{P}_1 \subseteq {}^\perp \mathcal{DI}$.

Each element of $\mathcal{P}_1$ is a union of a continuous chain of submodules such that all successive factors are finitely presented cyclic, by [**FS**, Corollary IV.4.7]. By [**ET1**, Lemma 1], we get $\mathcal{P}_1^\perp = D^\perp$ where D is the direct sum of a representative set of all finitely presented cyclic modules. By [**FS**, Proposition II.2.2], $D^\perp = \mathcal{DI}$. In particular, $\mathcal{P}_1 \subseteq {}^\perp \mathcal{DI}$.

If R is a Dedekind domain, then all modules have projective dimension ≤ 1. By [**X**, 2.2.6], or [**CE**, Proposition VII.5.1], every module has a $\mathcal{DI}$-envelope which is simply its injective envelope. Otherwise, we have $\mathrm{gl.dim}(R) \geq 2$. By Auslander lemma, $\mathrm{gl.dim}(R) = \sup \mathrm{proj.dim}(R/I)$, where I runs over all ideals of R. So there is an ideal I such that $R/I \notin \mathcal{P}_1$. Note that I is a direct limit of its finitely generated submodules, $(I_\alpha \mid \alpha \in F)$. Since R is Prüfer, each I_α is projective, hence $R/I_\alpha \in \mathcal{P}_1$. Finally, R/I is a direct limit of $(R/I_\alpha \mid \alpha \in F)$. □

The last assertion of Theorem 3.4.1 generalizes [**FS**, VI.3.10].

Recently, Enochs et al. [**AEJO**, Corollary 4.3] have proved that for any ring and any $n < \omega$, $(\mathcal{P}_n, \mathcal{P}_n^\perp)$ is a complete cotorsion theory. It appears to be open to determine when the $\mathcal{P}_n^\perp$-preenvelopes have minimal versions (= envelopes) for $n \geq 1$. In view of Theorem 3.4, our next result gives a negative answer for $n = 1$ in the particular case of Prüfer domains with $\mathrm{proj.dim}(Q) \geq 2$ (cf. Proposition 2.1.1):

THEOREM 3.5. *Let R be a domain. Then the following are equivalent:*

1. *Each free module has a $\mathcal{DI}$-envelope.*
2. *R has a $\mathcal{DI}$-envelope.*
3. *$\mathrm{proj.dim}(Q) \leq 1$.*

PROOF. That 1. implies 2. is clear. Assume 2. Let $0 \to R \xrightarrow{\mu} D \to D/R \to 0$ be exact, where $\mu : R \hookrightarrow D$ is a (special) $\mathcal{DI}$-envelope of R. Take $0 \neq r \in R$. Since

$Dr = D$, there exists $d \in D$ with $dr = 1$. Since μ is special, $\operatorname{Ext}_R(D/R, D) = 0$. It follows that there exists $\psi_r \in \operatorname{Hom}_R(D, D)$ such that $\psi_r(1) = d$. Denote by ϕ_r the endomorphism of D given by the multiplication by r. Then $\mu = \psi_r \phi_r \mu$. By assumption, $\psi_r \phi_r$ is an automorphism of D. This proves that ϕ_r is monic for all $0 \neq r \in R$, hence D is torsion free. By [**FS**, Theorem VI.4.1], $D \cong Q^{(\kappa)}$ for a cardinal κ. By [**X**, 1.2.3], we have $\kappa = 1$. Since $Q/qR \cong Q/R$ for any $0 \neq q \in Q$, and μ is special, we infer from Theorem 3.3 that $Q/R \in \mathcal{P}_1$. The latter is equivalent to $\operatorname{proj.dim}(Q) \leq 1$.

3. implies 1. Assume $\operatorname{proj.dim}(Q) \leq 1$. Let $F = R^{(\kappa)}$ be a free module of rank κ and put $E = Q^{(\kappa)}$. We will prove that the inclusion $\nu : F \hookrightarrow E$ is a $\mathcal{DI}$-envelope of F. Since Q is $\sum$-injective, we have $\operatorname{Ext}_R(Q/R, Q^{(\lambda)}) = 0$ for any λ. Since $\operatorname{proj.dim}(Q) \leq 1$, [**FS**, Theorem VI.1.3] implies that $\mathcal{DI} = \operatorname{Gen}(Q)$, and clearly $\operatorname{proj.dim}(Q/R) \leq 1$. So $\operatorname{Ext}_R(Q/R, D) = 0$ for any divisible module D, and ν is a special $\mathcal{DI}$-preenvelope of F.

Assume that φ is an endomorphism of E with $\varphi\nu = \nu$. Since F is essential in E, φ is monic. Since $\operatorname{Hom}_R(E, E) = \operatorname{Hom}_Q(E, E)$, $\varphi(E)$ is a Q-subspace of E containing F, hence φ is surjective. This proves that ν is a $\mathcal{DI}$-envelope of F. □

Henning Krause asked whether there always exist FP-injective envelopes for modules over coherent rings. The answer is negative (cf. Proposition 2.1.1):

COROLLARY 3.6. *Assume R is a Prüfer domain. Then $\mathcal{FI} = \mathcal{DI}$. Moreover, if $\operatorname{proj.dim}(Q) \geq 2$ then no free module has an $\mathcal{FI}$-envelope.*

PROOF. Since R is coherent, each finitely generated submodule of a finitely presented module is likewise finitely presented. So if F is finitely presented, then there exist $n < \omega$ and a chain of submodules $0 = F_0 \subset F_1 \subset \cdots \subset F_n = F$ such that F_{i+1}/F_i is cyclic and finitely presented for all $i < n$. Since R is Prüfer, each finitely generated ideal is projective. By [**FS**, Proposition II.2.2], we have $\mathcal{FI} = \{\oplus_{I \subseteq R, \operatorname{gen}(I) < \omega} R/I\}^\perp = \mathcal{DI}$, and Theorem 3.5 applies. □

Let R be an arbitrary ring and T be a tilting module. If T is finitely generated, then T is finitely presented [**CM**, Proposition 1.7], hence pure-projective. In particular, $\operatorname{Gen}(T) = T^\perp$ is closed under pure submodules. If T is not finitely generated then T may fail to be pure-projective. This occurs already for abelian groups:

Consider the particular case of $P = \mathbb{P}$ in Example 3.2. Then $\mathcal{T}_P$ is the class of all divisible groups and $T_P = \mathbb{Q} \oplus \mathbb{Q}/\mathbb{Z}$. Let E be the pure-injective hull of $\mathbb{Z}$. Then E is reduced and torsion-free and $E/\mathbb{Z}$ is divisible and torsion-free, [**F**, VII.41.8]. It follows that T is not projective w.r.t. the exact sequence $0 \to \mathbb{Z} \to E \to E/\mathbb{Z} \to 0$.

If T is a finitely generated tilting module, then $\operatorname{Gen}(T)$ is axiomatized (in the language of R-modules) by a single implication of pp-formulas, [**P**, Theorem 4.4 and Lemma 4.6]. On the other hand, the classes $\mathcal{T}_P$ ($P \subseteq \mathbb{P}$) from Example 3.2 are axiomatized by sets of pp-formulas. So is the class of all divisible modules over a domain from Lemma 3.4. In fact, this forces the following closure properties of $\operatorname{Gen}(T)$:

PROPOSITION 3.7. *Let R be a ring and T a tilting module. The following are equivalent:*

1. $\operatorname{Gen}(T)$ *is axiomatized by a set of implications of pp-formulas;*

2. $\mathrm{Gen}(T)$ *is axiomatized by a set of coherent formulas;*
3. $\mathrm{Gen}(T)$ *is closed under pure submodules.*

PROOF. 1. implies 2. By definition.

2. implies 3. By [**R**, Proposition 6.2].

3. implies 1. Since $\mathrm{Gen}(T) = T^\perp$ is closed under products, the result follows by [**R**, Corollary 6.4]. □

4. Cotilting modules and cotilting torsion-free classes

Let R be a ring. A module C is *cotilting* if $\mathrm{Cog}(C) = {}^\perp C$, [**CDT**]. The (torsion-free) class $\mathrm{Cog}(C)$ is then called a *cotilting torsion-free class*.

Equivalently, C is cotilting iff inj.dim$(C) \leq 1$, $\mathrm{Ext}_R(C^\kappa, C) = 0$ for all cardinals κ and there exist $C_1, C_2 \in \mathrm{Prod}(C)$ and an exact sequence $0 \to C_2 \to C_1 \to W \to 0$, where W is an injective cogenerator for Mod-R, [**CDT**], [**ATT**, Proposition 2.3].

By [**ATT**, Theorem 2.5]), if $\mathcal{F}$ is a torsion-free class of modules, then $\mathcal{F}$ is cotilting iff every module has a special $\mathcal{F}$-precover. In fact, all known examples of cotilting torsion-free classes are closed under direct limits, so every module has a $\mathcal{F}$-cover, cf. [**X**, 2.2.12].

The cotorsion theories generated by cotilting modules allow for a dual description to the one in Theorem 3.3:

THEOREM 4.1. *Let R be a ring and C be a cotilting module. Let $0 \to C_2 \to C_1 \to W \to 0$ be a short exact sequence with $C_1, C_2 \in \mathrm{Prod}(C)$ and W an injective cogenerator in Mod-R. Denote by $\mathcal{Y}_C$ the class of all direct summands of the modules Y such that there exist cardinals κ, λ, and an exact sequence $0 \to C_2^\lambda \to Y \to W^\kappa \to 0$.*

Then $\mathcal{Y}_C = ({}^\perp C)^\perp$, so the cotorsion theory generated by C is $\mathcal{G}_C = (\mathrm{Cog}(C), \mathcal{Y}_C)$. Moreover, $\mathrm{Ker}\mathcal{G}_C = \mathrm{Prod}(C) \subseteq \mathcal{Y}_C \subseteq \mathcal{I}_1 \cap \mathrm{Prod}(C)^\perp$.

PROOF. Dual to the proof of Theorem 3.3, using [**CDT**, Proposition 1.8] in place of [**CT**, Lemma 1.2]. □

If R is a finite dimensional k-algebra over a field k, then finitely generated cotilting modules are just the k-vector space duals of the finitely generated tilting modules. We have a similar result for arbitrary commutative rings:

LEMMA 4.2. *Let R be a commutative ring and T be a tilting module. Let T^* be a dual module of T. Then T^* is a cotilting module iff for each cardinal κ, $(T^{(\kappa)})^{**} \in \mathrm{Gen}(T)$. In this case, every module has a $\mathrm{Cog}(T^*)$-cover and a $\mathcal{Y}_{T^*}$-envelope.*

PROOF. Since proj.dim$(T) \leq 1$, we have inj.dim$(T^*) \leq 1$. From the short exact sequence $0 \to R \to T_1 \to T_2 \to 0$ with $T_1, T_2 \in \mathrm{Add}(T)$ we get $0 \to T_2^* \to T_1^* \to R^* \to 0$ with $T_1^*, T_2^* \in \mathrm{Prod}(T^*)$. Note that R^* is an injective cogenerator for Mod-R. It follows that T^* is cotilting iff $\mathrm{Ext}_R((T^*)^\kappa, T^*) = 0$ for all cardinals κ. Applying [**CE**, VI.5.1], we have

$$\mathrm{Ext}_R((T^*)^\kappa, T^*) = 0 \text{ iff } \mathrm{Tor}_R((T^*)^\kappa, T) = 0 \text{ iff } \mathrm{Tor}_R(T, (T^*)^\kappa) = 0$$
$$\text{iff } \mathrm{Tor}_R(T, (T^{(\kappa)})^*) = 0 \text{ iff } \mathrm{Ext}_R(T, (T^{(\kappa)})^{**}) = 0.$$

Since T^* is pure-injective, every module has a ${}^\perp T^*$-cover by Theorem 1.2.2. By [**X**, 2.2.6], each module has a $\mathcal{Y}_{T^*}$-envelope. □

There are futher relations between modules and their duals:

LEMMA 4.3. *Let R be an S-algebra and M be a module.*
1. *M is flat iff M^* is injective;*
2. *Let I be a right ideal. Let M be a left R-module. Then M is I-torsion-free iff M^* is I-divisible. In particular, $M \in \mathcal{TF}$ iff $M^* \in \mathcal{DI}$.*
3. *Let I be a right ideal of R such that I has a projective resolution consisting of finitely generated projective modules. Then M is I-divisible iff M^* is I-torsion-free. If R is a domain, then $M \in \mathcal{DI}$ iff $M^* \in \mathcal{TF}$.*
4. *Assume R is right coherent. Then M is I-divisible for all finitely generated ideals I iff M^* is flat.*

PROOF. 1. and 2. are well-known (cf. [**CE**, VI.5.1]).
3. and 4. M is I-divisible iff $\operatorname{Ext}_R(R/I, M) = 0$ iff $\operatorname{Tor}_R(R/I, M^*) = 0$, by [**CE**, Remark VI.5.3]. 4. now follows from the Flat Test Lemma. □

As an application, we produce tilting torsion, and cotilting torsion-free, classes over arbitrary commutative rings:

THEOREM 4.4. *Let R be a commutative ring. Let $\mathcal{I}$ be a set of finitely generated projective ideals of R. Denote by $\mathcal{T}_I$ the class of all modules which are I-divisible for all $I \in \mathcal{I}$, and by $\mathcal{F}_I$ the class of all modules which are I-torsion-free for all $I \in \mathcal{I}$. Then $M^* \in \mathcal{T}_I$ iff $M \in \mathcal{F}_I$, for each module M.*

Moreover, $\mathcal{T}_I$ is a tilting torsion class closed under pure submodules, and $\mathcal{F}_I$ is a cotilting torsion-free class closed under direct limits. Denote by T a tilting module generating $\mathcal{T}_I$.

Then each module has a special $\mathcal{T}_I$-preenvelope, a $\mathcal{Y}_{T^*}$-envelope, a special $\mathcal{X}_T$-precover, and an $\mathcal{F}_I$-cover.

PROOF. Put $N = \oplus_{I \in \mathcal{I}} R/I$. Then $\mathcal{T}_I = N^\perp$, so each module has a special $\mathcal{T}_I$-preenvelope. Since N has projective dimension ≤ 1, $N^\perp$ is closed under homomorphic images. Since each $I \in \mathcal{I}$ is finitely generated, $N^\perp$ is closed under direct sums. By Theorem 3.1, $N^\perp$ is a tilting torsion class. Proposition 3.7 provides for the closure properties of $\mathcal{T}_I$.

Let T be a tilting module with $\operatorname{Gen}(T) = \mathcal{T}_I$. By Lemma 4.2 and by parts 2. and 3. of Lemma 4.3, T^* is a cotilting module. By Theorem 1.2.2, every module has a $\operatorname{Cog}(T^*)$-cover. Finally, for each module M, $\operatorname{Ext}_R(T, M^*) = 0$ iff $\operatorname{Tor}_R(T, M) = 0$ iff $\operatorname{Tor}_R(M, T) = 0$ iff $\operatorname{Ext}_R(M, T^*) = 0$ by [**CE**, VI.5.1]. It follows that $M^* \in \mathcal{T}_I$ iff $M \in \operatorname{Cog}(T^*)$. So $\operatorname{Cog}(T^*) = \mathcal{F}_I$ by Lemma 4.3.2. □

COROLLARY 4.5. *Let R be a domain. Let δ be the Fuchs' divisible module. Let $\mathcal{I}$ be the set of all finitely generated projective ideals of R. Then δ^* is a cotilting torsion-free module, $\mathcal{T}_I = \operatorname{Gen}(\delta) = \mathcal{DI}$, $\mathcal{F}_I = \operatorname{Cog}(\delta^*) = \mathcal{TF}$ and $\mathcal{Y}_{\delta^*} = \mathcal{SC}$. The kernel $\operatorname{Ker}_{\mathcal{G}_{\delta^*}}$ is the class of all torsion-free pure-injective modules of injective dimension ≤ 1. If R is a Prüfer domain, then $\mathcal{X}_\delta = \mathcal{P}_1$.*

PROOF. By [**FS**, II.2.2], $\mathcal{T}_I$ is the class of all divisible modules. The rest follows by Theorems 3.4 and 4.4. □

If R is a Dedekind domain, we let $\mathcal{I}$ be a subset of the set of all maximal (= non-zero prime) ideals of R. Then we have (cf. Example 3.2, [**GT**, §2] and [**ET2**, Theorem 16]):

COROLLARY 4.6. *Let R be a Dedekind domain. Let P be a set of maximal ideals of R. Denote by $\mathcal{T}_P$ the class of all modules which are I-divisible for all $I \in P$, and by $\mathcal{F}_P$ the class of all modules which are I-torsion-free for all $I \in P$. Then $\mathcal{T}_P$ is a tilting torsion class, and $\mathcal{F}_P$ is a cotilting torsion-free class. Every module has a special $\mathcal{T}_P$-preenvelope and an $\mathcal{F}_P$-cover.*

References

[AEJO] S.T. Aldrich, E. Enochs, O.M.G. Jenda and L. Oyonarte, *Envelopes and covers by modules of finite injective and projective dimensions*, preprint.

[AC] L. Angeleri Hügel and F. Coelho, *Infinitely generated tilting modules of finite projective dimension*, Forum Mathematicum, to appear.

[ATT] L. Angeleri Hügel, A. Tonolo and J. Trlifaj, *Tilting preenvelopes and cotilting precovers*, Algebras and Representation Theory, to appear.

[AR] M. Auslander and I. Reiten, *Applications of contravariantly finite subcategories*, Adv. Math. **86** (1991), 111-152.

[AS] M. Auslander and S. Smalo, *Preprojective modules over artin algebras*, J. Algebra **66** (1980), 61-122.

[BBE] L. Bican, R. El Bashir and E. Enochs, *All modules have flat covers*, Bull. London Math. Soc, to appear.

[B] K. Bongartz, *Tilted algebras*, in Proc. ICRA III, LNM **903**, Springer (1981), 26-38.

[BB] S. Brenner and M. Butler, *Generalizations of the Bernstein-Gelfand-Ponomarev reflection functors*, in Proc. ICRA II, LNM **832**, Springer (1980), 103-169.

[CE] H. Cartan and S. Eilenberg, *Homological Algebra*, Princeton Univ. Press, Princeton 1956.

[CF] R.R. Colby and K.R. Fuller, *Tilting, cotilting, and serially tilted rings*, Comm. Algebra **18** (1990), 1585-1615.

[C] R. Colpi, *Tilting modules and ∗-modules*, Comm. Algebra **21** (1993), 1095-1102.

[CDT] R. Colpi, G.D'Este and A. Tonolo, *Quasi-tilting modules and counter equivalences*, J. Algebra **191** (1997), 461-494.

[CM] R. Colpi and C. Menini, *On the structure of ∗-modules*, J. Algebra **158** (1993), 492-510.

[CT] R. Colpi and J. Trlifaj, *Tilting modules and tilting torsion theories*, J. Algebra **178** (1995), 492-510.

[ET1] P. Eklof and J. Trlifaj, *How to make Ext vanish*, Bull. London Math. Soc., to appear.

[ET2] P. Eklof and J. Trlifaj, *Covers induced by Ext*, J. Algebra, to appear.

[E] E. Enochs, *Injective and flat covers, envelopes and resolvents*, Israel J. Math. **39** (1981), 33-38.

[Fa] A. Facchini, *A tilting module over commutative integral domains*, Comm. Algebra **15** (1987), 2235-2250.

[F] L. Fuchs, *Infinite Abelian Groups*, Vol. I, Academic Press, New York 1970.

[FS] L. Fuchs and L. Salce, *Modules over Valuation Domains*, Lecture Notes in Pure and Appl. Math., Vol. 96, M.Dekker, New York 1985.

[GS] R. Göbel and S. Shelah, *Cotorsion theories and splitters*, Trans. Amer. Math. Soc., to appear.

[GT] R. Göbel and J. Trlifaj, *Cotilting and a hierarchy of almost cotorsion groups*, J. Algebra 224(2000), 110-122.

[HR] D. Happel and C. M. Ringel, *Tilted algebras*, Trans. Amer. Math. Soc. 274(1982), 399-443.

[L] T. Y. Lam, *Lectures on Modules and Rings*, Graduate Texts in Math. 189, Springer, New York 1999.

[Ma] E. Matlis, *Cotorsion modules*, Memoirs Amer. Math. Soc. 49(1964).

[Mi] Y. Miyashita, *Tilting modules of finite projective dimension*, Math. Z. 193(1986), 113-146.

[P] M. Prest, *Interpreting modules in modules*, Annals of Pure and Applied Logic 88(1997), 193-215.

[R] P. Rothmaler, *Purity in model theory*, Advances in Algebra and Model Theory, Gordon & Breach, Amsterdam 1997, 445-469.

[S] L. Salce, *Cotorsion theories for abelian groups*, Symposia Math. XXIII(1979), 11-32.

[Sc] P. Schultz, *Self splitting groups*, preprint, Univ. Western Australia, Perth 1988.

[T] J. Trlifaj, *Every ∗-module is finitely generated*, J. Algebra 169(1994), 392-398.

[W] R. Wisbauer, *Foundations of Module and Ring Theory*, Gordon & Breach, Amsterdam 1991.
[X] J. Xu, *Flat Covers of Modules*, Lecture Notes in Mathematics No. 1634, Springer, New York 1996.

KATEDRA ALGEBRY MFF UK, SOKOLOVSKÁ 83, 186 75 PRAHA 8, ČESKÁ REPUBLIKA
E-mail address: `trlifaj@karlin.mff.cuni.cz`

Steadiness is tested by a single module

Jan Žemlička

DEDICATED TO PROFESSORS LÁSZLÓ FUCHS AND LADISLAV PROCHÁZKA IN HONOUR OF THEIR 75-TH AND 70-TH BIRTHDAY

ABSTRACT. The main goal of the paper is to provide a module-theoretic criterion of the steadiness of a ring. We show that the existence of an infinitely generated dually slender module depends on the existence of an infinitely generated dually slender submodule of a single module. In particular, steadiness is tested over a commutative regular ring R by the module R^*.

Dually slender modules are the modules M for which the covariant functor $\mathrm{Hom}(M, -)$ commutes with direct sums. In particular, finitely generated modules are dually slender. Countably generated modules are not dually slender. The notion of dually slender module dualizes the well known notions of slender modules and slim modules [**EM**]. Bass noticed that a module is dually slender if and only if it is not the union of a countable strictly increasing chain of submodules [**B**]. From this characterization it simply follows that the class of all dually slender modules is closed under homomorphic images. For some types of rings dually slender modules form a class much larger than the finitely generated ones. For instance, it is proved in [**CT**] that the class of all dually slender modules over the endomorphism ring of an infinitely generated free module contains the class of all injective modules.

Rings for which the dually slender modules coincide with the finitely generated modules are called *right steady* rings. Rings satisfying various finiteness conditions are known to be steady. For instance, right noetherian [**R, CT**], right perfect [**CT**], and semiartinian rings of countable socle length [**T2**] belong to the class of right steady rings. It is proved in [**RTZ**] that commutative semiartinian rings over which no cyclic module contains an infinitely generated dually slender submodule are steady as well.

Although a general ring-theoretic characterization of steady rings has not been found as yet, and so it is still an open problem, we present a characterization that uses a single module. Assume that R is a ring. It is shown here that a ring is right steady if and only if a single module of cardinality bounded by $2^{2^{card(R)}}$ contains no infinitely generated dually slender submodule. It is also proved that any representative class of dually slender modules over a commutative regular ring

---

Research supported by grants GAUK 254/2000/B MAT/MFF and MSM 113200007.

© 2001 American Mathematical Society

is a set, and we give an estimate of the cardinality of each dually slender module. Furthermore, we present a more precise characterization of steady commutative regular rings. A commutative regular ring R is steady if and only if R^* contains no infinitely generated dually slender submodule.

In the rest of the paper module means a right R-module. An ideal is a two-sided ideal. A regular ring means a von Neumann regular ring. A regular ring is called abelian regular provided all of its idempotents are central. A ring is said to be semilocal if its Jacobson radical is the intersection of finitely many maximal right ideals.

A representative set of simple modules over a ring will be denoted by $Simp$. R^* will denote the right R-module $R^* = (_RR)^* = \mathrm{Hom}_{\mathbf{Z}}(_RR, \mathbf{Q}/\mathbf{Z})$.

Let M be a module and I be a right ideal. The set of all maximal submodules of M will be denoted by $Max(M)$. $J(M)$ will denote the Jacobson radical of M and $E(M)$ the injective envelope of M. The minimal cardinality of sets of generators of M will be denoted by $gen(M)$. Finally, $Ann_I(M)$ will stand for the right annihilator of M in I, i.e. $Ann_I(M) = \{i \in I;\ Mi = 0\}$.

For further notation we refer to [**AF**] and [**G**].

1. General case

DEFINITION 1.1. Let R be a ring and $\kappa = card(R)^+$. Define the modules

$$T_1 = \prod_{S \in Simp} S^\kappa \quad \text{and} \quad T_2 = \prod_{S \in Simp} S^{(\omega)}.$$

Finite rings are noetherian, so they are right and left steady. Note that every dually slender module over a skew-field is a finitely generated semisimple module.

LEMMA 1.2. *Let R be a ring and M be an infinitely generated dually slender module such that $J(M) = 0$. Then there exists a homomorphism $\phi \colon M \to T_1$ such that $\phi(M)$ is infinitely generated. Moreover, if the ring R is commutative, then M is embeddable in T_2.*

PROOF. The ring R is not right steady, so it is infinite.

Let ρ_N denote the natural projection $M \to M/N$ for $N \in Max(M)$. Since $J(M) = 0$, the product $h : M \to \prod_{N \in Max(M)} M/N$ of the homomorphisms ρ_N is a monomorphism. Define a set $X \subset M$ by letting X be an arbitrary subset of M of cardinality κ if $card(M) \geq \kappa$, and $X = M$ if $card(M) < \kappa$ (recall that $\kappa = card(R)^+$). Clearly, the cardinality of the module XR ($\subseteq M$) is at most κ.

For each $a \in XR$, $a \neq 0$, fix a module $N_a \in Max(M)$ for which $a \in M \setminus N_a$. Put $U = \{N_a;\ a \in XR,\ a \neq 0\} \subseteq Max(M)$. Obviously, $card(U) \leq card(XR) \leq \kappa$. Let ρ be the natural projection of $\prod_{N \in Max(M)} M/N$ onto $\prod_{N \in U} M/N$. Then $\rho h\lceil_{(XR)}$ is a monomorphism. If $X = M$, ρh is a monomorphism as well. If $card(X) = \kappa$, then $card(\rho h(M)) \geq card(\rho h(XR)) \geq \kappa > card(R^{(\omega)}) = card(R)$, because R is an infinite ring. Hence $\rho h(M)$ is an infinitely generated module. As $card(U) \leq \kappa$ and $\rho h(M)$ is a submodule of $\prod_{N \in U} M/N$, $\rho h(M)$ is embeddable in T_1.

Assume now that R is commutative. Then for every $N \in Max(M)$ there exists an $I \in Max(R)$ such that $MI \subseteq N$. Since M is a dually slender R-module, M/MI is a dually slender R/I-module, and so $M/MI \cong (R/I)^{(n_I)}$ for each $I \in Max(R)$ and for some finite number n_I. As $\bigcap_{I \in Max(R)} MI = 0$, the product

$h\colon M \to \prod_{I\in Max(R)}(R/I)^{(n_I)}$ of the projections $\rho_I\colon M \to M/MI$, $I \in Max(R)$, is a monomorphism. Obviously, there is a natural injection $i : \prod_{I\in Max(R)}(R/I)^{(n_I)} \to T_2$, so ih embeds M into T_2. □

Note that in the commutative case all finitely generated modules with zero Jacobson radical are embeddable in T_2; the same holds for infinitely generated dually slender modules with zero Jacobson radical.

LEMMA 1.3. *Let R be a ring and M be a non-zero module such that $J(M) = M$. Then there exists a homomorphism $\phi\colon M \to E(S)$ for a simple module S such that $\phi(M)$ is infinitely generated.*

PROOF. Fix an arbitrary non-zero element $m \in M$. Then there exists a non-zero homomorphism $\phi^*\colon mR \to E(S)$ for a suitable $S \in Simp(R)$. As $E(S)$ is an injective module, we can extend ϕ^* to a homomorphism $\phi\colon M \to E(S)$. Since both M and $\phi(M)$ are non-zero and have no maximal submodule, $\phi(M)$ is not finitely generated. □

THEOREM 1.4. *Let R be a ring. If R is commutative, put $T = T_2$, otherwise put $T = T_1$. Then the following conditions are equivalent:*

(1) *R is right steady.*
(2) *Each dually slender submodule of T and each dually slender submodule of R^* is finitely generated.*
(3) *Each dually slender submodule of T and each dually slender submodule of every $E(S)$, $S \in Simp$, is finitely generated.*

PROOF. (1)→(2) Obvious.

(2)→(3) It is well known that R^* is an injective cogenerator, see for instance [**S**, Chapter I, Proposition 9.3]. Thus both S and $E(S)$ are embeddable in R^* for each $S \in Simp$.

(3)→(1) Let R be non-steady and M be an infinitely generated dually slender module. If $M/J(M)$ is infinitely generated, T contains an infinitely generated dually slender submodule by Lemma 1.2. On the other hand, assume that $M = F + J(M)$ for a finitely generated module F. Then $J(M/F) = M/F$. Applying Lemma 1.3 we get a simple module S such that $E(S)$ contains an infinitely generated dually slender submodule. □

COROLLARY 1.5. *Let R be a semilocal ring. Then R is right steady if and only if $E(R/J(R))$ contains no infinitely generated dually slender submodule.*

PROOF. Since R is semilocal, $R/J(R)$ is semisimple. So the module T is semisimple as well. Hence T contains no infinitely generated dually slender submodule and the assertion follows immediately from Theorem 1.4. □

PROPOSITION 1.6. *Let R be a commutative regular ring. Then each dually slender module is embeddable in T_2.*

PROOF. Let M be a dually slender module. It is well known that every module over a commutative abelian regular ring has zero Jacobson radical. Hence, from Lemma 1.2 it follows that M embeds in T_2. □

It is an immediate consequence of the last proposition that a representative class of dually slender modules over a commutative regular ring is a set.

PROPOSITION 1.7. *Let R be a commutative ring. If R^* contains no infinitely generated dually slender submodule, then a representative class of dually slender modules is a set.*

PROOF. Suppose that a representative class of dually slender modules is proper. Obviously, R is non-steady, so R is infinite. Thus there exists a dually slender module M such that $gen(M) > card(T_2)$. By Lemma 1.2 $M/J(M)$ embeds in T_2, so there is a submodule N of M such that $gen(N) \leq card(T_2)$ and $N + J(M) = M$. Hence M/N is an infinitely generated dually slender module containing no maximal submodule. Applying Lemma 1.3 we get an infinitely generated dually slender submodule of R^*, a contradiction. □

In [**T2**] it is shown that the class of all dually slender modules over the endomorphism ring of an infinitely generated free module contains the class of all injective modules. Thus any representative class of dually slender modules over this ring is a proper class.

2. Steadiness of commutative regular rings

DEFINITION 2.1. Let M be a module and I be a two-sided ideal which is maximal as a right ideal. Then define
$$d_I(M) = dim_{R/I}((M + MI)/MI).$$

Clearly, the notion is well defined because R/I is a skew-field.

Note that $d_I(N) = dim_{R/I}(N + MI/MI)$ for all modules M and N over an abelian regular ring, $N \subseteq M$. Indeed, $N \cap NI = N \cap MI$ over each abelian regular ring, so $(N + NI)/NI \cong N/(N \cap NI) = N/(N \cap MI) \cong (N + MI)/MI$.

If M is a dually slender module, $d_I(M) < \omega$ for each maximal ideal I.

LEMMA 2.2. *Let R be an abelian regular ring and M be a dually slender module.*
(1) *Assume that N is a submodule of M such that $d_I(N) = d_I(M)$ for each $I \in Max(R)$. Then $M = N$.*
(2) *Assume that J is an ideal satisfying $d_I(M) = 0$ for each $I \in Max(R)$ such that $J \subseteq I$. Then $M = MJ$.*

PROOF. Each module over an abelian regular ring has zero Jacobson radical. In particular, every non-zero dually slender module over an abelian regular ring contains a maximal submodule.

(1) Let I be an arbitrary maximal ideal. Since $(N + MI)/MI = M/MI$, we get following isomorphisms: $0 = M/(N + MI) \cong (M/N)/((N + MI)/N) \cong (M/N)/((M/N)I)$. Hence $d_I(M/N) = 0$ for each maximal ideal I, so the module M/N contains no maximal submodule. From the hypothesis it follows that M/N is the zero-module, and so $M = N$.

(2) Assume that $J \subseteq I$. Then $(MJ + MI)/MI = 0$. Therefore $d_I(MJ) = 0 = d_I(M)$.

On the other hand, assume that $J \not\subseteq I$. Since I is maximal, $J + I = R$. Thus $MJ + MI = M(J + I) = M$, and so $d_I(MJ) = d_I(M)$ for every maximal ideal I. Hence (1) implies the assertion. □

It follows from Proposition 1.6 that the cardinality of each dually slender module over an infinite commutative regular ring is bounded by $2^{2^{card(R)}}$. The following assertion improves on the estimate of the cardinality.

COROLLARY 2.3. *Let R be an infinite commutative regular ring. Let M be a dually slender module. Then $gen(M) \leq card(M) \leq 2^{card(R)}$.*

PROOF. Since M is a dually slender module, $d_I(M) < \omega$ for each $I \in Max(R)$. Hence there exist finitely generated modules $F_I \subseteq M$ such that $F_I + MI = M$. Thus, by Lemma 2.2 (1), $\sum_{I \in Max(R)} F_I = M$. As the cardinality of $Max(R)$ is bounded by $2^{card(R)}$ and the cardinality of each F_I is bounded by $card(R)$, $card(M) \leq 2^{card(R)}$. □

LEMMA 2.4. *Assume that M is a module and $M = \sum_{i<\omega} M_i$, where the M_i, $i < \omega$, are submodules of M such that no factor-module of M_i contains an infinitely generated dually slender submodule. Then no factor-module of M contains an infinitely generated dually slender submodule.*

PROOF. This is an easy generalization of [**ZT**, Lemma 5] where the statement is proved for $M_i = m_i R$, $i < \omega$ and $M = \sum_{i<\omega} m_i R$. □

Let R be an abelian regular ring. Fix a module P and an element $x \in R$. Note that $PxR = PeR = Pe$ for a suitable central idempotent $e \in R$ (moreover, $xR = eR$). Thus PxR is a homomorphic image of P (a homomorphism is defined as multiplication by the central element e). Consequently, PxR is dually slender, if P is a dually slender module.

LEMMA 2.5. *Let R be an abelian regular ring such that no cyclic module contains an infinitely generated dually slender submodule. Let N be an infinitely generated dually slender module. Define the set*

$$S = \{r \in R;\ gen(NrR) < \omega\}.$$

Then S is an ideal. In addition, either N/NS is an infinitely generated module or there exists a finitely generated submodule F such that $(N/F)S = N/F$ and $S/Ann_S(N/F)$ is infinitely generated.

PROOF. Since each factor of R contains no infinitely generated dually slender ideal, finitely generated modules contain no finitely generated dually slender modules (Lemma 2.4).

First, we will prove that S is an ideal. Since R is abelian regular, it is sufficient to show that S is a right ideal.

Fix two elements $r, s \in S$. We have observed that $N(r+s)R$ is a homomorphic image of N, hence $N(r+s)R$ is a dually slender module. Moreover, $N(r+s)R$ is a submodule of the finitely generated module $NrR + NsR$. It follows from Lemma 2.4 that $N(r+s)R$ is finitely generated. Thus $r + s \in S$. Now fix elements $i \in S$ and $r \in R$. Similarly, $NirR$ is a dually slender submodule of the finitely generated module NiR, hence $NirR$ is finitely generated as well. Therefore, $ir \in S$. This proves that S is a right ideal.

Assume that N/NS is finitely generated. Then there exists a finitely generated module $F \subseteq N$ for which $F + NS = N$. Thus $N/F = (N/F)S$. If $S/Ann_S(N/F)$ were finitely generated, there would exist a central idempotent $e \in S$ such that $S = eR + Ann_S(N/F)$. Obviously, $Ne + F = N$, so N would be finitely generated, a contradiction. Thus $S/Ann_S(N/F)$ is infinitely generated. □

LEMMA 2.6. *Let R be a non-steady abelian regular ring. Then:*

(1) There exist an infinitely generated dually slender module M and an ideal J such that $MJ = M$ and $gen(J/Ann_J(M)) \geq \omega$.
(2) There exist an infinitely generated dually slender module M and a strictly increasing chain of ideals $(J_\alpha;\ \alpha < \kappa)$ for an uncountable cardinal $\kappa = cf(\kappa)$ such that $M = \bigcup_{\alpha<\kappa} MJ_\alpha$ and $M \neq MJ_\alpha$ for each $\alpha < \kappa$.

PROOF. (1) If there exists a factor-ring of R which contains an infinitely generated dually slender ideal I, put $M = I$ and let J be a lifting of I to the ring R. Clearly, $MJ = II = I = M$ since R is a regular ring.

Assume that no factor of R contains any infinitely generated dually slender ideal. Since R is non-steady, there exists an infinitely generated dually slender module N. Now applying Lemma 2.5 we get an ideal S such that either N/NS is an infinitely generated module or there exists a finitely generated submodule F such that $(N/F)S = N/F$ and $S/Ann_S(N/F)$ are infinitely generated ideals.

Assume that N/NS is a finitely generated module. Put $M = N/F$ and $J = S$. Obviously, $MJ = M$ and $J/Ann_J(M)$ is infinitely generated.

On the other hand, let $\overline{N} = N/NS$ be an infinitely generated module.

Assume that there exists a central idempotent $e \in R \setminus S$ for which $gen(\overline{N}e) < \omega$. From the definition of S it follows that $gen(Ne) \geq \omega$. Moreover, there is a finitely generated module $F \subseteq Ne$ for which $Ne + NS = F + NS$. Since R is an abelian regular ring, we get that $Ne \subseteq NSe + Fe \subseteq NeS + F$. Hence $(Ne/F)S = Ne/F$. If $S/Ann_S(Ne/F)$ were finitely generated, there would exist a central idempotent $f \in S$ such that $Nef + F = Ne$, so Ne would be finitely generated, a contradiction. Now put $M = Ne/F$ and $J = S$. We have proved that $M = MJ$, and $J/Ann_J(M)$ is infinitely generated.

Finally, assume that there exists no idempotent $e \in R \setminus S$ such that $gen(\overline{N}e) < \omega$. If $\overline{N}x = 0$ for an element $x \in R$, $0 = \overline{N}xR = \overline{N}e$ for a central idempotent such that $xR = eR$. Then $e \in S$ and $xR \subseteq S$. Hence $\overline{N} = N/NS$ is faithful over $\overline{R} = R/S$. Since $\overline{R}$ is also non-steady (indeed, N/NS is an infinitely generated dually slender module over $\overline{R}$), it is not semisimple. Thus there exists an infinitely generated maximal ideal in $\overline{R}$; let us denote it by I. Moreover, there is a finitely generated module K for which $K + \overline{N}I = \overline{N}$. Now both $\overline{N}$ and $M = \overline{N}/K$ are faithful over $\overline{R}$, because finitely generated modules contain no infinitely generated dually slender module (Lemma 2.4). Let the ideal J be a lifting of the ideal I to the ring R. Obviously, $MJ = M$ and $gen(J/Ann_J(M)) \geq \omega$.

(2) Applying (1) we get an infinitely generated dually slender module M and an ideal J such that $MJ = M$ and $gen(J/Ann_J(M)) \geq \omega$. Fix M and J for which $gen(J/Ann_J(M))$ is minimal (but infinite). W.l.o.g. we can suppose that M is faithful (i.e. $Ann_J(M) = Ann_R(M) = 0$). Note that $MjR \neq M$ for each $j \in J \neq R$, since $jR = eR \neq R$ for a suitable central idempotent $1 \neq e \in jR$ and $Mjr = Me \neq Me \oplus M(1-e) = M$. Let $(J_\alpha;\ \alpha < \kappa)$ be an arbitrary filtration of J by submodules such that $cf(\kappa) = \kappa$ and $card(J_\alpha) < card(J)$ for each $\alpha < \kappa$. By the minimality of $gen(J)$, $MJ_\alpha \neq M$ for each $\alpha < \kappa$. Since M is a dually slender module (i.e. M is not the union of any countable chain of submodules), κ is not countable. $\square$

Now we are ready to characterize the steadiness of a commutative regular ring in terms of the module R^* (it is easy to generalize our result to abelian regular rings).

THEOREM 2.7. *Let R be a commutative regular ring. Then R is steady if and only if R^* contains no infinitely generated dually slender submodule.*

PROOF. Let R be a non-steady commutative regular ring. Applying Lemma 2.6 we get an infinitely generated dually slender module M and a strictly increasing chain of ideals $(J_\alpha; \alpha < \kappa)$ such that $M = \bigcup_{\alpha < \kappa} MJ_\alpha$ and $M \neq MJ_\alpha$ for each $\alpha < \kappa$. We will define a sequence of maximal ideals $(I_\beta; \beta < \kappa)$ and a strictly increasing sequence of ordinals $(\alpha_\beta; \beta < \kappa)$ such that $d_{I_\beta}(M) \neq 0$, $J_{\alpha_\beta} \subseteq I_\beta$ and $J_{\alpha_{\beta+1}} \not\subseteq I_\beta$ via transfinite induction.

Fix an arbitrary maximal ideal I_0 such that $d_{I_0}(M) \neq 0$ and let $\alpha_0 = 0$.

Assume that both I_β and α_β are defined. Since $d_{I_\beta}(M) \neq 0$, $M \not\subseteq MI_\beta$. Moreover, $M = \bigcup_{\alpha < \kappa} MJ_\alpha$, hence there exists $\alpha > \alpha_\beta$ such that $MJ_\alpha \not\subseteq MI_\beta$. Now put $\alpha_{\beta+1} = \alpha$. Obviously, $J_{\alpha_{\beta+1}} \not\subseteq I_\beta$. Applying Lemma 2.2(2) (for $J = J_{\alpha_{\beta+1}}$), we get $I_{\beta+1} \in Max(R)$ for which $J_{\alpha_{\beta+1}} \subseteq I_{\beta+1}$ and $d_{I_{\beta+1}}(M) \neq 0$.

If γ is a limit ordinal, put $\alpha_\gamma = sup(\alpha_\beta; \beta < \gamma)$ and define I_γ in the same way as in the non-limit step.

Since $\kappa = cf(\kappa)$ is an uncountable cardinal, there exists an $n < \omega$ such that there is a cofinal subset C of κ satisfying $d_{I_\beta}(M) = n$ for each $\beta \in C$. Hence w.l.o.g we can assume that $d_{I_\beta}(M) = n$ for each $\beta < \kappa$.

Now we are ready to find a suitable infinitely generated factor of M with the essential socle.

Denote by ρ_β the natural projection of M onto $M/(MI_\beta)$, and let $\rho: M \to \prod_{\beta < \kappa} M/(MI_\beta)$ be the product of the homomorphisms ρ_β, $\beta < \kappa$. Note that $\bigcap_{\beta < \kappa} \rho(M)I_\beta = \bigcap_{\beta < \kappa} \rho(MI_\beta) = \bigcap_{\beta < \kappa} \rho(ker\rho_\beta) = 0$. As the module M is dually slender and the module M/MI_β is semisimple, $M/MI_\beta \cong (R/I_\beta)^{(n)}$. Since $J_{\alpha_\beta} \subseteq I_\gamma$ for each $\gamma \geq \beta$, $MJ_{\alpha_\beta} \subseteq MI_\gamma$. From this it follows that $\rho_\gamma(MJ_{\alpha_\beta}) = 0$ and $\rho(MJ_{\alpha_\beta}) \neq \rho(M)$ for each $\gamma \geq \beta$. Moreover, $\rho(M) = \bigcup_{\alpha < \kappa} \rho(MJ_{\alpha_\beta}) = \bigcup_{\beta < \kappa} \rho(M)J_{\alpha_\beta}$, because $M = \bigcup_{\alpha < \kappa} MJ_\alpha$. So $\rho(M)$ is the union of a strictly increasing chain of submodules (a suitable subchain of $\rho(M)J_{\alpha_\beta}$). Consequently, $\rho(M)$ is an infinitely generated dually slender module.

Fix an arbitrary non-zero element $m \in \rho(M)$. Let β be the minimal ordinal such that $mR \not\subseteq \rho(M)I_\beta$ (it exists because $\bigcap_{\beta < \kappa} \rho(M)I_\beta = 0$). By the construction of sequences $(\alpha_\beta; \beta < \kappa)$ and $(I_\beta; \beta < \kappa)$ there exists a central idempotent $e \in J_{\alpha_{\beta+1}} \setminus I_\beta$. As I_β is a maximal ideal, $eR + I_\beta = R$, so $meR + mI_\beta = mR$. Hence $d_{I_\beta}(meR) = d_{I_\beta}(mR) \neq 0$. Since $\rho_\gamma(meR) \subseteq \rho_\gamma(mR) = 0$ for each $\gamma < \beta$ (minimality of β) and $\rho_\gamma(meR) = 0$ for each $\gamma > \beta$ (indeed, $eR \subseteq J_{\alpha_{\beta+1}} \subseteq I_\gamma$), $0 \neq meR \subseteq \rho_\beta(M)$. As $\rho_\beta(M)$, $\beta < \kappa$, is semisimple, $Soc(\rho(M))$ is essential in $\rho(M)$.

In addition, $Soc(\rho(M))$ is a submodule of $\bigoplus_{\beta < \kappa} M/MI_\beta \cong \bigoplus_{\beta < \kappa} (R/I_\beta)^{(n)}$. So $Soc(\rho(M))$ is embeddable in $(R^*)^{(n)}$. Since $(R^*)^{(n)} \cong (R^*)^n$ is an injective module, $\rho(M)$ is embeddable in $(R^*)^n$ as well. Hence, from Lemma 2.4 it follows that R^* contains an infinitely generated dually slender submodule. □

Theorem 2.7 shows that the implication of Proposition 1.7 cannot be reversed. Indeed, there exist examples of non-steady commutative regular rings [**EGT**]. From Proposition 1.6 it follows that any representative class of dually slender modules is a set, but the module R^* contains infinitely generated dually slender submodules.

References

[AF] F.W. Anderson and K.R. Fuller, *Rings and Categories of Modules*. 2nd edition, Springer, New York, 1992.

[B] H.Bass, *Algebraic K-theory*, Benjamin, New York, 1968.

[CM] R. Colpi and C. Menini, *On the structure of *-modules*, J. Algebra **158**, 1993, 400–419.

[CT] R. Colpi and J. Trlifaj, *Classes of generalized *-modules*, Comm. Algebra **22**, 1994, 3985–3995.

[EGT] P.C. Eklof, K.R. Goodearl and J. Trlifaj, *Dually slender modules and steady rings*, Forum Math., 1997, **9**, 61–74.

[EM] P.C. Eklof and A.H. Mekler, *Almost Free Modules*, North-Holland, New York, 1990.

[G] K. R. Goodearl, *Von Neumann Regular Rings*, London, 1979, Pitman, Second Ed., Melbourne, FL, 1991, Krieger.

[R] R. Rentschler, *Sur les modules M tels que* $\text{Hom}(M, -)$ *commute avec les sommes directes*, C.R. Acad. Sci. Paris, **268**, 1969, 930–933.

[RTZ] P.Růžička, J.Trlifaj and J.Žemlička, *Criteria of steadiness*, Proc. Conf. "Abelian Groups, Module Theory and Topology" (Padova 1997), Marcel Dekker, New York, 1998, 359–371.

[S] Bo Stenström, *Rings of Quotients*, Berlin, 1975, Springer-Verlag.

[T1] J. Trlifaj, *Strong incompactness for some non-perfect rings*, Proc. Amer. Math. Soc. **123**, 1995, 21–25.

[T2] J. Trlifaj, *Steady rings may contain large sets of orthogonal idempotents*, Proc. Conf. "Abelian Groups and Modules" (Padova 1994), Kluwer, Dordrecht, 1995, 467–473.

[ZT] J. Žemlička and J. Trlifaj, *Steady ideals and rings*, Rend. Sem. Mat. Univ. Padova, 1997, **98**, 161–172.

KATEDRA ALGEBRY, MFF UK, PRAHA 8, SOKOLOVSKÁ 83, 186 75, CZECH REPUBLIC
E-mail address: `zemlicka@karlin.mff.cuni.cz`